STEREOSELECTIVITY IN SYNTHESIS

STEREOSELECTIVITY IN SYNTHESIS

TSE-LOK HO
National Chiao Tung University
Hsinchu, Taiwan
Republic of China

A Wiley-Interscience Publication
JOHN WILEY & SONS, INC.
New York • Chichester • Weinheim • Brisbane • Singapore • Toronto

This book is printed on acid-free paper. ♾

Published simultaneously in Canada.

Library of Congress Cataloging in Publication Data

Ho, Tse-Lok
Stereoselectivity in synthesis / Tse-Lok Ho.
p. cm.
"A Wiley-Interscience publication."
Includes index.
ISBN 0-471-32922-3 (cloth : allk. paper)
1. Organic compounds—Synthesis. 2. Stereochemistry. I. Title
QD262.H616 1999
547' .2—dc21 98-38413

Printed in the United States of America.

10 9 8 7 6 5 4 3 2 1

Dedicated with admiration to
Professor D. Seebach
for his elegant contributions
Professor Z. Valenta
who initiated my fascination in stereochemistry
Professor G. A. Olah
who continues to support my efforts

CONTENTS

PREFACE

As organic synthesis enters the mature phase, reaction stereoselectivity becomes a focal point of research and practical consideration. The great advances in methodology related to this aspect attest to its importance. The subject matter, being so enormous in scope, has created a heavy burden on students and practicing chemists alike. Although flurries of reviews and texts ranging from introductory to comprehensive accounts have appeared in recent years, from a practical point of view they are not particularly easy to assimilate and exploit for synthetic design and execution, as most of these summaries describe reaction types and methods in general terms only. At the planning stage of a synthesis a chemist must face the choice of devices for solving associated stereochemical problems, while knowing only which transformations must be made between a pair of structures. The most urgent and relevant question would be: Under what conditions can one achieve the desired stereochemical outcome for a compound with a particular substitution pattern or conjectured conformation? The consultation of a compilation of transformations according to similar structural profiles would be helpful in finding the answer.

The need for such a monograph prompted my present effort. The writing based on examples mostly from syntheses of complex natural products in the recent literature to illustrate various themes is deemed appropriate as these examples are inspiring, interesting, and practical. Historically, the development of stereochemistry was largely connected with investigations in the natural product domain. In this volume certain syntheses are discussed in more than one place because different aspects and contexts are involved. Stereoselectivity can be induced by reagents and by substrates, and this book deals mainly with substrate-induced stereoselective transformations. As the theme differs from existing texts which are reaction-based, necessarily one finds dispersion of synthetic methods in various chapters and sections.

A restriction on the book size has forced me to omit discussion of many other areas, including the stereoselective formation and manipulation of alkenes. Moreover, limitations in my knowledge and judgment, and lack of sufficient time to prepare this work have contributed to the many defects, even if these are hardly exonerable pretexts.

Tse-Lok Ho

GENERAL ABBREVIATIONS

Ac	acetyl
acac	acetylacetonate
AIBN	2,2′-azobisisobutyronitrile
aq	aqueous
Ar	aryl
BBN	9-borabicyclo[3.3.1]nonane
binap	2,2′-bis(diphenylphosphino)-1,1′-binaphthyl
Bn	benzyl
Boc	*t*-butoxycarbonyl
Bu	*n*-butyl
Bz	benzoyl
18-c-6	18-crown-6
c-	cyclo
CAN	cerium(IV) ammonium nitrate
cat	catalytic
Cbz	benzyloxycarbonyl
COD	1,5-cyclooctadiene
Cp	cyclopentadienyl
CSA	10-camphorsulfonic acid
dba	dibenzylideneacetone
DBN	1,5-diazabicyclo[4.3.0]non-5-ene
DBU	1,8-diazabicyclo[5.4.0]undec-7-ene
DCC	*N,N*′-dicyclohexylcarbodiimide

DDQ	2,3-dichloro-5,6-dicyano-1,4-benzoquinone
de	diastereomer excess
DEAD	diethyl azodicarboxylate
DEIPS	diethylisopropylsilyl
DET	diethyl tartrate
DHP	dihydropyran
DIBAL	diisobutylaluminum hydride
DIPT	diisopropyl tartrate
DMAD	dimethyl acetylenedicarboxylate
DMAP	4-(dimethylamino)pyridine
DME	1,2-dimethoxyethane
DMF	dimethylformamide
DMSO	dimethyl sulfoxide
ee	enantiomer excess
EE	1-ethoxyethyl
EG	ethylene glycol
Et	ethyl
EVE	ethyl vinyl ether
EVK	ethyl vinyl ketone
Fu	furanyl
HMDS	hexamethyldisilazane
HMPA	hexamethylphosphoric triamide
hv	light
Im, imid	imidazole
Ipc	isopinocamphenyl
L	ligand
L.A.	Lewis Acid
LAH	lithium aluminum hydride
LDA	lithium diisopropylamide
LTA	lead tetraacetate
lut	2,6-lutidine
MCPBA	*m*-chloroperbenzoic acid
Me	methyl
MEM	(2-methoxyethoxy)methyl
Men	menthyl
Mes	mesityl
MOM	methoxymethyl
Ms	mesyl (methanesulfonyl)
MVK	methyl vinyl ketone
NBS	*N*-bromosuccinimide
NCS	*N*-chlorosuccinimide
NMO	*N*-methylmorpholine *N*-oxide
Nu	nucleophile
PCC	pyridinium chloroformate

PDC	pyridinium dichromate
Ph	phenyl
Pht, Phth	phthaloyl
Piv	pivaloyl
PPA	polyphosphoric acid
PPTS	pyridinium *p*-toluenesulfonate
Pr	*n*-propyl
py	pyridine
Red-Al	sodium dihydrobis(2-methoxyethoxy)aluminate
selectride	tri-*s*-butylborohydride
Sia	siamyl (3-methyl-2-butyl)
TBAF	tetrabutylammonium fluoride
TBDMS	*t*-butyldimethylsilyl
TBDPS	*t*-butyldiphenylsilyl
TBS	TBDMS
TEA	triethylamine
TES	triethylsilyl
Tf	triflyl (trifluoromethanesulfonyl)
TFA	trifluoroacetic acid
TFAA	trifluoroacetic anhydride
Thex	*t*-hexyl
THF	tetrahydrofuran
THP	tetrahydropyranyl
Thx	thexyl (2,3-dimethyl-2-butyl)
TIPS	triisopropylsilyl
TMEDA	*N,N,N′,N′*-tetramethylethylenediamine
TMS	trimethylsilyl
Tol	*p*-tolyl
TPAP	tetrapropylammonium perruthenate
TPS	TBDPS
Tr	trityl (triphenylmethyl)
Troc	2,2,2-trichloroethoxycarbonyl
Ts	tosyl (*p*-toluenesulfonyl)
TTN	thallium(III) nitrate
X_c	chiral auxiliary
Vn	vinyl
Z	Cbz
Δ	heat

STEREOSELECTIVITY IN SYNTHESIS

1

SOME FUNDAMENTAL CONCEPTS

A stereoselective reaction is one which creates a new stereogenic center and gives rise to one stereoisomer in excess. Stereoselectivity is explainable on the basis of the free-energy difference between two transition states leading to diastereomeric products:

$$\Delta\Delta G^{\ddagger} = \Delta\Delta H^{\ddagger} - T\Delta\Delta S^{\ddagger}$$

As seen in Table 1-1, a very small free-energy difference has a profound effect on the diastereomer excess (de) which relates two diastereomers in amounts A and B by the equation: de = $100(|A - B|)/(A + B)$. The product ratio is temperature dependent.

TABLE 1-1 Relationship of Diastereomer Ratio, Diastereomer Excess with $\Delta\Delta G^{\ddagger}$ at 25°C

Diastereomer Ratio	Diastereomer Excess	$\Delta\Delta G^{\ddagger}$ (kcal mol^{-1})
1.0	0.0	0.00
3.0	50.0	0.65
9.0	80.0	1.30
19.0	90.0	1.74
99.0	98.0	2.72
99.9	99.8	4.09

1.1. CONFIGURATIONS

Different compounds of the same composition are isomers. Three kinds of isomers are known: constitutional isomers, diastereomers, and enantiomers, which possess different properties in terms of topology, geometry, and topography, respectively. The latter two are stereoisomers. A reaction which gives rise to unequal amounts of stereoisomers is stereoselective. Generally it is desirable to obtain one stereoisomer as much as possible.

In identification of stereoisomers it is important to describe unambiguously the configuration of a stereogenic center (stereocenter). By the universally adopted Cahn–Ingold–Prelog (CIP) convention[1] such a center can be assigned as *R* or *S*. The most fundamental part of this system is the *sequence rule* which specifies the hierarchy of substituents or ligands bound to the stereocenter. The rank of a substituent is determined in relation to the others by its mass (the higher mass of the directly attached atom has a higher rank), and among the most common atoms the rule indicates the serial priority: I > Br > Cl > S > P > F > O > N > C > H. A lone electron-pair is considered as a phantom atom of zero mass. Isotopic substituents are similarly arranged, e.g., T > D > H.

When viewing at a chirality center X which is bonded to four substituents with ranking of a > b > c > d such that the substituent of the lowest rank d is at its back (i.e., along the X–d bond), one is faced with a triangle abc. If the a–b–c sequence describes a clockwise course, the stereogenic center X has an (*R*)-configuration (**1-1**). If the a–b–c sequence has a counterclockwise order, the configuration is (*S*) (**1-2**). [Latin *rectus*, right; *sinister*, left.]

If two or more atoms attaching to the stereogenic center are the same, the substitution patterns of these atoms must be evaluated in terms of the other substituents. If ordering is still not possible, the sphere of atoms is extended outwards until a decision can be made. Double and triple bonds are regarded as two and three single bonds, carrying two and three atoms of the same nature, respectively.

The transformation of a nonstereogenic center to a stereogenic center can be achieved by replacing one or two identical ligands (e.g., the two hydrogen atoms of CH_2) with a different one. In that circumstance the nonstereogenic center is a prochiral center, and the ligands are *heterotopic*. Heterotopic ligands may be *enantiotopic* or *diastereotopic*, their classification being made on the basis of symmetry criteria. To assign these ligands on X(aabc), the ligand {a} is *pro-R* if the other identical ligand is transformed into {a′} of lower rank to create an (*R*)-configured X(aa′bc). Accordingly, in ethanol (**1-3**) the enantiotopic H_A and H_A' are *pro-R* and *pro-S*, respectively, as shown.

Specification of heterotopic faces is also important because stereocenters can also be formed by addition reactions, e.g., of alkenes and carbonyl compounds. In facing a stereomodel X(abc) with an a > b > c ranking (**1-4**), one finds a clockwise sequence when the face is *Re*. The opposite face would be *Si*. For example, the

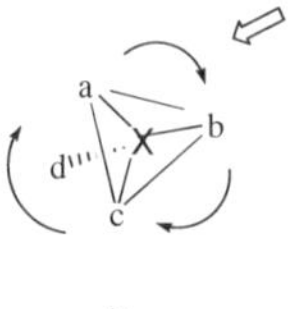

R

(1-1)

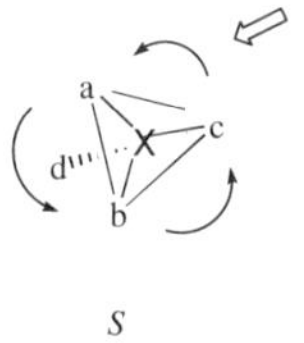

S

(1-2)

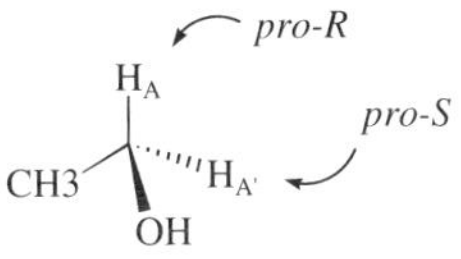

(1-3)

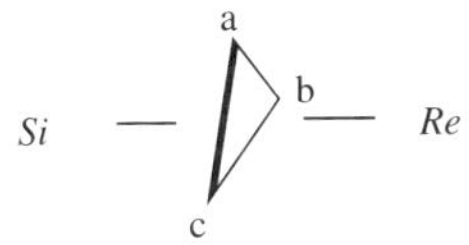

(1-4)

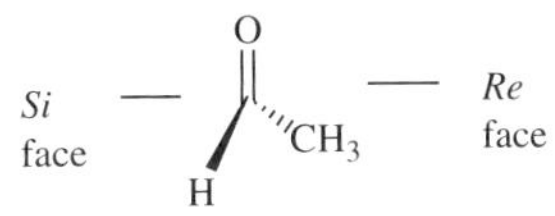

(1-5)

nucleophilic addition to acetaldehyde (**1-5**) is discriminated in the sense of *Re*-face attack and *Si*-face attack.

The description of relative configuration of diastereomers is more difficult, but the CIP system can be used. To specify the relative configuration of racemic diastereomers, both enantiomers are stated, e.g., *R,R/S,S* and *R,S/S,R* which are equivalent to *R**,*R** and *R**,*S**, respectively. Seebach and Prelog[2] proposed the *l* (=like) and *u* (=unlike) descriptors to indicate the relationship of each pair, whereas Masamune et al.[3] preferred a *syn/anti* nomenclature on a zigzag conformation of a carbon backbone (**1-6**). There is also the carbohydrate-based *c-erythro/c-threo* designation.

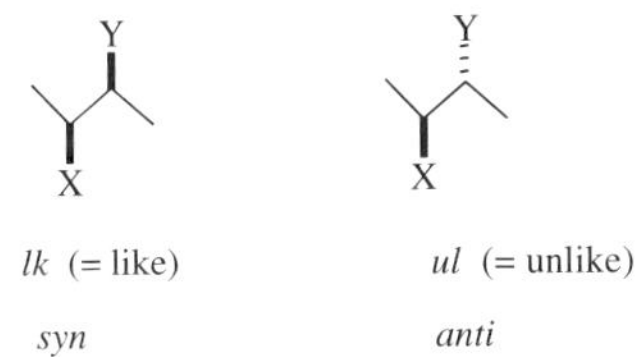

(1-6)

1.2. CONFORMATIONS

Conformation is the spatial arrangement of atoms in a molecule. Even with fixed atomic connectedness and configuration, a molecule may assume different conformations by rotation around single bonds.

According to Klyne and Prelog,[4] local conformation can be defined on the basis of torsional angles. In a stereomodel of a four-atom segment A–X–Y–B, the torsional angle (τ) about the stereogenic axis X–Y is related to the terms:

synperiplanar	(*sp*)	$0° \pm 30°$
+synclinal	(+*sc*)	$+60° \pm 30°$
+anticlinal	(+*ac*)	$+120° \pm 30°$
antiperiplanar	(*ap*)	$180° \pm 30°$
–anticlinal	(–*ac*)	$-120° \pm 30°$
–synclinal	(–*sc*)	$-60° \pm 30°$

Antiperiplanar and synperiplanar conformations are sometimes denoted as *trans* and *cis*, respectively. Such terms appear in many other contexts and it is ill-advised to use

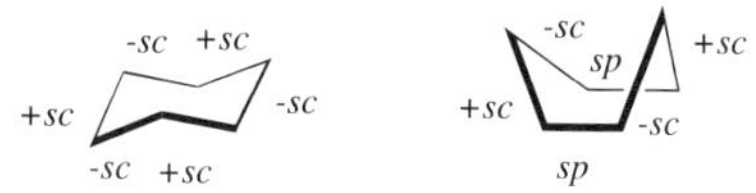

(1-7)

them here again. The description of τ = 30° or 90° is also ambiguous. However, rings (e.g., **1-7**) can be characterized by this method.

In the literature one finds conformational terms such as *eclipsed*, *staggered*, and so forth, which are represented by Newman projections shown below. For planar conjugated systems, conformations are designated *s-cis* and *s-trans*.

Conformation is an important factor in determining the stereochemical outcome of a reaction. On such a premise many models have been proposed for the nucleophilic addition to carbonyl compounds which bear a stereocenter at the α-position (1,2-stereoinduction). In the Cram open-chain model (**1-8**)[5] it is assumed that the largest substituent (L) of the stereocenter is antiperiplanar to the carbonyl group and perpendicular attack of the reagent comes predominantly from the less hindered side. The diastereoselectivity is dependent on the relative severity of the R′<->S and R′<->M interactions. This model is similar to Prelog's rule[6] for predicting the Grignard reaction on chiral esters of α-keto acids (**1-9**). (Note the more distant separation of the reaction site and the stereocenter.)

A cyclic or chelated model (**1-10**)[7] complements the above. The involvement of an α-substituent capable of chelation with an organometallic reagent fixes the conformation and allows the nucleophile to add from the opposite side of the large substituent.

The presence of the α-heteroatom leads to lower stereoselectivity as the polarity of the solvent is increased. This behavior can be ascribed to the competition between chelation and the extended conformation (dipolar model)[8] in which the carbonyl group and the polar substituent lie as far apart as possible. It is favored in polar solvents.

Karabatsos took into account the conformational stability of the carbonyl compound and furnished a reactant-like model (**1-11**) for the transition state.[9] The

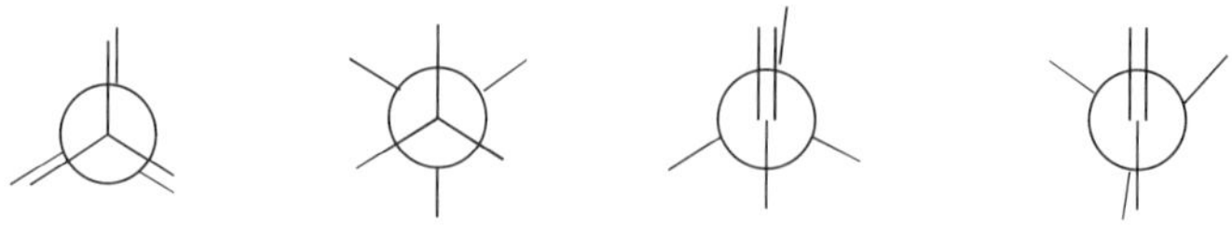

eclipsed staggered eclipsing bisecting

s-cis s-trans

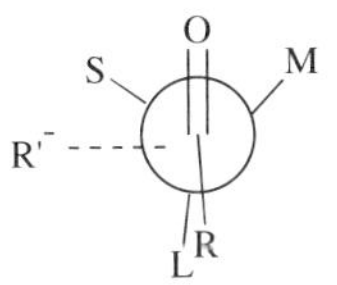

L, M, S =large, medium, small

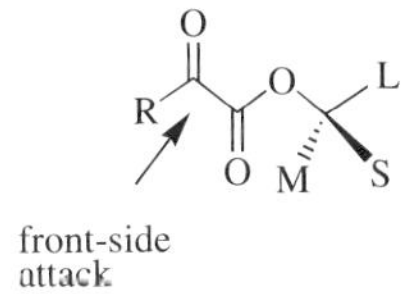

Cram model (**1-8**)

(**1-9**)

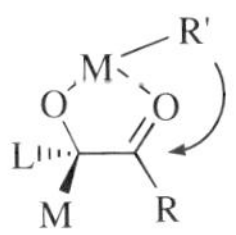

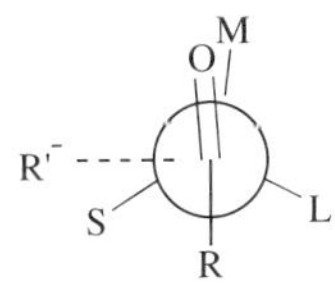

Cram cyclic model
(**1-10**)

Karabatsos model
(**1-11**)

origin of diastereoselectivity is considered to be the difference between O<->M and O<->L interactions. However, all of the models have difficulties in explaining the enhanced stereoselectivity with increasing bulk of R and R′, which prompted Felkin to hypothesize a staggered transition state (**1-12**).[10] Any other arrangement would be destabilized by larger R and R′. According to this model, an electronegative n-donor, if present, should occupy the L position irrespective of its size (antiperiplanar effect)[11] to avoid interaction with the nucleophile and utilize its σ-orbital to interact with the π*-orbital of the carbonyl group.

All of the above models suppose an orthogonal approach of the nucleophile to the carbonyl group. With modification on a trajectory[12,13] along an angle of 100–110° the Felkin model can account for most of the observed stereoselectivity.

The Felkin model can be applied to cyclic ketones. Axial attack (**1-13**) on unhindered ketones encounters little steric interference but equatorial approach (**1-14**) incurs torsional strain. The trend is reversed in 3-substituted cyclohexanones.[13]

For addition to a *trans* double bond the Felkin–Ahn model (**1-15**) is valid when the attack is from an obtuse angle, otherwise the Houk model (**1-16**)[14] fits better. When

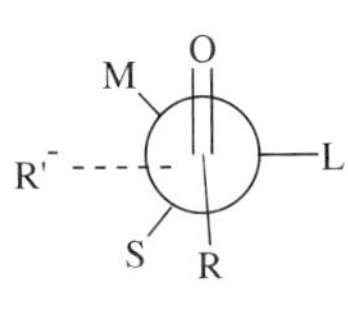

Felkin model

(**1-12**)

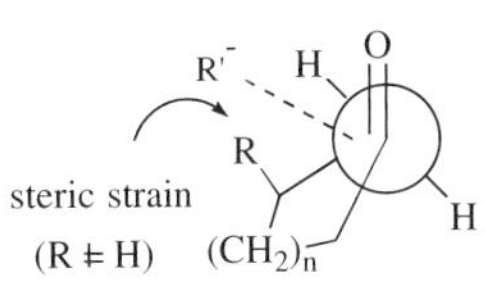

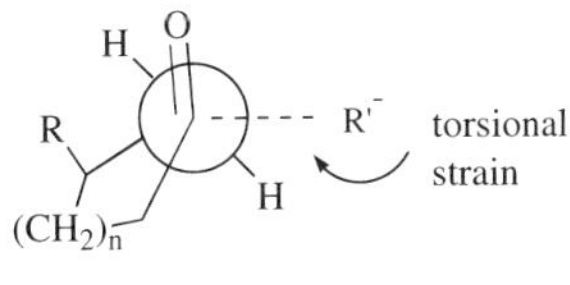

(**1-13**)

(**1-14**)

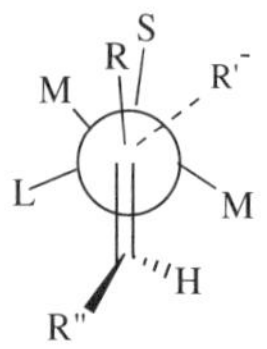

Felkin-Ahn model

(1-15)

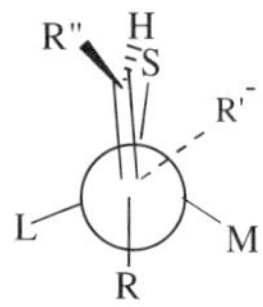

Houk model

(1-16)

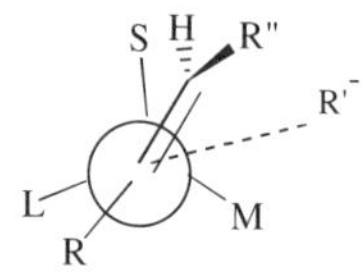

eclipsed model

(1-17)

the double bond contains a *cis* substituent (pointing toward the stereocenter), an eclipsed model (**1-17**) gives the best prediction, due to the influence of allylic 1,3-strain.[15]

1.3. THRUSTS OF SUBSTRATE-INDUCED STEREOSELECTIVITY

Processes of asymmetric synthesis can be classified in mechanistic terms as additions to prostereogenic atoms and substitutions (aliphatic, mainly of S_N2 type), but it is also useful to categorize them according to the sources of chirality or diastereoselection. This book discusses mainly substrate-induced stereoselectivity because this aspect has not been systematically or extensively summarized. In a substrate-induced asymmetric reaction the stereochemical information must be effectively communicated to the site of stereogenesis. The information comes from structural elements present in the substrate. For those substrates lacking transmissible information one must install proper auxiliaries to serve the purpose. Major results of such an approach have been summarized,[16] therefore an effort in an exhaustive survey is not to be repeated, although cursory examples are still included throughout the present volume.

The other strategies for achieving asymmetric synthesis pertain to reagent direction and the use of chiral catalysts.[17] When the element of chirality resides in the reagent, its effects are exerted in the selected transition state. Maximization of the energy differences in diastereomeric transition states would give rise to optimal results. With chiral reagents enantioselective processes emerge. Chiral catalysts provide restrictive environments to reactants, thus enzymes enforce the generation of one set of enantiomers/diastereomers.

The availability of organic compounds from chiral carbon pools (amino acids, carbohydrates, terpenes, etc.) underscores the popularity of their employment as building blocks (substrates).[18–20]

1.4. DOUBLE AND TRIPLE STEREODIFFERENTIATIONS

Stereoselection can be induced by the substrate and/or the reagent. Reagent-induced stereoinductions are divided into auxiliary-induced and additive-induced categories,[21] with *additive* meaning reactant or catalyst which is not covalently linked to the substrate. When both substrate and reagent contain stereogenic elements, their reaction is subject to mutual influence in the steric sense. A simple relationship exists between $\Delta G^{\ddagger}$ values of reference reactions, $\Delta G_1^{\ddagger}$ and $\Delta G_2^{\ddagger}$, and those of the so-called matched and mismatched pairs[22]:

$$\Delta G^{\ddagger}(\text{matched}) = \Delta G_1^{\ddagger} + \Delta G_2^{\ddagger} + \Delta G_{12}^{\ddagger}$$

$$\Delta G^{\ddagger}(\text{mismatched}) = \Delta G_1^{\ddagger} - \Delta G_2^{\ddagger} + \Delta G_{12}^{\prime\ddagger}$$

Triple stereodifferentiation in the union of two reactants which is promoted by an enantiopure reagent is shown with an example of the aldol condensation (Scheme 1-1).[23]

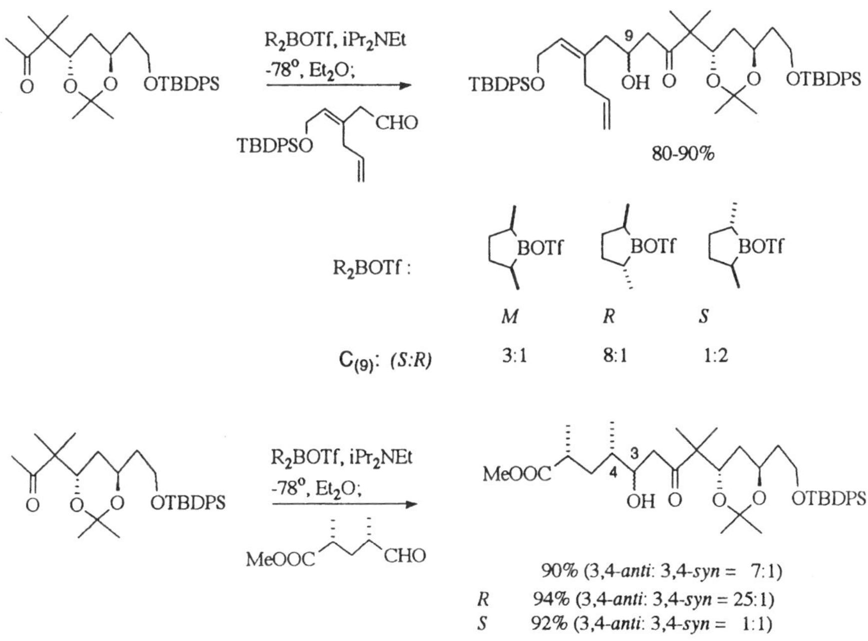

Scheme 1-1. Triple asymmetric aldol.

REFERENCES

1. Cahn, R.S.; Ingold, C.; Prelog, V. *Angew. Chem. Int. Ed.* **1966**, *5*, 385.
2. Seebach, D.; Prelog, V. *Angew. Chem. Int. Ed.* **1982**, *21*, 654; Brewster, J.H. *J. Org. Chem.* **1986**, *51*, 4751.
3. Masamune, S.; Ali, S.A.; Snitman, D.L.; Garvey, D.S. *Angew. Chem. Int. Ed.* **1980**, *19*, 557.

4. Klyne, W.; Prelog, V. *Experientia* **1960**, *16*, 521.
5. Cram, D.J.; Abd Elhafez, F.A. *J. Am. Chem. Soc.* **1952**, *74*, 5828.
6. Prelog, V. *Helv. Chim. Acta* **1953**, *39*, 308.
7. Cram, D.J.; Kopecky, K.R. *J. Am. Chem. Soc.* **1959**, *81*, 2748.
8. Cornforth, J.W.; Cornforth, R.H.; Mathew, K.K. *J. Chem. Soc.* **1959**, 112.
9. Karabatsos, G.J. *J. Am. Chem. Soc.* **1967**, *89*, 1367.
10. Chérest, M.; Felkin, H.; Prudent, N. *Tetrahedron Lett.* **1968**, 2199; Anh, N.T. *Top. Curr. Chem.* **1980**, *88*, 145.
11. Caramella, P.; Rondan, N.G.; Paddon-Row, M.N.; Houk, K.N. *J. Am. Chem. Soc.* **1981**, *103*, 2438.
12. Bürgi, H.B.; Dunitz, J.D.; Lehn, J.M.; Wipf, G. *Tetrahedron* **1974**, *30*, 1563.
13. Chérest, M.; Felkin, H. *Tetrahedron Lett.* **1968**, 2205; *Tetrahedron Lett.* **1971**, 379.
14. Houk, K.N.; Paddon-Row, M.N.; Rondan, N.G.; Wu, Y.-D.; Brown, F.K.; Spellmeyer, D.C.; Metz, J.T.; Li, Y.; Loncharich, R.J. *Science* **1986**, *231*, 1108.
15. Johnson, F. *Chem. Rev.* **1968**, *68*, 375; Hoffmann, R.W. *Chem. Rev.* **1989**, *89*, 1841.
16. Seyden-Penne, J. *Chiral Auxiliaries and Ligands in Asymmetric Synthesis*; Wiley: New York, 1995.
17. Noyori, R. *Asymmetric Catalysts in Organic Synthesis*; 1994, Wiley: New York, 1994.
18. Hanessian, S. *Total Synthesis of Natural Products: The Chiron Approach*; Pergamon Press: New York, 1983.
19. Coppola, G.M.; Schuster, H.F. *Asymmetric Synthesis: Construction of Chiral Molecules Using Amino Acids*; Wiley: New York, 1987.
20. Ho, T.-L. *Enantioselective Synthesis: Natural Products from Chiral Terpenes*; Wiley: New York, 1992.
21. Helmchen, G. *Houben-Weyl Stereoselective Synthesis* **1996**, *E21/1*, 56.
22. Masamune, S.; Choy, W.; Petersen, J.S.; Sita, L.R. *Angew. Chem. Int. Ed. Engl.* **1985**, *24*, 1.
23. Duplantier, A.J.; Nantz, M.H.; Roberts, J.C.; Short, R.P.; Somfai, P.; Masamune, S. *Tetrahedron Lett.* **1989**, *30*, 7357.

2

1,2-STEREOINDUCTION

Stereoinduction is often due to the influence exerted by an existing stereocenter on the steric course of reagent approach. Such an influence can be overwhelming or barely sufficient to show some difference in the creation of a new stereocenter(s), but the effect is most substantial from groups in proximity. Accordingly, 1,2- and 1,3-stereoinductions are most frequently invoked to set the course for gainful progress in a synthetic endeavor.

In this and later chapters I collate examples mostly showing stereoinduction at a site quite close to a stereocenter. In other words, the reactions are substrate-controlled. It must be emphasized that *incipient* stereoinduction is also possible. When two diastereomeric transition states of different energy can be formed from a prochiral substrate, the reaction will preferably proceed from the one with a lower activation energy, with the product distribution profile depending on the energy difference. For example, chiral reagents or catalysts such as chiral organoboranes[1] in hydroboration and reduction and the titanium alkoxide – tartrate ester – *t*-butyl hydroperoxide ensemble[2] in the epoxidation of alkenes are used in those circumstances.

In addition, it is also very popular to employ chiral auxiliaries to form a cleavable covalent bond with achiral compounds, thereby bringing the reaction site under its influence.

2.1. ALKYLATIONS

Alkylation reactions include reaction of nucleophiles with carbonyl compounds (also Michael additions for conjugated carbonyl compounds and related substances), and reaction of enolates with electrophiles.

Although precapnelladiene (**2-1**) possesses only two stereocenters which are 1,3-related, the eight-membered ring is the dominant feature. Retrosynthetic analysis

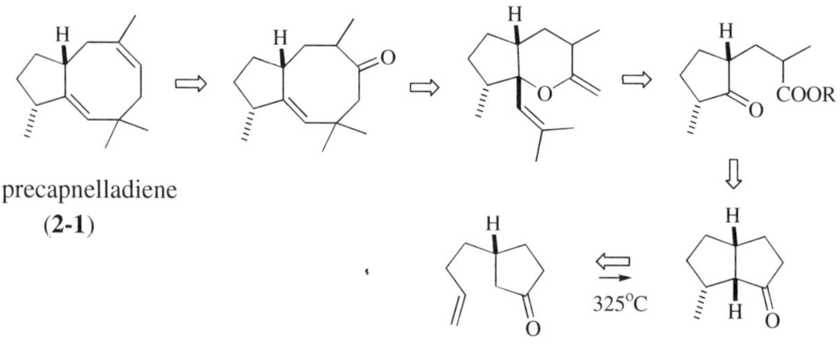

precapnelladiene
(**2-1**)

indicates the advantage of employing a *cis*-2,5-disubstituted cyclopentanone to construct the molecular skeleton, as all stereochemical issues would have been settled. A route that involved Grignard reaction of the cyclopentanone and Claisen rearrangement of a derived enol ether with an exocyclic methylene group to elaborate the eight-membered ring[3] was probably facilitated by the formation of a more stable *cis*-5/6-fused lactone.

1,2-Stereoinduction by a substituent attached to an adjacent carbon in cyclohexane derivatives is well appreciated. Thus, the planning of sesquiterpene synthesis including those targeted at fukinone (**2-2**)[4] and bakkenolide-A (**2-3**)[5] was amply rewarded.

Of course the stereoinduction was not limited to simple alkylation: Michael addition such as used in an approach to isopetasol (**2-4**)[6] followed the same trend. Ironically, the last step of borohydride reduction of the saturated ketone proceeded predominantly from the equatorial side (1,2- and 1,3-stereoinduction were in operation).

Although there was no overall increase in the number of stereocenters, the introduction of an alkyl group to C-2 of *cis*-2-cyclopentene-1,4-diol (**2-5**)[7] proceeded with complete stereoselectivity. The alkylation step was apparently subjected to stereoinduction by the hydroxyl group, otherwise the return of the secondary alcohol

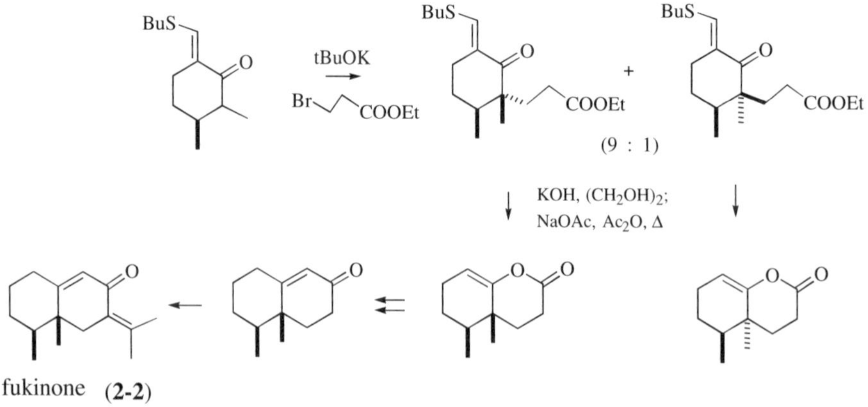

fukinone (**2-2**)

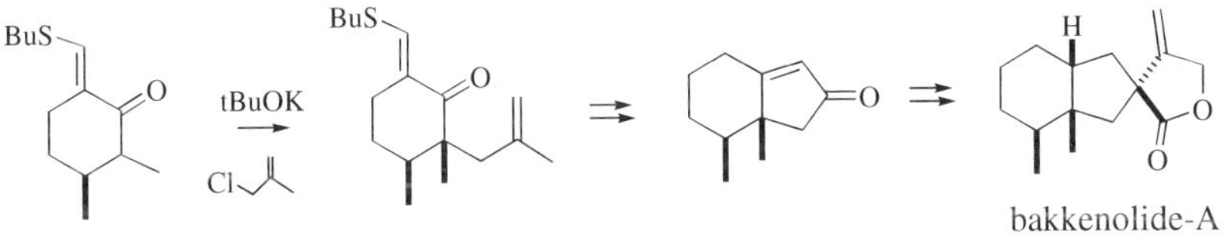

bakkenolide-A
(**2-3**)

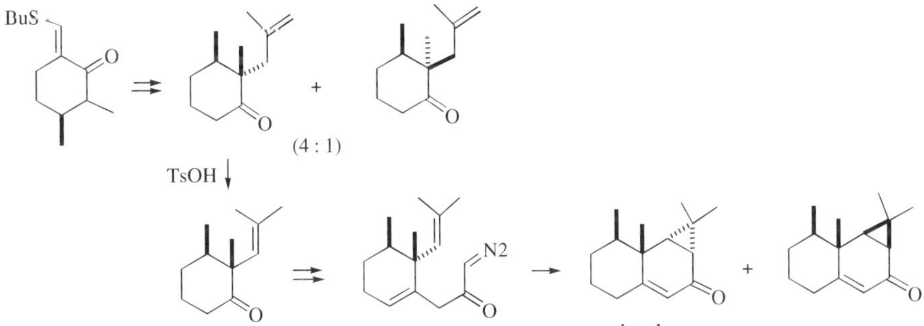

aristolone

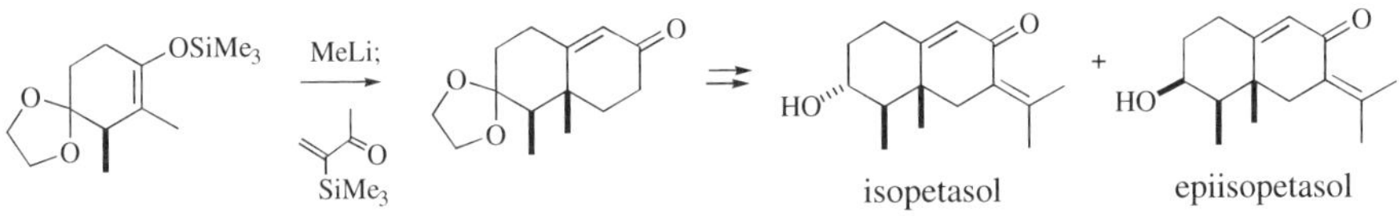

isopetasol
(**2-4**)

epiisopetasol

to the original position and configuration by a [2,3]-sigmatropic rearrangement would not have been possible.

The ready availability of a 3-aryl-4-nitro-1-methylenecyclopentane from [3+2]-cycloaddition opened a route to cephalotaxine (**2-6**).[8] The nitro group served an activator role (to empower the α-carbon as Michael donor) in the elaboration of the pyrrolidine ring prior to reduction. The base-catalyzed reaction with methyl acrylate gave an adduct with stereochemistry in agreement with the target. The Michael addition took place away from the aryl group.

Many methods for synthesis of estrone and its methyl ether (**2-7**) exist. One route featuring conjugate addition of 6-methoxy-α-tetralone to a chiral 2-*p*-toluenesulfinyl-2-cyclopentenone[9] is quite efficient. The stereochemistry of $C_{(13)}$ was established by prenylation of the enolate which was generated by desulfinylation. The prenyl group entered from the side opposite the bulky tetralone.

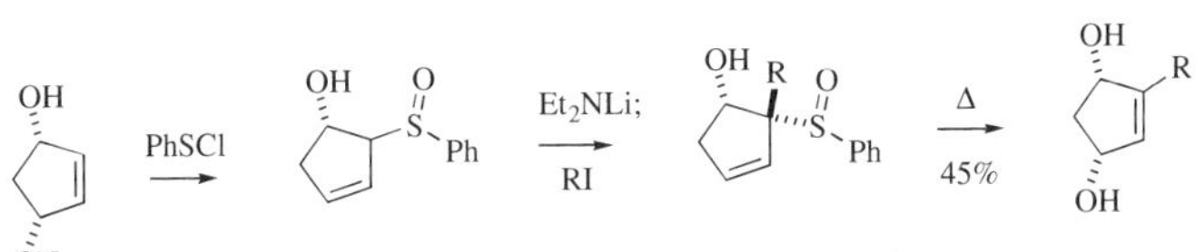

R = $(CH_2)_6COOBu^t$

(**2-5**)

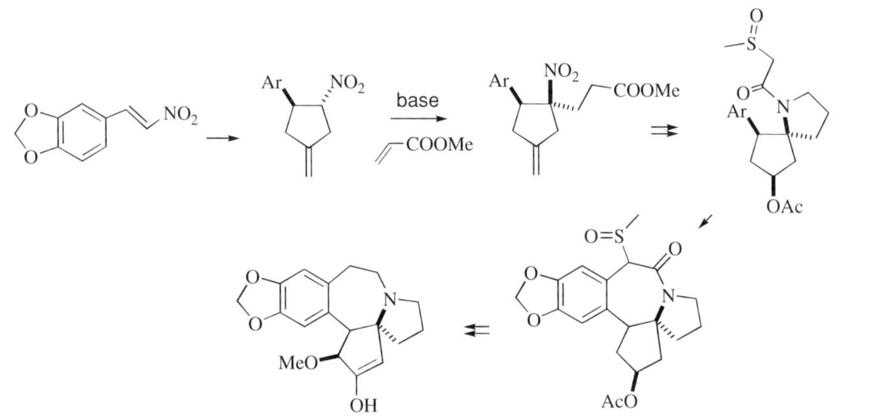

cephalotaxine (**2-6**)

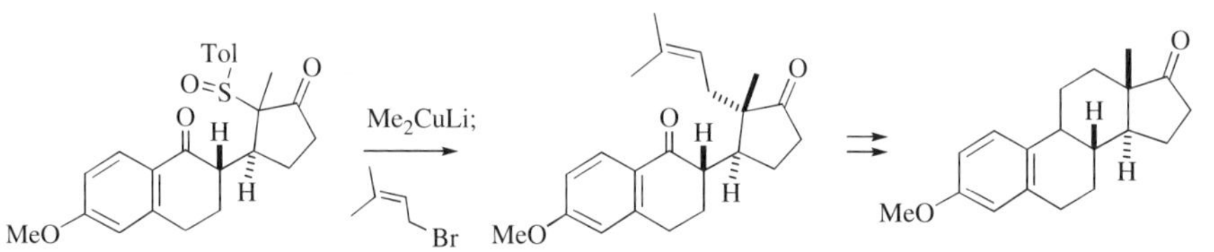

estrone methyl ether (**2-7**)

Of interest is the observation that reaction of 3-isopropenyl-2-methylcyclopentanone with methyl acrylate gave a 98:2 mixture of diastereomers favoring entry of the ester chain from the side distal to the $C_{(3)}$ substituent. This result boded well for a synthesis of confertin (**2-8**).[10] However, when the isopropenyl group was changed to a simpler allyl group, the stereoselectivity diminished drastically to 68:28.

The highly stereoselective reaction of 2-methyl-3-phenyldimethylsilyl-1-tributylstannyloxycyclopentene with the tricarbonyliron complex of protonated

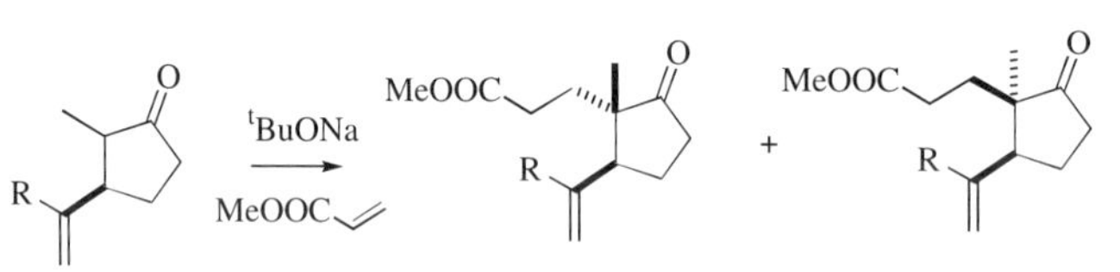

R = Me	98%	2%
R = H	68%	28%

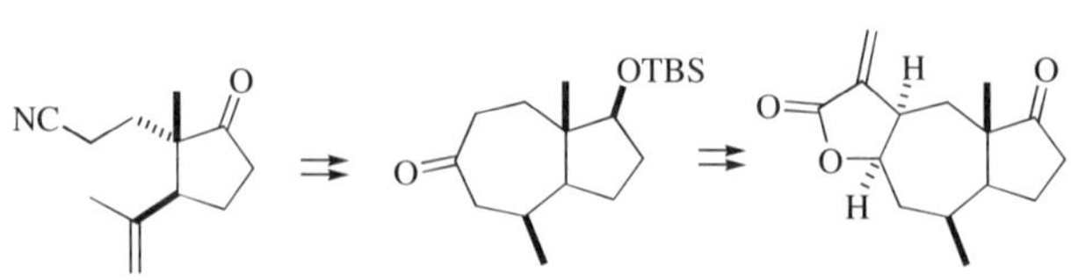

(+)-confertin
(**2-8**)

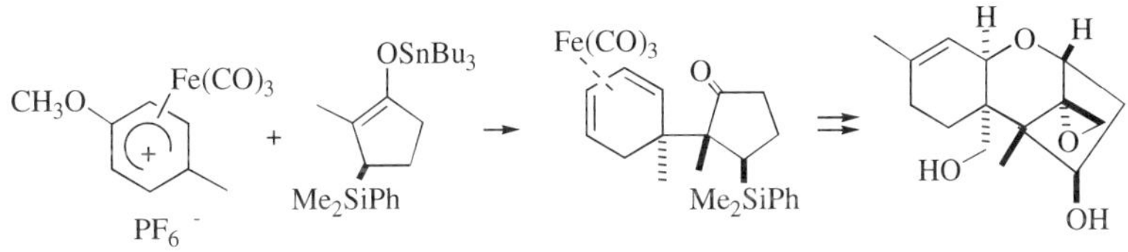

trichodermol
(**2-9**)

p-cresyl methyl ether was a good start for a synthesis of trichodermol (**2-9**).[11] The bulky silyl group, which directed the C–C bond formation, was itself to be transformed into the secondary alcohol with retention of configuration. Interestingly, the newly introduced cyclohexadienyliron complex was larger than the silyl residue and it exerted dominance over the course of α-hydroxylation of the cyclopentanone which was next in the synthetic agenda.

The spiroannulation route to β-vetivone (**2-10**)[12,13] was characterized by chemoselective intermolecular alkylation (with an allylic halide) followed by an intramolecular version. The final result reflected the effect of the secondary methyl group. Another method[14] featuring a homo-Michael reaction and internal Wittig reaction revealed stereoinduction in the first step.

An interesting phenomenon was discovered when dilithiated mono-(+)-menthyl ester of (*R,R*)-1,2-cyclopentanecarboxylic acid (**2-11**) was allylated. With allylic

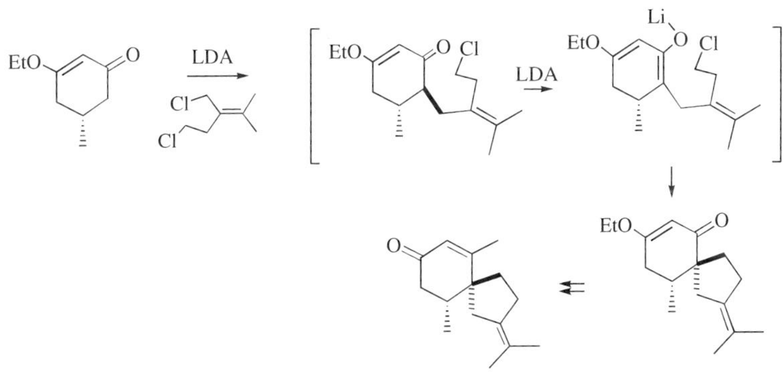

β-vetivone (**2-10**)

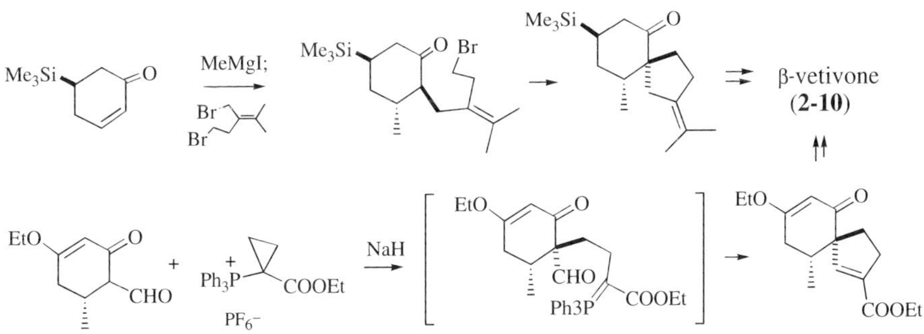

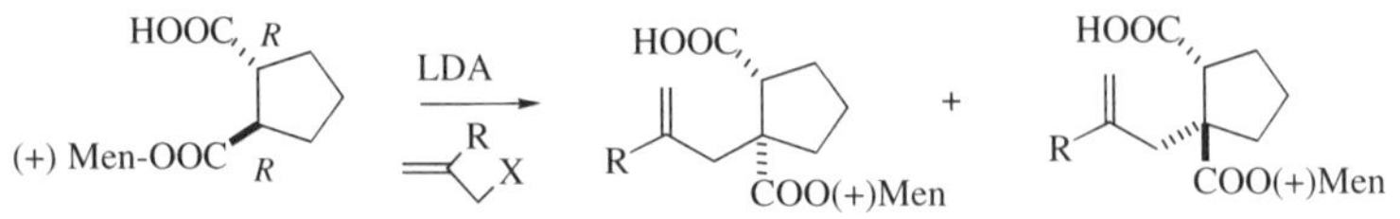

(**2-11**)

R = Me	X = Cl	1 : 4
	X = Br	3.4 : 1
R = H	X = Cl	1 : 4.9
	X = Br [LDA]	1.7 : 1
	[NaN(SiMe$_3$)$_2$]	11.2 : 1

bromides the major products were normal, but the change to allylic chlorides led predominantly to formation of contrasteric alkylation.[15] The increase in the proportion of the abnormal product with hardness of the electrophile can be explained by postulating a tighter coordination with the lithium ion. The normal product was useful for a synthesis of (–)-ptaquilosin (**2-12**).

Annulation to give a *cis*-5/6-fused dienone was accomplished in a single step by a similar reaction. The ketone product was converted to bilobalide (**2-13**)[16] and it already contained all of the skeletal carbon atoms of the target molecule.

If the Mannich reaction is considered to be an alkylation process, mention may be made of a case of 1,2-stereoinduction in the synthesis of dasycarpidone (**2-14**).[17] Protonation of the Δ^2-piperideine led to the more stable tetrahydropyridinium ion which then underwent cyclization.

Intramolecular interception of cyclic *N*-(3-butenyl)acyliminium ions forms izidinones. If the β-position of the alkenyl chain is fully substituted or bonded to a cation-stabilizing group (e.g., phenyl, methoxy), then the pyrrolizidinones are the major products. Trapping from the rearranged side chain gave rise to more stable intermediates. Combined with 1,2-stereoinduction by an existing substituent, this phenomenon allowed access to alkaloids such as (–)-hastanecine (**2-15**).[18]

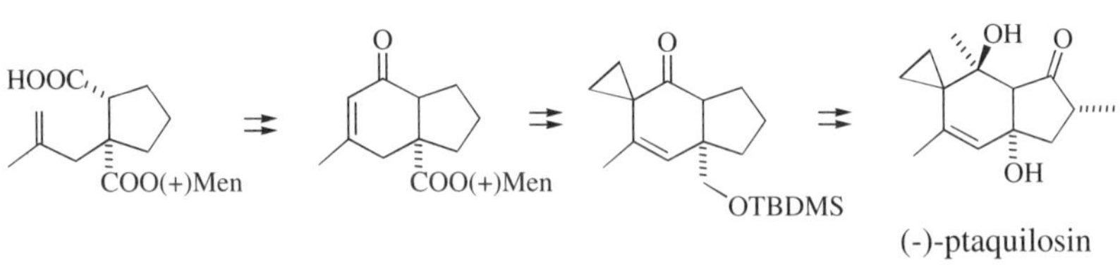

(-)-ptaquilosin
(**2-12**)

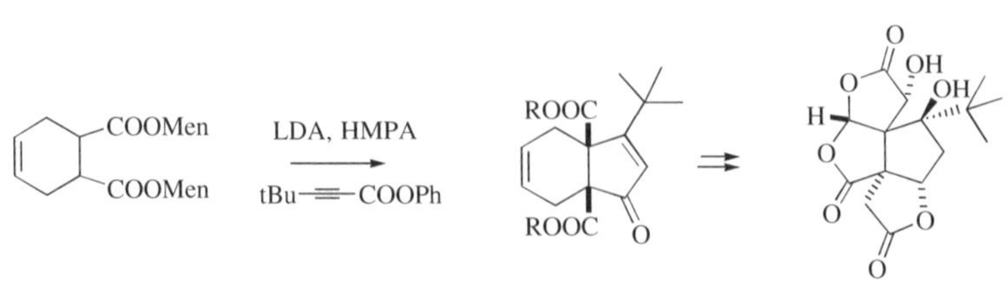

bilobalide (**2-13**)

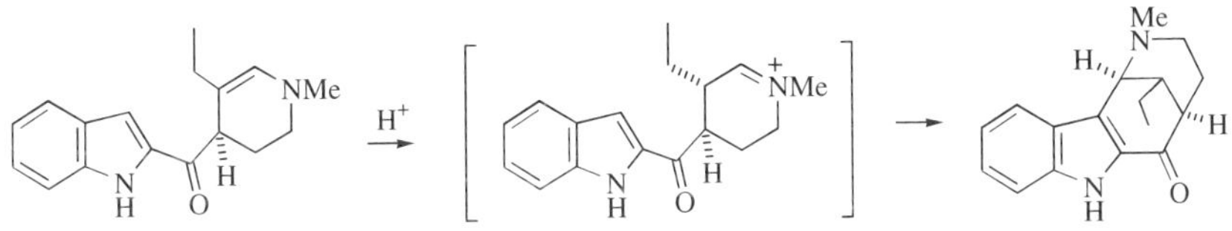

dasycarpidone
(**2-14**)

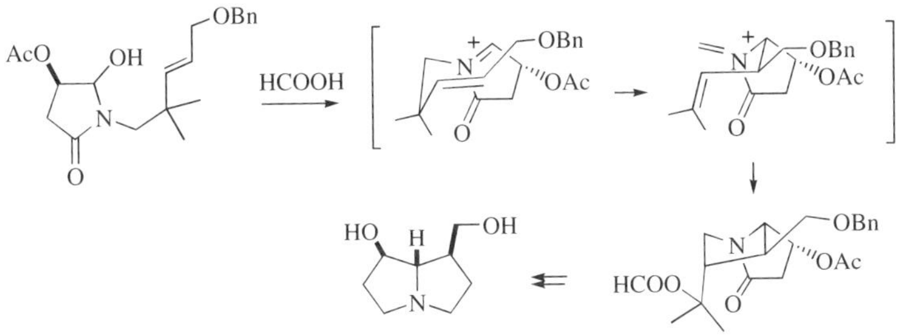

(-)-hastanecine
(**2-15**)

2.2. ADDITIONS

When an acceptor for conjugate addition presents a reaction center whose two sides are of different steric arrangements, stereoselectivity would be displayed. Thus, in emetine (**2-16**) the C-ring is vicinally *trans*-disubstituted with a tetrahydro-isoquinolinylmethyl residue and an ethyl group. Considering a Bischler–Napieralski cyclization–reduction sequence for ring closure from a B-seco precursor, deletion of the large substituent simplifies the synthesis of emetine to an unsaturated lactam subtarget. Apparently the guidance of the Michael addition by the ethyl group would lead to a suitable intermediate. Such expectation is well grounded and rendered highly

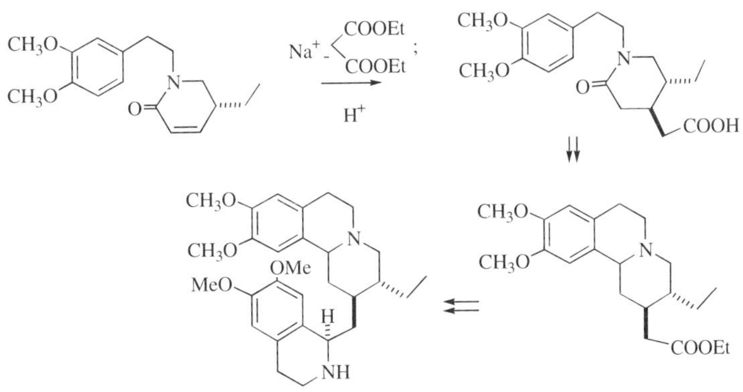

emetine (**2-16**)

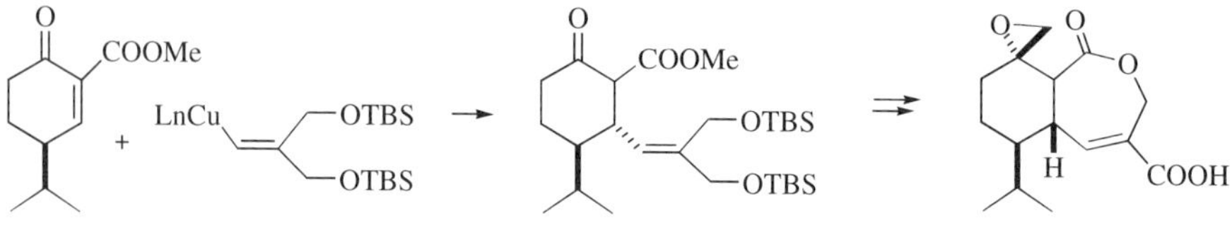

heptelidic acid
(**2-17**)

plausible using a Michael donor like malonate anion.[19] By virtue of reversibility the generation of a more stable product is ensured.

When heptelidic acid (**2-17**) is disconnected at the epoxide site, an all-*trans* 2,3,4-trisubstituted cyclohexanone is unveiled. Thus, a scheme for its synthesis from cryptone or the α-carboxylic ester presents itself.[20,21]

Preparation of the diene substrate for cationic cyclization, which generated an intermediate for a synthesis of nagilactone-F (**2-18**),[22] was delegated to a conjugate addition to an unsaturated δ-lactone. Regardless of the major reason for 1,2-stereoinduction, whether it be a stereoelectronic effect or steric effect which operated individually on reaction through two different conformations of the lactone (i.e., with the acetal side chain pseudoaxial and pseudoequatorial, respectively), the result was beneficial. Interestingly, the allylic substituent (acetal chain) provided the steric steering until the AB-ring system was complete, and its chirality was subsequently destroyed.

γ-Hydroxymethyl-2-butenolide is a widely used building block for synthesis. When derivatized with a bulky group (e.g., trityl, [*t*-butyl]dimethylsilyl), the outstanding suppression of reactions on the same face as the pendant gives functionalization at both the α- and β-positions of the lactone with clearcut results. A recent example describes the photoinitiated aminoalkylation[23] which furnished a product correlative to lindelofidene (**2-19**).

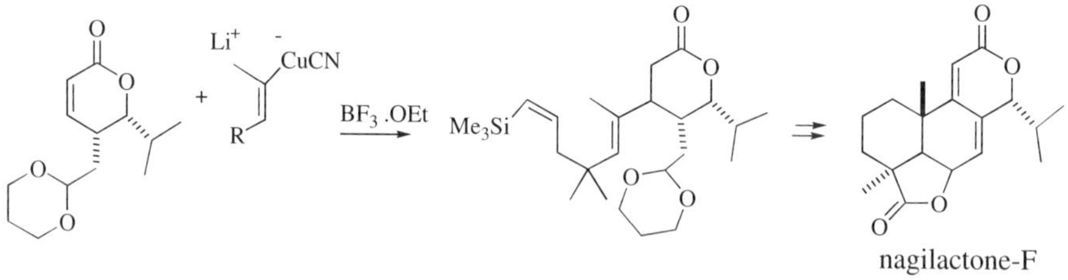

nagilactone-F
(**2-18**)

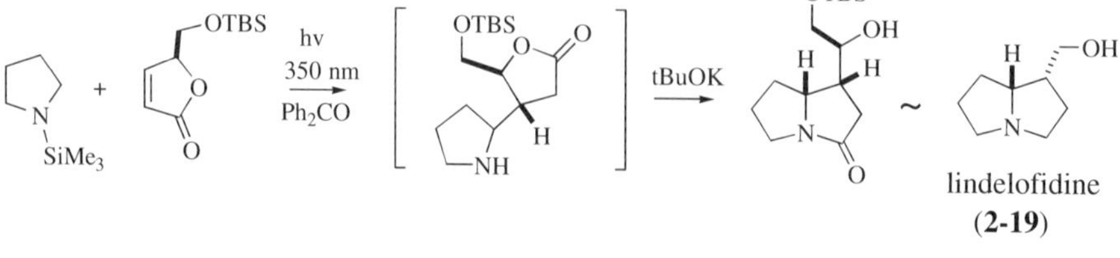

lindelofidine
(**2-19**)

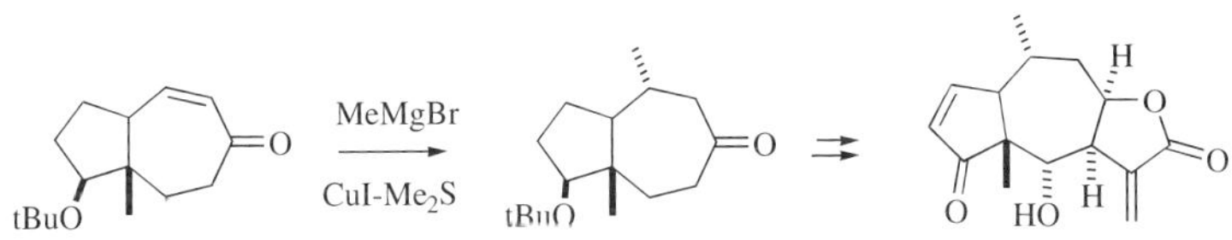

87%

helenalin
(**2-20**)

Michael acceptors embedded in more rigid systems behave even better stereochemically. The 10α-methyl group of helenalin (**2-20**) was readily introduced by a Kharasch reaction on a hydrazulenone.[24] The small methine hydrogen adjacent to the incident site was α-oriented.

11-Oxygenated steroids such as cortisone (**2-21**) are very important therapeutic agents and their chemical synthesis has attracted much attention. Modern strategic thinking declares the preference of building blocks containing the functionality and uses it to direct the construction process. The proper *trans*-hydrindenone fulfills the need[25,26] to evolve plans which are based on conjugate addition to attach structural elements for completing the AB-ring component. The conjugate addition serves to establish the correct configuration of $C_{(8)}$ as the approach of the nucleophilic reagent from the same side of the angular methine hydrogen is sterically favored. Of course the steric course also avoids incipient 1,3-diaxial interaction between the nucleophile and the angular methyl group. But it should also be noted that promotion by chlorotrimethylsilane was counterproductive in terms of stereoselectivity, as it led to a 1:1 stereoisomeric mixture.

An interesting finding which can be readily rationalized is the formation of a racemic oxopyrrolizidine (**2-22**) nitrile with retention of configuration on the reaction of a chiral *exo*-mesylate with sodium cyanide.[27] A pathway through elimination and Michael addition is probably uncontestable. At higher reaction temperature the cyano group was translocated to the angular position.

Reaction at an sp^2-carbon adjacent to a fully substituted ring junction often forms a new bond from the side opposite the existing angular substituent. Such a course was critical for synthesis of ceroplastol-I[28] because the new stereocenter could not be modified without undue manipulations.

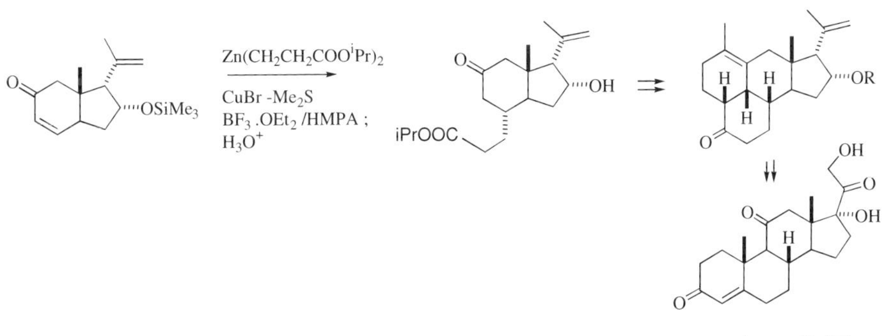

cortisone (**2-21**)

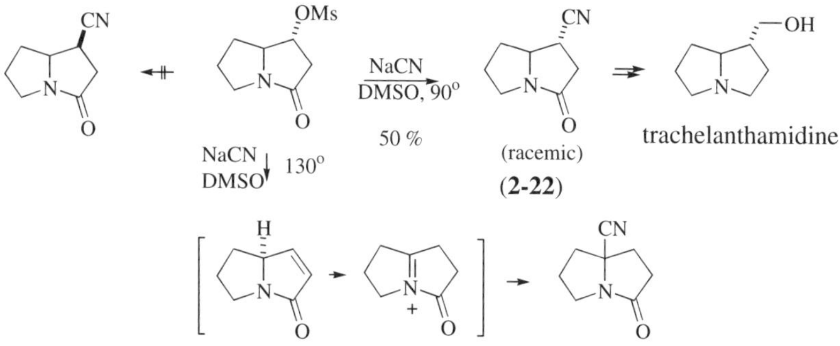

trachelanthamidine

(racemic)
(**2-22**)

A relatively simple solution for access to a stereohomogeneous intermediate of zizaene (**2-23**)[29] involved introduction of the bridging cyclopentane unit by conjugate vinylation of a hydrindenone in which the secondary methyl was already present. In fact, this subangular methyl group determined the entry of the vinyl group and hence the elaboration of the entire skeleton.

A similar type of reaction initiated a synthesis of acorenone-B (**2-24**).[30] Interestingly, the cyclic ketone group was to be devolved as an ester and eventually converted to an isopropyl pendant with the correct configuration.

Vicinal difunctionalization of enones such as tandem conjugated addition–electrophilation is a synthetically valuable operation.[31] For cycloalkenones the appended groups tend to be *trans*, due to 1,2-stereoinduction from the first entrant. This stereochemical outcome is favored by both kinetic and thermodynamic considerations. Preparation of the diquinane enone intermediate of quadrone (**2-25**)[32] was initiated by this protocol, and among the numerous other examples that can be

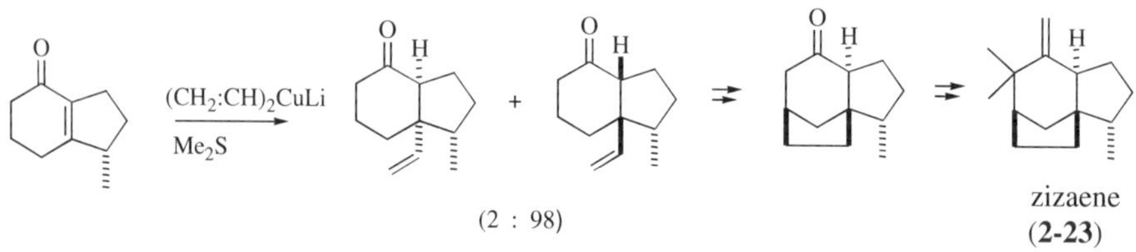

zizaene
(**2-23**)

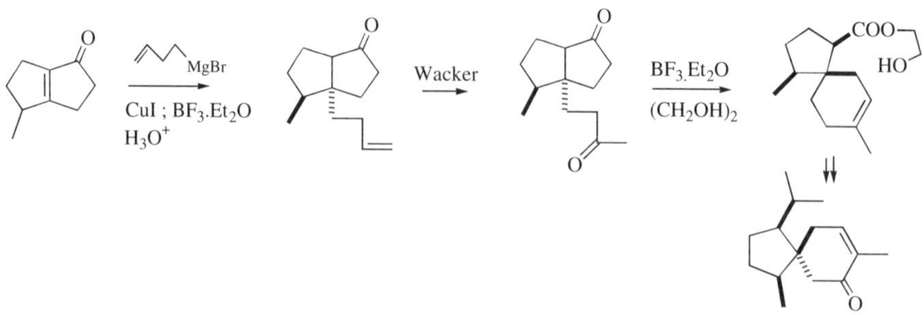

acorenone-B (**2-24**)

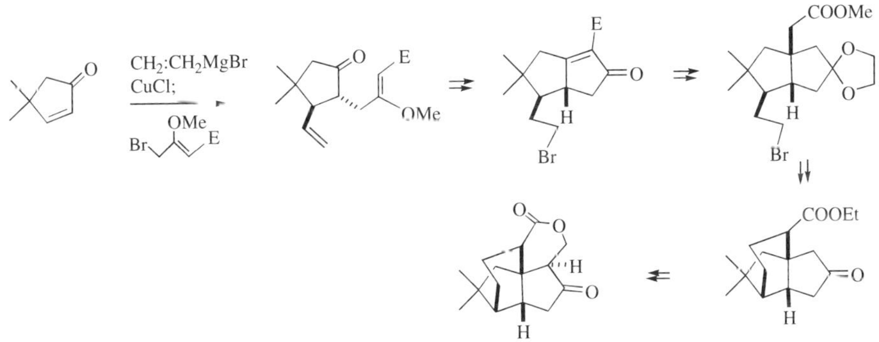

quadrone (**2-25**)

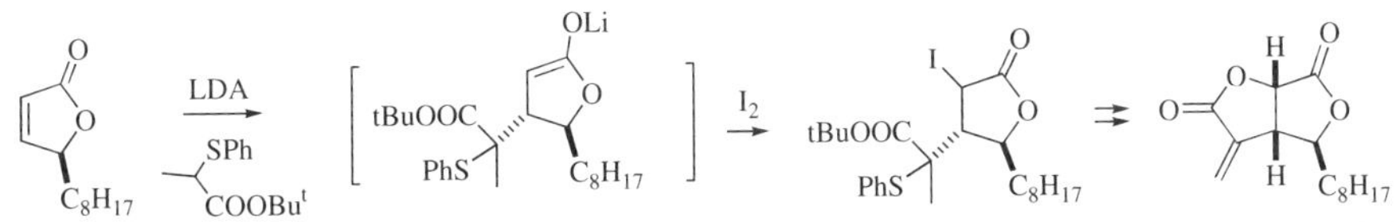

avenaciolide (**2-26**)

found in the chemical literature one might arbitrarily indicate those involved in the synthesis of avenaciolide (**2-26**)[33] and ethyl *trans*-jasmonate (**2-27**).[34]

Internal delivery of borane to a tetrasubstituted double bond resolved the regiochemical as well as stereoselectivity problems concerning the synthesis of curcumanolide-A (**2-28**).[35] The isopropylol group at the allylic position exerted the required 1,2-stereoinduction.

Epoxidation of a tricyclic olefin was en route to presilphiperfolan-9-ol (**2-29**).[36] This reaction occurred *cis* to the methine hydrogen of an adjacent angular position. Rearrangement of the epoxide was next performed to install the last stereocenter at the ring junctions. The ketone is less stable than the diastereomer with a *cis*-5/6 ring fusion and basic conditions must be avoided during the subsequent olefination.

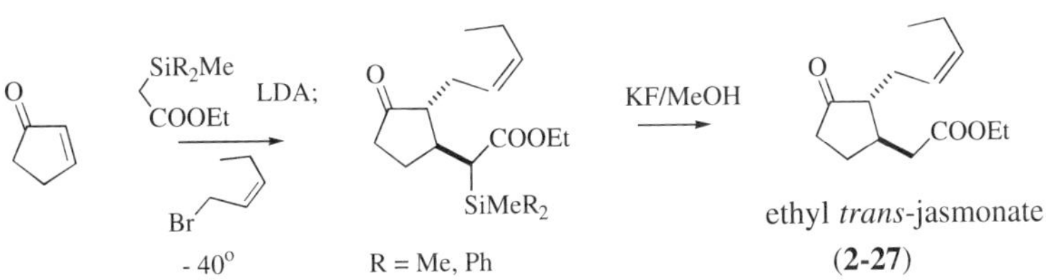

ethyl *trans*-jasmonate (**2-27**)

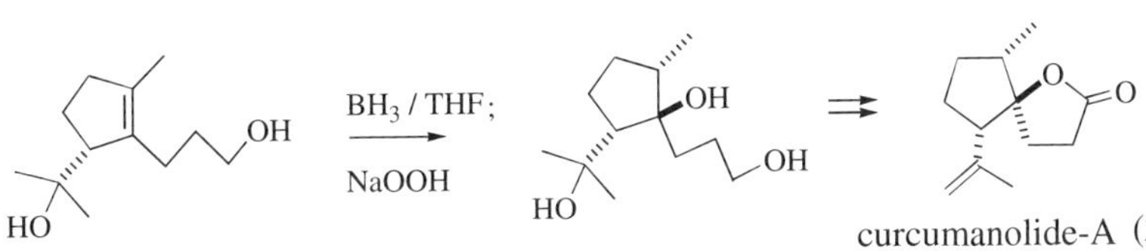

curcumanolide-A (**2-28**)

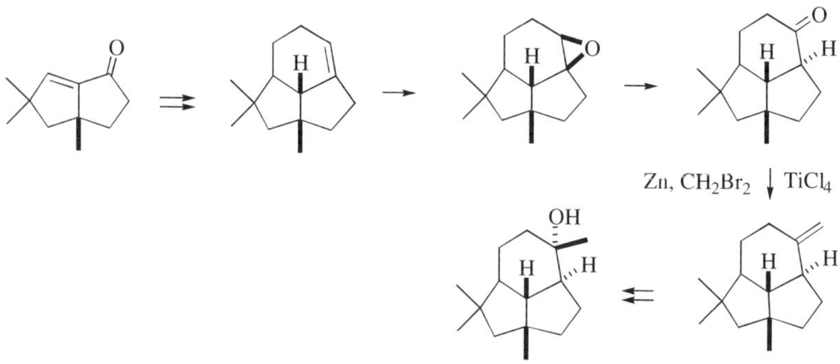

presilphiperfolan-9-ol
(**2-29**)

An allylic silyl group is also effective for inducing stereoselective reactions of the double bond. Thus, dihydroxylation with osmium tetroxide can be so executed, and the use of Sharpless reagents leads to chiral *cis*-1,2-diols. One application of this method is in the synthesis of (+)-conduritol-E (**2-30**).[37]

Naturally a *t*-butyldimethylsiloxy group has the same effect and a *cis,trans*-1,2,3-triol array such as that present in shikimic acid is available from the allyl silyl ether.[38,39] Methods for generating the precursor of methyl shikimate (**2-31**) can be quite diversified.

Dibromocarbene addition to a similar diether followed by conversion of the bromine atoms to methyl groups led to a useful precursor (**2-32**) of *cis*-pyrethroid insecticides.[40] Incorporation of the fused *gem*-dimethylcyclopropane unit to a more complex system as required for synthesis of phorbol[41] was also successfully performed owing to the characteristic fusion mode of the cyclohexene with the

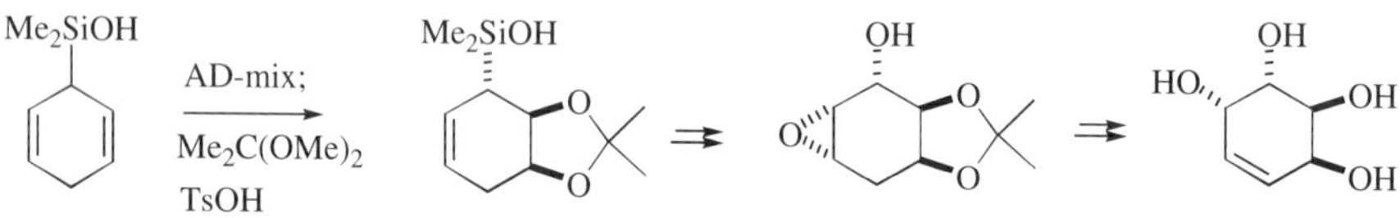

(+)-conduritol-E
(**2-30**)

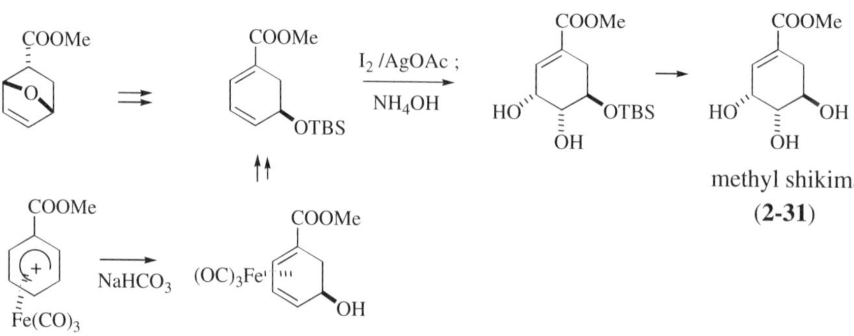

methyl shikimate
(**2-31**)

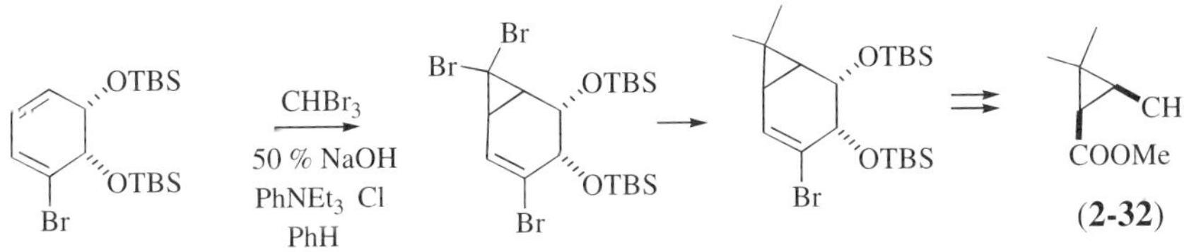

(**2-32**)

oxabicyclooctane. It is reasonable to expect that bond formation was selectively *syn* to the angular hydrogen but *anti* to the oxygen bridge. Because the C–O bond is axial to the cyclohexene, it is an impediment to external reagents approaching from its direction.

For the conversion of a 2-(*p*-anisyl)methyl-3-pyrroline derivative to anisomycin (**2-33**), direct epoxidation is unwarrantable. The *cis*-epoxide must be obtained in another manner such as via the iodohydrin.[42] Iodohydrination, like epoxidation, was initiated by attack of the electrophilic species on the opposite side of the anisylmethyl group.

Catalytic hydrogenation of a double bond adjacent to a stereocenter generally occurs on a surface distant from the largest group. On this premise, a *cis*-3,4-disubstituted *N*-methylpiperidine suitable for elaboration of tubifoline (**2-34**)[43] was acquired by reduction of the 4-alkylidene-3-ethyl derivative.

The rhodium(I)-catalyzed intramolecular addition of a formyl group across a double bond to form a cyclopentanone has been shown to exhibit good

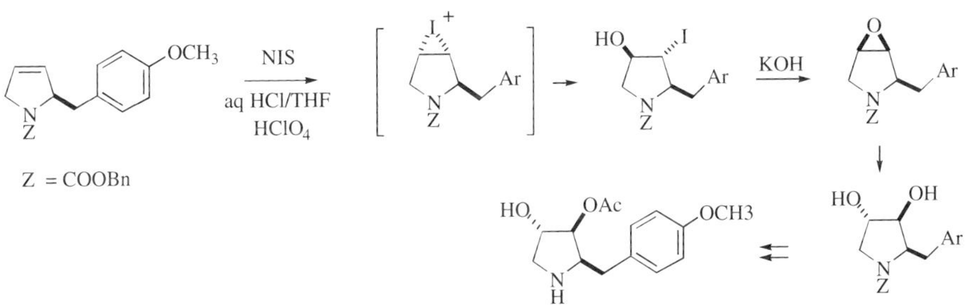

anisomycin (**2-33**)

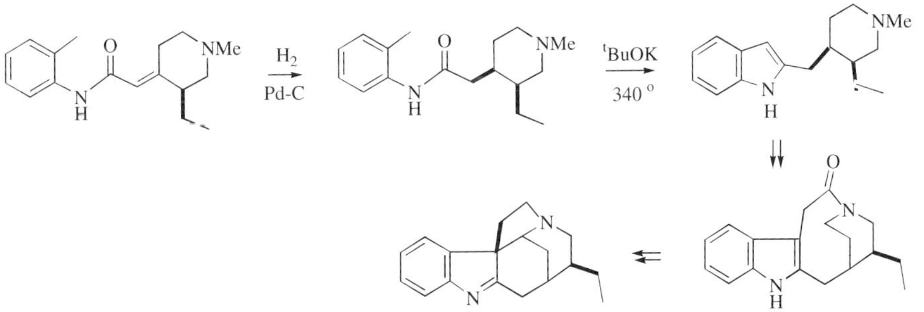

tubifoline (**2-34**)

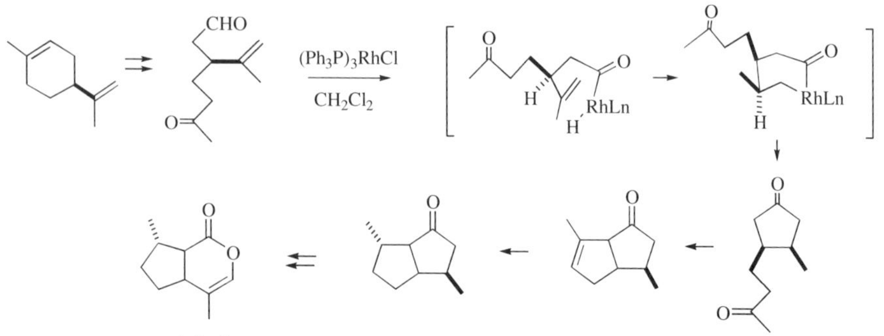

nepetalactone
(**2-35**)

stereoselectivity. This reaction has been applied to the ketoaldehyde derived from limonene to afford *cis*-3-methyl-4-(3-oxobutyl)cyclopentanone.[44] The stereochemistry was confirmed by conversion of the product to nepetalactone (**2-35**).

The preferred transition state can be explained on the basis of a torsional effect. To form the *trans*-isomer the double bond of the acylrhodium hydride must rotate to force an eclipse of the methyl group with the four-carbon chain. On the other hand, the observed product can arise without suffering from such adverse interaction in the transition state leading to the metallocycle intermediate.

2.3. CYCLOADDITIONS

The Diels–Alder reaction of a 4-substituted 2-cyclohexenone is subject to stereoinduction so that the diene tends to approach from the opposite side of the 4-substituent. This reaction characteristic is well suited for the assembly of bicyclic intermediates of sesquiterpenes such as dehydrofukinone (**2-36**).[45] Actually, to compromise on reactivity an ester group was placed at $C_{(3)}$ of the dienophile as surrogate to a methyl group. Because of this the overall effectiveness of the strategy was diminished by the requirement of maneuvers to readjust the oxidation level of the angular substituent.

Formation of a bridged ring system while under the steric influence of an allylic substituent in the diene was predictably consistent. A very concise approach to nootkatone (**2-37**)[46] was based on this principle to set up the *vic-cis*-dimethyl pattern.

In a synthesis of quassin (**2-38**),[47] three new stereocenters were created in the formation of the tricarbocyclic intermediate. The inappropriately oriented angular

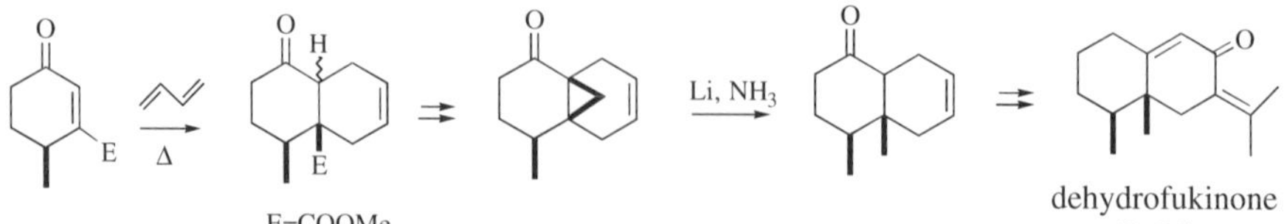

dehydrofukinone
(**2-36**)

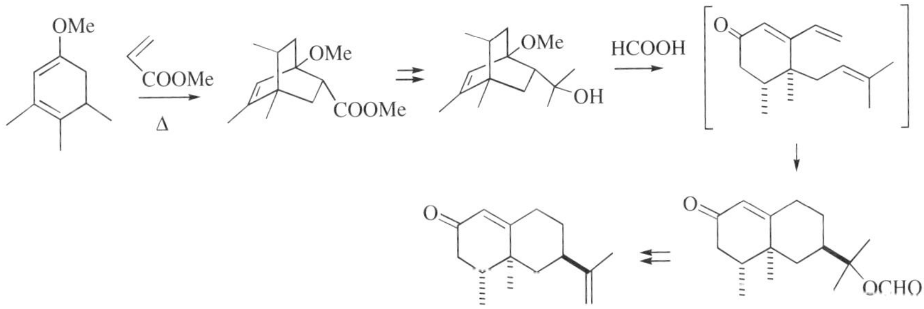

nootkatone
(**2-37**)

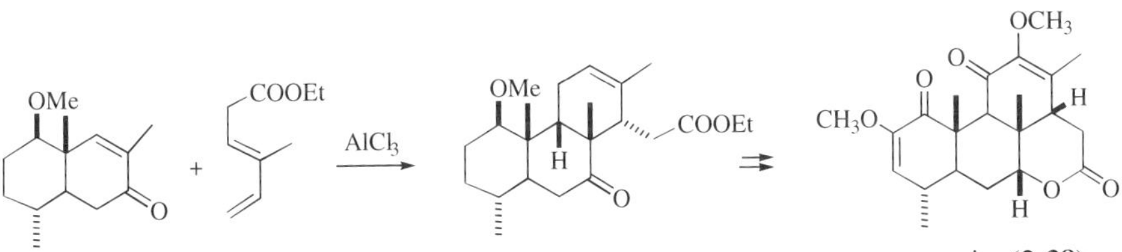

quassin (**2-38**)

methine hydrogen was to be corrected automatically on introduction of a ketone group at its adjacency. In this case the concerted and suprafacial nature of the Diels–Alder reaction was actually used to establish a new quaternary carbon site. In other words, the angular methyl group of the octalone steered the rehybridization of the vinylic methyl to its own side.

An intramolecular Diels–Alder process led to an intermediate of forskolin (**2-39**).[48,49] As expected, *cis*-5/6 ring fusion was heavily favored.

Himgravine (**2-40**) has a structure which is suggestive of an intramolecular Diels–Alder approach.[50] The adoption of a less crowded *endo* transition state, in

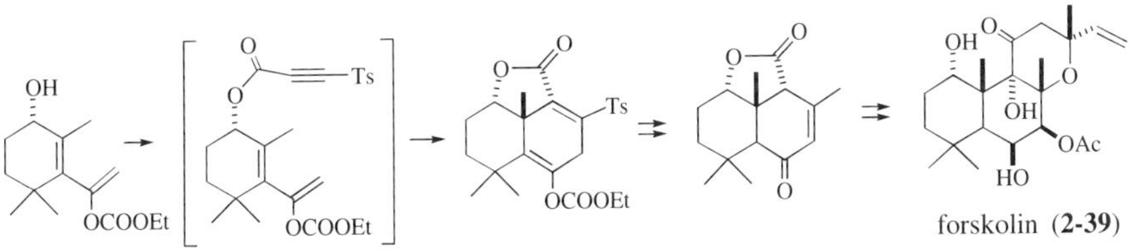

forskolin (**2-39**)

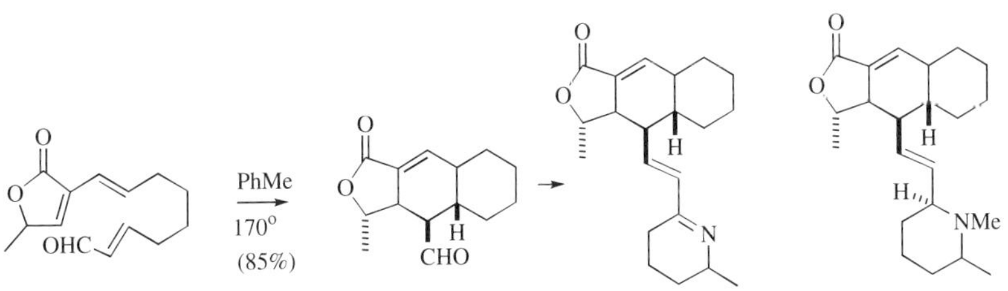

major

himgravine (**2-40**)

which the chain containing the (*E*)-dienophile folds over the diene moiety from the face away from the secondary methyl group of the butenolide, would give rise to an ideal precursor. Thus, a single point of asymmetry decrees generation of a correct array of four contiguous stereocenters.

The cytochalasans are challenging synthetic targets. The perhydroisoindole moiety which contains the most stereocenters deservedly attracts attention of the highest order. Fortunately, this molecular portion is also very amenable to elaboration by a Diels–Alder reaction between a diene and an unsaturated γ-lactam.[51] The benzyl group of cytochalasan-B (**2-41**) is *cis* to the angular methine hydrogen and *trans* to the methyl three bonds away, forming a stereochemical array which can be generated by an *endo*-selective Diels–Alder reaction, if simultaneous 1,2- and 1,3-stereoinductions operate. The Diels–Alder approach is even more justifiable by the fact that the configuration of the carbon atom common to both the macrocyclic ring and the cyclohexane correlates to a more readily accessible (*E,E*)-diene.

The convenient generation of a pure cyclopenta[*d*]pyridazine with two benzylic methyl groups in a *cis* disposition is very useful for completion of a *cis*-trikentrin (**2-42**) synthesis.[52] An enamine of 2,4-dimethylcyclopentanone was found to be enriched (3:1) in the *cis* form, yet the pyridazine prepared from the mixture was sterically homogeneous. There must be a kinetic preference for the Diels–Alder reaction with the *cis*-enamine, and the *trans*-isomer was depleted by continuing equilibration. The selection can be attributed to hindrance to approach of the diene by the allylic methyl group.

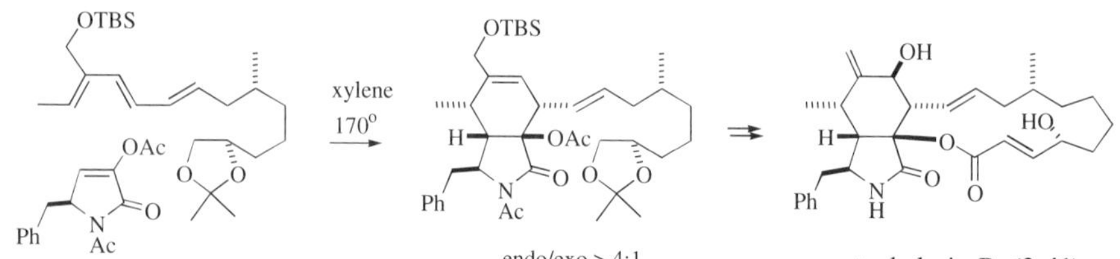

cytochalasin-B (**2-41**)

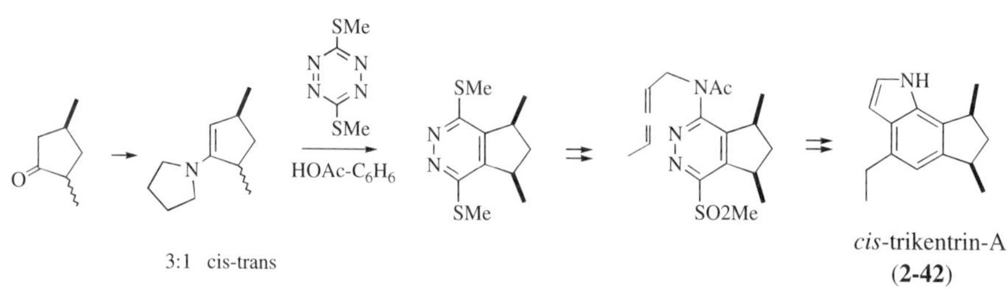

cis-trikentrin-A
(**2-42**)

The 1,2-/1,2′-stereoinduction of 5-substituted 1,3-cyclopentadienes in Diels–Alder reactions has been implicitly recognized for a long time. A most significant exploitation of this phenomenon may be in prostaglandin synthesis.[53] A chiral norbornene adduct (**2-43**) can be prepared[54,55] by the addition of a C_2-symmetric diazaaluminolidine catalyst.

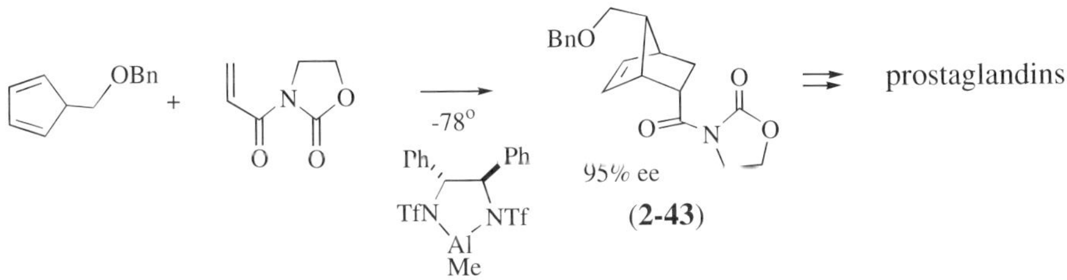

(2-43)

Generally the [2+2]-photocycloaddition is quite stereoselective and it manifests the effect of proximal substituents in the addend(s). Thus, when the method was applied to a synthesis of α- and β-bourbonene,[56] the only defect was of regiochemical origin. The allylic isopropyl group determined the approach of 2-cyclopentenone.

Synthetic design incorporating an intramolecular [2+2]-photocycloaddition step has been very popular because many poorly accessible ring systems can be obtained and delicate stereochemical issues may be settled. Whereas the angular triquinane sesquiterpene isocomene (**2-44**) contains four stereocenters, a photochemical route[57] employed a cyclohexenone with only one such center.

Another related sesquiterpene is pentalenic acid (**2-45**). A synthesis which featured cleavage of the photocycloadduct but not a molecular rearrangement has been developed.[58] Excellent 1,2-stereoinduction was observed.

In another case, the spirocyclic structure was actually present in the target, α-acoradiene (**2-46**).[59] Thus, the photocycloaddition and fragmentation combination was very accommodating.

Sometimes the major photocycloadduct is not the desired diastereomer for conversion into the target molecule. For example, the transition state leading to the *endo*-methyl isomer was disfavored due to interaction with an oxy functionality present in the enone moiety. Such epimers correspond to bulnesene and precapnelladiene, respectively, and the routes pursued thus led to epibulnesene (**2-47**)[60] and epiprecapnelladiene[61] predominantly or practically exclusively.

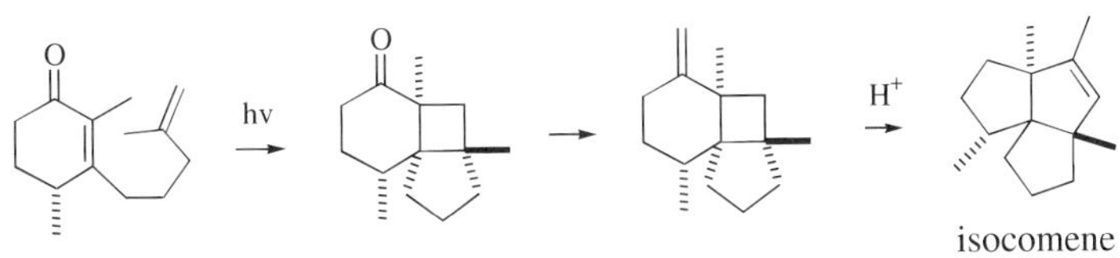

isocomene
(**2-44**)

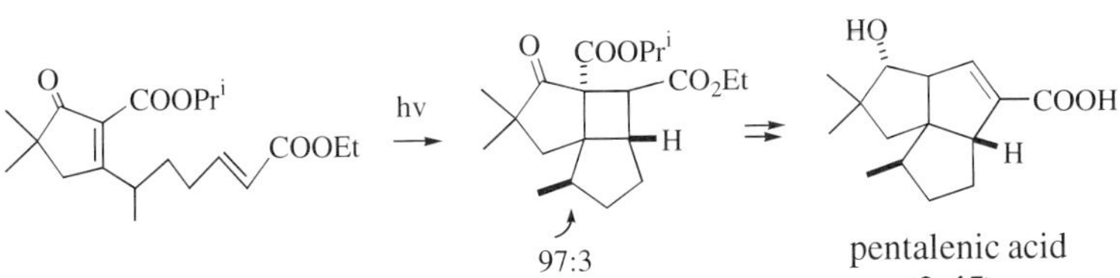

pentalenic acid
(**2-45**)

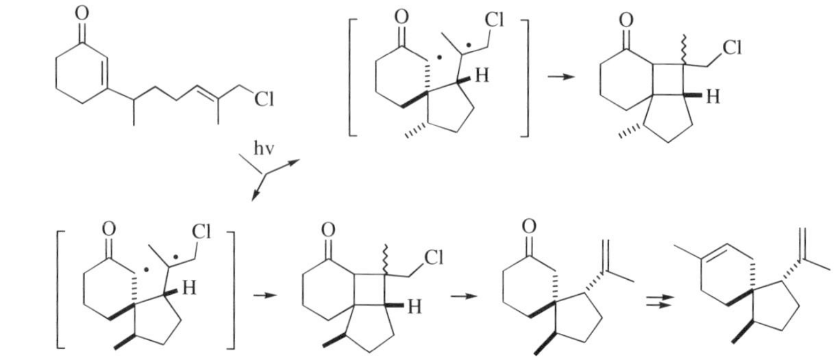

α-acoradiene
(**2-46**)

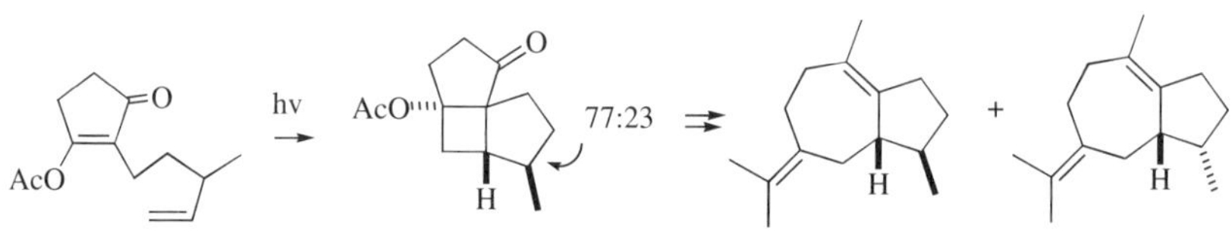

epibulnesene (**2-47**) bulnesene

Thermally induced [2+2]-cycloadditions involving ketenes are complementary to the photochemical method for the formation of cyclobutanes. If cyclobutanones are desirable, the ketene route may be more direct. A tricyclic compound suitable for the elaboration of (–)-silphiperfol-6-ene (**2-48**) has been acquired[62] in a rather straightforward manner. All four stereocenters were set in place in the cycloaddition step.

Nickel-catalyzed intramolecular [4+4]-cycloaddition gives 1,5-cyclooctadienes. To extrapolate this process to a synthesis of (+)-asteriscanolide (**2-49**),[63] one of the diene units was incorporated as an α-vinylbutenolide while the other was linked to the

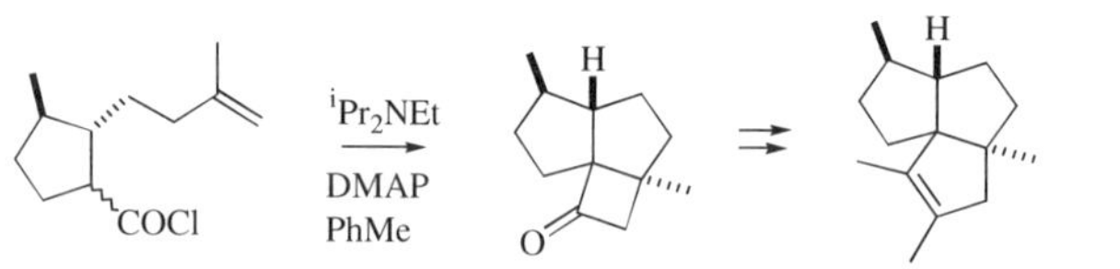

(-)-silphiperfol-6-ene
(**2-48**)

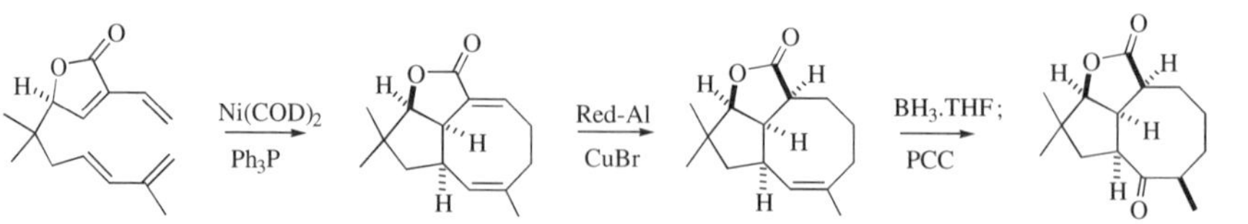

(+)-asteriscanolide
(**2-49**)

γ-position of the heterocycle. The stereochemistry at the ring junctions was completely established during the cycloaddition: *cis*-5/5 fusion and further extension to the next carbon atom as a result of mechanistic requirement.

2.4. OTHER REACTIONS

The preparation of an intermediate for the synthesis of trehazolin (**2-50**) was attempted by reduction of a symmetrical oxime ether with diborane.[64] The undesirable epimer was formed predominantly (ratio 12:1), probably due to the combined factor of precapture of the reagent by the adjacent hydroxyl groups and the overwhelming favoritism for *exo* attack on the diquinane nucleus. The situation was reversed (to a ratio of 1:6) by changing the two hydroxyl groups to the corresponding *t*-butyldimethylsilyl ethers.

On two occasions, 1,2-stereoinduction guided a successful synthetic venture in gloiosiphone-A dimethyl ether (**2-51**).[65] First the reduction of the enedione system was rendered regioselective by the difference in the hybridization state of the two carbon atoms adjacent to the spirocyclic center. In return the desymmetrized ring containing the enone conferred stereoselectivity in the dihydroxylation step.

The [2,3]-Wittig rearrangement of allyloxy carbanions has been developed into a valuable method for C–C bond formation.[66] In principle the steric course can be influenced by a stereocenter at the other (nonoxygenated) allylic position and certain diastereomer may be formed preferentially. Due to adoption of a Felkin–Ahn-type transition state, chiral products (**2-52, 2-53**) for a projected synthesis of amphotericin-B become readily available.[67]

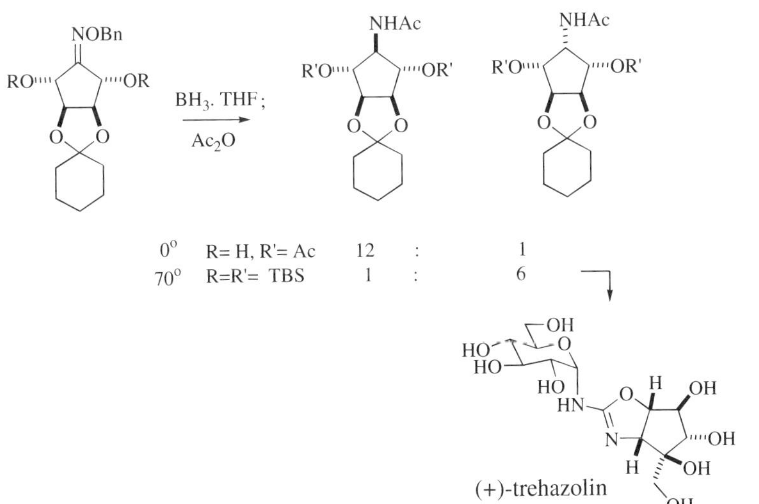

(+)-trehazolin
(**2-50**)

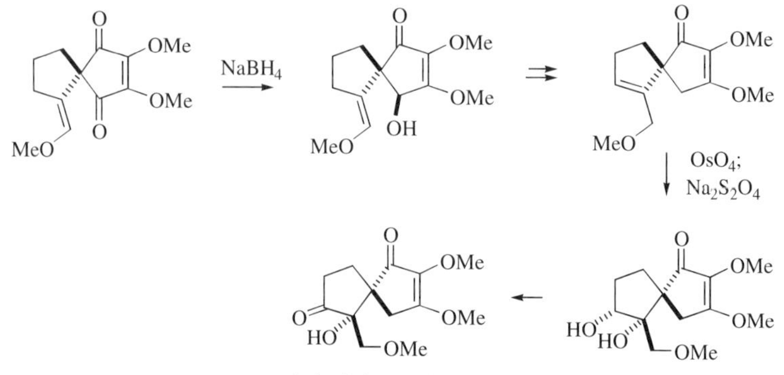

gloiosiphone-A dimethyl ether (**2-51**)

An alternative method leading to Stork's prostaglandin intermediate (**2-54**) employed a [2,3]-Wittig rearrangement of a bis(*t*-butyldimethylsiloxy)cyclopentenylmethyl propargyl ether.[68] A single isomer was obtained in that step.

The Claisen rearrangement and its variants are strongly disposed to generate (*E*)-alkenes because the large carbinyl substituent prefers to occupy a pseudoequatorial position in the transition state. It is interesting to note that bulky Lewis acids such as methylaluminum bis(4-bromo-2,6-di-*t*-butylphenoxide) promote the formation of the (*Z*)-alkenes.[69] A complexed allyl vinyl ether is forced to place its more sterically demanding group in a pseudoaxial orientation to avoid excessive congestion.

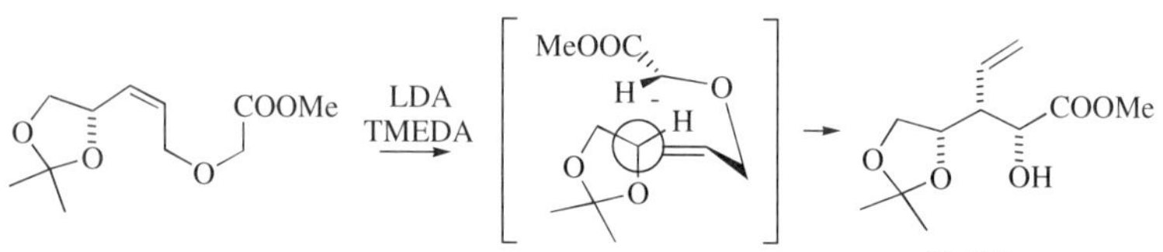

(**2-52**)

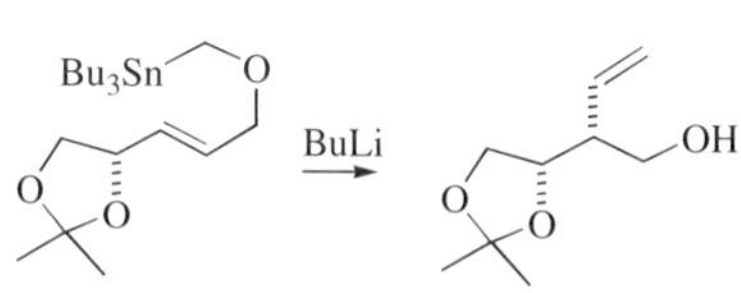

(**2-53**)

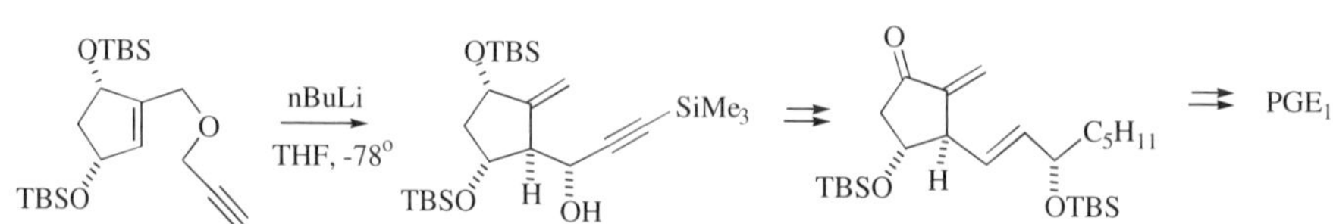

(**2-54**)

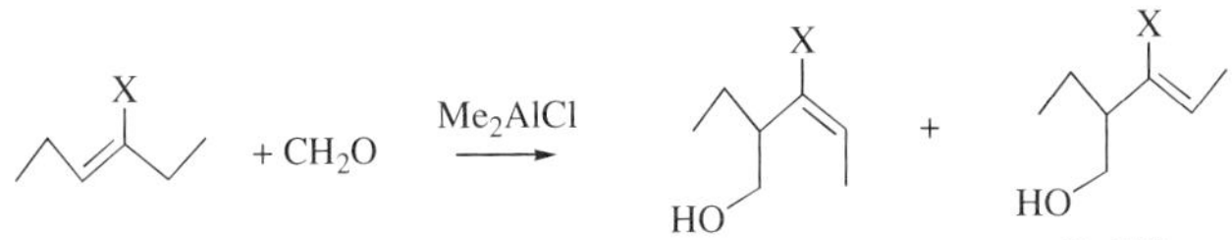

(**2-55**)

X			
X = H	15	:	85
X = $SiMe_3$	98	:	2

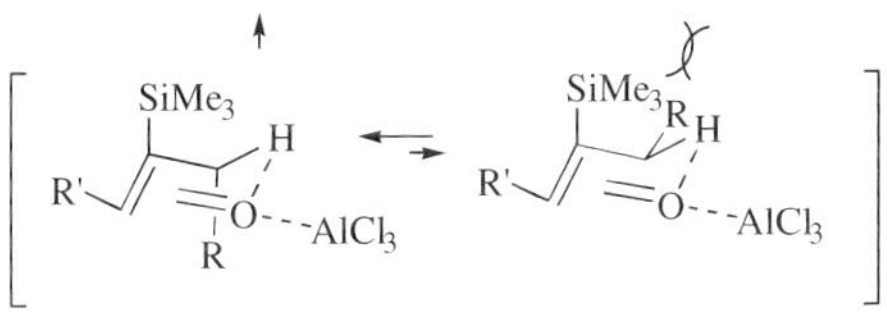

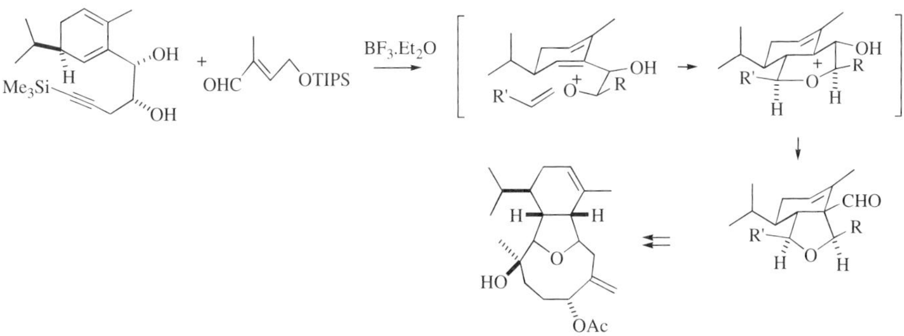

(-)-7-deacetoxyalcyonin acetate (**2-56**)

The double bond of the major ene reaction product (**2-55**) from an aldehyde and alkene is *trans*. However, a changeover to *cis*-selectivity can be made by using a silylalkene.[70] Synclinal arrangement of the silyl group and the main chain causes allylic 1,2-strain.

An important stereochemical issue in the annexation of the tetrahydrofuran ring to the *p*-menthene nucleus for a synthesis of (–)-7-deacetoxyalcyonin acetate (**2-56**)[71]

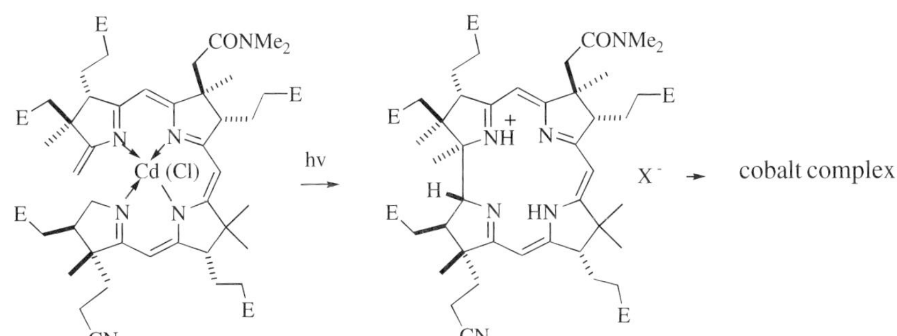

E = COOMe

(**2-57**)

was settled by the direction of the isopropyl group. An intramolecularized Prins reaction proceeded accordingly, and with the succeeding rearrangement the first stage assembly was satisfactorily concluded.

Perhaps the most remarkable 1,2-stereoinduction is that found in the photoinduced cyclization (sigmatropic rearrangement) of a seco-corrin (**2-57**).[72] Of the two possibilities the cyclization led to a product with the natural configuration. The acetic ester chain in the D-ring is thought to be responsible for this result.

2.5. SELF-IMPOSED 1,2-STEREOINDUCTIONS

The stereochemical consequences of intramolecular reactions which result in ring formation are particularly easy to unravel. Generally, *cis*-fused products arise from a small or common sized cyclic compound with a short chain separating the reaction site and the adjacent atom. The steric constraint can be exploited in many situations.

A popular strategy for assembly of both linear and angular triquinanes involves intramolecular alkylation, aldol condensation, ene reaction, or Nazarov cyclization of diquinane intermediates in which the cyclizing chain is *exo*. Preparation of the proper intermediates is always facilitated by topographical factors, and the various cyclization protocols indicated above invariably led to the desired products. For details the synthesis of isocomene,[73] pentalenene,[74] $\Delta^{9(12)}$-capnellene (**2-58**),[75] and silphinene (**2-59**)[76,77] may be consulted.

An intramolecular [2+2]-photocycloaddition gave a fenestrane compound from which silphinene (**2-59**) could be unraveled.[78] Opening of the cyclobutane unit

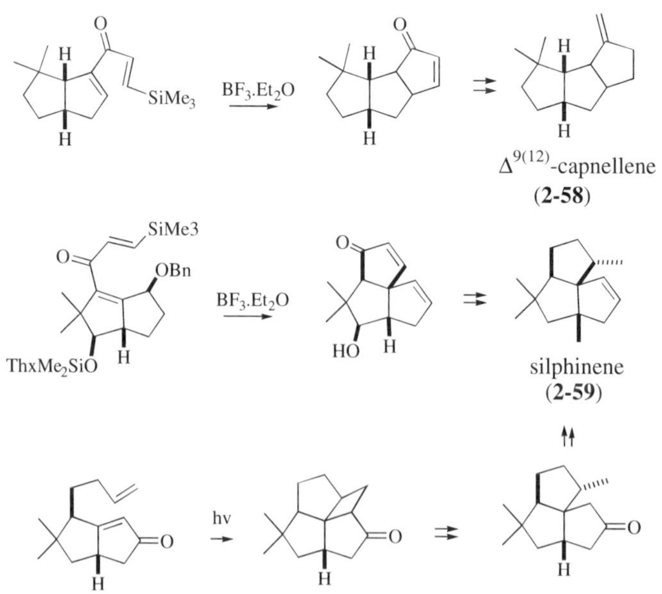

$\Delta^{9(12)}$-capnellene (**2-58**)

silphinene (**2-59**)

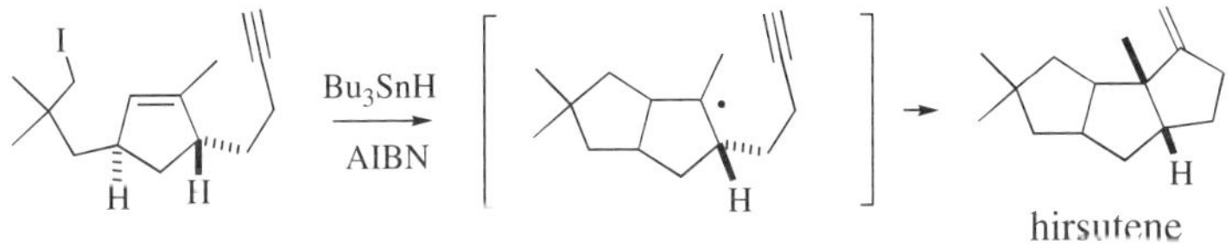

hirsutene
(**2-60**)

liberated an iodomethyl group which was promptly reduced. Thus, two contiguous stereocenters were created unambiguously and simultaneously.

Very rapid construction of triquinane motifs such as those related to hirsutene (**2-60**) and silphiperfolene (**2-61**) by relayed free radical reactions has been developed.[79–81] Each new ring was formed in *cis*-fused fashion.

The convex shape of the hydrophenanthrene nucleus of morphine (**2-62**) is accentuated by the furanoid adjunct. A tandem addition scheme can be deployed to advance a synthesis of morphine considerably. With an aminoethyl chain present in the procyclic isomer in some protected form, the cyclization would provide an intermediate of even higher attactiveness.[82,83] The premise is that these tandem processes, either ionic or free radical in nature, automatically generate the *cis*-fused products.

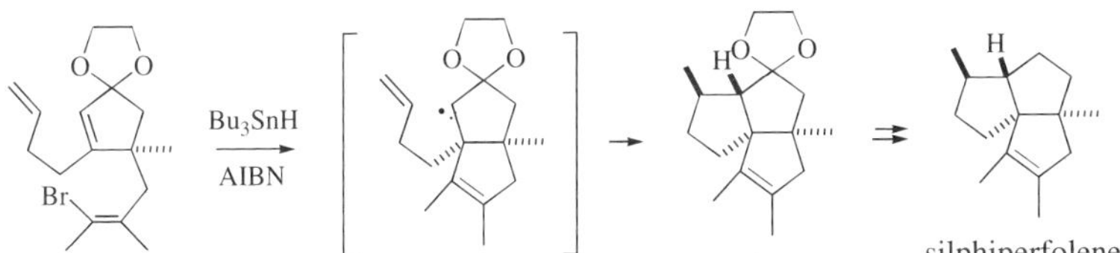

major

silphiperfolene
(**2-61**)

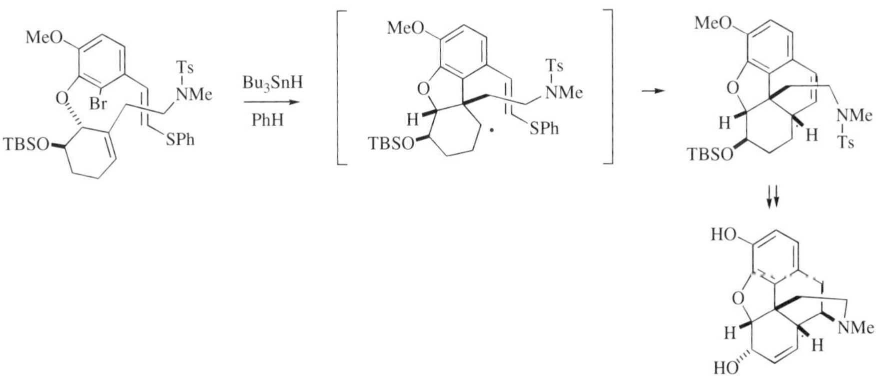

morphine
(**2-62**)

REFERENCES

1. Brown, H.C.; Ramachandran, P.V. *Adv. Asymm. Synth.* **1995**, *1*, 147.
2. Sharpless, K.B. *Tetrahedron* **1994**, *50*, 4235.
3. Kenney, W.A.; Coghlan, M.J.; Paquette, L.A. *J. Am. Chem. Soc.* **1985**, *107*, 7352.
4. Piers, E.; Smillie, R.D. *J. Org. Chem.* **1970**, *35*, 3997.
5. Evans, D.A.; Sims, C.L.; Andrews, G.C. *J. Am. Chem. Soc.* **1977**, *99*, 5453.
6. Bohlmann, F.; Otto, W. *Liebigs Ann. Chem.* **1982**, 186.
7. Evans, D.A.; Crawford, T.C.; Fujimoto, T.T.; Thomas, R.C. *J. Org. Chem.* **1974**, *39*, 3176.
8. Ishibashi, H.; Okano, M.; Tamaki, H.; Maruyama, K.; Yakura, T.; Ikeda, M. *Chem. Commun.* **1990**, 1436.
9. Posner, G.H.; Asirvatham, E. *J. Org. Chem.* **1985**, *50*, 2589.
10. Quinkert, G.; Schmalz, H.-G.; Walzer, E.; Gross, S.; Kowalczyk-Przewloka, T.; Schierloh, C.; Dürner, G.; Bats, J.W.; Kessler, H. *Liebigs Ann. Chem.* **1988**, 283.
11. O'Brien, M.K.; Pearson, A.J.; Pinkerton, A.A.; Schmidt, W.; Willman, K. *J. Am. Chem. Soc.* **1989**, *111*, 1499.
12. Stork, G.; Danheiser, R.L.; Ganem, B. *J. Am. Chem. Soc.* **1973**, *95*, 3414.
13. Asaoka, M.; Takenouchi, K.; Takei, H. *Tetrahedron* Lett. **1988**, *29*, 325.
14. Dauben, W.G.; Hart, D.J. *J. Am. Chem. Soc.* **1975**, *97*, 1622.
15. Kigoshi, H.; Imamura, Y.; Mizuta, K.; Niwa, H.; Yamada, K. *J. Am. Chem. Soc.* **1993**, *115*, 3056.
16. Corey, E.J.; Su, W.-g. *J. Am. Chem. Soc.* **1987**, *109*, 7534; Corey, E.J.; Su, W.-g. *Tetrahedron Lett.* **1988**, *29*, 3423.
17. Jackson, A.; Gaskell, A.J.; Wilson, N.D.V.; Joule, J.A. *Chem. Commun.* **1968**, 364.
18. Hart, D.J.; Yang, T.-K. *J. Org. Chem.* **1985**, *50*, 235.
19. Battersby, A.R.; Turner, J.C. *J. Chem. Soc.* **1960**, 717.
20. Danishefsky, S.J.; Mantlo, N. *J. Am. Chem. Soc.* **1988**, *110*, 8129.
21. Riehs, G.; Urban, E. *Tetrahedron Lett.* **1995**, *36*, 7233.
22. Burke, S.; Kort, M.E.; Strickland, S.M.; Organ, H.M.; Silk, L.A., III. *Tetrahedron Lett.* **1994**, *35*, 1503.
23. de Alvarenga, E.S.; Mann, J. *J. Chem. Soc. Perkin Trans. 1* **1993**, 2141.
24. Roberts, M.R.; Schlessinger, R.H. *J. Am. Chem. Soc.* **1979**, *101*, 7626.
25. Snider, B.B.; Kirk, T.C. *J. Am. Chem. Soc.* **1983**, *105*, 2364.
26. Horiguchi, Y.; Nakamura, E.; Kuwajima, I. *J. Am. Chem. Soc.* **1989**, *111*, 6257.
27. Murray, A.; Proctor, G.R.; Murray, P.J. *Tetrahedron Lett.* **1995**, *36*, 291.
28. Boeckman, R.K.; Arvanitis, A.; Voss, M.E. *J. Am. Chem. Soc.* **1989**, *111*, 2737.
29. Piers, E.; Banville, J.; Lau, C.K.; Nagakura, I. *Can. J. Chem.* **1982**, *60*, 2965.
30. Nagumo, S.; Suemune, H.; Sakai, K. *Chem. Commun.* **1990**, 1778.
31. Chapdelaine, M.J.; Hulce, M. *Org. React.* **1990**, *38*, 225.
32. Danishefsky, S.; Vaughn, K.; Gadwood, R.; Tsuzuki, K. *J. Am. Chem. Soc.* **1981**, *103*, 4136.
33. Herrmann, J.L.; Berger, M.H.; Schlessinger, R.H. *J. Am. Chem. Soc.* **1979**, *101*, 1544.

34. Oppolzer, W.; Guo, M.; Bättig, K. *Helv. Chim. Acta* **1983**, *66*, 2140.
35. Hirukawa, T.; Oguchi, M.; Yoshikawa, N.; Kato, T. *Chem. Lett.* **1992**, 2343.
36. Weyerstahl, P.; Marschall, H.; Schulze, M.; Schwope, I. *Liebigs Ann. Chem.* **1996**, 799.
37. Angelaud, R.; Landais, Y. *J. Org. Chem.* **1996**, *61*, 5202.
38. Campbell, M.M.; Sainsbury, M.; Yavarzadeh, R. *Tetrahedron* **1984**, *40*, 5063.
39. Birch, A.J.; Kelly, L.F.; Weerasuria, D.V. *J. Org. Chem.* **1988**, *53*, 278.
40. Banwell, M.G.; Forman, G.S. *J. Chem. Soc. Perkin Trans. 1* **1996**, 2565.
41. Wender, P.A.; Lee, H.Y.; Wilhelm, R.S.; Williams, P.D. *J. Am. Chem. Soc.* **1989**, *111*, 8954; Wender, P.A.; Kogen, H.; Lee, H.Y.; Munger, J.D.; Wilhelm, R.S.; Williams, P.D. *J. Am. Chem. Soc.* **1989**, *111*, 8957.
42. Schumacher, D.P.; Hall, S.S. *J. Am. Chem. Soc.* **1982**, *104*, 6076.
43. Wu, A.; Snieckus, V. *Tetrahedron Lett.* **1975**, 2057.
44. Suemune, H.; Oda, K.; Saeki, S.; Sakai, K. *Chem. Pharm. Bull.* **1988**, *36*, 172.
45. Torii, S.; Inokuchi, T.; Yamafuji, T. *Bull. Chem. Soc. Jpn.* **1979**, *52*, 2640.
46. Dastur, K.P. *J. Am. Chem. Soc.* **1974**, *96*, 2605.
47. Vidari, G.; Ferrino, S.; Grieco, P.A. *J. Am. Chem. Soc.* **1984**, *106*, 3539.
48. Corey, E.J.; Da Silva Jardine, P.; Rohloff, J.C. *J. Am. Chem. Soc.* **1988**, *110*, 3672.
49. Corey, E.J.; Da Silva Jardine, P.; Mohri, T. *Tetrahedron Lett.* **1988**, *29*, 6409.
50. De Baecke, G.; De Clercq, P.J. *Tetrahedron Lett.* **1995**, *36*, 7515.
51. Stork, G.; Nakahara, Y.; Nakahara, Y.; Greenlee, W.G. *J. Am. Chem. Soc.* **1978**, *100*, 7775.
52. Boger, D.L.; Zhang, M. *J. Am. Chem. Soc.* **1991**, *113*, 4230.
53. Corey, E.J.; Weinshenker, N.M.; Schaaf, T.K.; Huber, W. *J. Am. Chem. Soc.* **1969**, *91*, 5675.
54. Corey, E.J.; Imwinkelried, R.; Pikul, S.; Xiang, Y.B. *J. Am. Chem. Soc.* **1989**, *111*, 5493.
55. Corey, E.J.; Imai, N.; Pikul, S. *Tetrahedron Lett.* **1991**, *32*, 7517.
56. White, J.D.; Gupta, D.N. *J. Am. Chem. Soc.* **1968**, *90*, 6171.
57. Pirrung, M.C. *J. Am. Chem. Soc.* **1981**, *103*, 82.
58. Crimmins, M.T.; DeLoach, J.A. *J. Am. Chem. Soc.* **1986**, *108*, 800.
59. Oppolzer, W.; Zutterman, F.; Bättig, K. *Helv. Chim. Acta* **1983**, *66*, 522.
60. Oppolzer, W.; Wylie, R.D. *Helv. Chim. Acta* **1980**, *63*, 1198.
61. Birch, A.M.; Pattenden, G. *J. Chem. Soc. Perkin Trans. 1* **1983**, 1913.
62. Vo, N.H.; Snider, B.B. *J. Org. Chem.* **1994**, *59*, 5419.
63. Wender, P.A.; Ihle, N.C.; Correia, C.R.D. *J. Am. Chem. Soc.* **1988**, *110*, 5904.
64. Goering, B.K.; Li, J.; Ganem, B. *Tetrahedron Lett.* **1995**, *36*, 8905.
65. Paquette, L.A.; Sturino, C.F.; Doussot, P. *J. Am. Chem. Soc.* **1996**, *118*, 9456.
66. Nakai, T.; Mikami, K. *Chem. Rev.* **1986**, *86*, 885.
67. Brückner, R.; Priepke, H. *Angew. Chem. Int. Ed. Engl.* **1988**, *27*, 278.
68. Nakazawa, M.; Sakamoto, Y.; Takahashi, T.; Tomooka, K.; Ishikawa, K.; Nakai, T. *Tetrahedron Lett.* **1993**, *34*, 5923.
69. Nonoshita, K.; Banno, H.; Maruoka, K.; Yamamoto, H. *J. Am. Chem. Soc.* **1990**, *112*, 316.

70. Mikami, K.; Loh, T.-P.; Nakai, T. *J. Am. Chem. Soc.* **1990**, *112*, 6737.
71. MacMillan, D.W.C.; Overman, L.E. *J. Am. Chem. Soc.* **1995**, *117*, 10391.
72. Eschenmoser, A.; Wintner, C.E. *Science* **1977**, *196*, 1410.
73. Paquette, L.A.; Han, Y.-K. *J. Am. Chem. Soc.* **1981**, *103*, 1835.
74. Annis, G.D.; Paquette, L.A. *J. Am. Chem. Soc.* **1983**, *105*, 7358.
75. Crisp, G.T.; Scott, W.J.; Stille, J.K. *J. Am. Chem. Soc.* **1984**, *106*, 7500.
76. Paquette, L.A.; Leone-Bay, A. *J. Am. Chem. Soc.* **1983**, *105*, 7352.
77. Franck-Neumann, M.; Miesch, M.; Lacroix, E. *Tetrahedron Lett.* **1989**, *30*, 3533.
78. Crimmins, M.T.; Mascarella, S.W. *J. Am. Chem. Soc.* **1986**, *108*, 3435.
79. Curran, D.P.; Chen, M.-H. *Tetrahedron Lett.* **1985**, *26*, 4991.
80. Curran, D.P.; Rakiewicz, D.M. *J. Am. Chem. Soc.* **1985**, *107*, 1448.
81. Curran, D.P.; Kuo, S.-C. *Tetrahedron* **1987**, *43*, 5653.
82. Toth, J.E.; Fuchs, P.L. *J. Org. Chem.* **1987**, *52*, 473.
83. Parker, K.A.; Fokas, D. *J. Am. Chem. Soc.* **1992**, *114*, 9688.

3

1,3-STEREOINDUCTION

3.1. CONSEQUENCES OF 1,3-INTERACTIONS IN A SIX-CENTERED ARRAY

In a chairlike conformation the presence of an axial substituent hinders the approach of reagents from that direction. Reactions usually take place seemingly from the equatorial side, but may actually involve the twist boat transition states. This is an old problem which is related to synthesis of resin acids such as podocarpic acid (**3-1**)[1] and dehydroabietic acid (**3-2**).[2]

A more spectacular achievement concerned the total synthesis of cortisone acetate (**3-3**) in which the tricarbocyclic ketone underwent methylation and methallylation sequentially[3] in the process of D-ring construction. There was an axial

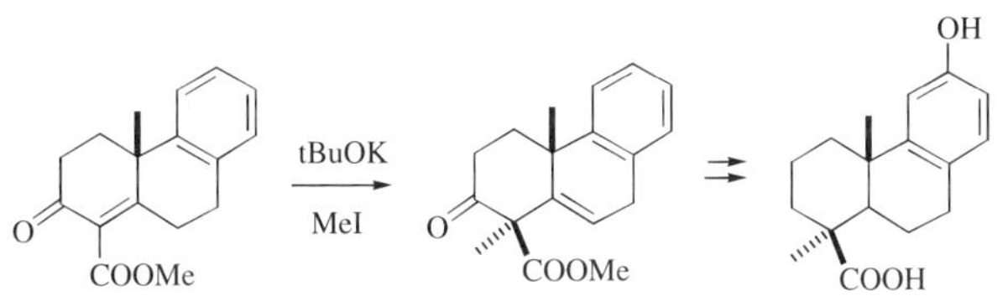

podocarpic acid
(**3-1**)

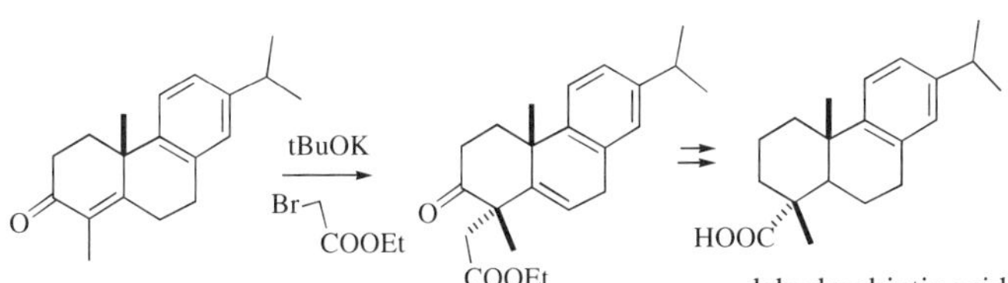

dehydroabietic acid
(**3-2**)

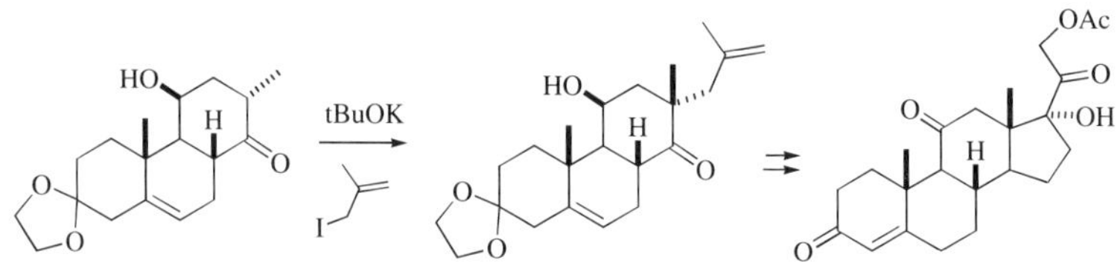

cortisone acetate
(**3-3**)

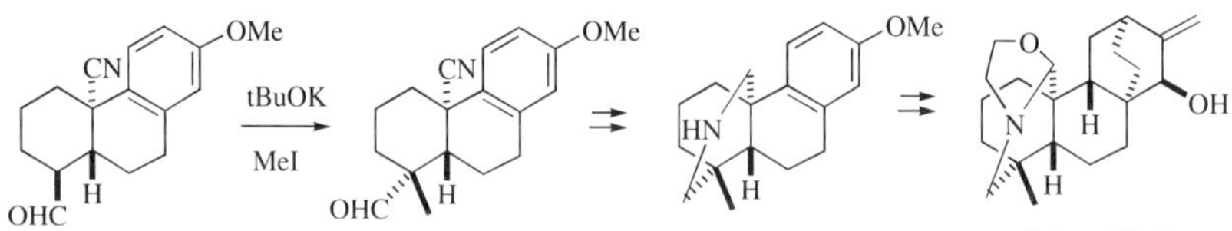

atisine (**3-4**)

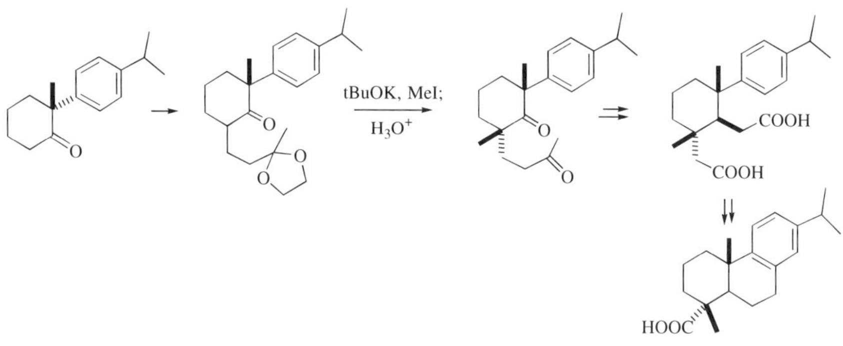

dehydroabietic acid
(**3-2**)

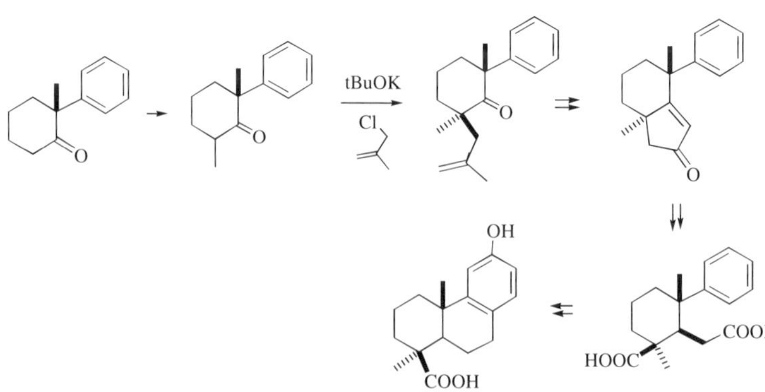

podocarpic acid
(**3-1**)

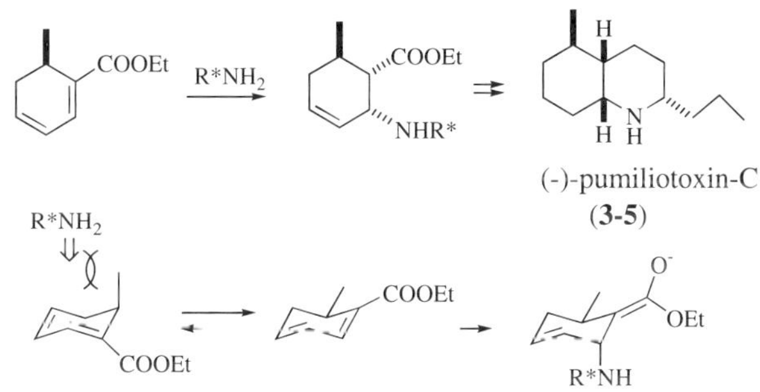

hydroxyl group (eventually to become the ketone at $C_{(11)}$) to steer each of the alkylation steps to the α-side of the molecule.

The activating group for alkylation may be situated outside the ring, as was observed in a synthesis of atisine (**3-4**).[4] However, stereoelectronic effects are absent in this case and equatorial entry of the methyl group proceeded unopposed.

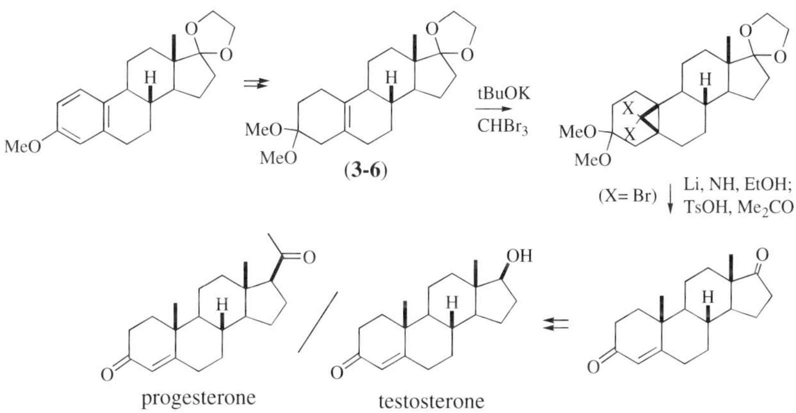

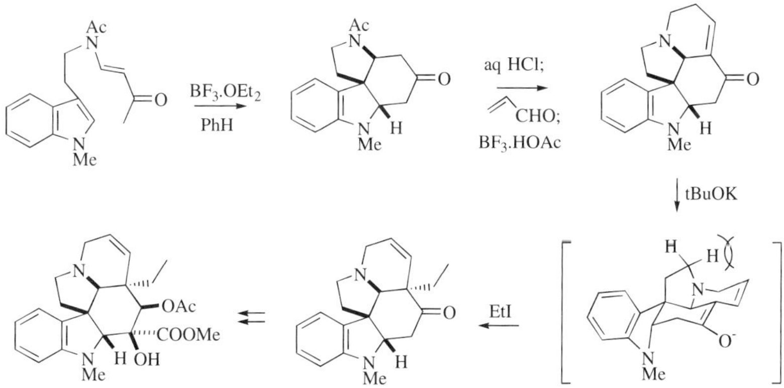

vindorosine (**3-7**)

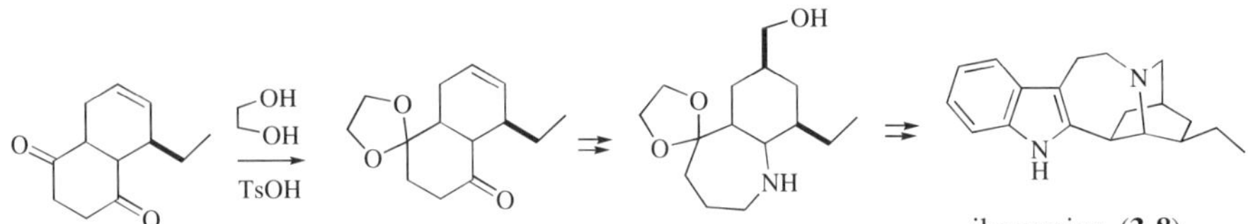

ibogamine (**3-8**)

Consideration of conformational stabilization by a large substituent (e.g., aryl group) in combination with stereoelectronic effects led to the development of a stereoselective route to resin acids such as dehydroabietic acid (**3-2**)[5] and podocarpic acid (**3-1**).[6] Thus, alkylation of a 2-aryl-2-methylcyclohexanone in succession always places the last substituent *trans* to the aryl group.

Operation of both 1,3- and 1,2-stereoinduction is evident during the preparation of an intermediate of (–)-pumiliotoxin-C (**3-5**).[7] After the large nitrogen nucleophile was introduced, it in turn determined the protonation.

The development of the Birch reduction of aromatic compounds has enabled the conversion of estrone methyl ether into other steroids.[8] Introduction of the 19-methyl group through dibromocarbene addition to the $\Delta^{5(10)}$-3-ketone dimethyl acetal (**3-6**) was managed by steric impediment from the α side by the axial methoxy group at $C_{(3)}$.

In a vindorosine (**3-7**) synthesis,[9] the ring system was rapidly assembled from N^{a}-methyltryptamine using an intramolecular Michael–Mannich reaction tandem, followed by Michael and aldol reactions. Note that three contiguous stereocenters so created correspond to the most stable configuration and also those of the target molecule, all 5/6-ring junctions being *cis*. The next step of C-ethylation also furnished a product with a *cis*-octahydroquinoline unit, owing to blocking of the β-face by the N-CH_2 group to allow approach of the reagent.

The ibogamine (**3-8**) synthesis[10] starting from the Diels–Alder reaction of 1,3-hexadiene and *p*-benzoquinone is characterized by excellent chemoselectivity, regioselectivity, and stereoselectivity. After reduction of the conjugated double bond the diketone was converted to a monoacetal. The controlling element was the ethyl

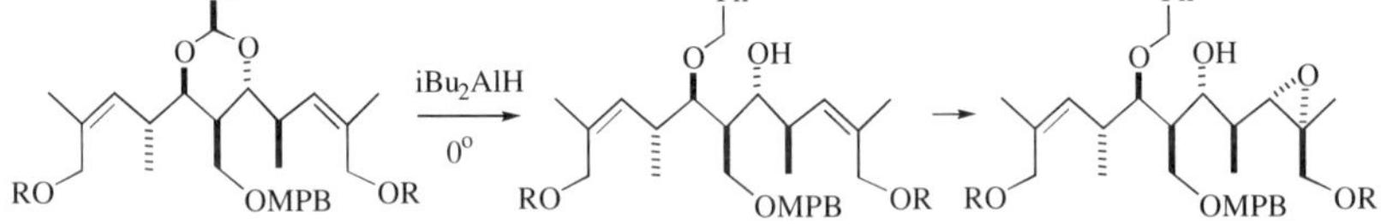

(**3-9**) R = TBDPS 72% (sel. >20:1)

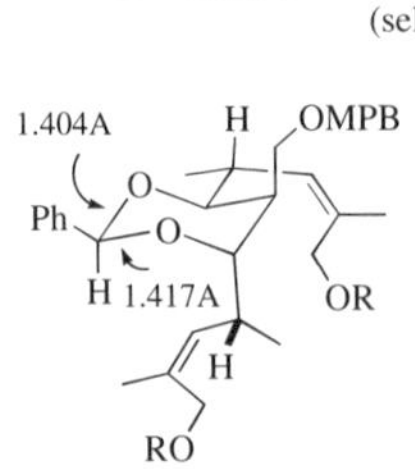

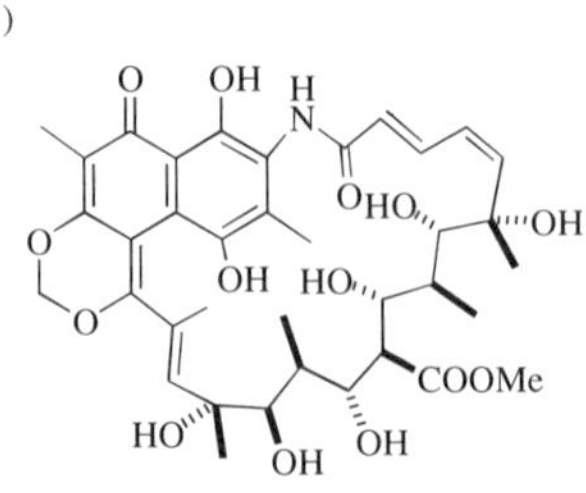

streptovaricin-G

group which occupied an equatorial position and thereby inhibited acetalization of the proximal carbonyl. The diacetal would experience at least one set of 1,3-diaxial interaction, whatever conformation is adopted.

A C_2-symmetric 1,3-dioxane (**3-9**) was found to undergo regioselective cleavage on reduction with diisobutylaluminum hydride.[11] The cause for a stretched and hence weaker C–O bond of the acetal unit which differed from its geminal bond in length (shown by x-ray diffraction) was likely to be the presence of an axial group which must interact with the benzylic hydrogen. It should be noted that there was another axial substituent but only the lone electron-pairs of two oxygen atoms of the heterocycle were 1,3-diaxially related to it.

A very important result which contributed to the success of a picrotoxinin (**3-10**) synthesis[12] is that a bicyclo[3.3.1]nonenone reacted with lithium acetylide from the side of the saturated cyclohexane ring, probably because the *exo*-oriented isopropyl group offered hindrance to the incoming nucleophile (even if it is quite small). Note that in a bulnesol (**3-11**) synthesis[13] the key step of rearrangement was set up by the stereoselective reduction of a ketone from the unsaturated branch of a bicyclo[4.3.1]decene.

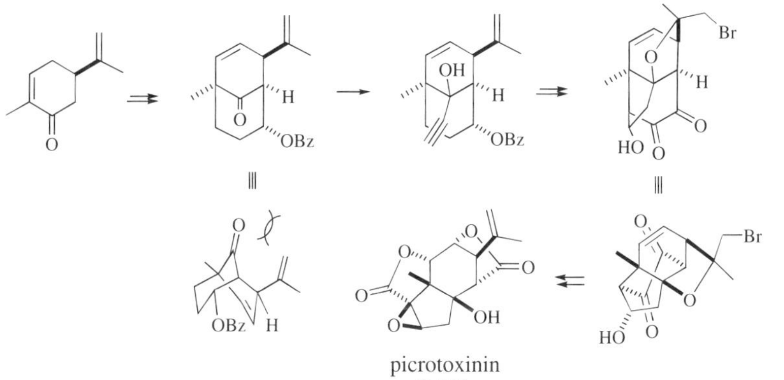

picrotoxinin
(**3-10**)

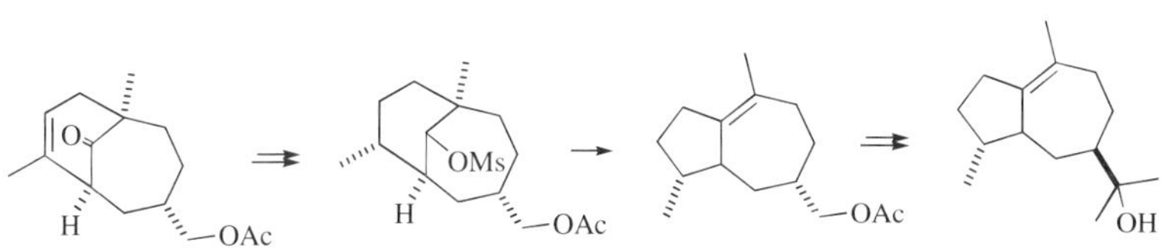

bulnesol (**3-11**)

A large number of synthetic applications of the chiral 5-trimethylsilyl-2-cyclohexenone[14] are based on the silyl substituent to consolidate the conformation, directing *anti* face reactions such as reactions on the enone system in the 1,2- or 1,4-mode which happen to be sterically and stereoelectronically favored. Representative are routes to (–)-*O*-methyljoubertiamine (**3-12**)[15] and (+)-ptilocaulin (**3-13**).[16]

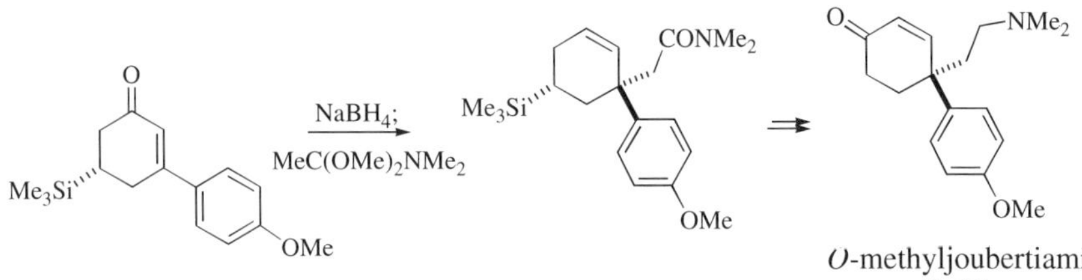

O-methyljoubertiamine
(**3-12**)

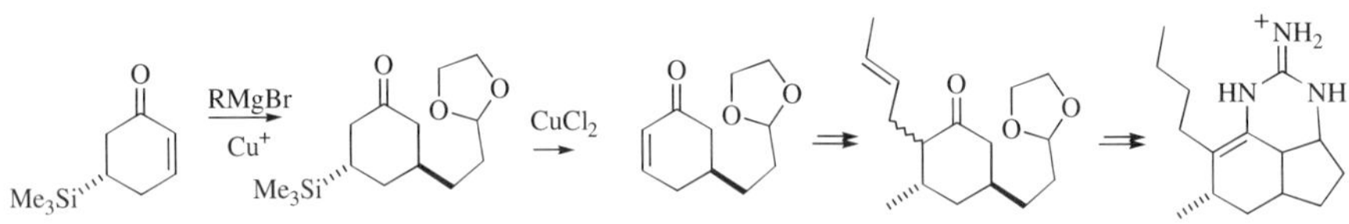

ptilocaulin
(**3-13**)

During deiodination of a bridged lactone[17] the axial (with respect to the cyclohexane ring) iodine atom was replaced with inversion of configuration and the product was converted to (+)-epianastrephin (**3-14**). The tin hydride reagent submitted its hydrogen atom from the equatorial side because the approach encountered less hindrance.

The synthesis of β-multistriatin (**3-15**) cannot be accomplished through an acetalization step because it is prone to epimerize to give δ-multistriatin (**3-16**). Advantage can be taken from the mutual hindrance of the two axial methyl groups such that when one is replaced with an unsaturated unit the hydrogenation would be forced to occur from the equatorial direction.[18]

Bridged ring systems prepared from the Diels–Alder reaction containing a bulky *endo* group are sterically more favorable. When the double bond is removed, the situation can be different. Thus, one such diketone was epimerized after

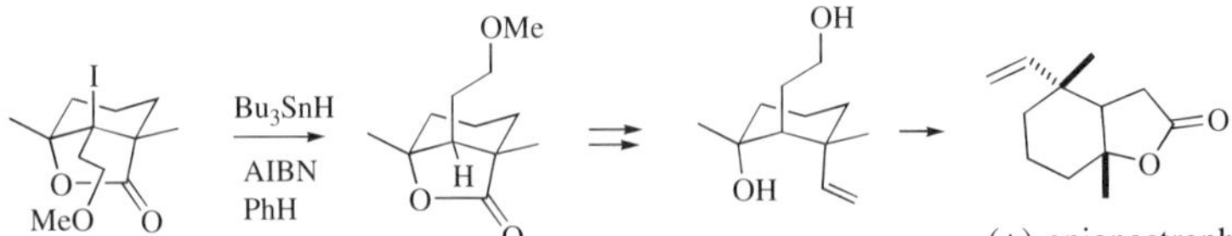

(+)-epianastrephin
(**3-14**)

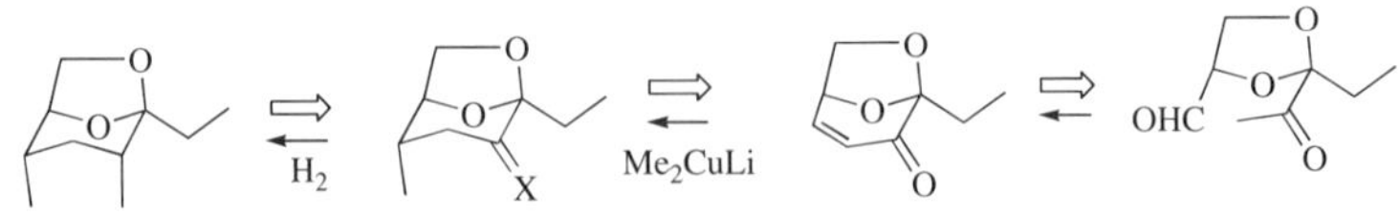

β-multistriatin
(**3-15**)

X = O
X = CH$_2$

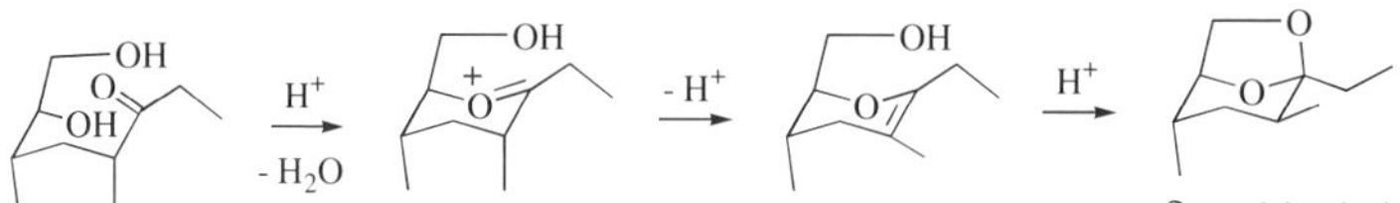

δ-multistriatin
(**3-16**)

hydrogenation and a synthesis of patchouli alcohol (**3-17**)[19] was completed by proper manipulation.

Another synthesis of patchouli alcohol (**3-17**)[20] featured an intramolecular Diels–Alder reaction. Only one of two diastereomers reacted to give the dehydro precursor because the other isomer encountered incipient 1,3-diaxial interactions between the methyl group attached to the double bond and that located in the side chain.

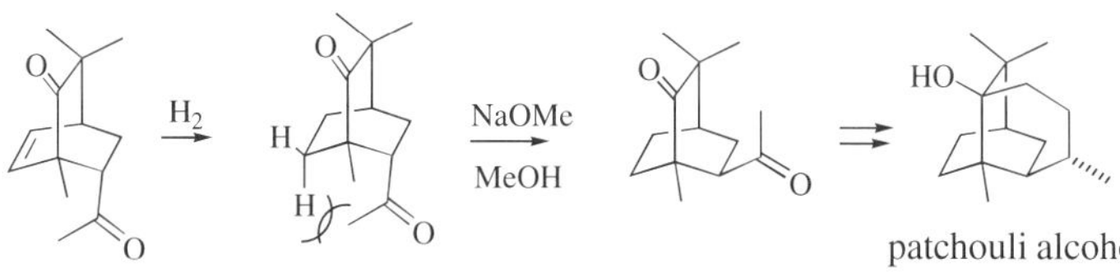

patchouli alcohol
(**3-17**)

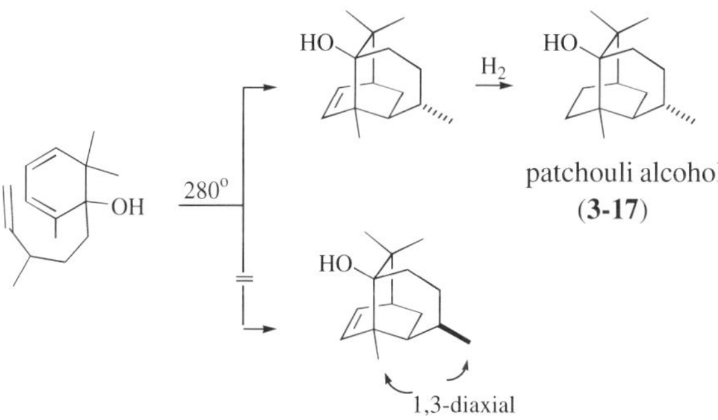

patchouli alcohol
(**3-17**)

The more familiar 1,3-stereoinduction by an axial angular methyl group can be extended to an intramolecular Diels–Alder reaction.[21] With a detachable tether of sufficient length, an advanced tricarbocyclic intermediate for adrenosterone (**3-18**) was readily prepared.

The structural pattern of mamanuthaquinone (**3-19**) presents an irresistible suggestion of a Diels–Alder approach. However, the possible transition states must be scrutinized, and only a product from the *exo* transition is suitable as there is no simple recourse for configurational amendment. When the *gem*-dimethyl group of the diene component plays a decisive role by offering a much more severe incipient 1,3-diaxial interaction with the aryl group in the *endo* transition state, the chance for success in adopting the desired *exo* approach increases greatly. Indeed, based on this analysis a Lewis acid catalyzed Diels–Alder reaction was explored and a short route to mamanuthaquinone was developed.[22]

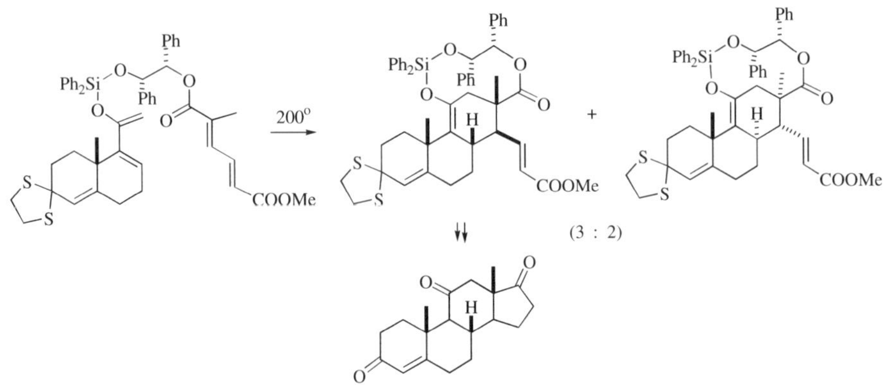

(+)-adrenosterone
(**3-18**)

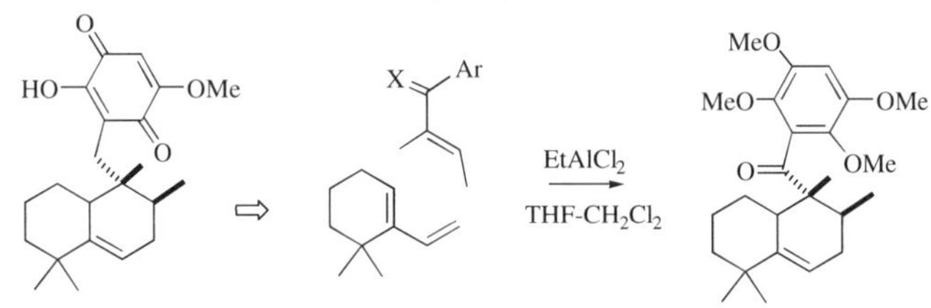

mamanuthaquinone
(**3-19**)

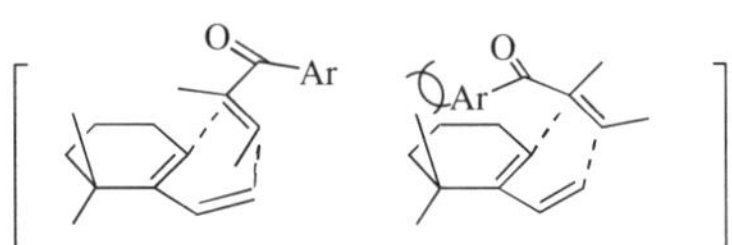

favored TS disfavored TS

Similarly, 1,3-stereoinduction due to presence of a 5-substituent in a dienophilic cyclohexenone was instrumental in acquisition of the desired isomer for synthesis of luciduline (**3-20**)[23] and fawcettimine (**3-21**).[24]

The diastereofacial selection, which was determined by avoidance of 1,3-interactions in a [2+2]-photocycloaddition between piperitone and cyclobutene

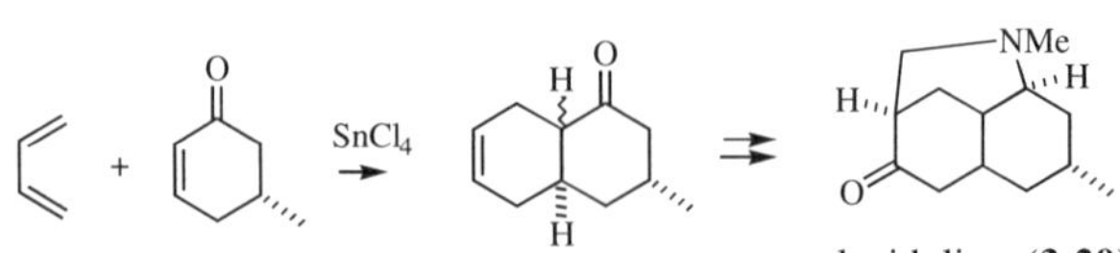

luciduline (**3-20**)

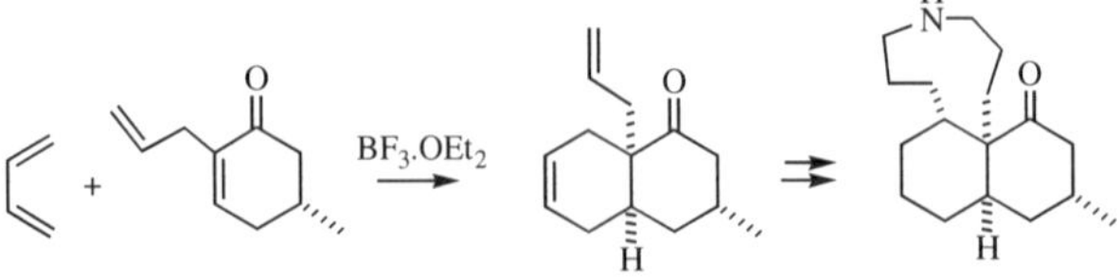

fawcettimine (**3-21**)

derivatives,[25,26] was exploited for yielding properly constituted intermediates for shyobunone (**3-22**) and 10-epijunenol (**3-23**).

In a much less obvious route to subergorgic acid (**3-24**),[27] a spiro[4.5]decane system became the platform for the elaboration of the third ring. The important stereoselective attachment of a two-carbon chain was delegated to a [2+2]-photocycloaddition which was controlled by the secondary methyl group in the same ring.

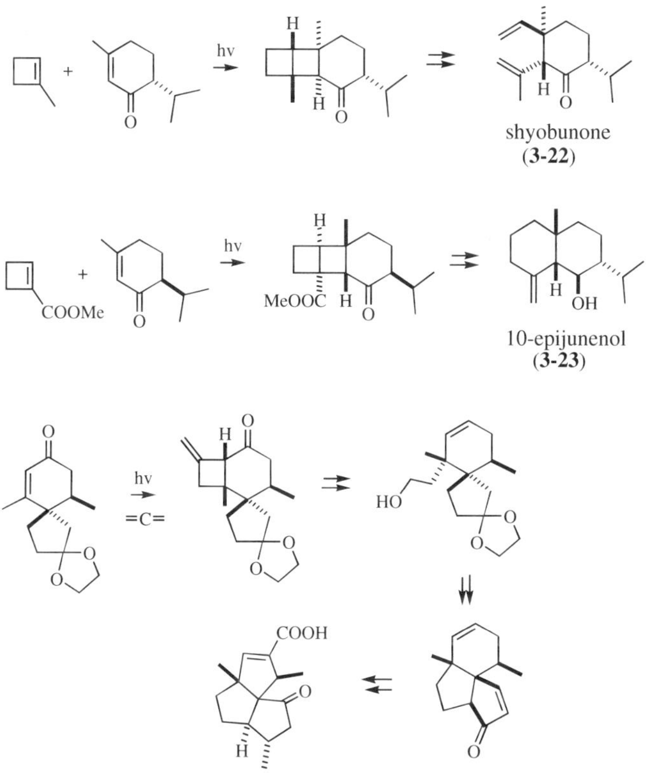

shyobunone (**3-22**)

10-epijunenol (**3-23**)

subergorgic acid (**3-24**)

Calcimycin is an ionophore antibiotic featuring a spiroacetal unit in which the anomeric center is subject to thermodynamic control. By a two-bond disconnection of the spiroacetal (**3-25**) such as to take away a methyl group and the ether linkage from the more highly substituted tetrahydropyran ring to yield a trisubstituted dihydropyran, and entertaining the regeneration of the whole unit by a Simmons–Smith reaction and acid-catalyzed cyclopropane cleavage, a synthetic route is simplified.[28] 1,3-Stereoinduction is involved in the process, but it is important that the two vicinal alkyl groups remaining in the heterocycle are *cis* to one another. The Simmons–Smith reaction took place from the *anti* face to avoid incipient 1,3-diaxial interaction.

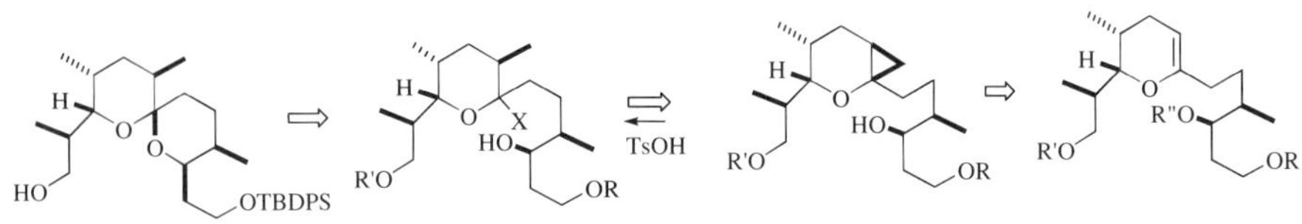

(**3-25**)

A synthesis of africanol (**3-26**)[29] employed an orthoester Claisen rearrangement to introduce a two-carbon chain for eventual incorporation into the five-membered ring. It also created a new stereocenter. Two key aspects were involved: stereoselective reduction of the enone (selectivity 88:12) was due to preference of the *s-cisoid* conformation, and the transition state selection that minimized steric interactions (potential 1,3-diaxial interactions of the alternative form).

The C_2-symmetrical *trans*-3,4-diphenyl-*N,N′*-bisarenesulfonyl-2,5- diazaborolidine is a very useful controller for asymmetric synthesis. Its application to promoting aldol condensation is illustrated in a synthesis of sitophilure (**3-27**).[30] There is correspondence of chirality of the product with the transition state, and in the alternative transition state the ethyl group of the enolate suffers a severe 1,3-diaxial interaction with one of the arene residues.

If a molecule is allowed to assume different transition states, it usually selects the one with the least steric interactions. Often one can find a different chain folding mode

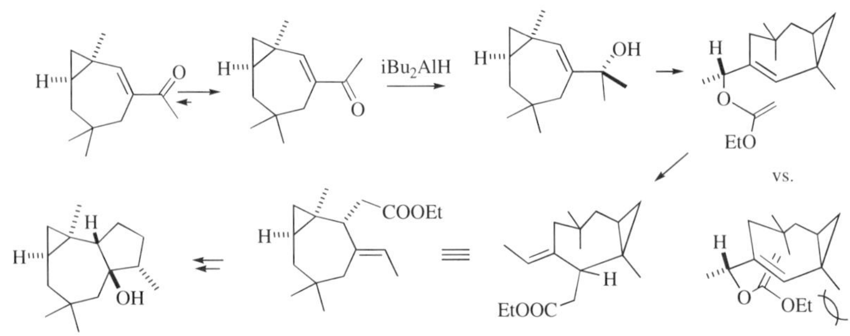

africanol (**3-26**)

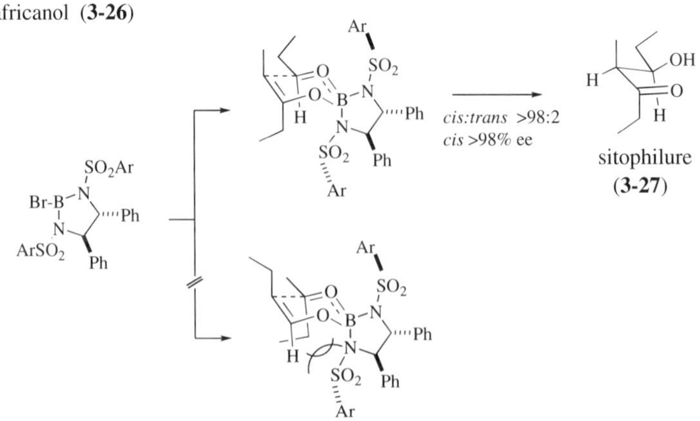

sitophilure (**3-27**)

that can avoid a conformation which presents bulky group(s) in an axial orientation. Thus, the intramolecular 2-azaallyl anion cycloaddition to an extended styrene led to a hydroindole with the allylic substituent *trans* to the aryl group (and the methine hydrogen) at the ring junction. Fortunately, a synthesis of crinine (**3-28**)[31] could be achieved by epimerization of the allylic alcohol at a later stage.

A large orthoester of N^a-protected tryptophan is an excellent stereocontroller. A derivative thus equipped readily underwent enantioselective transformation to give a useful intermediate of vindorosine (**3-29**).[32] In the sequence of intramolecular β-aminoenone-alkene [2+2]-photocycloaddition, retro-Mannich cleavage, and alterpositional Mannich reaction, the first step was most crucial because the stereochemistry was set by the cyclic orthoester which demanded an equatorial position in an envelope conformation defined by the three-atom chain and two proximal sp^2 termini. When occupying the pseudoaxial position, it would have butted against the benzene ring.

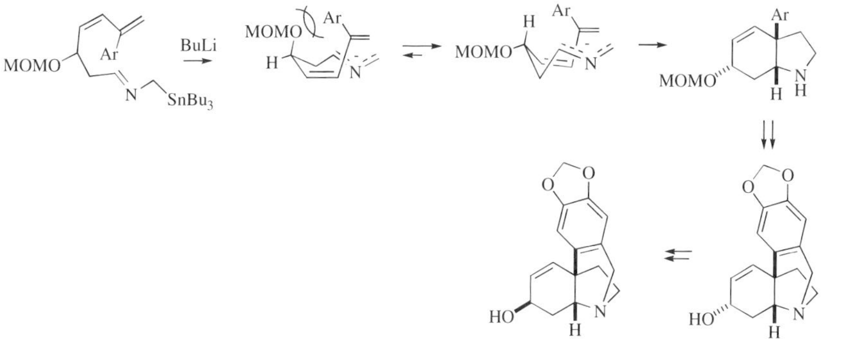

crinine (**3-28**)

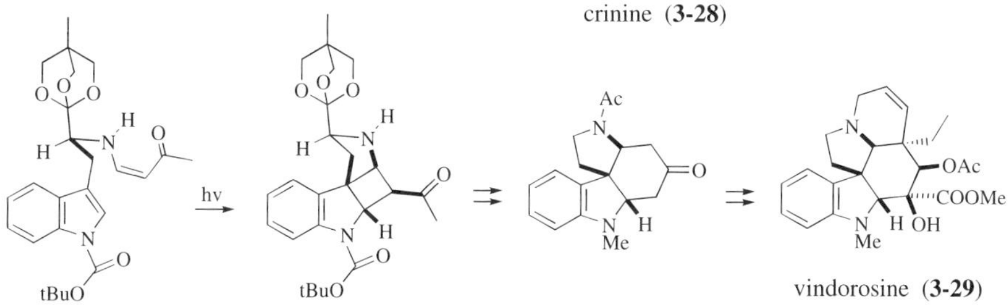

vindorosine (**3-29**)

Stereoselective addition of H/CH_2OH to the double bond of an allylic alcohol to give the 1,4-diol product can be effected by using a free radical located in a silyl tether. The configuration of the original double bond is very important, as manifested by the different reactivities of pregn-17(20)-enes (**3-30, 3-31**).[33] The (*E*)-alkenes are unreactive due to inevitable formation of products with methyl groups in 1,3-diaxial relationship. The transition states leading to such compounds must be highly destabilized.

For potential uses in polyether and ionophore synthesis, the stereochemically defined polysubstituted tetrahydropyrans are valuable. One method for elaboration of

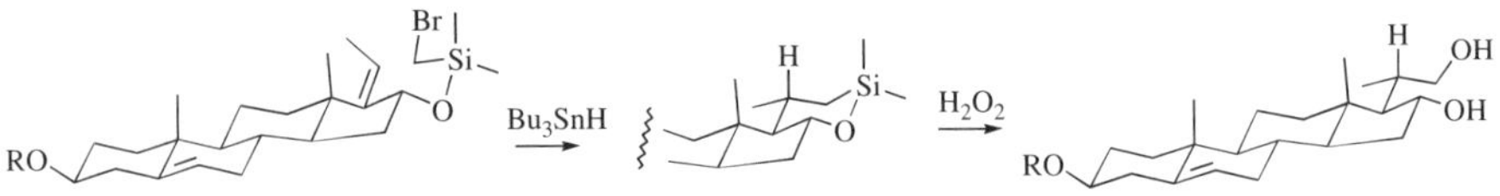

(3-30) R = THP

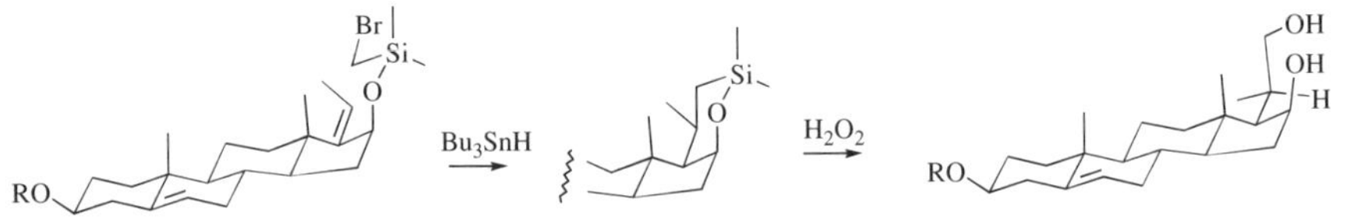

(3-31)

such compounds is by a free radical cyclization of vinylsilanes (e.g., **3-32**).[34] Equatorial disposition of the acceptor unit becomes a determining factor.

Cyclization by intramolecular alkylation of an ester enolate with epoxide proceeds from a chairlike transition state (**3-33**) in which both the enolate and the epoxide lie equatorially.[35] Because the enolate resembles a *t*-butyl group, 1,3-diaxial interactions with hydrogen atoms can be severe. This observation served as the basis of a design to unravel the three contiguous stereocenters en route to (–)-histrionicotoxin (**3-34**).[36]

Scopadulciol (**3-35**) and scopadulcic acid were successfully acquired by carrying a BC → BCD → ABCD ring construction scheme.[37] It was found that Li/NH_3 reduction of the octalone intermediate was intimately dependent on the configuration of a methyl group at the α′-position. Protonation at the ring junction successfully avoided the form that sustained a 1,3-diaxial interaction of the methyl group with the other ring.

There is a clear preference for the transition state leading to the zizaane skeleton by a tandem free radical process. The alternative mode of bicyclization would suffer severe interaction of the 1,3-diaxial type. Thus zizaene and khusimone (**3-35a**) can be rapidly synthesized.[37a]

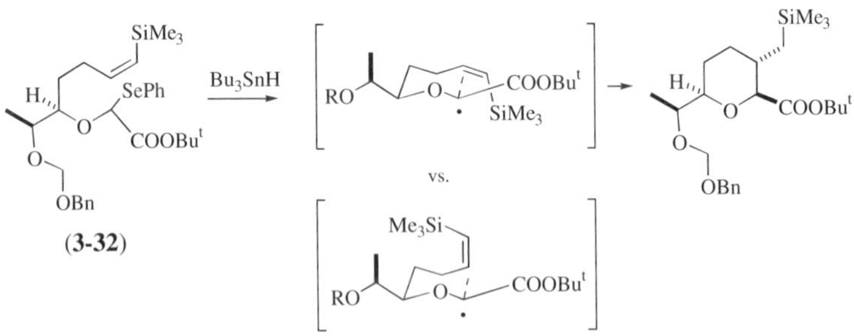

(3-32)

(3-33)

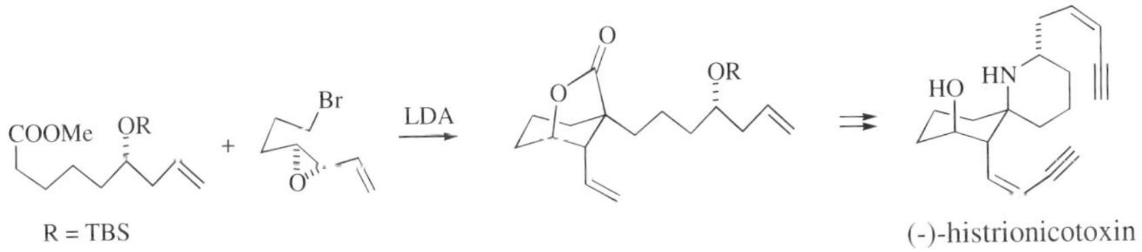

(-)-histrionicotoxin
(**3-34**)

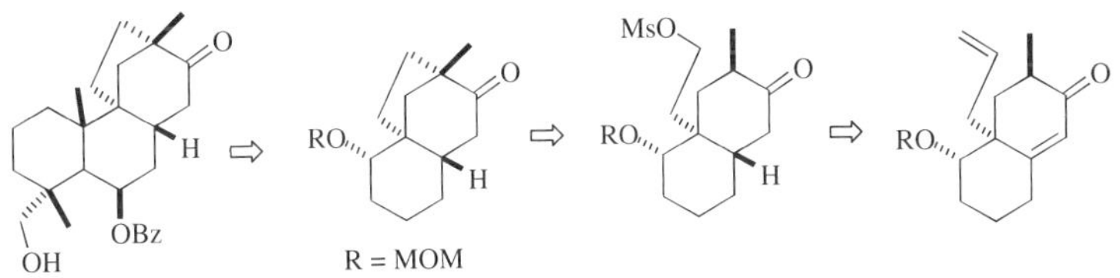

scopadulciol
(**3-35**)

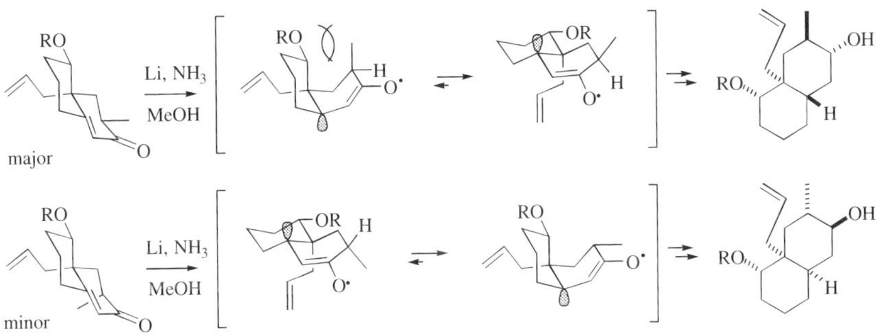

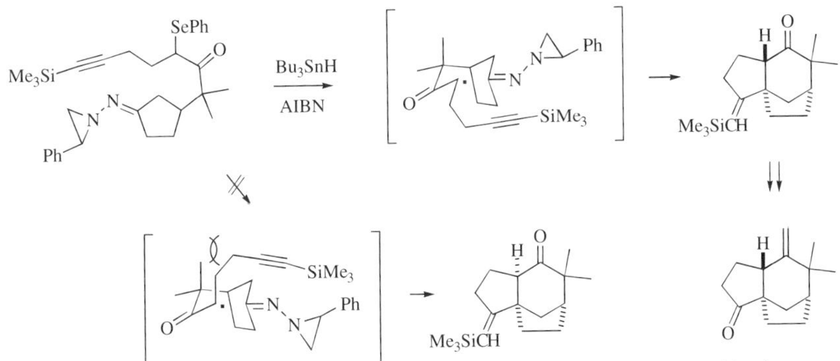

disfavored TS

khusimone
(**3-35a**)

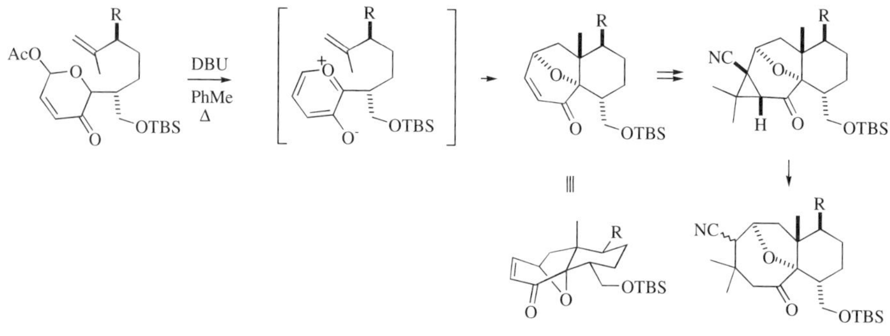

(3-36)

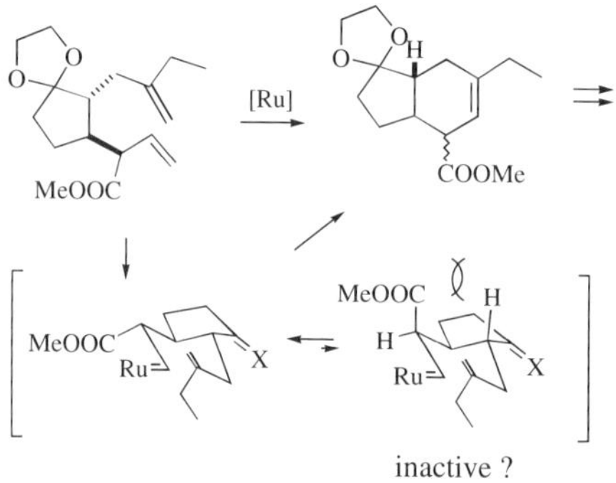

coronafacic acid
(3-37)

A novel approach to the BC ring system (**3-36**) of taxane diterpenes[38] involved intramolecular cycloaddition to form an oxabicyclo[3.2.1]octenone and ring expansion. Using a *t*-butyldimethylsiloxymethyl group to anchor a chairlike conformation, the relative configuration of quaternary carbon centers were established.

As a method to effect ring closure to access the skeleton of coronafacic acid (**3-37**),[39] metathesis of side chains at the α- and β-positions of a cyclopentanone derivative was shown to be selective for one diastereomer (equatorial ester?) when a ruthenium catalyst was employed.

Incipient CH_3/CH_2 interactions in one folding mode preceding Claisen rearrangement of a cyclopentenylmethyl vinyl ether disfavored one form (to the extent

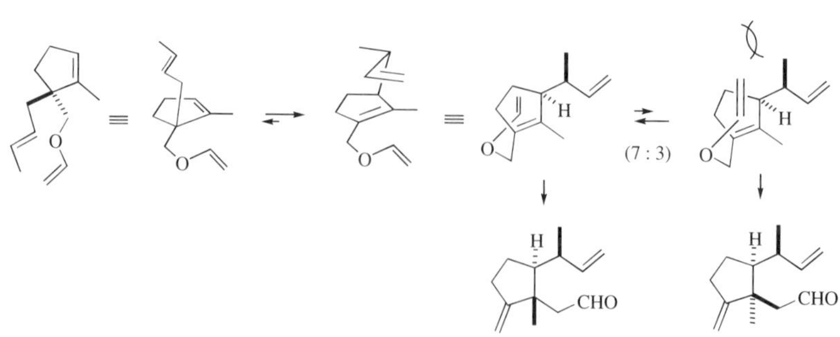

(3-38)

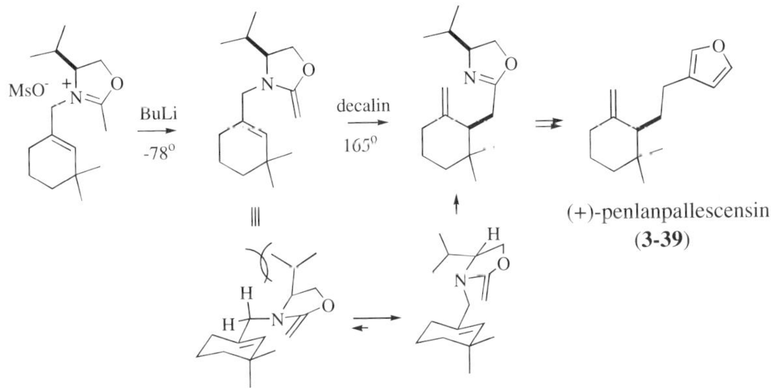

(+)-penlanpallescensin (**3-39**)

of 30 : 70%).[40,41] The major product (**3-38**) was useful for synthesis of pseudoguaianolide sesquiterpenes.

The aza-Claisen rearrangement has been successful in the elaboration of (+)-penlanpallescensin (**3-39**).[42] The 1,5-stereoinduction was operated through one of the chairlike transition states. The preferred conformation apparently was devoid of interaction between the isopropyl group and the pseudoaxial hydrogen atom adjacent to the nitrogen atom. Note that the two conformations differ in the orientation of the lone electron-pair.

1,3-Stereoinduction frequently appears in reactions with allylmetals. With two chiral components the Hiyama reaction served to assemble an intermediate of albolic acid (**3-40**) and ceroplastol-II (**3-41**).[43] There was a double differentiation.[44]

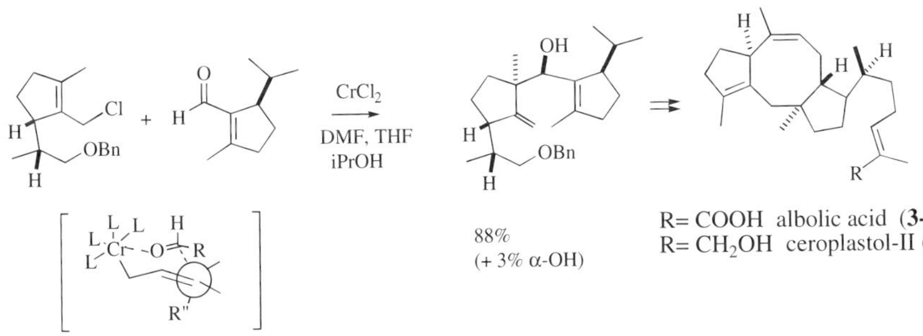

R= COOH albolic acid (**3-40**)
R= CH_2OH ceroplastol-II (**3-41**)

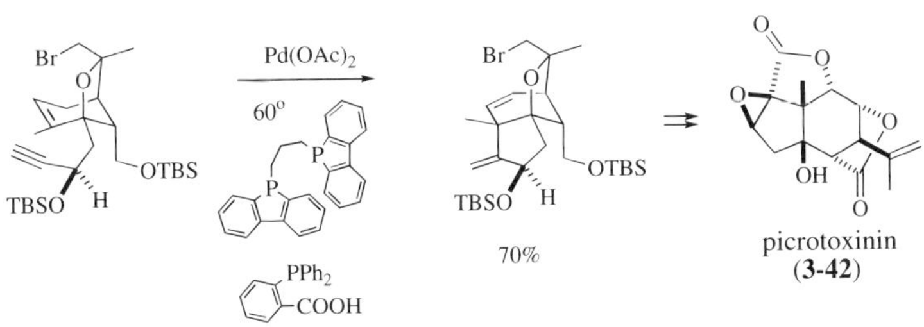

picrotoxinin (**3-42**)

It is possible to defy the adverse effects of 1,3-diaxial interactions. However, its achievement requires rigid systems which do not allow the reacting molecules to assume alternative conformations. For example, with carefully designed ligands a Pd-catalyzed cycloisomerization was accomplished in good yield and the product was parlayed into picrotoxinin (**3-42**).[45]

3.2. OTHER 1,3-STEREOINDUCTIONS

A classic synthesis of cedrol (**3-43**)[46] began with the alkylation of a cyclopentanonedicarboxylic ester. In this step the quaternary carbon common to all three rings was affixed. The reaction resulted in a *cis*-diester which was important for correlation with a degradation product of the natural compound and the final reconstitution. The distal ester group caused the alkylation from the opposite face.

Consider the preparation of a 2,2,3-trisubstituted cyclopentanone that contains three contiguous stereocenters as precursor for damsin. Correlation of this precursor with a bicyclo[2.2.1]heptan-2-one simplifies the stereochemical manipulation because of *endo/exo* biased reactivity and the possibility of discrimination during generation of 7-*syn* and 7-*anti* substituents. The availability of a bicyclo[2.2.1] hept-5-en-2-one from the inexpensive norbornadiene prompted an extensive investigation of the potential of the compound, and a route to damsin (**3-44**)[47] was part of the reward. One critical aspect was the *endo*-methylation at C-3 to avoid 1,3-interaction with the existing *syn*-7-methyl substituent. [Note that the secondary methyl group thus introduced underwent epimerization when the bridged ring was opened and parlayed into a condensed lactone, rendering accessibility to helenalin (**3-45**).[48]]

In a longipinene (**3-46a, 3-46b**) synthesis,[49] ring expansion of a cyclopentanone subunit was effected via a spiroepoxide and a Tiffeneau–Demjanov reaction of the derived amino alcohol. This sequence accommodated the facial selectivity during homologation. The alternative, consisting of Wittig reaction and epoxidation, would not be suitable. The steric determinator appears to be the bridgehead methyl group.

trans-2,5-Disubstituted tetrahydrofurans and pyrrolidines are quite easily accessible by displacement reactions which are generally participated by the neighboring heteroatom. In other words, the addition of nucleophiles to an incipient sp^2-hybridized carbon is stereoselective, and directed trans to the existing group at the α'-position. Complex molecular systems assembled in this manner include synthetic

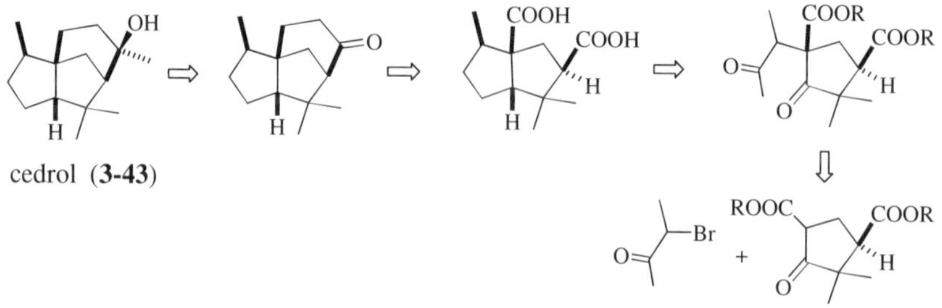

cedrol (**3-43**)

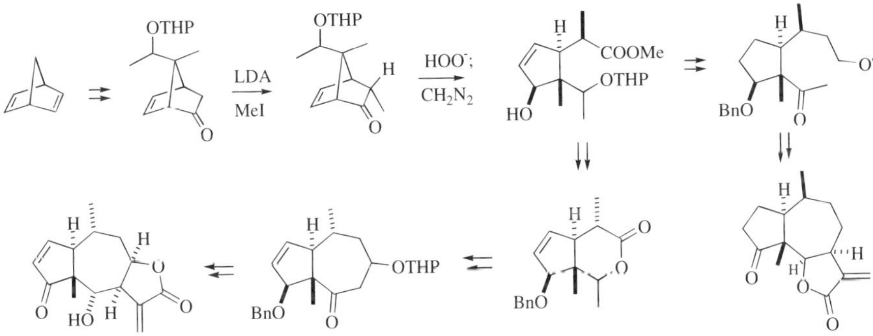

helenalin (**3-45**)

damsin (**3-44**)

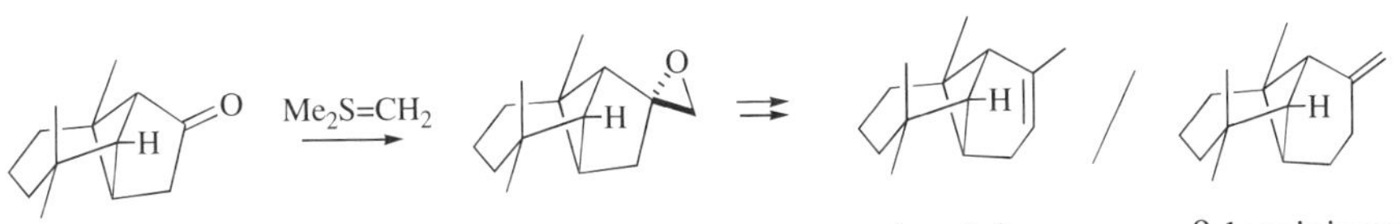

α-longipinene (**3-46a**)

β-longipinene (**3-46b**)

intermediates of (+)-tetronomycin (**3-47**)[50] and (+)-anatoxin-*a* (**3-48**).[51] A 5-propylproline derivative similarly acquired is useful for elaboration of (–)-pseudoconhydrine (**3-48a**).[51a] Hydrogenation of an exocyclic double bond is also subject to the same stereoinduction, as shown by a preparation of nonactic acid derivative (**3-49**).[52]

The preferred steric course for 1,3-dipolar cycloadditions involving tetrahydropyridine *N*-oxides is from the opposite face of the α′-substituent. Accordingly, a route to solenopsin-A (**3-50**)[53] was thus formulated.

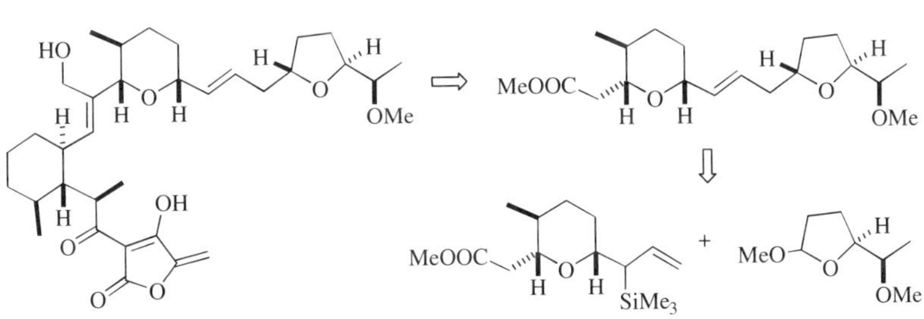

(+)-tetronomycin (**3-47**)

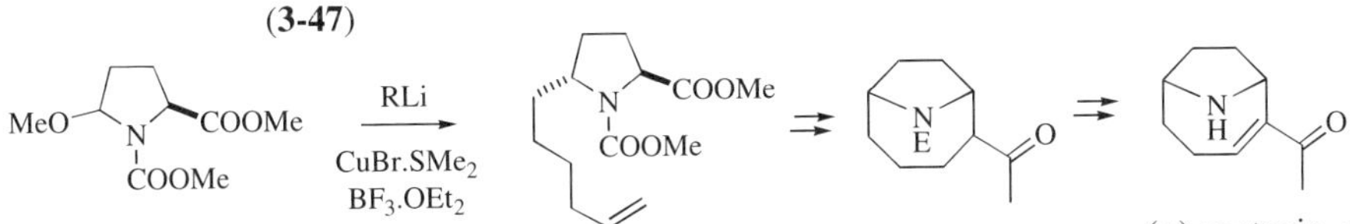

(+)-anatoxin-*a* (**3-48**)

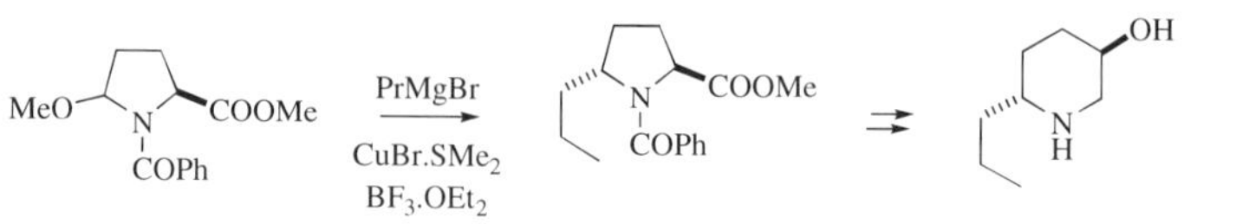

(*trans:cis* 9:1)

(-)-pseudoconhydrine (**3-48a**)

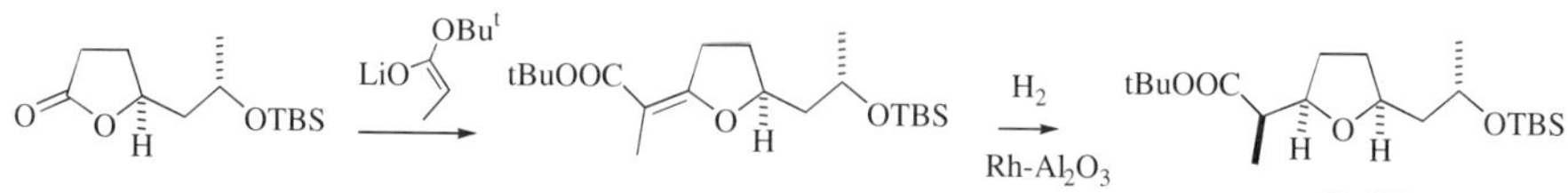

(**3-49**)

The inner tetrahydrofuran unit (**3-51**) of ionomycin during its synthesis was formed by an intramolecular oxymercuration.[54] The (*Z*)-double bond derived from a Wittig reaction exposed the *re*-face for reaction with the mercury(II) ion, the *si*-face being guarded by the secondary methyl group of the 1,3-dioxane unit.

The key to the establishment of the $C_{(14)}$ configuration in a synthesis of (+)-estradiol (**3-52**)[55] from a 2*S*-(*p*-toluenesulfinyl)-2-cyclopentenone was that the Michael donor attacked from the β-face as dictated by the chirality at sulfur. The introduced α-tetralone unit then took the responsibility for 1,2-stereoinduction in the subsequent alkylation steps. Note that metal chelation fixed the sulfinyl group in one conformation in relation to the enone unit, thereby exposing the face of the sulfur lone electron-pair to attack, as the other face is patrolled by the large tolyl substituent.

In situ trapping of the dioxolenium ions generated from Beckmann fragmentation of 2,3-isopropylidene-dioxycyclohexanone oxime esters with organoalanes created a new C–C bond.[56] The high stereoselectivity makes the

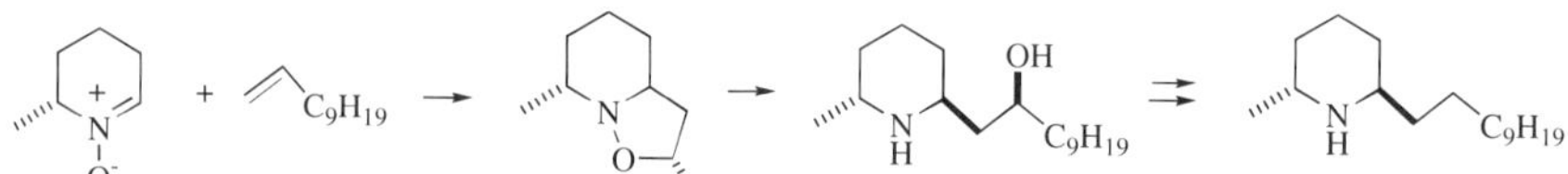

solenopsin-A (**3-50**)

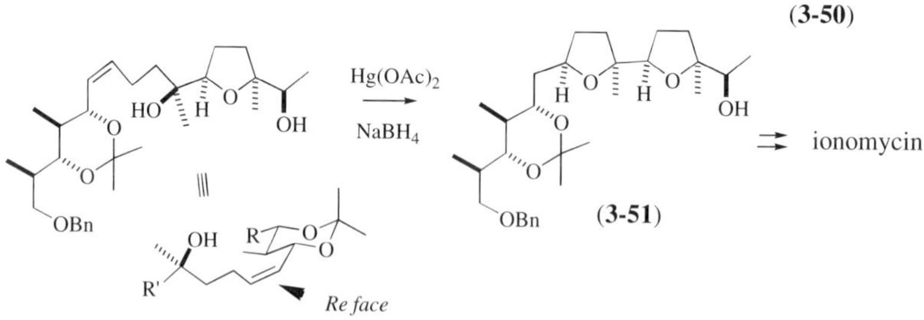

(**3-51**)

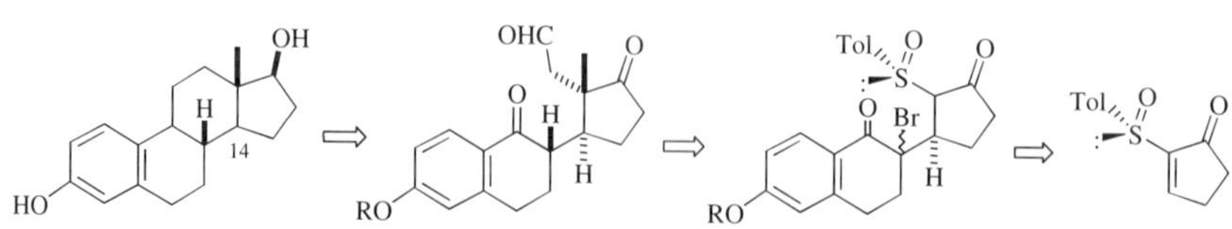

estradiol (**3-52**)

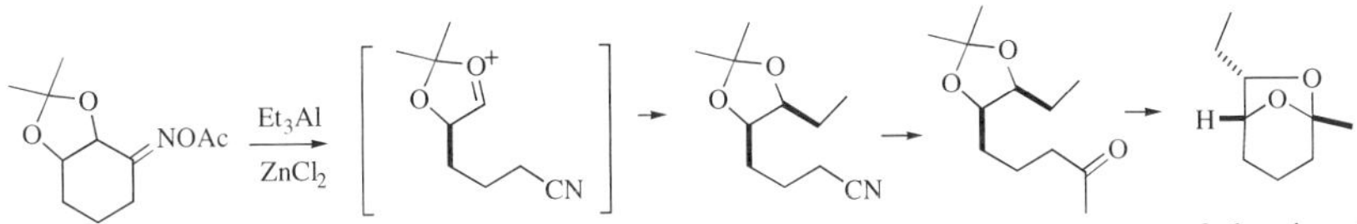

endo-brevicomin
(**3-53**)

protocol very useful in syntheses such as that of *endo*-brevicomin (**3-53**). The nucleophilic residue furnished by the aluminum reagent tends to add to the cationic intermediate from the same side as the emerging chain. It is conjectured that 1,3-diaxial interaction of the approaching nucleophile with one branch of the *gem*-dimethyl group was too severe that it would tolerate repulsion from the neighboring chain.

A building block for ophiobolin-C (**3-54**) was obtained by a Brook–Claisen rearrangement sequence of a cyclopentenol.[57] Thus, the configuration of the allylic alcohol was important. Although the stereoselectivity (3:1) for reduction of the enone was moderate, it nevertheless gave the desired compound as the major product. The only inducer of any significance was the alkoxyethyl group.

Thus, it is seen that a carbon chain may be inserted into a ring structure by arranging an allylic alcohol system in its precursor even if the desired product is devoid of the double bond. A particularly ingenious tactic was deployed in the assembly of the $C_{(28)}$–$C_{(49)}$ segment **(3-54a)** of rapamycin[58] by an Ireland–Claisen rearrangement followed by a decarboxylative maneuver. A cyclohexenol prepared from 2-deoxy-D-glucose was merged with the other building block represented by an acid. The method solved a difficult and awkward synthetic problem in connecting two stereogenic centers to a methylene group.

Transpositional reduction of cycloalkenones in a stereoselective manner was very critical to a synthesis of pachydictyol-A (**3-55**)[59] and was accomplished by reduction of the tosylhydrazone with catecholborane. In this case the reducing agent

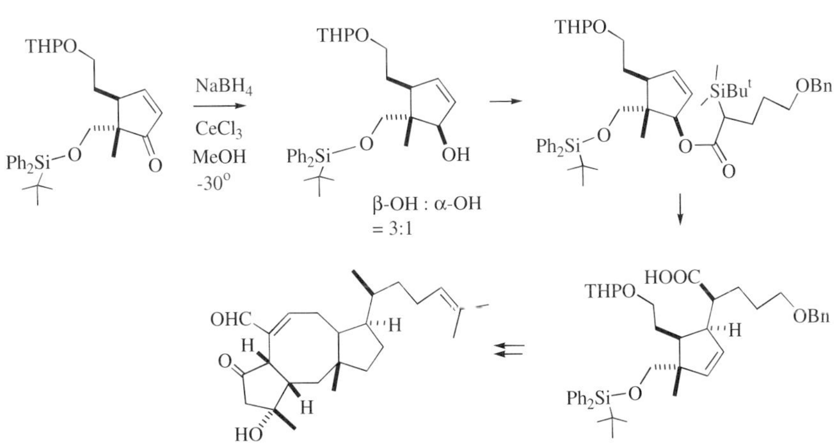

ophiobolin-C (**3-54**)

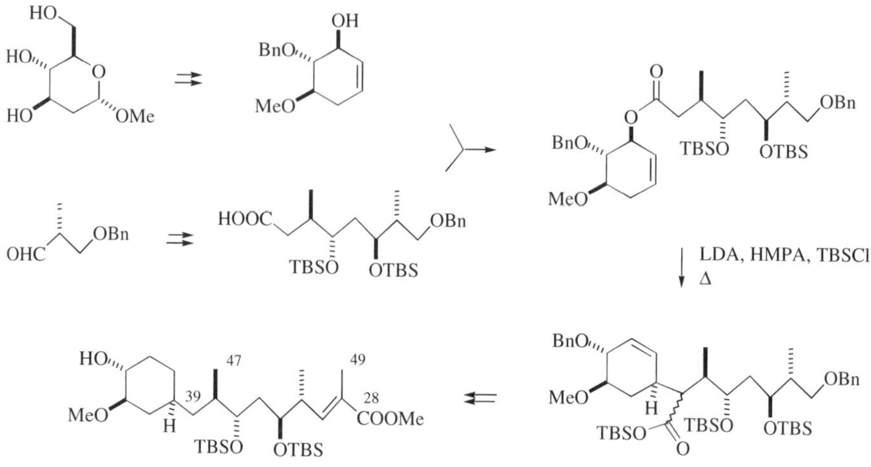

(**3-54a**)

approached from the same side as the angular hydrogen atom and the tosylhydrazine decomposed with hydrogen transfer to the β-position of the original enone system.

Carbonylative stitching of diene by hydroboration with thexylborane and succeeding manipulation furnished a B-seco A-aromatic steroid intermediate (**3-56**).[60] Further extension to 2,3-dialkenylcyclopentanol derivatives led to hydrazulenones which are known precursors (**3-57, 3-58**) of confertin and helenalin.[61,62] All of these hydroboration steps were diastereoselective due to the presence of stereocenters at allylic sites.

During a synthetic pursuit of the bistetrahydrofuran unit (**3-59**) of asteltoxin,[63] a monocyclic intermediate was subjected to dihydroxylation (OsO_4–NMO). Agreeable diastereofacial selectivity was obtained and was attributed to shielding of one face of

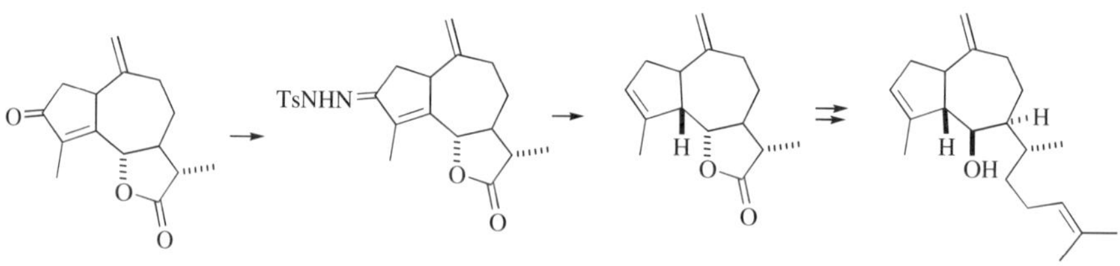

pachydictyol-A
(**3-55**)

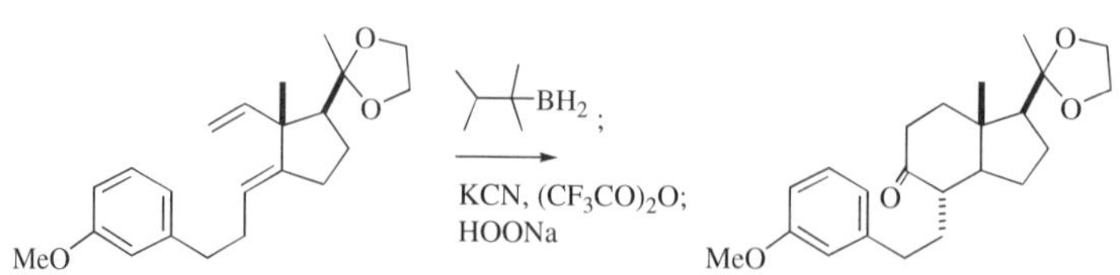

(**3-56**)

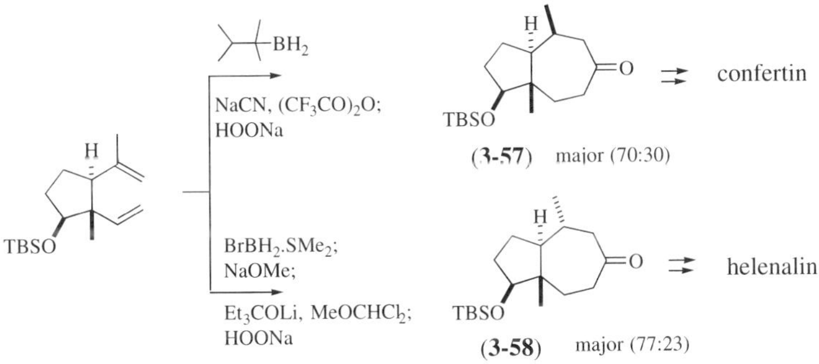

(3-57) (3-58)

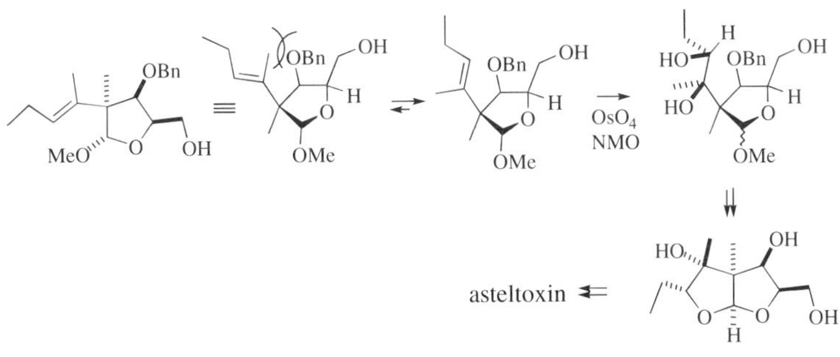

(3-59)

the double bond by the benzyloxy group. This conformation was apparently preferred as the alternative featured a synperiplanar relationship.

The Pauson–Khand reaction for cyclopentenone synthesis is subject to steric interactions in the formation of the intermediates containing C–Co σ-bonds. The attendant stereoselectivity was very beneficial to the formation of a precursor of pentalenene (**3-60**).[64] In reaching the alternative complex, the transition state must tolerate mutual interference of a methyl group and a methylene unit in spatial proximity. There was involvement of a detachable silyl group to control the reaction pathway in a synthesis of coriolin (**3-61**).[65]

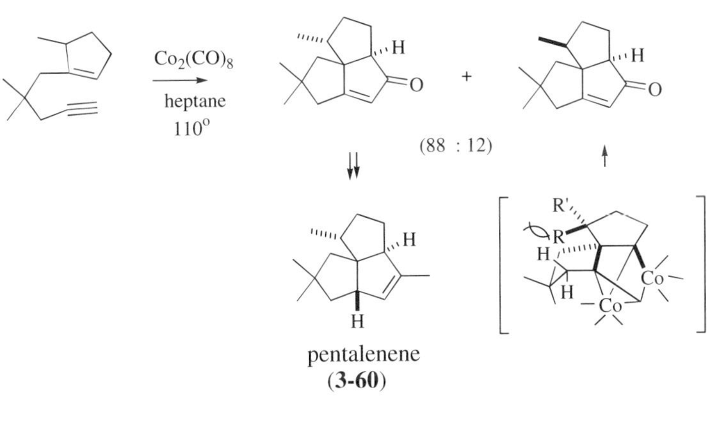

pentalenene
(**3-60**)

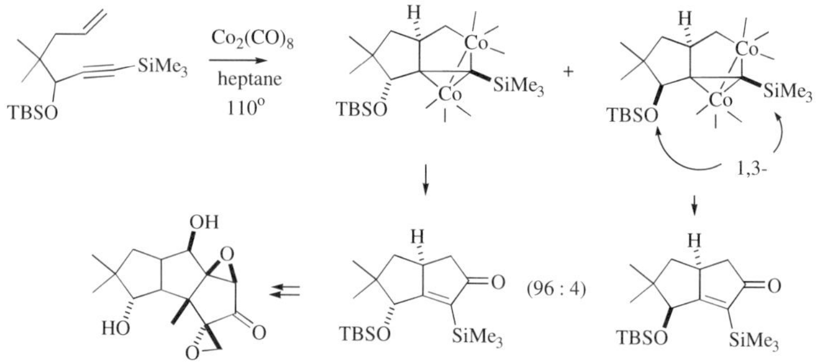

coriolin (**3-61**)

The synthesis of pentalenic acid (**3-62**),[66] based on an intramolecular [2+2]-photocycloaddition process, was smoothed by abstention of the allylic methyl group from incipient 1,3-interaction with the α-ester. This step set up three stereocenters of the target compound.

A significant contribution to modern synthesis concerned the elaboration of ginkgolide-B (**3-63**).[67] The γ-lactone interstitial between the two cyclopentane units evolved from a cyclobutanone. The guidepost for stereoselective construction of this array that comprises the central core of the molecule was a *t*-butyl group. By virtue of 1,3-induction to ensure the intramolecular ketene–alkene cycloaddition from the opposite side, stereochemistry at the ring junctions was established on the firmest footing.

An example in which an allylic substituent played a crucial role in creating a new stereocenter by sacrificing itself is found in the cyclization step leading to daucene (**3-64**).[68]

The Mukaiyama version of a directed aldol reaction involves an oxonium ion species as the acceptor. If the acetal is derived from a chiral diol, asymmetric induction becomes possible. The closest stereocenter to the reaction site is carbinolic and a bulky group at that position can give rise to stereoinduction. For best manipulation a C_2-symmetric chiral auxiliary appears to be needed. Generally, inversion is observed

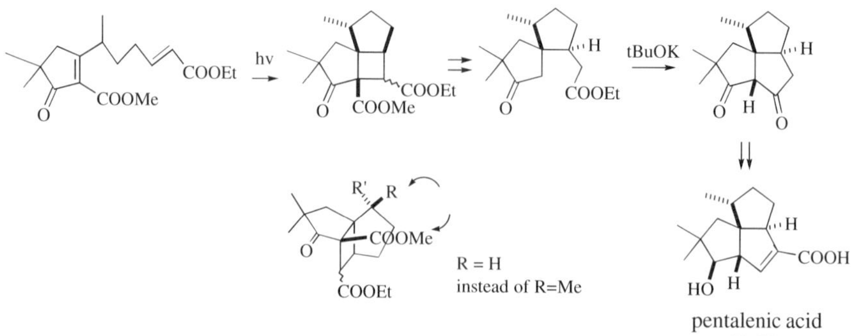

pentalenic acid
(**3-62**)

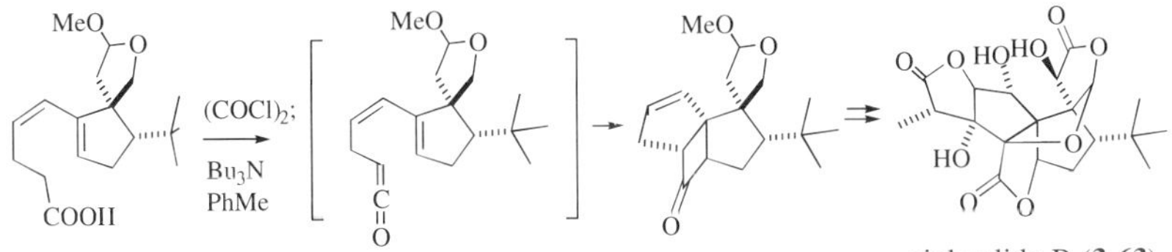

ginkgolide-B (**3-63**)

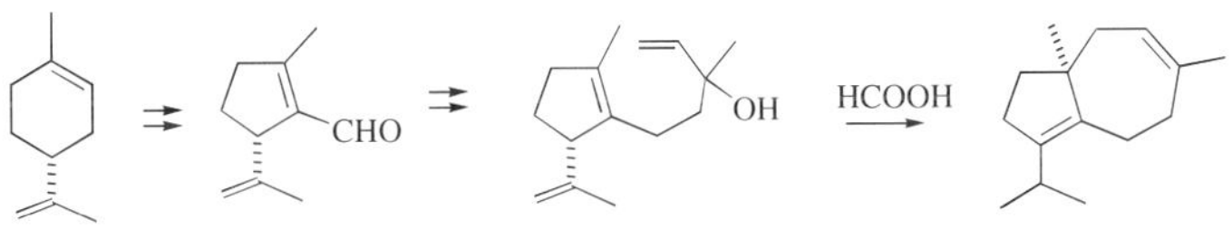

(-)-daucene (**3-64**)

with sterically unhindered aliphatic acetals and weak Lewis acids which tend to form intimate ion pairs.[69]

When aklavinone (**3-65**) was considered as a synthetic target,[70] an A-seco aldehyde was recognized as possible candidate for assembly by an aldol reaction. With an acetal derived from (*2R,3R*)-2,3-butanediol the reaction afforded two isomers in a 10:1 ratio, the major compound being the desired product.

A chiral AB-synthon (**3-66**) for the anthracycline antibiotics is quite readily accessible from a dienyl ether which was obtained from ring opening of a substituted 1,3-dioxane.[71]

By virtue of the steric bulk of the substituent, 1,3-stereoinduction is commonplace during alkylation of γ-trityloxymethyl-γ-butyrolactone and it has been practiced routinely. More interestingly, it is possible to change certain reaction sequences to elaborate such compounds into enantiomeric natural products: (+)-quebrachamine (**3-67**)[72] and (–)-quebrachamine (**3-67′**).[73]

The configuration of $C_{(20)}$ in the steroid side chain is a center of contention for many syntheses. Several methods are now available and those involving reactions with 17-ethylideneandrostane derivatives (**3-68**) are well developed. If the double

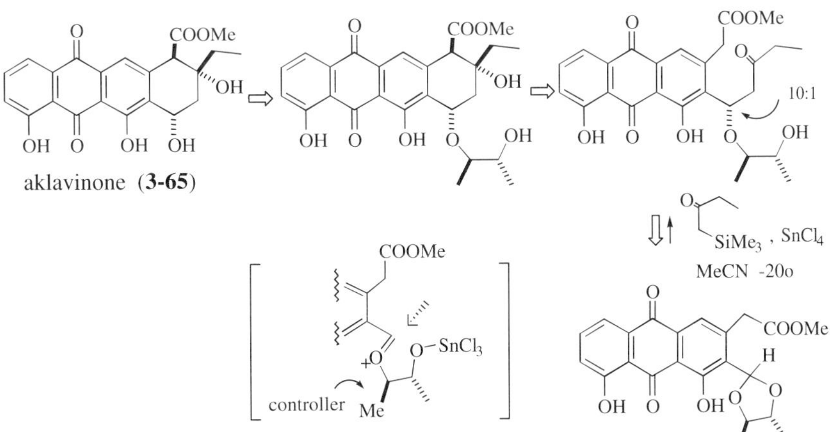

aklavinone (**3-65**)

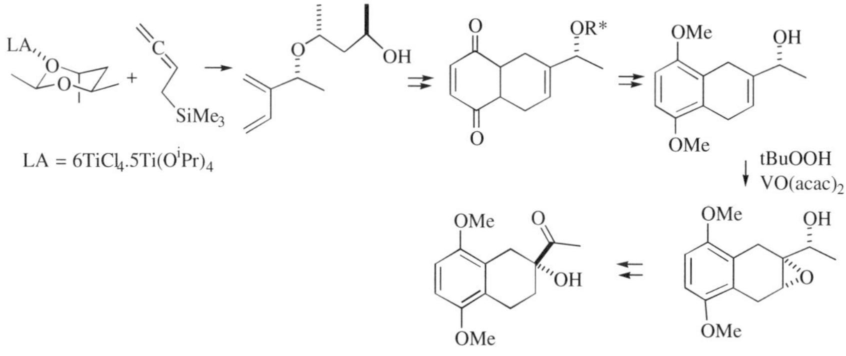

(3-66)

bond is (Z), the addition occurring from underside as dictated by the 18-methyl group uniformly gives rise to the natural configuration. Thus, hydroboration with subsequent alkylation of the derived boranes,[74] ene reaction,[75] conjugate addition or allylic displacement[76] with cuprate reagents, and displacement of π-allylpalladium complexes with nucleophiles[77] follow the same steric course. (Regiochemistry may be different for the last set of reactions due to attack on the metal atom prior to ligand transfer.)

The biomimetic synthesis of indole alkaloids based on secodine cyclization is very effective. By installing a stereocenter at the α-position of the N^b-atom, either

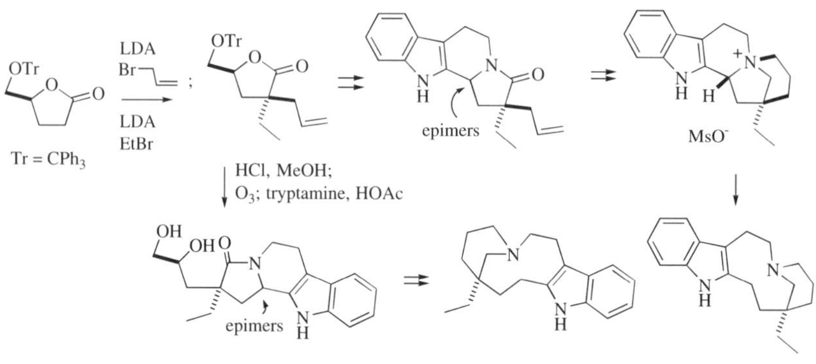

(-)-quebrachamine **(3-67')** (+)-quebrachamine **(3-67)**

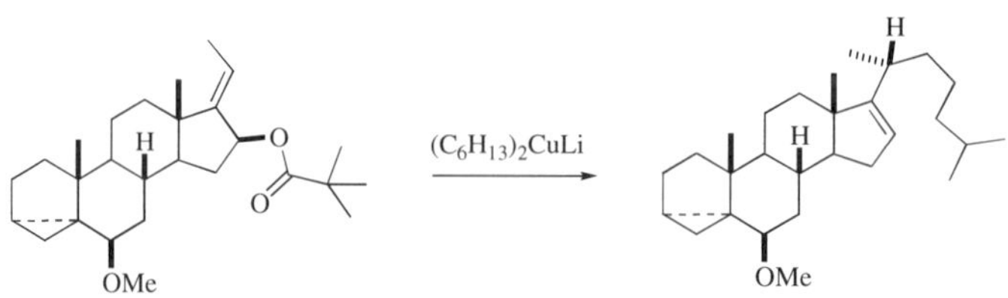

(3-68)

inside the chain or in a removable substituent, to achieve 1,3-induction, it has been possible to modify the routes to generate chiral products such as tubotaiwine (**3-69, 3-69′**).[78]

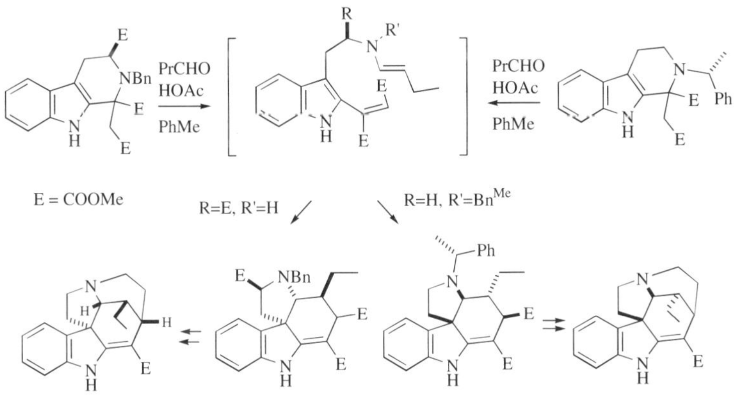

(+)-tubotaiwine
(**3-69**)

(-)-tubotaiwine
(**3-69'**)

The *trans*-1,3-disubstituted tetrahydroquinoline pattern of (+)-corlumine (**3-70**) gives a direct hint to its synthesis.[79] Reversing the roles of the two components, with a huge auxiliary group on the nitrogen atom to direct the coupling of chiral benzylideneimines with nucleophiles represents an alternative method for synthesis of compounds such as corynoline (**3-71**).[80]

Reinforcement of 1,3-stereoinduction by 1,4-induction has been demonstrated in the formation of 1,3-disubstituted tetrahydro-β-carbolines (**3-72**).[81]

An interesting pressure effect on chemoselectivity of [4+2]-cycloaddition has been observed during a synthetic study on (+)-erysotrine (**3-73**).[82] However, 1,3-diaxial-like interference from an ester group cannot be overcome, although the opposition from a benzylic hydrogen was overwhelmed under high pressure.

Using the general method developed by Wenkert, a synthesis of (+)-vallesiachotamine (**3-74**)[83] was accomplished. In the reaction of an *N*-tryptophylpyridinium salt with a cyclic nucleophile to establish the side chain at $C_{(15)}$, combined 1,2- and 1,4-stereoinductions operated. The large group immediately served to influence (1,3-stereoinduction) the C-ring cyclization.

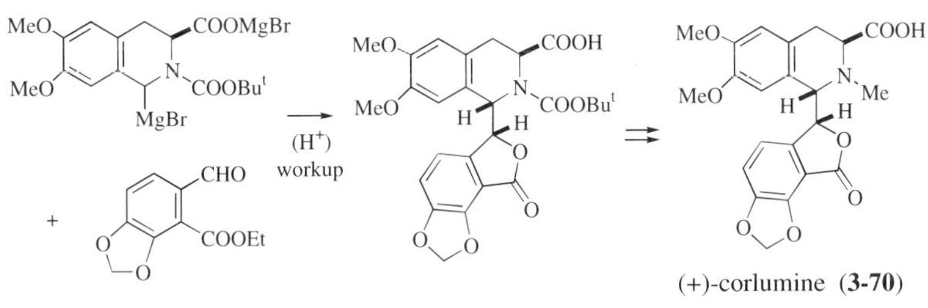

(+)-corlumine (**3-70**)

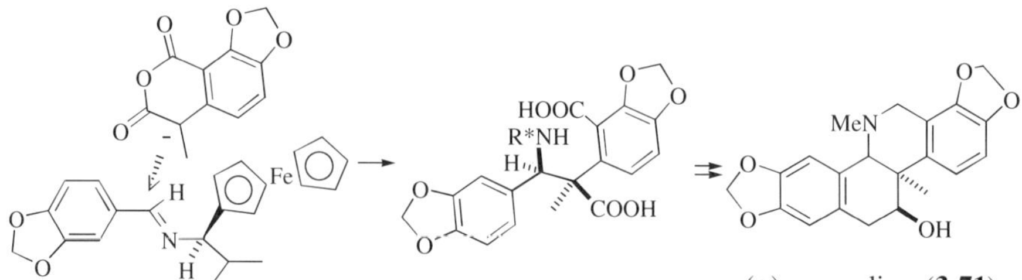

(+)-corynoline (**3-71**)

To assemble the saturated portion of vallesamidine (**3-75**) from a bicyclic imine (the reactive form is the enamine) the angular ethyl group ensured positioning of the cinnamic acid in its opposite face. The two components combined in the fashion that minimized nonbonding interactions.[84] The first bond that was formed eventually became the C_α–C_β bond of the indoline moiety.

Intramolecular desilylative capture of iminium salts by an allylsilane is expected to proceed antarafacially. However, the imine must have a well-defined geometry for the generation of two contiguous stereocenters. To apply this method to a morphine (**3-76**) synthesis[85] requires a "*cis*-imine" portion. Fortunately this can be easily managed by attaching a large group to the nitrogen atom.

Spirocyclization by a Pd-catalyzed intramolecular zinco-ene reaction as the key step for the synthesis of (–)-erythrodiene (**3-76a**)[85a] relied on a stereoselective C–C bond formation from the side opposite to the bulky isopropyl group.

An intramolecular free radical ring closure to form hydrindenes, which involved hydrogen abstraction from the ring junction atom, was found to be greatly affected by a substituent three atoms away.[86] The degree of steric hindrance in substrates such as (**3-77**) and (**3-78**) to the tin hydride reagent was directly related to the *cis:trans* ratio of the products.

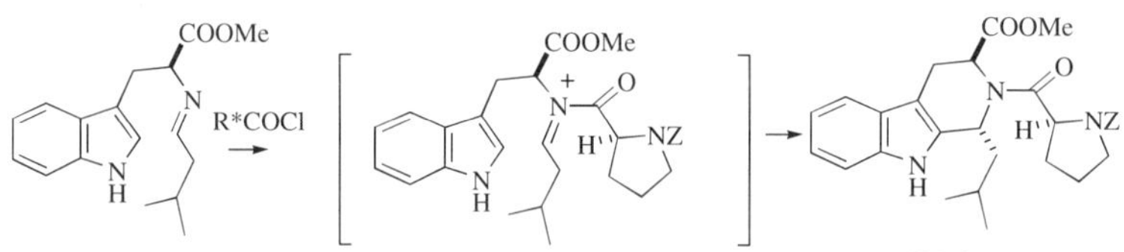

(**3-72**)

Z = COOBn

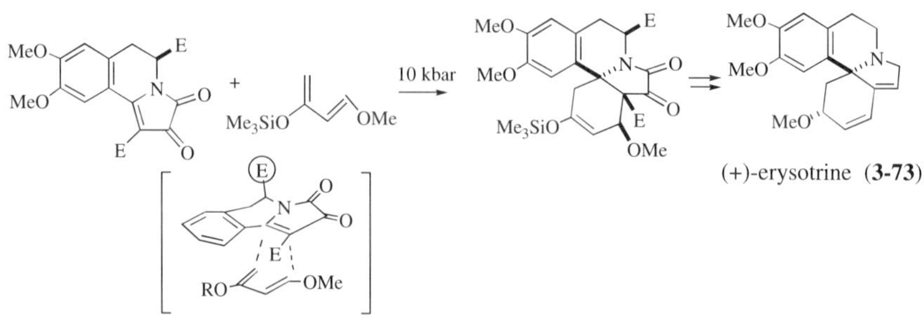

(+)-erysotrine (**3-73**)

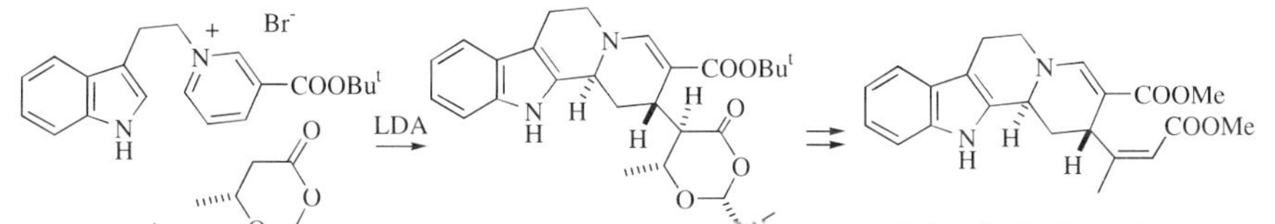

(+)-vallesiachotamine
(**3-74**)

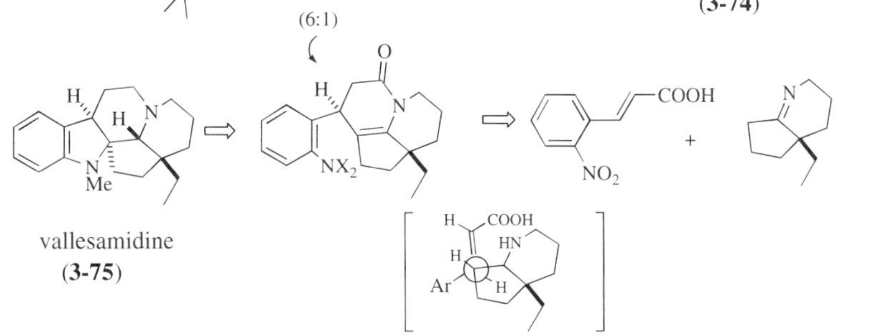

vallesamidine
(**3-75**)

favored TS

An intriguing protocol for the generation of asymmetric quaternary centers involves formation and alkylation of a chiral 4-substituted 2-*t*-butyl-1,3-dioxolan-4-one.[87] On condensation of pivalaldehyde with a chiral α-hydroxyl acid the chirality is transferred to the lactolic carbon atom, while forming the cyclic product having *cis* substituents. Although enolization destroys the original chirality but the direction of alkylation (*anti*) is determined by the *t*-butyl group. The great many syntheses of chiral natural products attest to the usefulness of this method: frontalin (**3-79**),[88]

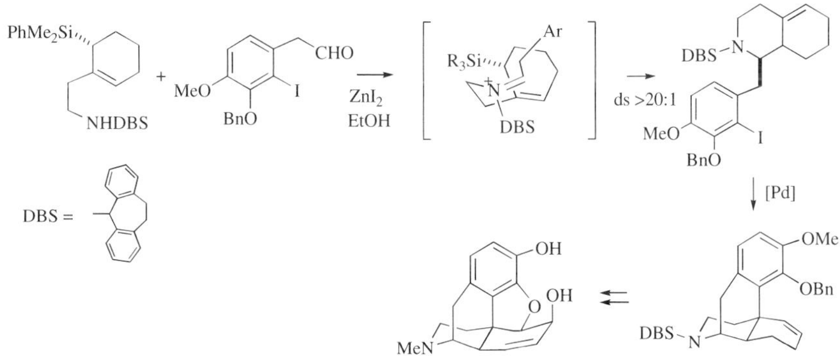

(-)-morphine
(**3-76**)

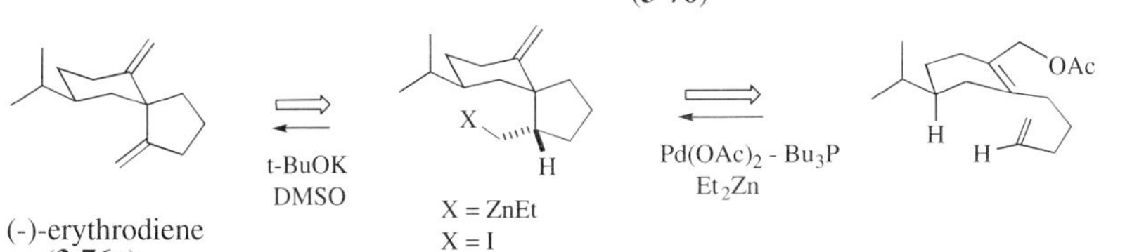

(-)-erythrodiene
(**3-76a**)

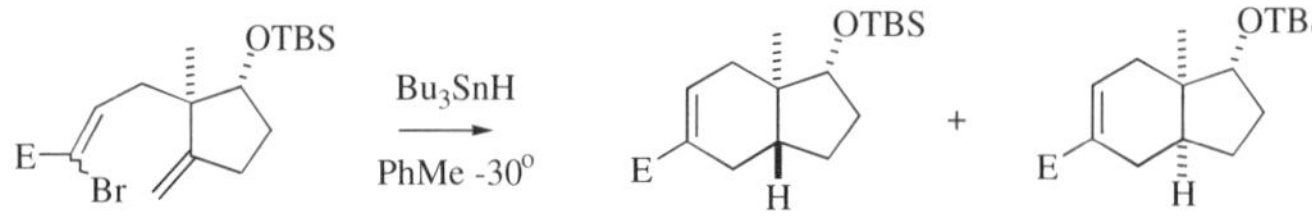

(3-77) (95 : 5)

E = COOMe

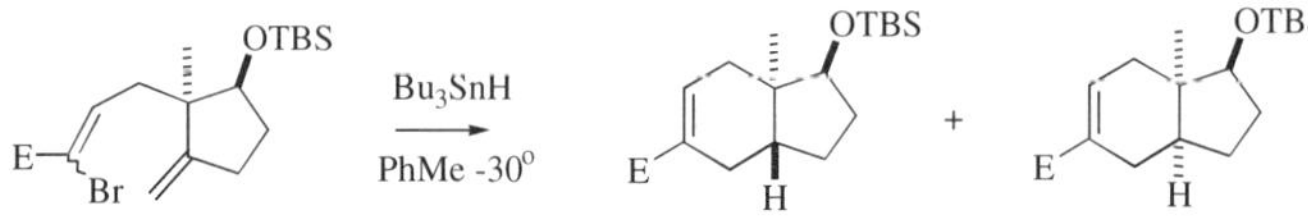

(3-78) (57 : 43)

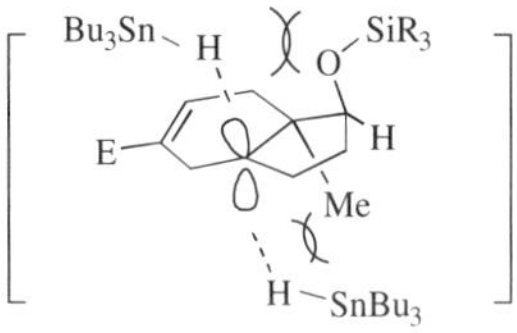

β-rhodomycinone (**3-80**),[89] eremantholide-A (**3-81**),[90] ipomeamarone (**3-82**),[91] camptothecin (**3-83**)[92] are only a few examples.

A related system consists of the enantiomeric 2-*t*-butyl-1-carbomethoxy-2,3-dihydropyrimidin-4(1*H*)-ones (**3-84, 3-84′**).[93]

A sequence of noteworthy maneuver developed three stereocenters of the dynemicin-A (**3-85**) system.[94] First, the benzylic center induced the addition of an acetylide anion to the quinolinium ion, then a [2,3]-Wittig rearrangement followed by dehydration furnished an (*E,E*)-diene moiety for an intramolecular Diels–Alder

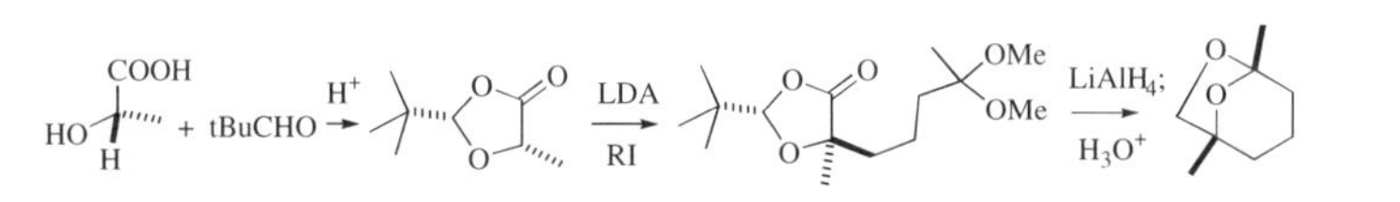

(*S*)(+)-lactic acid (+)-frontalin (**3-79**)

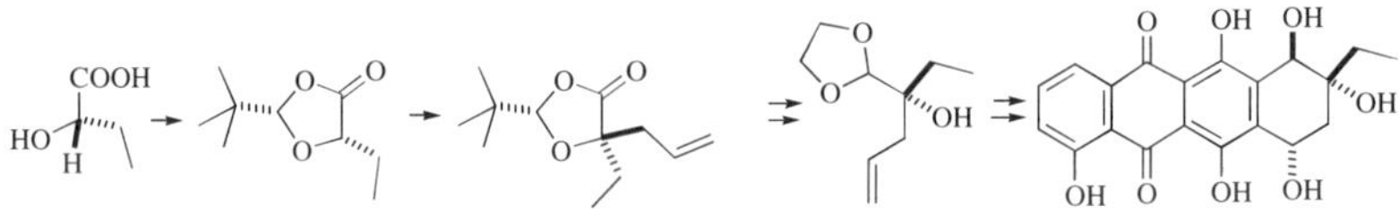

(*R*)-β-rhodomycinone
(**3-80**)

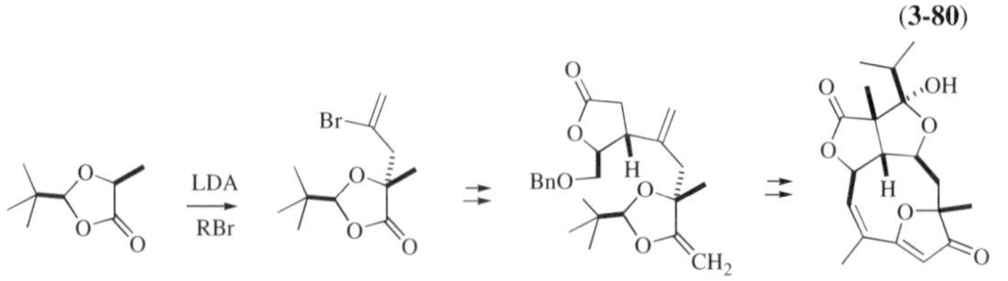

(+)-eremantholide-A
(**3-81**)

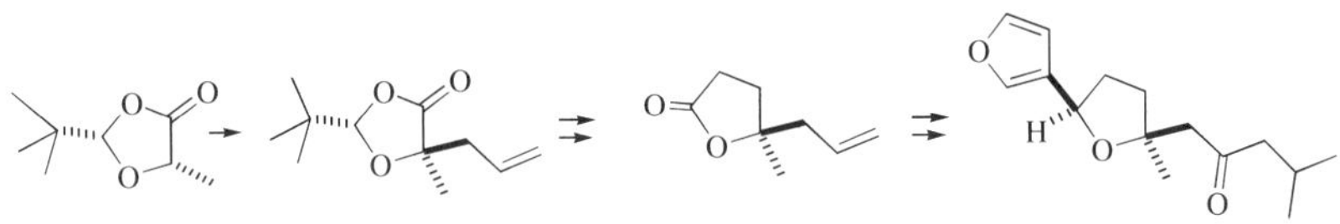

(+) ipomeamarone
(**3-82**)

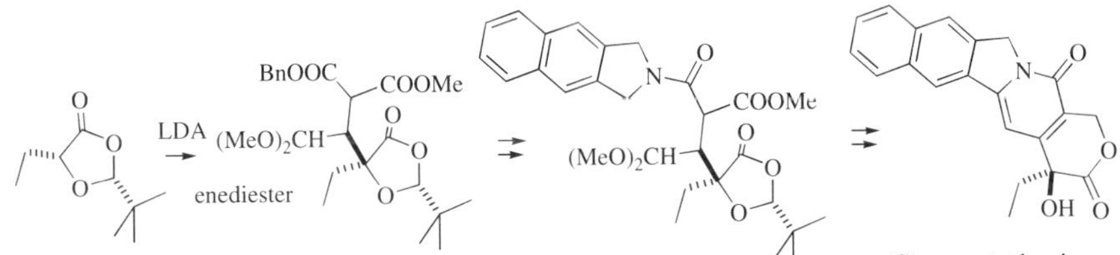

(*S*)-camptothecin
(**3-83**)

(*S*)- (**3-84**) (*R*)- (**3-84'**)

reaction that generated the correct stereochemistry for closure of the cyclodecenediyne circuit.

Sigmatropic rearrangements which are stereochemically influenced by elements outside the pericycle are well known. For the sake of illustration the Claisen rearrangement of an iridenol derived from limonene is mentioned. The isopropyl group dominated the steric course and thereby rendered the product useful for the elaboration of daucene (**3-86**)[95] and isoamijiol (**3-87**).[96] Similarly, rearrangement of a chiral amide enolate was found to give excellent results with respect to asymmetric

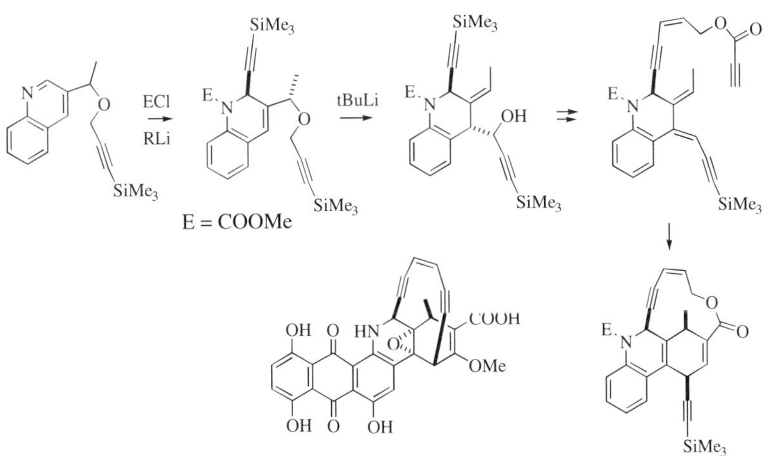

dynemicin-A (**3-85**)

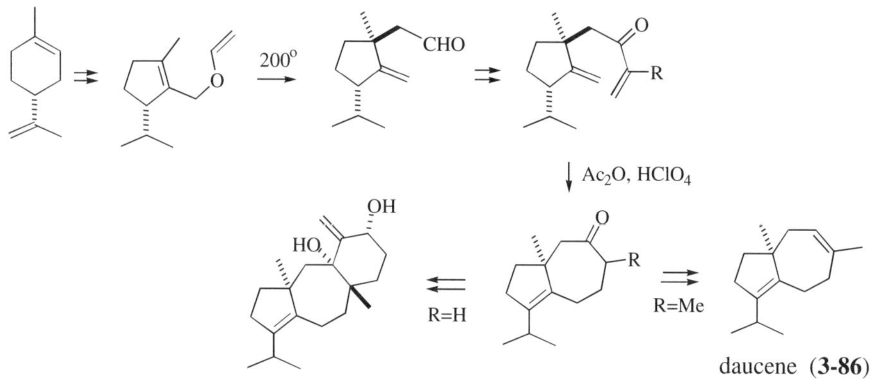

daucene (**3-86**)

isoamijiol (**3-87**)

induction.[97] Here two stereocenters outside of the pericycle cooperated (by 1,3- and 1,4-induction) to favor one particular transition state, enabling a highly effective synthesis of (–)-isoiridomyrmecin (**3-88**) to materialize.

It is not always possible to predict the degree of stereoinduction by a stereocenter in a freely rotating carbon chain. Thus, it was a useful observation in connection with an approach to isocaespitol (**3-89**) that a homoallylic bromine atom directed epoxidation of a (*Z*)-double bond to the extent of 8:1 in diastereoselectivity.[98]

On the other hand, an adverse 1,3-stereoinduction which unexpectedly surpassed 1,2-induction distressed a synthesis of (–)-furaquinocin-C (**3-90**).[99] Contrary to general consensus, an undesirable conformation of the butenolide which was designated to undergo conjugate addition has the lowest energy.

Certain circumstances may arise such that a more expedient performance of net 1,3-induction is by simultaneous 1,2- and 1,3-induction followed by destruction of

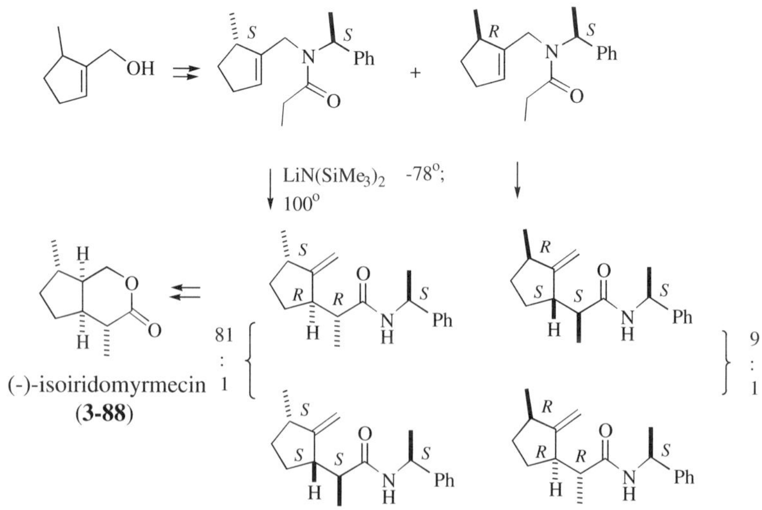

(-)-isoiridomyrmecin
(**3-88**)

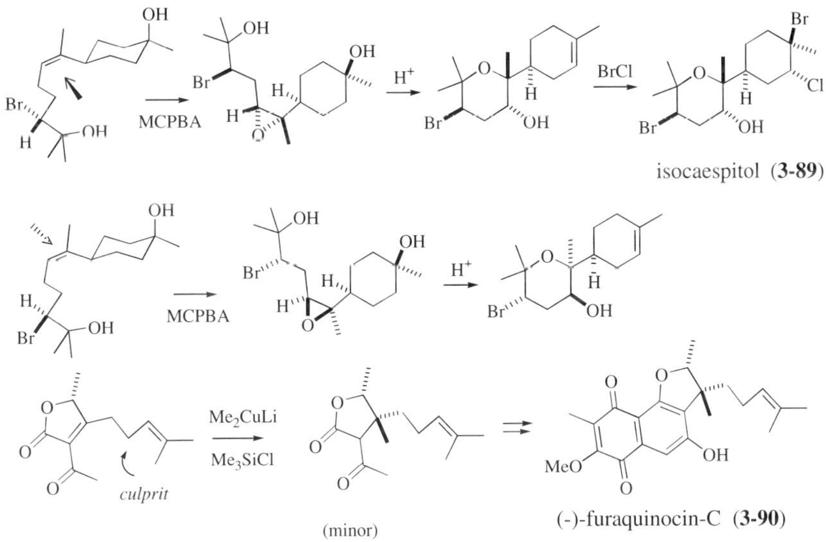

isocaespitol (**3-89**)

(minor)

(-)-furaquinocin-C (**3-90**)

one stereocenter. A stereoselective reaction on a double bond presents such a scenario, as demonstrated in a synthesis of coriolin (**3-61**).[100]

1,3-Transpositional reactions on allylic systems generally follow well-defined stereochemical courses. In addition to sigmatropic rearrangements, S_N2' reactions have been exploited to introduce molecular fragments or functionalities. 1,3-Chirality transfer is very readily accomplished by displacement reactions involving copper and palladium reagents or catalysts.[101] Thus, it permitted the establishment of the $C_{(7)}$–$C_{(13)}$ fragment of **(3-90a)** rapamycin[102] from either D-glucal or D-galactal, by merely adjusting the double bond configuration of each intermediate.

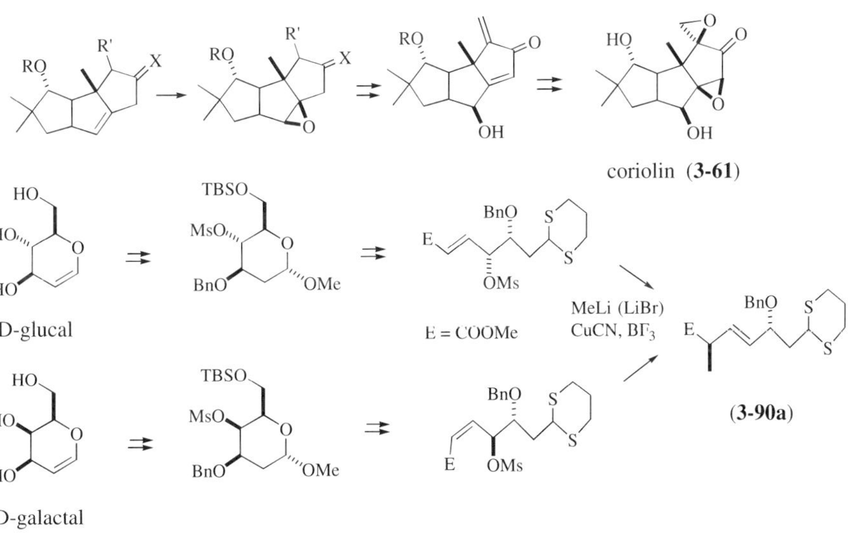

coriolin (**3-61**)

D-glucal

D-galactal

(**3-90a**)

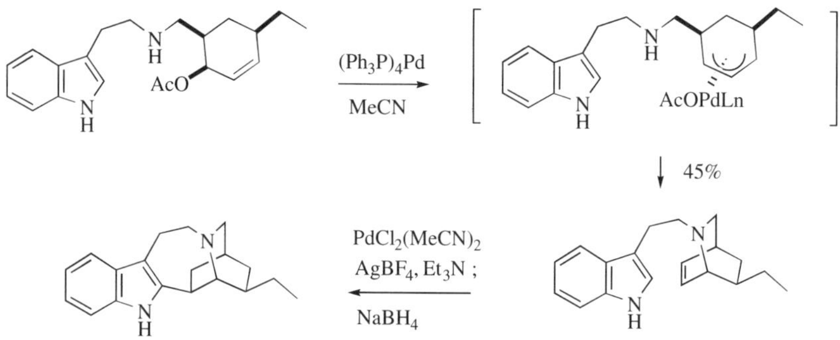

ibogamine (**3-90b**)

The palladium-catalyzed allylic substitution is a particularly popular synthetic process.[103] Double inversion of configuration results from π-allylpalladium complex formation and attack by the nucleophile. A concise synthesis of ibogamine (**3-90b**)[104] demonstrates the power of the reaction. The substrate was obtained from a Diels–Alder route, and interestingly, the two subsequent cyclization steps were mediated by Pd species.

3.3. $A^{1,3}$-STRAIN AND REACTIVITY

The conversion of a double bond into sp^3 centers is greatly affected by groups at the allylic position. From a purely steric standpoint, the allylic system that contains a (*Z*)-substituent (e.g., R=Me...) would not abide in one particular conformation in which that substituent is coplanar with an allylic group other than a hydrogen atom. Supporting evidence is numerous for the existence of this effect, which is now known as allylic 1,3-strain ($A^{1,3}$-strain).[105]

Generally, when the allylic position is not fully substituted, the conformation (**3-91**) which has an eclipsing hydrogen atom with the (*Z*)-substituent is an energy minimum. However, deviation of ±30° from the eclipsed form by rotation of the intervening $C_{(sp2)}$–$C_{(sp3)}$ bond is allowed because the cost in energy is < 1 kcal/mol. Such a rotation may accommodate stereoelectronic demands from reactions at the double bond.

When the C–X bond (X being the allylic substituent) contributes to a reaction stereoelectronically, its optimal effect would be felt when aligned parallel to the π-orbital. Two such conformations (**3-92, 3-93**) differing in energy of > 2 kcal/mol can

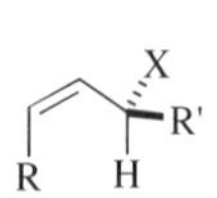

(**3-91**)

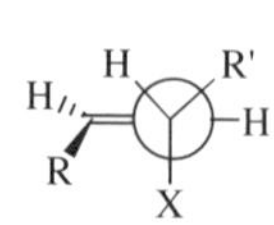

(**3-92**)

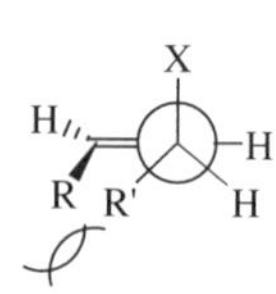

(**3-93**)

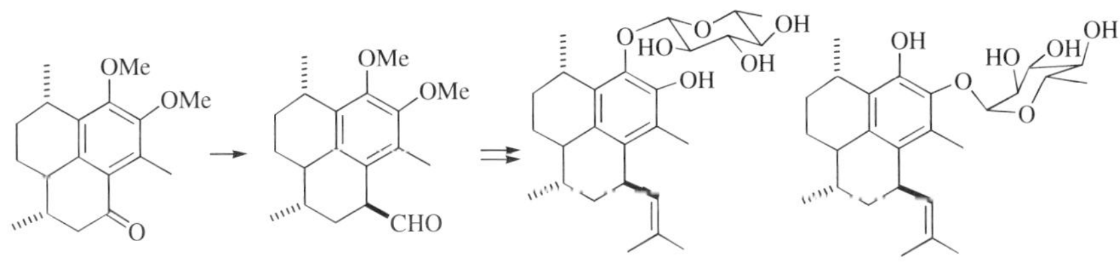

pseudopterosin-A (**3-94**) pseudopterosin-E (**3-95**)

be identified, one of which sustains allylic 1,3-strain. Consequently, a higher stereoinduction results when the system embodies a larger (*Z*)-substituent.

Peri-interaction in substituted tetralins makes a pseudoaxial benzylic substituent more stable. Thus, generation of the isobutenyl side chain of pseudopterosin-A (**3-94**) and pseudopterosin-E (**3-95**) by Wittig reaction of an aldehyde did not create any stereochemical problem.[106] Actually, the aldehyde was prepared by homologation of the ketone and it already had the same configuration.

A tetrahydrophenanthridine derivative (**3-96**) which was used in a synthesis of dynemicin-A underwent addition stereoselectively.[107] In this case the magnesium chelate formation involving the alcohol and a methoxy group locked the substrate into

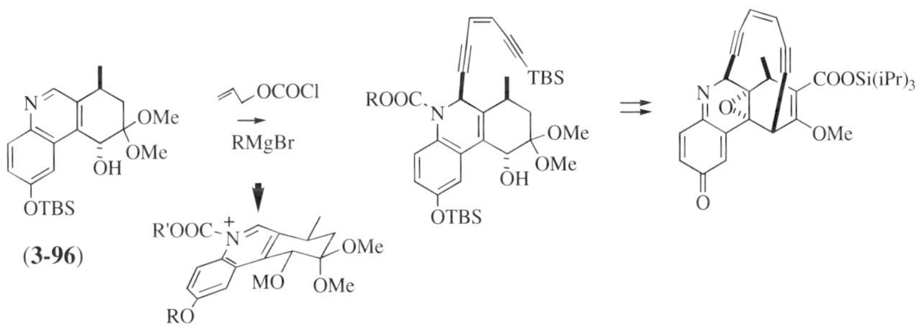

(**3-96**)

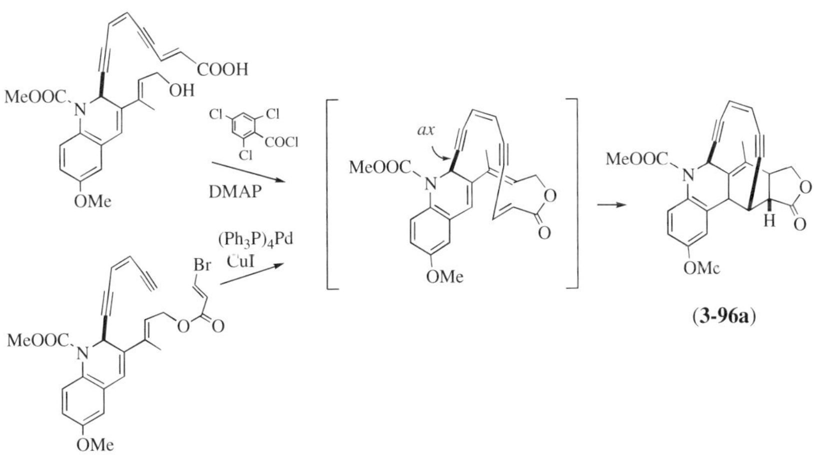

(**3-96a**)

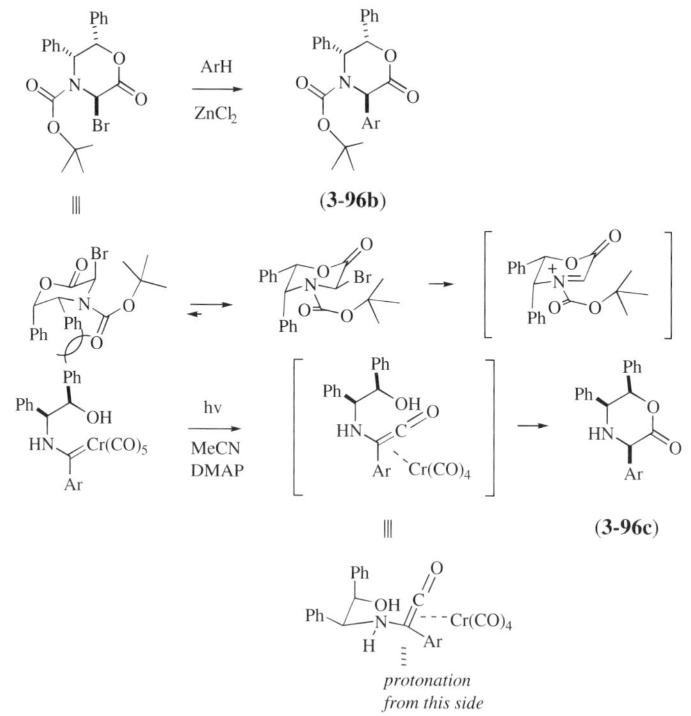

(3-96b)

(3-96c)

a conformation that oriented the benzylic methyl pseudoequatorially. The attack by the Grignard reagent followed an axial course. Accordingly, the product disposed the methyl and the enediyne chain in a *cis* relationship as required.

A transannular Diels–Alder reaction[108] to furnish the dynemicin-A system was apparently promoted by the allylic 1,3-strain factor. The disposition of the acetylenic girdle in an axial orientation favored the cycloaddition. Actually, macrocyclization either by lactonization or by organometallic coupling between the terminal alkyne and a bromoalkene at room temperature led directly to the intricate ring system (**3-96a**).

The observation that Friedel–Crafts alkylation[109] with *N*-(*t*-butoxycarbonyl)-5(*R*)-bromo-2(*R*),3(*S*)-diphenylperhydro-1,4-oxazin-6-one resulted in the displacement with retention of configuration at $C_{(5)}$ to give (**3-96b**) is a relevant phenomenon. It seems likely that the preferred conformation had an axial phenyl substituent on the carbon adjacent to the nitrogen atom, and the aryl group avoided the 1,3-diaxial interaction with the same. On the other hand, the situation must be different in a photocyclization route[110] to analogous perhydrooxazinones (**3-96c**), due to the absence of the ester substituent on the amine.

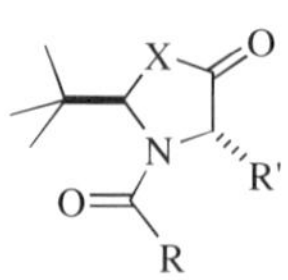

(3-96d)

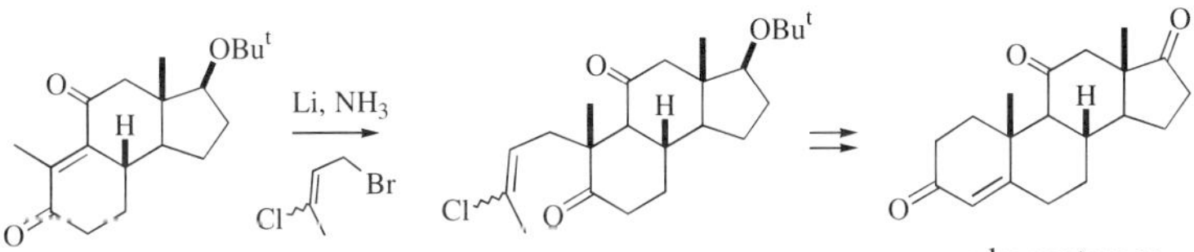

adrenosterone
(**3-97**)

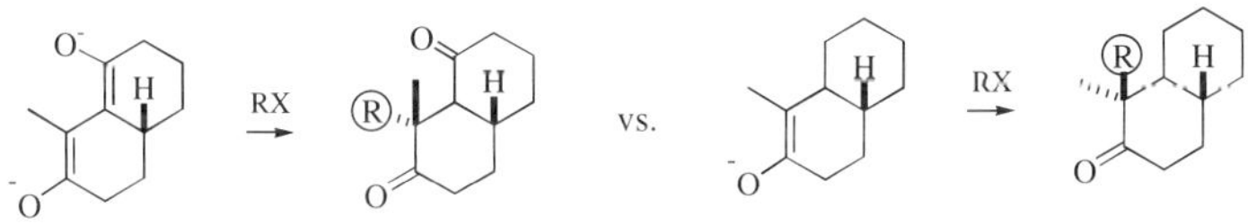

The above examples illustrate that the ester group of a planar carbamate exerted the strain because the amide C–N bond has considerable double bond character. It is interesting to note that for some molecules such as **3-96d** and their enolates, decrease of severe $A^{1,3}$-strain can occur by pyramidalization of the nitrogen atom.[111]

A proven method for synthesizing 11-ketosteroids [e.g., adrenosterone (**3-97**)] involves reductive alkylation of tricyclic enediones with an electrophilic chain for subsequent elaboration of the A-ring.[112] The methyl group which is squeezed between two oxygen atoms in the dienediolate species would enjoy some relief by tilting toward the angular hydrogen (pyramidalization) and thereby favoring the reaction with the electrophile from the other side. Without the *peri*-oxygen atom the stereoelectronic effect dominates, which leads to opposite stereoselection.

Cyclobutanecarboxylic esters can be prepared by intramolecular alkylation. When further substituted with methyl and isopropenyl groups at $C_{(2)}$ and $C_{(3)}$, respectively, ethyl 5-tosyloxypentanoate underwent cyclization to afford predominantly a *cis*-dialkyl ester (ratio 97:3).[113] The allylic 1,3-strain minimization caused the enolate to align the methine hydrogen coplanarly with the other substituents of the fully substituted double bond. This consequence also enabled the utility of the major product for a synthesis of fragranol (**3-98**). The spirocyclization step following reaction of a fulvene (**3-99**) with lithium dimethylcuprate[114] appeared

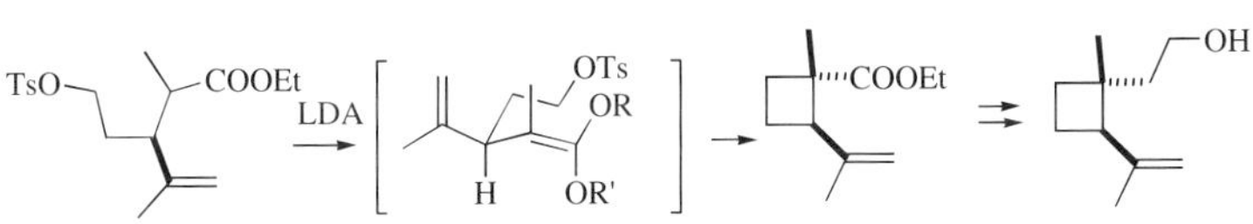

R= Li, R'= Et
or R= Et, R'= Li

fragranol
(**3-98**)

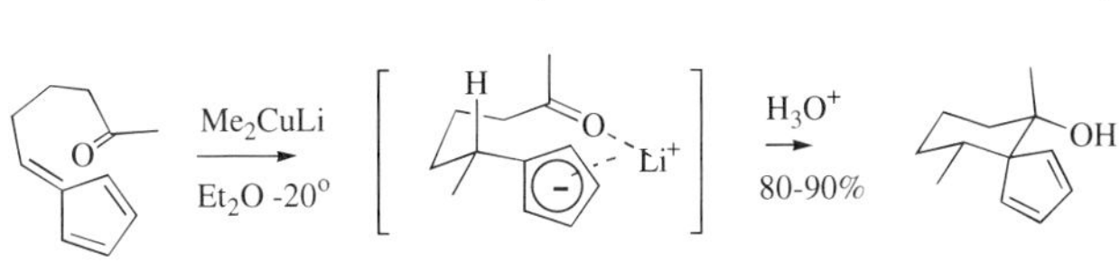

(**3-99**)

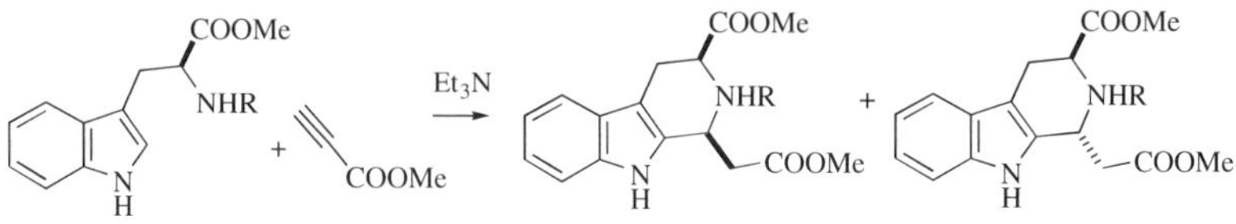

(3-100) (3-101)

	(3-100)	(3-101)
R = H	71	29
R = Bn	27	73

to be governed by $A^{1,3}$-strain factor as well as chelation and minimization of 1,3-diaxial interaction.

The major condensation product from a tryptophan methyl ester with methyl propynoate is the *cis*-1,3-disubstituted tetrahydro-β-carboline (**3-100**), whereas when the nitrogen atom is monobenzylated the *trans*-isomer (**3-101**) becomes predominant.[115] The equatorial *N*-benzyl group reinforces the $A^{1,3}$-strain due to interaction of the acetic ester chain with the indole moiety.

Conjugate addition of a methyl group to a Δ^2-piperideine ester (**3-102**) was found to be affected by a group at $C_{(6)}$.[116] The steric course was attributed to $A^{1,3}$-strain but it seems more likely that $A^{1,2}$-interaction between the esters on $C_{(2)}$ and the nitrogen atom caused the *N*-substituent to be pseudoaxial and also to inhibit the nucleophile to attack from the same side. A boatlike transition state was adopted.

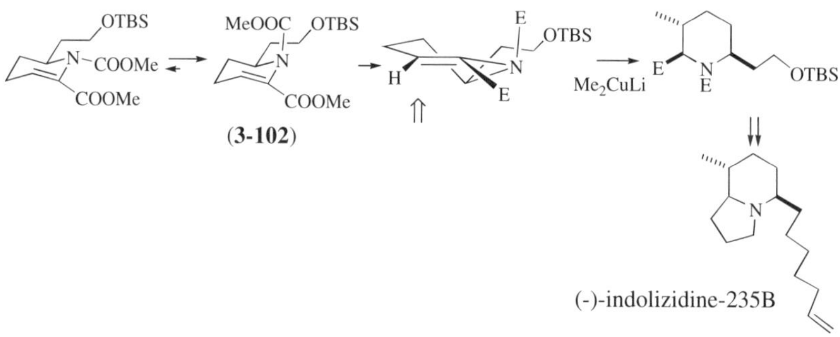

(3-102)

(-)-indolizidine-235B

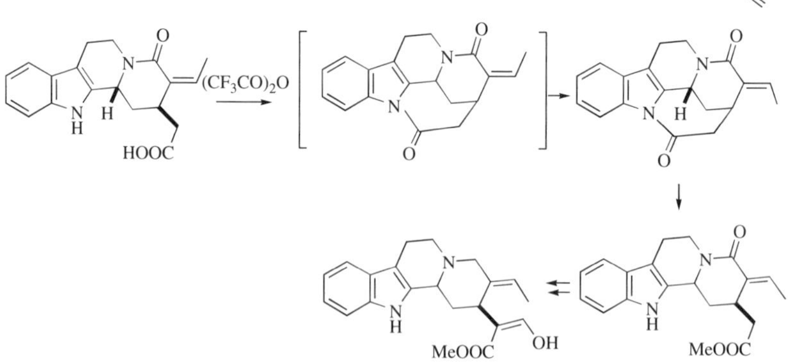

geissoschizine (**3-103**)

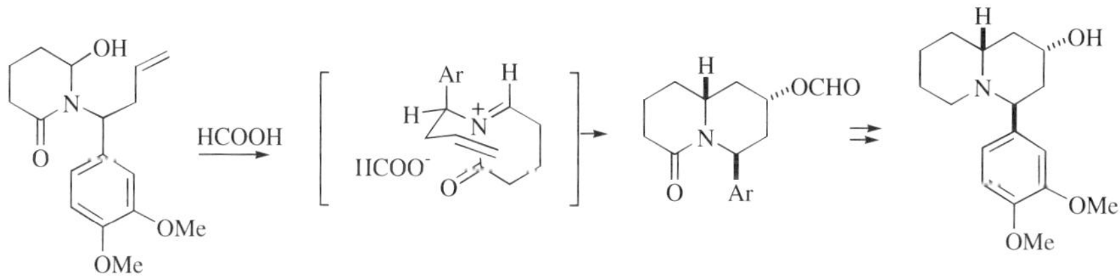

lasubine-I (**3-104**)

Because of severe $A^{1,3}$-strain, establishment of the (*E*)-double bond of geissoschizine (**3-103**) requires special attention. In one synthetic approach,[117] this double bond made its appearance in the (*Z*)-form in the early intermediates. Only when a seven-membered lactam was created to relieve the strain was isomerization possible.

The formolysis of an *N*-(α′-aryl-3-butenyl)-δ-hydroxy-δ-valerolactam led to a product with an axial aryl group and hence a perfect precursor of lasubine-I (**3-104**)[118] (cf. synthesis of vertaline[119]). The alternative conformation would have occasioned an Ar : CO interaction similar to $A^{1,3}$-strain because the acyliminium segment is planar.

An analogous situation is in the epimerizability of an aldehyde to an axial orientation, which was essential to bring about the ring closure that eventually completed a synthesis of *O*-methylpallidinine (**3-105**).[120]

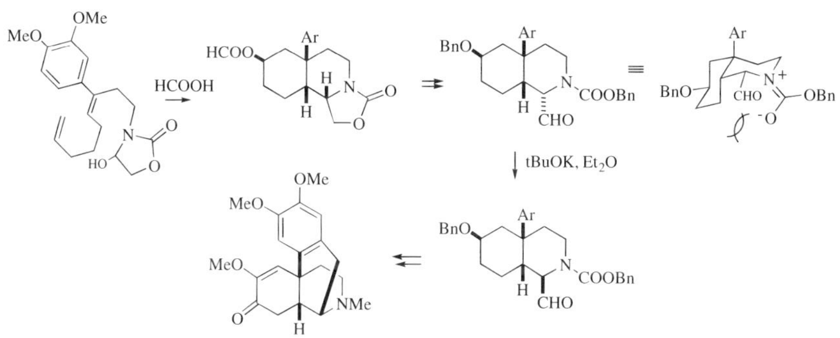

O-methylpallidinine
(**3-105**)

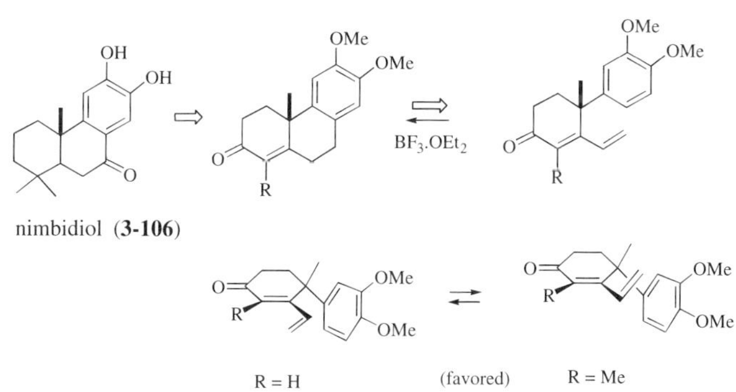

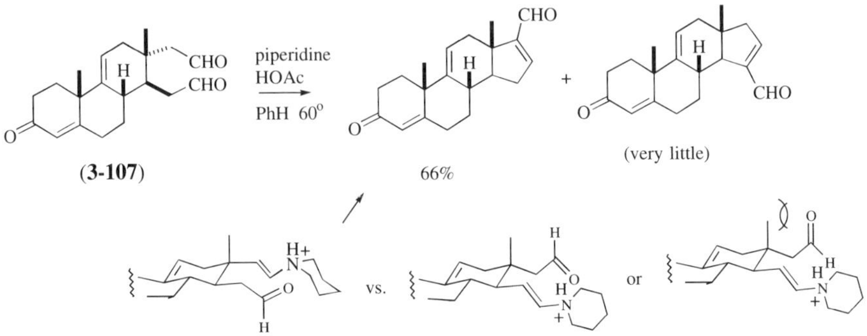

When a synthesis of nimbidiol (**3-106**) was conducted through B-ring formation by an intramolecular conjugate Friedel–Crafts alkylation process,[121] the importance of conformational factors was revealed. The dienone in which the α-position was substituted was a sluggish substrate and a large amount of catalyst was required to bring about a modest yield of the tricyclic product. The reason is likely to be a preference for the *s-cis* conformation in the absence of an α-substituent. To shift the equilibrium toward the active *s-trans* conformer, insertion of a methyl group to introduce $A^{1,3}$-strain indeed had a decisive effect in giving the product in higher yield and under milder conditions.

The regioselective D-ring closure by an aldol condensation of the dialdehyde (**3-107**) in the synthesis of cholesterol[122] was interesting. A rationale that the less hindered aldehyde acted as the acceptor was presented. However, why that was so was not particularly clear. Another explanation offered here is based on the importance of $A^{1,3}$-strain factor. If we assume the cyclization proceeded by simultaneous proton transfer from the nitrogen atom to the carbonyl oxygen, there arose an interaction of the β-vinylic hydrogen with the methylene group adjacent to the nitrogen atom in the reverse donor. The $A^{1,3}$-strain can be eliminated by rotation of the protonated enamine and the carbonyl acceptor, but in so doing the carbonyl would incur 1,3-diaxial interaction with the 18-methyl group.

The stereochemistry of the product from a Sakurai-Hosomi reaction which involves attack of an allylsilane on a carbonyl compound, whether in 1,2- or 1,4-fashion (the latter to a conjugated ketone), is affected by a transition state conformation, with the $A^{1,3}$-strain factor playing an important role. It proved to be a very beneficial basis for designing synthetic routes to ageline-A (**3-108**)[123] and linaridial (**3-109**).[124]

An intramolecular Hiyama reaction was applied to close the six-membered ring in a synthesis of (+)-allopumiliotoxin-267A (**3-110**).[125] Generation of the correct

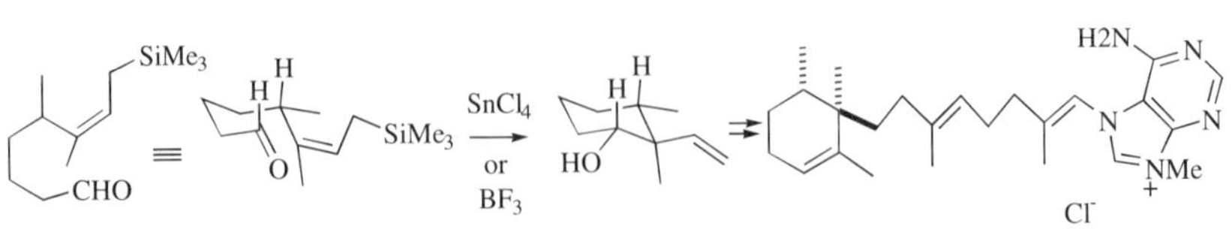

ageline-A (**3-108**)

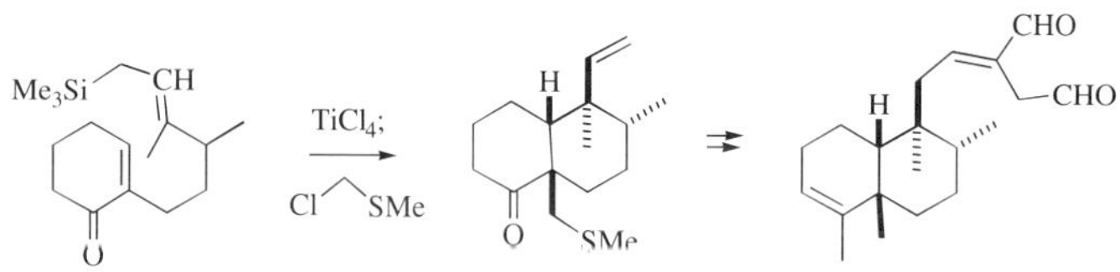

linaridial (**3-109**)

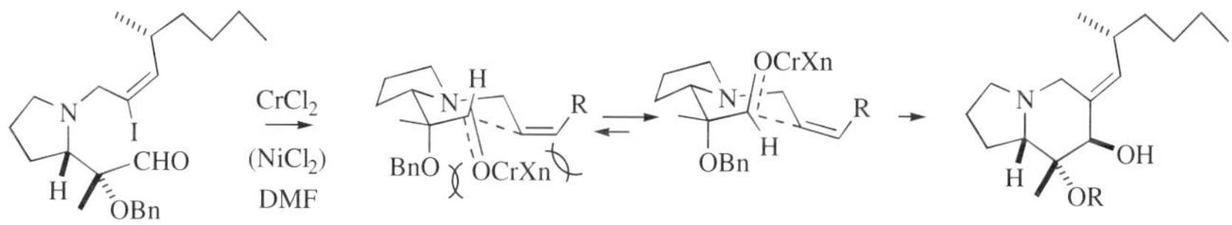

R = H allopumiliotoxin-267A
(**3-110**)

alcohol was due to two factors: to avoid dipole–dipole interactions between two oxygen functions and $A^{1,3}$-strain which must have been felt in the transition state.

An efficient construction of the DE-ring component of (–)-ajmalicine (**3-111**) has been developed.[126] In the process a double bond was consumed to trap an iminium ion while being involved in a simultaneous lactonization. The double bond must be (*Z*) to set the carbinyl center in the correct configuration, as the (*E*)-isomer gave rise to the precursor of 19-epiajmalicine. The *trans* ring junction followed the rule of $A^{1,3}$-strain minimization in the transition state.

Manipulation of the allylic strain phenomena in addition reactions to double bonds can be fruitful. For example, a critical step in the elaboration of the right-half portion (**3-112**) of monensin was epoxidation of a homoallylic alcohol.[127] The diastereofacial selectivity of the OH-directed reaction was possible due to conformational preference as dictated by $A^{1,3}$-strain.

$A^{1,3}$-Strain was an aid to the dihydroxylation of an alkenyl side chain in a δ-lactam.[128] The stereoselectivity was beneficial to an access to (–)-swainsonine (**3-113**).

Conjugate addition to alkylidene derivatives (**3-114**) of Meldrum's acid which contain a stereocenter at the allylic site is affected by the nature of the substituent. One of the two conformers is disfavored due to the occasioned $A^{1,3}$-strain. Whether the

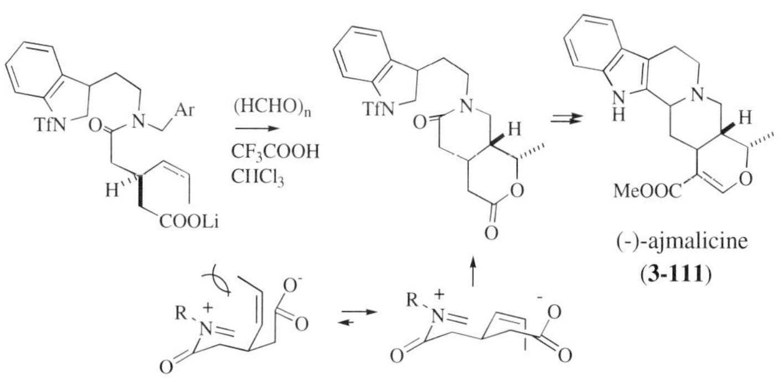

(-)-ajmalicine
(**3-111**)

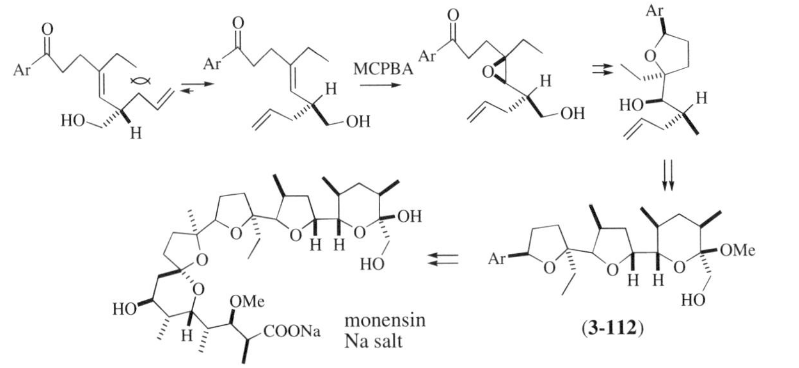

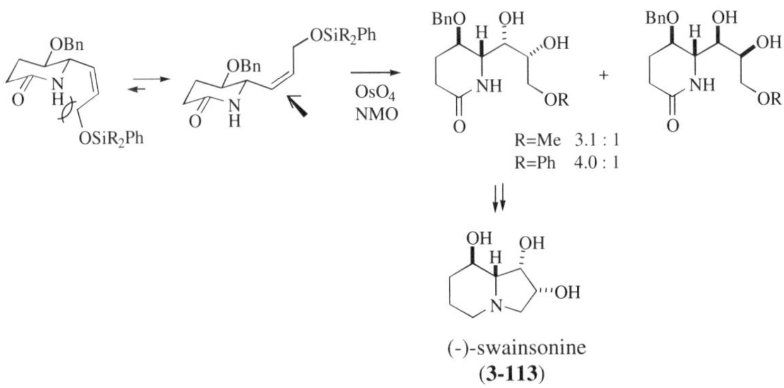

addition gives *syn* or *anti* product then depends on $A^{1,2}$-induction or chelated transition state. For example, Grignard reaction of a substrate bearing an allylic MEM-ether is *syn*-selective. On the other hand, a bulky, nonchelating group reverses the trend to favor the *anti* adduct. Clearly, the objective of eldanolide (**3-115**)[129] can be fulfilled by a proper choice of substrate structure and reaction conditions.

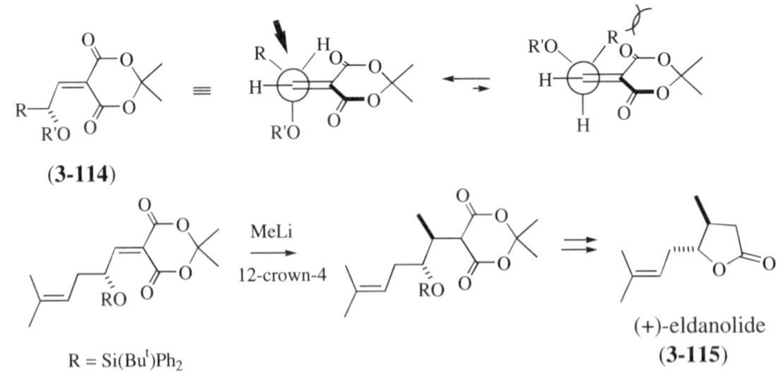

1-Substituted 3,4-dihydroisoquinolinium salts (**3-116**) equipped with a chiral *N*-substituent have been found to undergo reaction with nucleophiles from the side proximal to the smaller group of the *N*-substituent.[130] The conformation of these salts

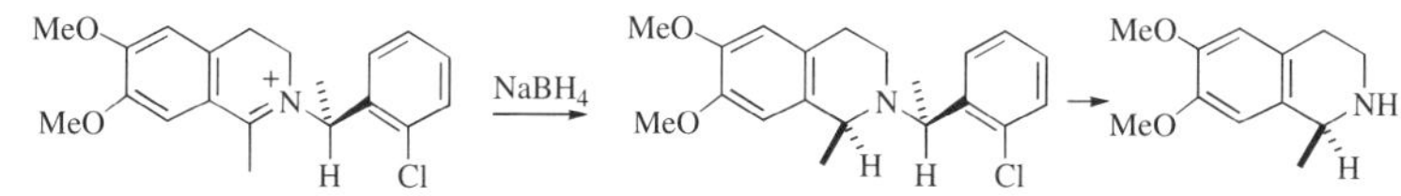

(**3-116**) ds 90% (-)-salsolidine

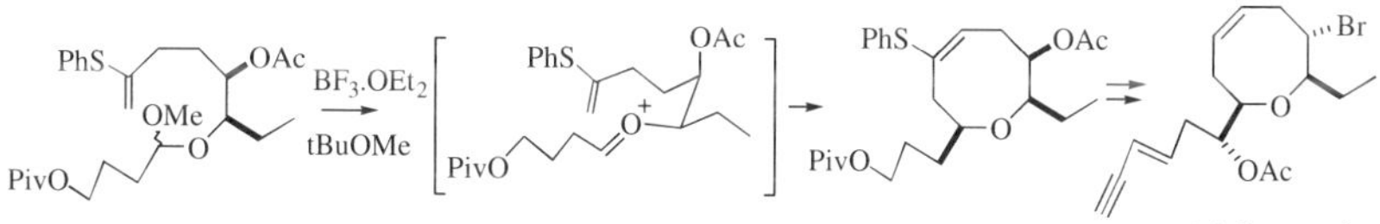

(+)-laurencin
(**3-117**)

is easy to determine and the smallest group (e.g., H) must be synperiplanar to the 1-substituent.

The occurrence of many *cis*-2,8-disubstituted oxocanes in nature has stimulated development of related synthetic methods. One particularly effective way to establish this stereochemistry is by intramolecular ene cyclization of oxonium ions, as illustrated in a synthesis of (+)-laurencin (**3-117**).[131] Ionization of the mixed acetal invariably gives rise to the oxonium which disposes *cis*-dipseudoequatorial groups at the two α-positions of the oxygen atom, to minimize $A^{1,3}$-strain. Actually, this phenomenon is not limited to the formation of the eight-membered ring.

In kainic acid (and α-allokainic acid) the 1,3-relationship of the two carboxyl groups is exploitable synthetically. Actually, an intramolecular free radical addition to a (*Z*)-unsaturated ester for ring closure is predicated on the minimized $A^{1,3}$-strain and nonbonding interactions.[131a] A conformation which places the allylic hydrogen closest to the ester the radical would prefer attack from the *Si*-face of the double bond to avoid steric interference with the CH_2OR group.

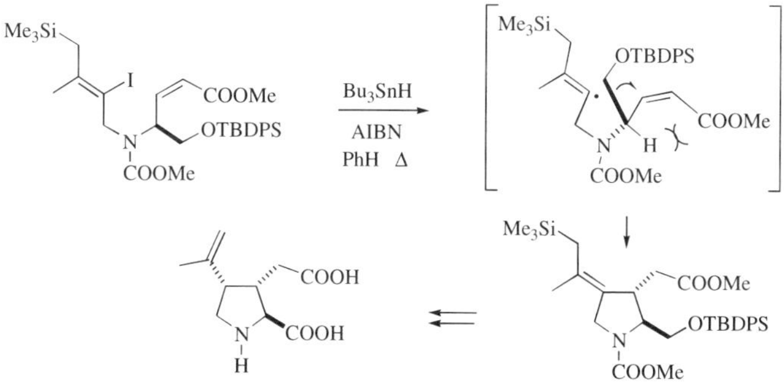

(-)-kainic acid
(**3-117a**)

The Grignard reaction of allylic sulfoxides shifts the double bond into the α,β′-position while accomplishing conjugate addition and deoxygenation at the sulfur atom. This chemistry can be exploited in a synthesis of (−)-sibirine (**3-118**).[132] When the sulfinyl function and an acetal unit of the substrate were *cis*-locked, the least steric

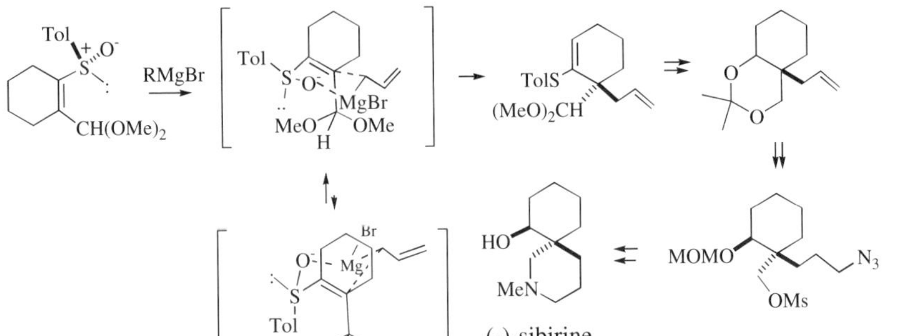

(**3-118**)

compression experienced by the molecule would have steered the tolyl group away from the acetal while offering its oxygen to direct the reagent.

Extracyclic elements can exert profound stereochemical effects on cycloaddition reactions, and therefore their synthetic value is increased. An allylic stereocenter (designated to locate at $C_{(17)}$) was shown to favor the formation of a desired diastereomer which represents a useful CD-ring building block (**3-119**) for steroids.[133] The 3:1 preference might indicate that in the other transition state the large group did not eclipse the vinylic hydrogen completely, and when bond formation started the torsional strain decreased.

It should be noted that a much better selectivity was observed in a Lewis acid catalyzed process which led the precursor (**3-120**) of the antibiotic X-14547A.[134,135] Perhaps complexation of the dienophile had a long range effect on spatial arrangement of the allylic substituent.

The *trans*-octalin sector of tetronolide and kijanolide is also provocative of a Diels–Alder approach for synthesis from an acyclic intermediate (**3-121**), although more attention must be paid to the elaboration of the more complicated interconnecting unit between the diene and the dienophile. Of the two *endo*/chair transition states one would be strongly disfavored on account of the $A^{1,3}$- interaction between two methyl groups.[136–138] In such reactions the use of a relatively weak Lewis acid may be advantageous.[139] Furthermore, it was observed that the two secondary methyl groups played a crucial role in determining the stereoselectivity. Without them

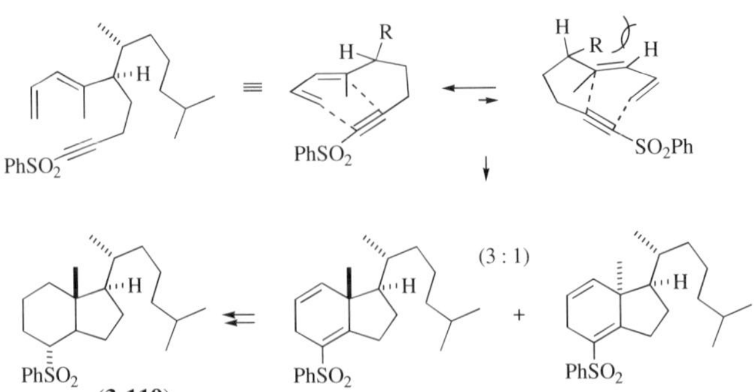

(**3-119**)

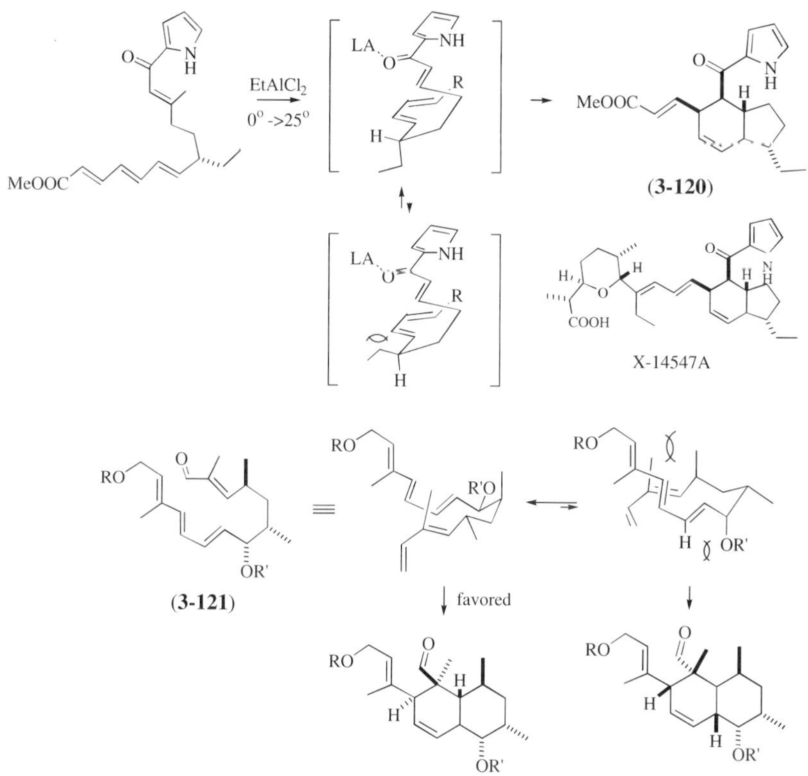

(3-120) X-14547A (3-121)

(as in a substrate for a similar approach to chlorothricolide), either very poor diastereoselection resulted, or the wrong diastereomer was obtained.

A synthetic route to (+)-himbacine (**3-122**)[140] can be adumbrated based on an intramolecular Diels–Alder reaction, since the angular hydrogen adjacent to the lactone is epimerizable and a *cis*-locked lactone ring is able to impart steric bias during hydrogenation of the endocyclic double bond. It is important to assign the diene unit conjugated to the ester carbonyl and the (*Z*)-form dienophile as an allyloxy functionality, to elaborate the three nonepimerizable stereocenters in the

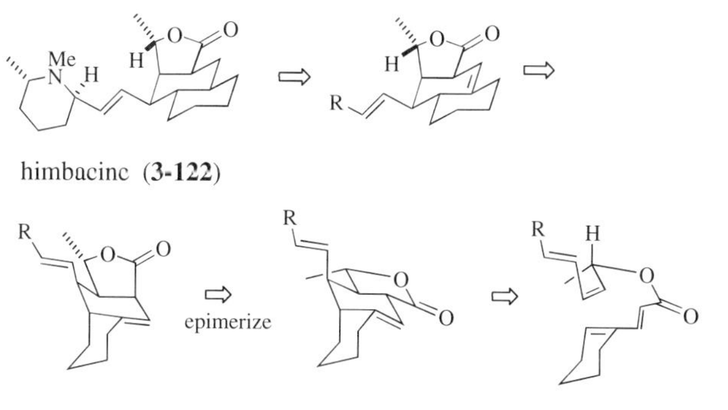

himbacine (**3-122**)

cycloaddition step. Then the carbinolic center can be relied on to direct the genesis of the skeleton. $A^{1,3}$-Strain favors the correct conformation.

An all-*cis* pattern of substituents at the ring junctions is required if the cycloadduct from an intramolecular Diels–Alder reaction is to be transformed into dendrobine (**3-123**).[141] Two different transition states differed by the presence of $A^{1,2}$- and $A^{1,3}$-strains, both involving the *N*-substituent. A bulky substituent favored the *endo* conformation because molecules generally tolerate $A^{1,2}$-strain better than $A^{1,3}$-strain.

The alkene/arene *meta*-photocycloaddition, particularly the intramolecular version, has been developed into one of the most powerful protocols for erecting complex structures. Thus, unlike all other syntheses of cedrene which are longer and plagued by generation of diastereomeric pairs about the secondary methyl group, the intramolecular *meta*-photocycloaddition method is unique. Three hypothetical exciplexes of different stability according to the degree of orbital overlap and severity of nonbonded interactions may be considered. The exciplex leading to C–C bond formation at the methoxylated site is disfavored by steric repulsion with the side chain, leaving only the option of cycloaddition involving the ipso carbon (where the side chain originates). Orbital overlap appears most efficient in the *exo* conformation (contrary to the intermolecular cycloaddition). The final decision is contributed by $A^{1,3}$-strain minimization between the benzylic methyl and the methoxy groups, and it

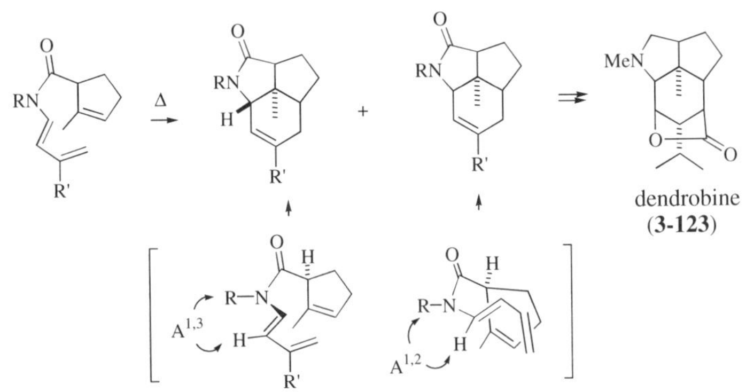

dendrobine (**3-123**)

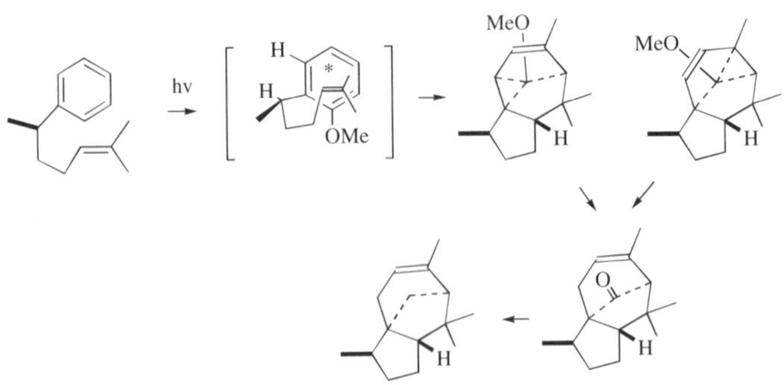

cedrene (**3-124**)

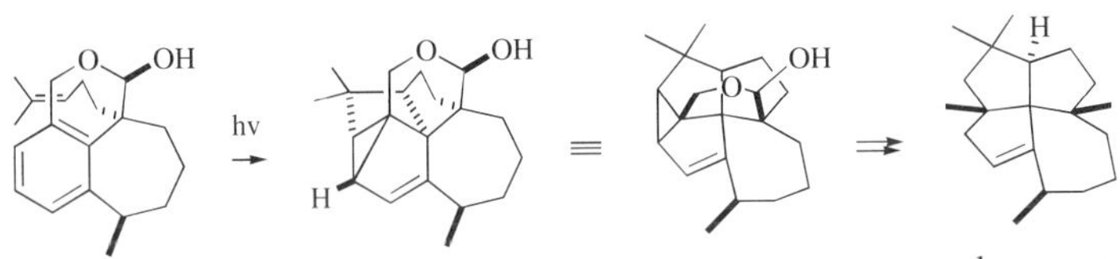

laurenene
(**3-125**)

happens that this transition state gives rise to tetracyclic structures embedding all of the critical features of cedrene (**3-124**).[142]

Besides the stream of success in providing unique solutions to isocomene,[143] silphinene,[144] silphiperfol-6-ene,[145] hirsutene,[146] subergorgic acid,[147] retigeranic acid-A,[148] and related triquinane terpenes, rudmollin[149] and laurenene (**3-125**)[150] are two other types that *meta*-photocycloaddition contributed mightily. Significantly, conventional synthetic routes to laurenene encountered much grief in that its precursors resisted insertion of the secondary methyl group in a stereochemically proper sense in the fenestrane framework,[151,152] the photoreaction of a sophisticated benzosuberane removed the problem posthaste, once benzylic alkylation of the δ-lactone was achieved. The alkylation was influenced by the methyl group at the other benzylic position, and the important implication is that this methyl group remained pseudoaxial throughout in order to avoid A^{1,3}-strain.

A^{1,3}-Strain must be responsible for the behavior of several substrates of intramolecular [2+2]-photocycloaddition. These include cyclohexenones that were used to synthesize acoradiene (**3-126**)[153] and epiprecapnelladiene.[154] A disagreeable result was apparent in the latter instance because steric effect worked against the wish to obtain the isomeric cycloadduct leading to the natural precapnelladiene.

A 3,3,4-trisubstituted cyclohexanone has been used to created a CD-block for cortisone (**3-127**) synthesis.[155] An ene reaction served to close the five-membered ring

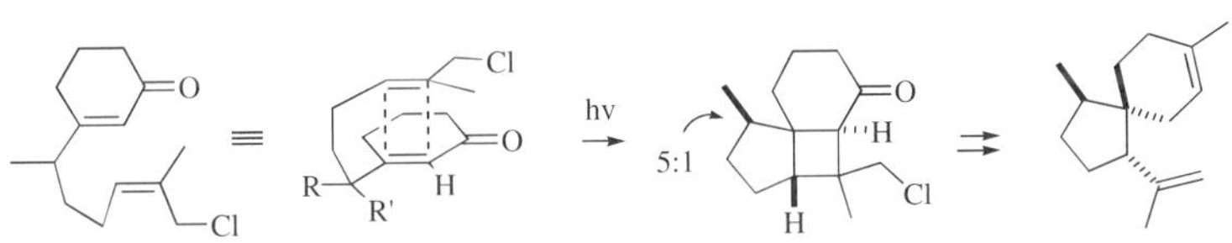

R=Me, R'=H favored

acoradiene (**3-126**)

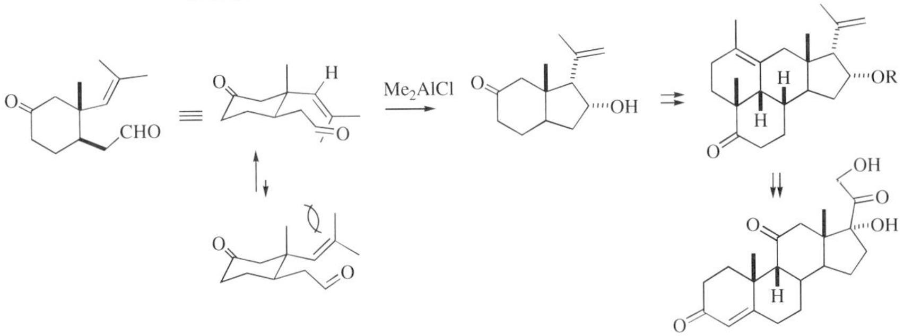

cortisone (**3-127**)

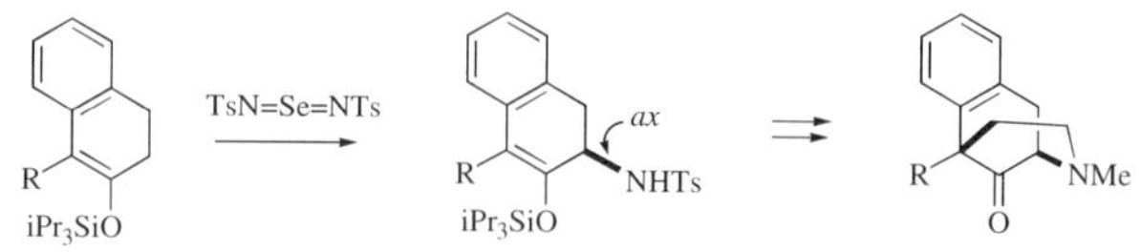

(**3-127a**)

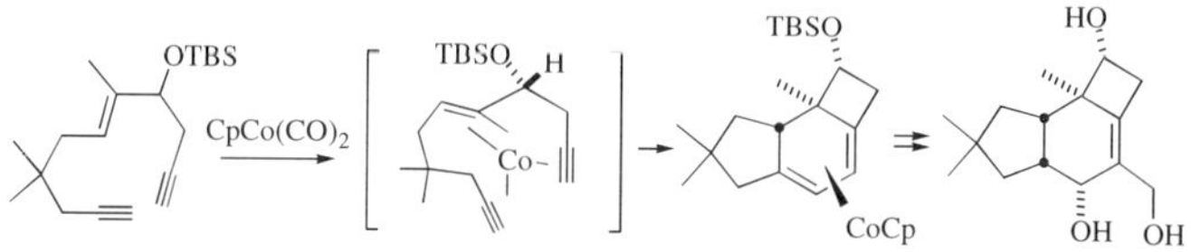

illudol (**3-128**)

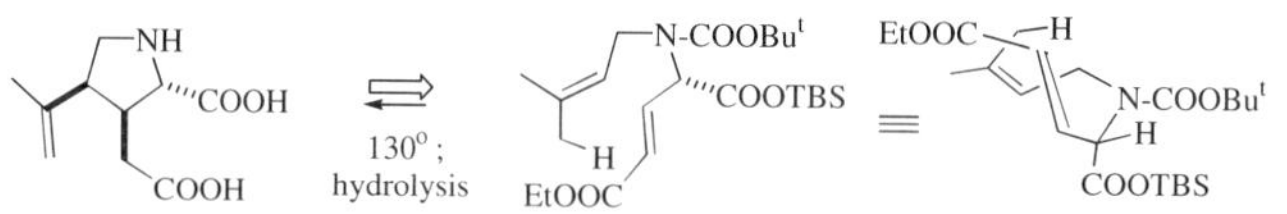

kainic acid
(**3-129**)

and it proceeded stereoselectively. $A^{1,3}$-Strain prevented the isopropenyl group from unfolding correctly, but the configuration could be readily amended at the end.

The allylic amination (ene-type reaction) of trialkylsilyl enol ethers places the nitrogen functionality in a pseudoaxial conformation, probably due to the combined effect of π–σ* stabilization and $A^{1,3}$-strain.[156] Such a conformation is most suitable for elaboration of azacycle such as the benzomorphanone system (**3-127a**).

The skeleton of illudol (**3-128**) can be assembled from an acyclic enediyne in one step.[157] Evolvement of three contiguous stereocenters as shown can be ascribed to $A^{1,3}$-strain of the TBSO group.

The intramolecular ene reaction[158] is suitable for construction of cyclic compounds with a *cis*-1,2-disubstitution pattern, and ideally that of the kainic acid (**3-129**) series. Actually, an *N*-prenyldehydroglutamic ester is easily identified as the substrate.[159] The α-ester provided instruction to the two reacting chains regarding their folding so as to minimize $A^{1,3}$-strain during the cyclization.

A similar situation is found in a zirconium-mediated ring contraction reaction.[160] Only in this case a medium ring was involved. Its synthetic potential was tested in a synthesis of (–)-macronecine (**3-130**).

The inurement of the Claisen rearrangement to synthesis is evident. Its significance is particularly prominent in the light of a very efficient route to the CD-ring component (**3-131**) of steroids.[161] Conformational selection on the basis of $A^{1,3}$-strain was again essential, and it was reinforced by nonbonded interaction with the existing ester chain.

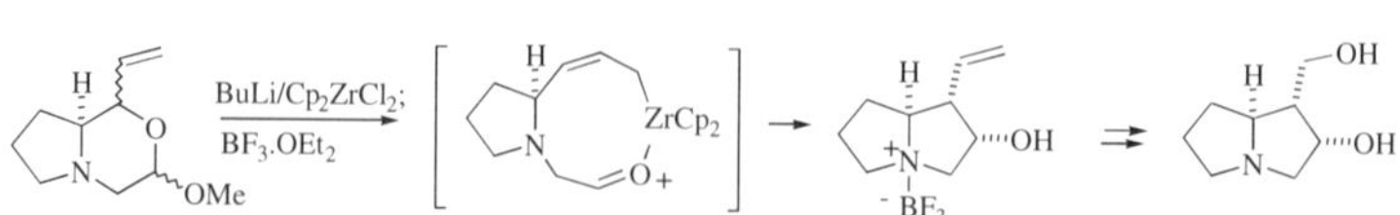

(-)-macronecine
(**3-130**)

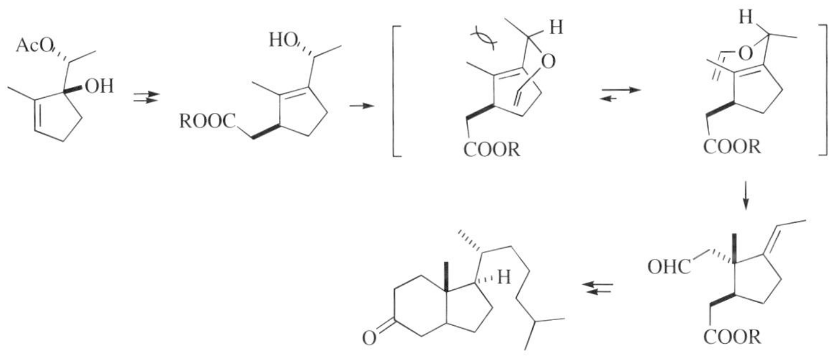

(**3-131**)

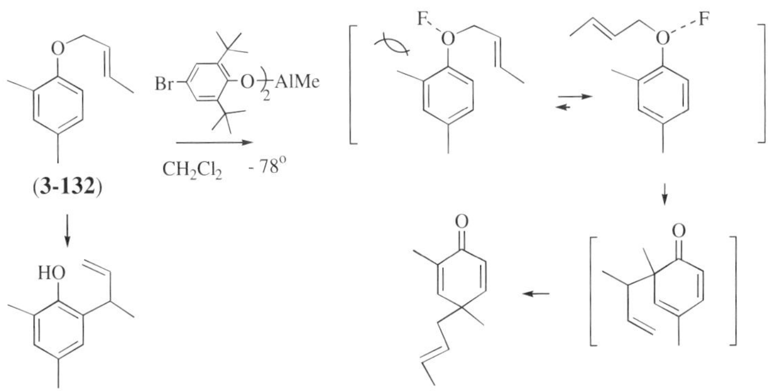

(**3-132**)

The effect of bulky Lewis acids to Claisen rearrangements has been demonstrated.[162] A 2,4-dimethylphenyl ether (**3-132**) behaved very differently in the presence or absence of a designed diaryloxyalane. The forced orientation of the smaller allyl chain toward the side of the *ortho*-methyl group led to a 4,4,6-trisubstituted 2,4-cyclohexadienone, which continued via a Cope rearrangement to give the 2,4,4-trisubstituted 2,5-cyclohexadienone.

Enhanced diastereoselectivity of aliphatic Claisen rearrangements by the same tactic has also been reported.[163]

A method for stereoselective formation of alkenes from conjugated aldehyde tosylhydrazones[164] involves reaction of the *N*-silylated derivatives (e.g., **3-133**) with an organolithium followed by treatment with acetic acid. Allylic diazene

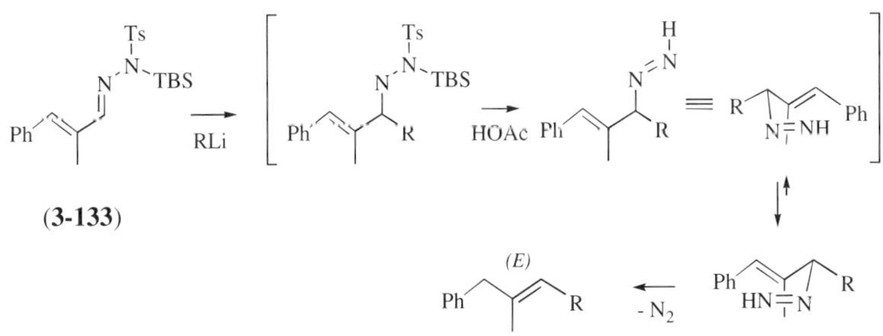

(**3-133**)

intermediates are implicated, and their decomposition via the transition state free of $A^{1,3}$-strain is apparent.

REFERENCES

1. Wenkert, E.; Afonso, A.; Bredenberg, J.B.; Kaneko, C.; Tahara, A. *J. Am. Chem. Soc.* **1964**, *86*, 2038.
2. Stork, G.; Schulenberg, J.W. *J. Am. Chem. Soc.* **1962**, *84*, 284.
3. Sarett, L.H.; Johns, W.F.; Beyler, R.E.; Lukes, R.M.; Poos, G.I.; Arth, G.E. *J. Am. Chem. Soc.* **1953**, *75*, 2112.
4. Nagata, W.; Sugasawa, T.; Narisada, M.; Wakabayashi, T.; Hayase, Y. *J. Am. Chem. Soc.* **1967**, *89*, 1483.
5. Ireland, R.E.; Kierstead, R.C. *J. Org. Chem.* **1966**, *31*, 2543.
6. Giarrusso, F.; Ireland, R.E. *J. Org. Chem.* **1968**, *33*, 3560.
7. Davies, S.G.; Bhalay, G. *Tetrahedron: Asymmetry* **1996**, *7*, 1595.
8. Birch, A.J.; Brown, J.M.; Subba Rao, G.S.R. *J. Chem. Soc.* **1964**, 3309.
9. Büchi, G.; Matsumoto, K.E.; Nishimura, H. *J. Am. Chem. Soc.* **1971**, *93*, 3299.
10. Salley, S.I. *J. Am. Chem. Soc.* **1967**, *89*, 6762.
11. Schreiber, S.L.; Wang, Z. *Tetrahedron Lett.* **1988**, *29*, 4085.
12. Corey, E.J.; Pearce, H.L. *J. Am. Chem. Soc.* **1979**, *101*, 5841.
13. Marshall, J.A.; Partridge, J.J. *Tetrahedron* **1969**, *25*, 2159.
14. Asaoka, M.; Takei, H. *J. Synth. Org. Chem. Jpn.* **1990**, *48*, 216.
15. Asaoka, M.; Fujii, N.; Takei, H. *Chem. Lett.* **1988**, 1655.
16. Asaoka, M.; Sakurai, M.; Takei, H. *Tetrahedron Lett.* **1990**, *31*, 4759.
17. Schultz, A.G.; Kirincich, S.J. *J. Org. Chem.* **1996**, *61*, 5626.
18. Berens, U.; Scharf, H.-D. *J. Org. Chem.* **1965**, *60*, 5127.
19. Bertrand, M.; Teisseire, P.; Pelerin, G. *Tetrahedron Lett.* **1980**, *21*, 2051, 2055.
20. Näf, F.; Ohloff, G. *Helv. Chim. Acta* **1974**, *57*, 1868.
21. Dzierba, C.D.; Zandi, K.S.; Möllers, T.; Shea, K.J. *J. Am. Chem. Soc.* **1996**, *118*, 4711.
22. Yoon, T.; Danishefsky, S.J.; de Gala, S. *Angew. Chem. Int. Ed.* **1994**, *33*, 853.
23. Oppolzer, W.; Petrzilka, M. *Helv. Chim. Acta* **1978**, *61*, 2755.
24. Harayama, T.; Takatani, M.; Inubushi, Y. *Chem. Pharm. Bull.* **1980**, *28*, 2394.
25. Wender, P.A.; Lechleiter, J.C. *J. Am. Chem. Soc.* **1978**, *100*, 4321.
26. Williams, J.R.; Callahan, J.F. *Chem. Commun.* **1979**, 404.
27. Iwata, C.; Takemoto, Y.; Doi, M.; Imanishi, T. *J. Org. Chem.* **1988**, *53*, 1623.
28. Boeckman, R.K.; Charette, A.B.; Asberom, T.; Johnston, B.H. *J. Am. Chem. Soc.* **1987**, *109*, 7553.
29. Paquette, L.A.; Ham, W.H. *Tetrahedron Lett.* **1986**, *27*, 2341.
30. Corey, E.J.; Imwinkelried, R.; Pikul, S.; Xiang, Y.B. *J. Am. Chem. Soc.* **1989**, *111*, 5493.
31. Pearson, W.H.; Lovering, F.E. *Tetrahedron Lett.* **1994**, *35*, 9173.
32. Winkler, J.D.; Scott, R.D.; Williard, P.G. *J. Am. Chem. Soc.* **1990**, *112*, 8971.

33. Koreeda, M.; George, I.A. *J. Am. Chem. Soc.* **1986**, *108*, 8098.
34. Burke, S.D.; Rancourt, J. *J. Am. Chem. Soc.* **1991**, *113*, 2335.
35. Stork, G.; Kobayashi, Y.; Suzuki, T.; Zhao, K. *J. Am. Chem. Soc.* **1990**, *112*, 1661.
36. Stork, G.; Zhao, K. *J. Am. Chem. Soc.* **1990**, *112*, 5875.
37. Ziegler, F.E.; Wallace, O.B. *J. Org. Chem.* **1995**, *60*, 3626.
37a. Kim, S.; Cheong, J.H. *Synlett* **1997**, 947.
38. Bauta, W.; Booth, J.; Bos, M.E.; De Luca, M.; Diorazio, L.; Donahoe, T.; Magnus, N.; Magnus, P.; Mendoza, J.; Pye, P.; Tarrant, J.; Thom, S.; Ujjainwalla, F. *Tetrahedron Lett.* **1995**, *36*, 5327.
39. Hölder, S.; Blechert, S. *Synlett* **1996**, 505.
40. Ziegler, F.E.; Piwinski, J.J. *J. Am. Chem. Soc.* **1980**, *102*, 6576.
41. Ziegler, F.E.; Piwinski, J.J. *J. Am. Chem. Soc.* **1982**, *104*, 7181.
42. Kurth, M.J.; Soares, C.J. *Tetrahedron Lett.* **1987**, *28*, 1031.
43. Kato, N.; Kataoka, H.; Ohbuchi, S.; Tanaka, S.; Takeshita, H. *Chem. Commun.* **1988**, 354.
44. Masamune, S.; Choy, W.; Petersen, J.S.; Sita, L.R. *Angew. Chem. Int. Ed. Engl.* **1985**, *24*, 1.
45. Trost, B.M.; Krische, M.J. *J. Am. Chem. Soc.* **1996**, *118*, 233.
46. Stork, G.; Clarke, F.H. *J. Am. Chem. Soc.* **1961**, *83*, 3114.
47. Grieco, P.A.; Ohfune, Y.; Majetich, G. *J. Am. Chem. Soc.* **1977**, *99*, 7393.
48. Ohfune, Y.; Grieco, P.A.; Wang, C.-L.; Majetich, G. *J. Am. Chem. Soc.* **1978**, *100*, 5946.
49. Miyashita, M.; Yoshikoshi, A. *J. Am. Chem. Soc.* **1974**, *96*, 1917.
50. Hori, K.; Hikage, N.; Inagaki, A.; Mori, S.; Nomura, K.; Yoshii, E. *J. Org. Chem.* **1992**, *57*, 2888.
51. Skrinjar, M.; Nilsson, C.; Wistrand, L.-G. *Tetrahedron: Asymmetry* **1992**, *3*, 1263.
51a. Cossy, J.; Dumas, C.; Pardo, D.G. *Synlett* **1997**, 905.
52. Barrett, A.G.M.; Sheth, H.G. *Chem. Commun.* **1982**, 170.
53. Carruthers, W.; Williams, M.J. *Chem. Commun.* **1986**, 1287.
54. Evans, D.A.; Dow, R.L.; Shih, T.L.; Takacs, J.M.; Zahler, R. *J. Am. Chem. Soc.* **1990**, *112*, 5290.
55. Posner, G.H.; Switzer, C. *J. Am. Chem. Soc.* **1986**, *108*, 1239.
56. Fujioka, H.; Miyazaki, M.; Kitagawa, H.; Yamanaka, T.; Yamamoto, H.; Takuma, K.; Kita, Y. *Chem. Commun.* **1993**, 1634.
57. Rowley, M.; Tsukamoto, M.; Kishi, Y. *J. Am. Chem. Soc.* **1989**, *111*, 2735.
58. Fisher, M.J.; Myers, C.D.; Joglar, J.; Chen, S.-H.; Danishefsky, S.J. *J. Org. Chem.* **1991**, *56*, 5826.
59. Greene, A.E. *J. Am. Chem. Soc.* **1980**, *102*, 5337.
60. Bryson, T.A.; Reichel, C.J. *Tetrahedron Lett.* **1980**, *21*, 2381.
61. Welch, M.C.; Bryson, T.A. *Tetrahedron Lett.* **1988**, *29*, 521.
62. Welch, M.C.; Bryson, T.A. *Tetrahedron Lett.* **1989**, *30*, 523.
63. Mulzer, J.; Mohr, J.-T. *J. Org. Chem.* **1994**, *59*, 1160.
64. Schore, N.E.; Rowley, E.G. *J. Am. Chem. Soc.* **1988**, *110*, 5224.

65. Exon, C.; Magnus, P. *J. Am. Chem. Soc.* **1983**, *105*, 2477.
66. Crimmins, M.T.; De Loach, J.A. *J. Org. Chem.* **1984**, *49*, 2076.
67. Corey, E.J.; Kang, M.-c.; Desai, M.C.; Ghosh, A.K.; Houpis, I.N. *J. Am. Chem. Soc.* **1988**, *110*, 649.
68. Yamazaki, M. *Chem. Commun.* **1972**, 606.
69. Denmark, S.E.; Almstead, N.G. *J. Org. Chem.* **1991**, *56*, 6458.
70. McNamara, J.M.; Kishi, Y. *J. Am. Chem. Soc.* **1982**, *104*, 7371.
71. Hatakeyama, S.; Sugawara, K.; Takano, S. *Chem. Commun.* **1991**, 1533.
72. Takano, S.; Chiba, K.; Yonaga, M.; Ogasawara, K. *Chem. Commun.* **1980**, 616.
73. Takano, S.; Yonaga, M.; Ogasawara, K. *Chem. Commun.* **1980**, 1153.
74. Midland, M.M.; Kwon, Y.C. *J. Org. Chem.* **1981**, *46*, 229; *Tetrahedron Lett.* **1982**, *23*, 2077.
75. Batcho, A.D.; Berger, D.E.; Uskokovic, M.R. *J. Am. Chem. Soc.* **1981**, *103*, 1293.
76. Schmuff, N.R.; Trost, B.M. *J. Org. Chem.* **1983**, *48*, 1404.
77. Trost, B.M.; Verhoeven, T.R. *J. Am. Chem. Soc.* **1978**, *100*, 3435.
78. Vercauteren, J.; Lavaud, C.; Levy, J.; Massiot, G. *J. Org. Chem.* **1984**, *49*, 2278.
79. Huber, I.M.P.; Seebach, D. *Helv. Chim. Acta* **1987**, *70*, 1944.
80. Cushman, M.; Chen, J.-k. *J. Org. Chem.* **1987**, *52*, 1517.
81. Nakagawa, M.; Fukushima, H.; Kawate, T.; Hongu, M.; Kodato, S.; Une, T.; Taniguchi, M.; Hino, T. *Tetrahedron Lett.* **1986**, *27*, 3235.
82. Tsuda, Y.; Hosoi, S.; Katagiri, N.; Kaneko, C.; Sano, T. *Chem. Pharm. Bull.* **1993**, *41*, 2087.
83. Amann, R.; Arnold, K.; Spitzner, D.; Majer, Z.; Snatzke, G. *Liebigs Ann. Chem.* **1996**, 349.
84. Heathcock, C.H.; Norman, M.H.; Dickman, D.A. *J. Org. Chem.* **1990**, *55*, 798.
85. Hong, C.Y.; Kado, N.; Overman, L.E. *J. Am. Chem. Soc.* **1993**, *115*, 11028.
85a. Oppolzer, W., Flachsmann, F. *Tetrahedron Lett.* **1998**, *39*, 5019.
86. Satoh, S.; Sodeoka, M.; Sasai, H.; Shibasaki, M. *J. Org. Chem.* **1991**, *56*, 2278.
87. Seebach, D.; Naef, R.; Calderari, G. *Tetrahedron* **1984**, *40*, 1313; Seebach, D.; Sting, A.R.; Hoffmann, M. *Angew. Chem. Int. Ed.* **1996**, *35*, 2709.
88. Naef, R.; Seebach, D. *Liebigs Ann. Chem.* **1983**, 1930.
89. Krohn, K.; Hamann, I. *Liebigs Ann. Chem.* **1988**, 949.
90. Boeckman, R.K.; Yoon, S.K.; Heckendorn, D.K. *J. Am. Chem. Soc.* **1991**, *113*, 9682.
91. Matsuo, K.; Arase, T. *Chem. Pharm. Bull.* **1995**, *43*, 890.
92. Fortunak, J.M.D.; Kitteringham, J.; Mastrocola, A.R.; Mellinger, M.; Sisti, N.J.; Wood, J.L.; Zhuang, Z.-P. *Tetrahedron Lett.* **1996**, *37*, 5683.
93. Chu, K.S.; Negrete, G.R.; Konopelski, J.P.; Lakner, F.J.; Woo, N.-T.; Olmstead, M.M. *J. Am. Chem. Soc.* **1992**, *114*, 1800.
94. Sakamoto, Y.; Takahashi, T. *Synlett* **1995**, 513.
95. Mehta, G.; Krishnamurthy, N. *Synth. Commun.* **1988**, *18*, 1267.
96. Mehta, G.; Krishnamurthy, N. *Tetrahedron Lett.* **1987**, *28*, 5945.
97. Tsunoda, T.; Tatsuki, S.; Kataoka, K.; Ito, S. *Chem. Lett.* **1994**, 543.

98. Gonzalez, A.O.; Martin, J.D.; Melian, M.A. *Tetrahedron Lett.* **1976**, 2279.

99. Smith, A.B., III; Sestelo, J.P.; Dormer, P.G. *J. Am. Chem. Soc.* **1995**, *117*, 10755.

100. Danishefsky, S.; Zomboni, R.; Kahn, M.; Etheredge, S.J. *J. Am. Chem. Soc.* **1980**, *102*, 2097.

101. Marshall, J.A. *Chem. Rev.* **1989**, *89*, 1503.

102. Chen, S.-H.; Horvath, R.F.; Joglar, J.; Fisher, M.J.; Danishefsky, S.J. *J. Org. Chem.* **1991**, *56*, 5834.

103. Tsuji, J. *Palladium Reagents and Catalysts*; Wiley: New York, 1995; pp 290–422.

104. Trost, B.M.; Godleski, S.A.; Genet, J.-P. *J. Am. Chem. Soc.* **1978**, *100*, 3930.

105. Johnson, F. *Chem. Rev.* **1968**, *68*, 375; Hoffmann, R.W. *Chem. Rev.* **1989**, *89*, 1841.

106. Corey, E.J.; Carpino, P. *J. Am. Chem. Soc.* **1989**, *111*, 5472.

107. Myers, A.G.; Fraley, M.E.; Tom, N.J. *J. Am. Chem. Soc.* **1994**, *116*, 11556.

108. Porco, J.A.; Schoenen, F.J.; Stout, T.J.; Clardy, J.; Schreiber, S.L. *J. Am. Chem. Soc.* **1990**, *112*, 7410.

109. Williams, R.M.; Hendrix, J.A. *J. Org. Chem.* **1990**, *55*, 3723.

110. Vernier, J.-M.; Hegedus, L.S.; Miller, D.B. *J. Org. Chem.* **1992**, *57*, 6914.

111. Seebach, D.; Lamatsch, B.; Amstutz, R.; Beck, A.K.; Dobler, M.; Egli, M.; Fitzi, R.; Gautschi, M.; Herradon, B.; Hidber, P.C.; Irwin, J.J.; Locher, R.; Maestro, M.; Maetzke, T.; Mourino, A.; Pfammatter, E.; Plattner, D.A.; Schickli, C.; Schweizer, W.B.; Seiler, P.; Stucky, G.; Petter, W.; Escalante, J.; Juaristi, E.; Quintana, D.; Miravilles, C.; Molins, E. *Helv. Chim. Acta* **1992**, *75*, 913.

112. Stork, G.; Logusch, E.W. *J. Am. Chem. Soc.* **1980**, *102*, 1219.

113. Kim, D.; Lee, Y.K.; Jang, Y.M.; Kim, I.O.; Park, S.W. *J. Chem. Soc. Perkin Trans. 1* **1990**, 3221.

114. Büchi, G.; Berthet, D.; Decorzant, R.; Grieder, A.; Hauser, A. *J. Org. Chem.* **1976**, *41*, 3208.

115. Bailey, P.D.; Hollinshead, S.P.; Dauter, Z. *Chem. Commun.* **1985**, 1507.

116. Momose, T.; Toyooka, N. *J. Org. Chem.* **1994**, *59*, 943.

117. Bohlmann, C.; Bohlmann, R.; Rivera, E.G.; Vogel, C.; Manandhar, M.D.; Winterfeldt, E. *Liebigs Ann. Chem.* **1985**, 1752.

118. Ent, H.; de Koning, H.; Speckamp, W.N. *Heterocycles* **1988**, *27*, 237.

119. Hart, D.J.; Kanai, K. *J. Org. Chem.* **1982**, *47*, 1555.

120. Kano, S.; Yokomatsu, T.; Nemoto, H.; Shibuya, S. *J. Am. Chem. Soc.* **1986**, *108*, 6746.

121. Majetich, G.; Siesel, D. *Synlett* **1995**, 559.

122. Woodward, R.B.; Sondheimer, F.; Taub, D.; Heusler, K.; McLamore, W.M. *J. Am. Chem. Soc.* **1952**, *74*, 4223.

123. Asao, K.; Iio, H.; Tokoroyama, T. *Tetrahedron Lett.* **1989**, *30*, 6401.

124. Tokoroyama, T.; Tsukamoto, M., Asada, T.; Iio, H. *Tetrahedron Lett.* **1987**, *28*, 6645.

125. Aoyagi, S.; Wang, T.-C.; Kibayashi, C. *J. Am. Chem. Soc.* **1993**, *115*, 11393.

126. Lögers, M.; Overman, L.E.; Welmaker, G.S. *J. Am. Chem. Soc.* **1995**, *117*, 9139.

127. Fukuyama, T.; Wang, C.-L.J.; Kishi, Y. *J. Am. Chem. Soc.* **1979**, *101*, 260.

128. Naruse, M.; Aoyagi, S.; Kibayashi, C. *J. Org. Chem.* **1994**, *59*, 1358.

129. Larcheveque, M.; Tamagnan, G.; Petit, Y. *Chem. Commun.* **1989**, 31.

130. Polniaszek, R.P.; Kaufman, C.R. *J. Am. Chem. Soc.* **1989**, *111*, 4859.

131. Bratz, M.; Bullock, W.H.; Overman, L.E.; Takemoto, T. *J. Am. Chem. Soc.* **1995**, *117*, 5958.

132. Imanishi, T.; Kurumada, T.; Maezaki, N.; Sugiyama, K.; Iwata, C. *Chem. Commun.* **1991**, 1409.

133. Clasby, M.C.; Craig, D.; Marsh, A. *Angew. Chem. Int. Ed.* **1993**, *32*, 1444.

134. Roush, W.R.; Myers, A.G. *J. Org. Chem.* **1981**, *46*, 1509.

135. Nicolaou, K.C.; Magolda, R.L. *J. Org. Chem.* **1981**, *46*, 1506.

136. Marshall, J.A.; Audia, J.E.; Grote, J.; Shearer, B.G. *Tetrahedron* **1986**, *42*, 2893.

137. Roush, W.R.; Brown, B.B.; Drozda, S.E. *Tetrahedron Lett.* **1988**, *29*, 3541.

138. Boeckman, R.K.; Barta, T.E.; Nelson, S.G. *Tetrahedron Lett.* **1991**, *32*, 4091.

139. Takeda, K.; Kawanishi, E.; Nakamura, H.; Yoshii, E. *Tetrahedron Lett.* **1991**, *32*, 4925.

140. Chackalamannil, S.; Davies, R.J.; Asberom, T.; Doller, D.; Leone, D. *J. Am. Chem. Soc.* **1996**, *118*, 9812.

141. Martin, S.F.; Li, W. *J. Org. Chem.* **1991**, *56*, 642.

142. Wender, P.A.; Howbert, J.J. *J. Am. Chem. Soc.* **1981**, *103*, 688.

143. Wender, P.A.; Dreyer, G.B. *Tetrahedron* **1981**, *37*, 4445.

144. Wender, P.A.; Ternansky, R.J. *Tetrahedron Lett.* **1985**, *26*, 2625.

145. Wender, P.A.; Singh, S.K. *Tetrahedron Lett.* **1985**, *26*, 5987.

146. Wender, P.A.; Howbert, J.J. *Tetrahedron Lett.* **1982**, *23*, 3983.

147. Wender, P.A.; De Long, M.A. *Tetrahedron Lett.* **1990**, *31*, 5429.

148. Wender, P.A.; Singh, S.K. *Tetrahedron Lett.* **1990**, *31*, 2517.

149. Wender, P.A.; Fisher, K. *Tetrahedron Lett.* **1986**, *27*, 1857.

150. Wender, P.A.; von Geldern, T.W.; Levine, B.H. *J. Am. Chem. Soc.* **1988**, *110*, 4858.

151. Crimmins, M.T.; Gould, L.D. *J. Am. Chem. Soc.* **1987**, *109*, 6199.

152. Tsunoda, T.; Amaike, M.; Tambunan, U.S.F.; Fujise, Y.; Ito, S.; Kodama, M. *Tetrahedron Lett.* **1987**, *28*, 2537.

153. Oppolzer, W.; Zutterman, F.; Bättig, K. *Helv. Chim. Acta* **1983**, *66*, 522.

154. Birch, A.M.; Pattenden, G. *J. Chem. Soc. Perkin Trans. 1* **1983**, 1913.

155. Horiguchi, Y.; Nakamura, E.; Kuwajima, I. *J. Am. Chem. Soc.* **1989**, *111*, 6257.

156. Magnus, P.; Coldham, I. *J. Am. Chem. Soc.* **1991**, *113*, 672.

157. Johnson, E.P.; Vollhardt, K.P.C. *J. Am. Chem. Soc.* **1991**, *113*, 381.

158. Oppolzer, W.; Snieckus, V. *Angew. Chem. Int. Ed.* **1978**, *17*, 476.

159. Oppolzer, W.; Thirring, K. *J. Am. Chem. Soc.* **1982**, *104*, 4978.

160. Ito, H.; Ikeuchi, Y.; Taguchi, T.; Hanzawa, Y. *J. Am. Chem. Soc.* **1994**, *116*, 5469.

161. Takahashi, T.; Yamada, H.; Tsuji, J. *J. Am. Chem. Soc.* **1981**, *103*, 5259.

162. Maruoka, K.; Sato, J.; Banno, H.; Yamamoto, H. *Tetrahedron Lett.* **1990**, *31*, 377.

163. Boeckman, R.K.; Neeb, M.J.; Gaul, M.D. *Tetrahedron Lett.* **1995**, *36*, 803.

164. Myers, A.G.; Kukkola, P.J. *J. Am. Chem. Soc.* **1990**, *112*, 8208.

4

STEREOINDUCTION AT LONG DISTANCES

Stereoinduction at relatively long range is observable, but unless in specially constituted compounds, the limit for effective manifestation of the phenomenon may be 1,5. In this chapter, examples of 1,4-induction and higher are presented in the context of synthetic applications. It should be noted that these effects are often conferred to very simple substances by the attachment of chiral auxiliaries. Actually, many techniques of ab initio asymmetric synthesis are based on such tactics.

4.1. 1,4-STEREOINDUCTION

Intramolecular transfer of chirality is quite general in this category. The Claisen rearrangement (Ireland variant) of enolizable ω-alkenyl lactones is a very valuable method for construction of carbocycles of different sizes. Because of steric constraints the chirality of the emerging stereogenic center at the α-position of the resulting carboxylic acid is related to the ω-carbinyl center of the substrate. After this reaction was applied to a properly equipped lactone with the intention of synthesizing widdrol (**4-1**),[1] it only remained to degrade the carboxyl group to give the tertiary alcohol.

Cyclic enol ethers bearing an alkenyl substituent at the α′-position undergo Claisen rearrangement to give cycloalkenes containing a carbonyl side chain. Interestingly, if the double bond of the enol ether is exocyclic, the rearrangement

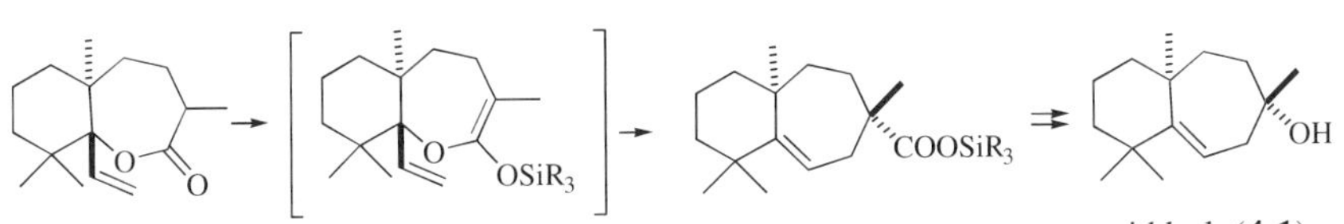

widdrol (**4-1**)

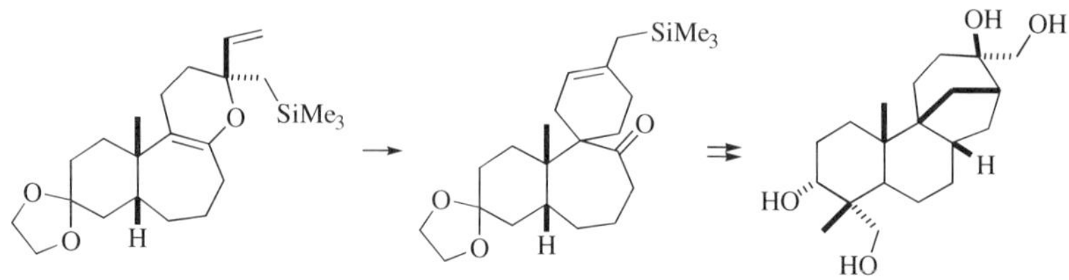

aphidicolin (**4-2**)

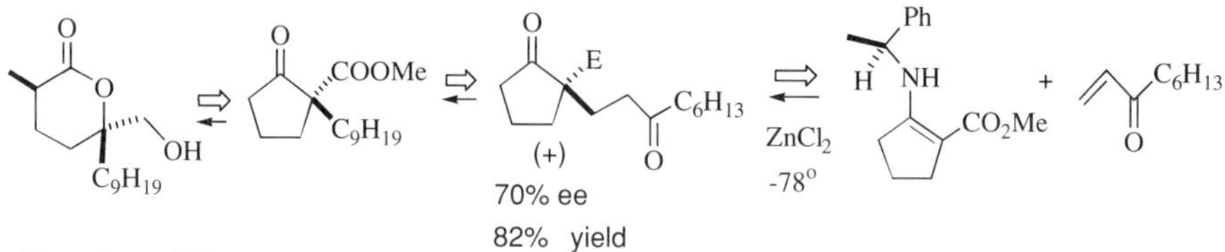

(-)-malyngolide
(**4-3**)

would lead to a ring-enlarged carbocycle. A novel route to aphidicolin (**4-2**),[2] which was formulated on the basis of this process, is instructive.

Many chiral auxiliaries display their capability of asymmetric induction in the 1,4-sense. Chiral enamines are versatile reactants and one such species derived from methyl 2-oxocyclopentanecarboxylate has been converted to (–)-malyngolide (**4-3**).[3]

In the assembly of Vinca alkaloids, the need for the establishment of the quaternary angular center ranks high in priority. With a readily available 3-ethyl-4-oxo-2-piperidone, an asymmetric Michael addition (to an acrylic ester) is a logical process to pursue. Fruition of the concept which delivered (+)-vincamine (**4-4**)[4] involved conversion of the β-ketolactam into a chiral enamine.

Intramolecular alkylation of a cyclic amide bearing an (*S*)-1-phenethyl group on the nitrogen atom was investigated for asymmetric induction in the formation of spirocyclic systems.[5] The most favorable transition state appears to have the

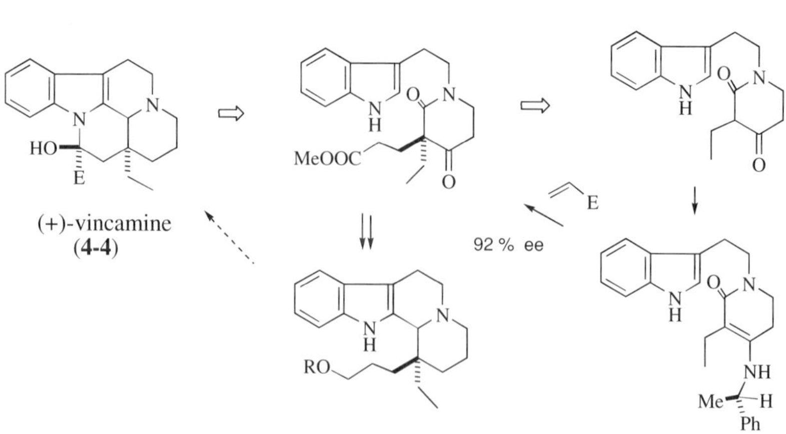

(+)-vincamine
(**4-4**)

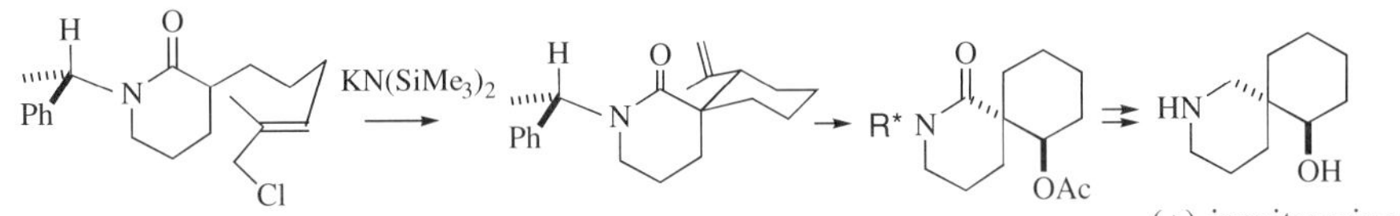

(+) isonitramine
(**4-5**)

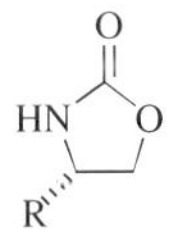

R= iPr, Ph, Bn
(**4-6**)

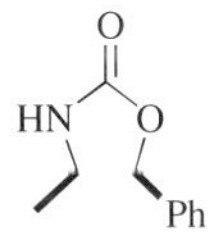

(**4-7**)

N-substituent of the amide enolate fixed in the H:O-eclipsed form and the electrophilic chain folded to the side of the secondary methyl group of the chiral auxiliary, with the carbon segment intervening the bond-forming atoms in a chairlike conformation. Thus, 1,4-induction by the asymmetric center is expected during rehybridization at the α-carbon atom of the amide enolate. This alkylation gave an intermediate of (+)-isonitramine (**4-5**).

1,3-Oxazolidin-2-ones (**4-6, 4-7**) derived from (*S*)-valinol, (*S*)-phenylglycinol, (*1S,2R*)-norephedrine are valuable chiral auxiliaries.[6] After *N*-acylation they induce asymmetric reactions at the α-position of the acyl group. Various metal (Li, Zn, S, Ti, B) enolates have been used to suit particular situations. The scope is wide, including alkylation, acylation, aldol condensation, Darzens reaction, Michael addition, and the ene reaction. Products can be cleaved to chiral acids with LiOH–H_2O_2, or to esters and amides by transesterification and transamidation protocols, with recovery of the chiral auxiliaries.

Taken as a recent example of common practice, a route to (–)-roccellaric acid (**4-8**)[7] that required only four steps is shown. Effective 1,4-asymmetric induction in the aldol condensation was evident.

An interesting variant is the use of a deprotonated Fischer carbene complex (**4-9**) incorporating a chiral imidazolidinone as a latent acetyl unit.[8] Its highly stereoselective reaction with isobutyraldehyde was demonstrated.

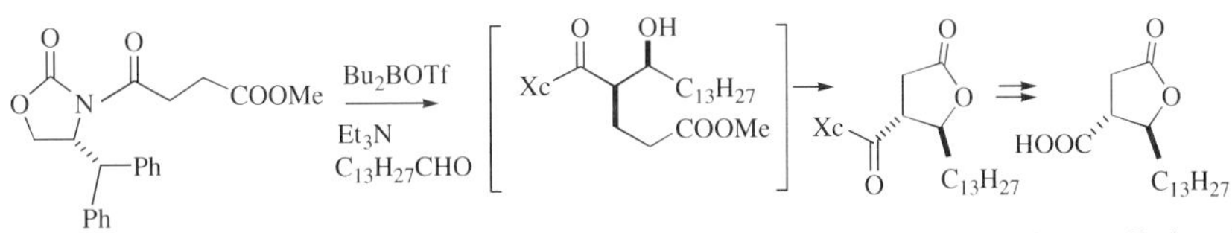

(-)-roccellaric acid
(**4-8**)

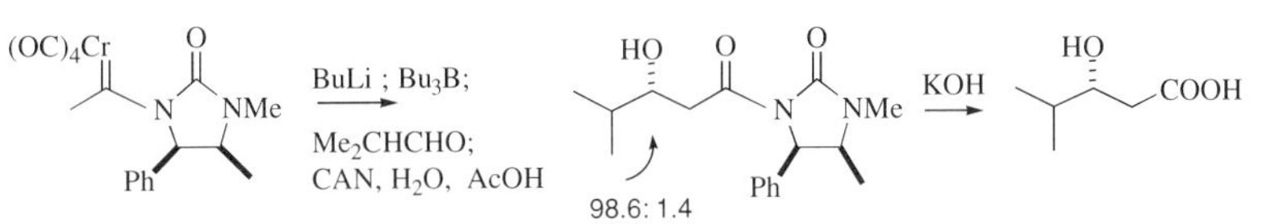

(**4-9**)

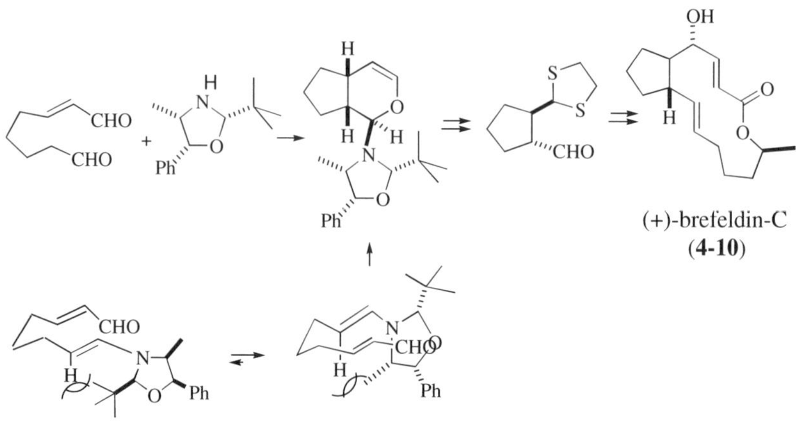

Reaction of 2-octenedial with a chiral all-*cis* 2-*t*-butyl-4-methyl-5-phenyl-1,3-oxazolidine led to a *cis*-fused aminodihydropyran. Cyclization of the enamine intermediate from one rotamer was favored. The other rotamer suffered from a strong 1,4-interaction between the β-H of the enamine and the *t*-butyl group. The product was eventually converted to (+)-brefeldin–C (**4-10**).[9]

Radical macrocyclization of (**4-11**) which can form 15- or 14-membered ring ketones was found to favor (by 8:1) the larger ring (*endo* mode).[10] The C_2-symmetric pyrrolidine is effective in exerting a diastereoselection to the extent of 97:3 in the *endo* mode of cyclization, yet the *exo* mode led to a 1:1 mixture of the 14-membered ketone isomers. There was an enormous difference between 1,4- and 1,5-stereoinduction.

Reactions of electrophilic hydrazonium ions derived from prolinol with nucleophiles are subject to influence of the side chain. On this basis an access to chiral salsolidine (**4-12**) is readily developed.[11]

1,4-Stereoinduction in the reduction of tricyclic hydrazones was the crux for a chiral synthesis of α-amino acids (**4-13**).[12,13] Interestingly, the dominant effect comes

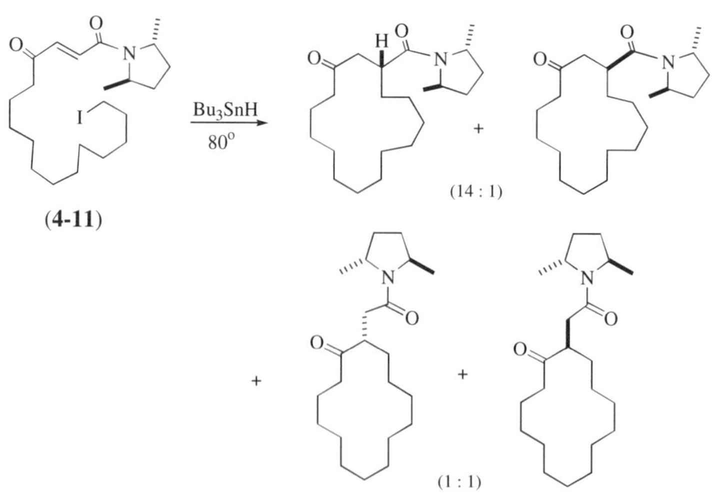

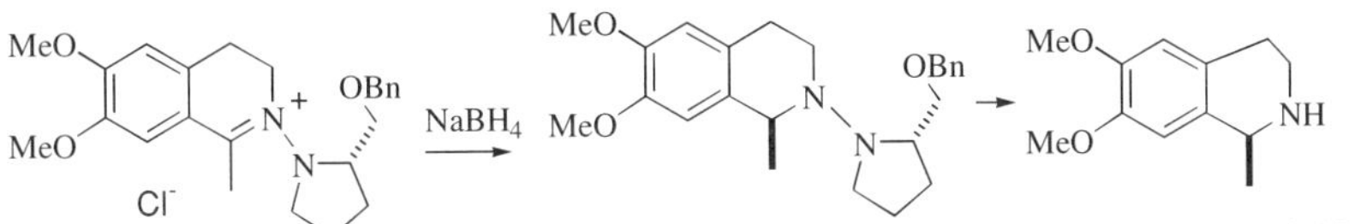

(-)-salsolidine (**4-12**)

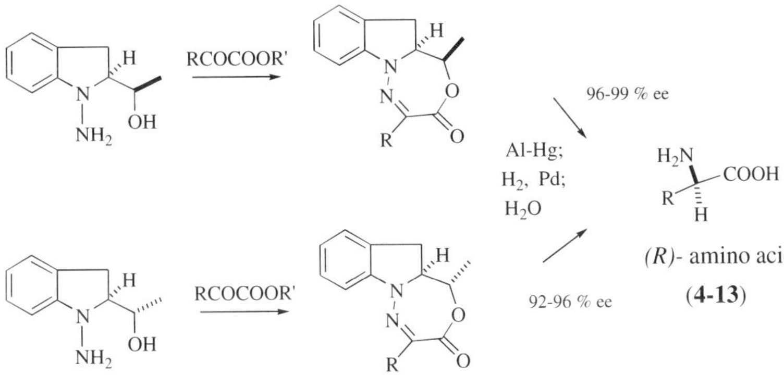

(*R*)- amino acids (**4-13**)

from the ring junction as hydrazones epimeric at the carbinolic center gave amino acids of the same optical series.

In the context of method development and evaluation of asymmetric induction during Birch reduction, a synthesis of chiral pumiliotoxin-C (**4-14**) was pursued.[14] The plan involved the incorporation of L-proline into a tricyclic substrate for the Birch reduction. The first new stereocenter was created on protonation at the same side as the existing methine hydrogen (1,4-stereoinduction).

In situ alkylation of the Birch reduction product appears to be stereoselective. It is significant that the steric courses are different for substrates with the chiral auxiliaries singly or doubly bound to the aromatic ring (cf. **6-88** and **6-89**).[15]

In certain Diels–Alder reactions involving furans, diastereoselection can be observed when a substituent bearing a chiral moiety is installed at $C_{(3)}$. The evolution of an extremely concise method for assembling a potential precursor of zaragozic acids (**4-15**)[16] attests to the maneuverability.

Diastereofacial selection was expected and demonstrated in nopinone derivatives. Thus, the operation of 1,5- and 1,4-asymmetric inductions by the *gem*-dimethylated bridge could be relied on to provide a useful intermediate of (+)-nootkatone (**4-16**).[17] The pertinent reactions are Michael addition and methylation.

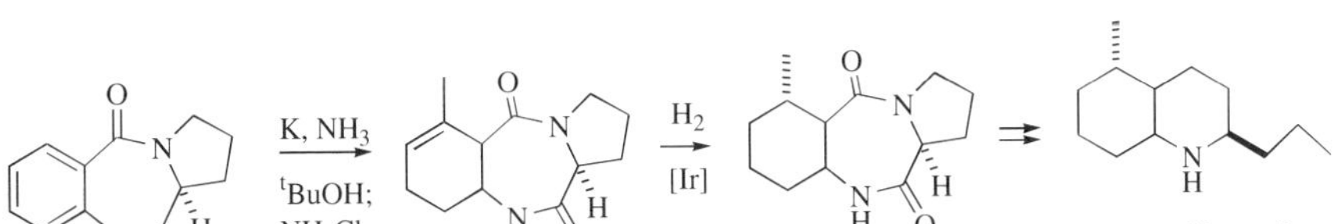

(+)-pumiliotoxin
(**4-14**)

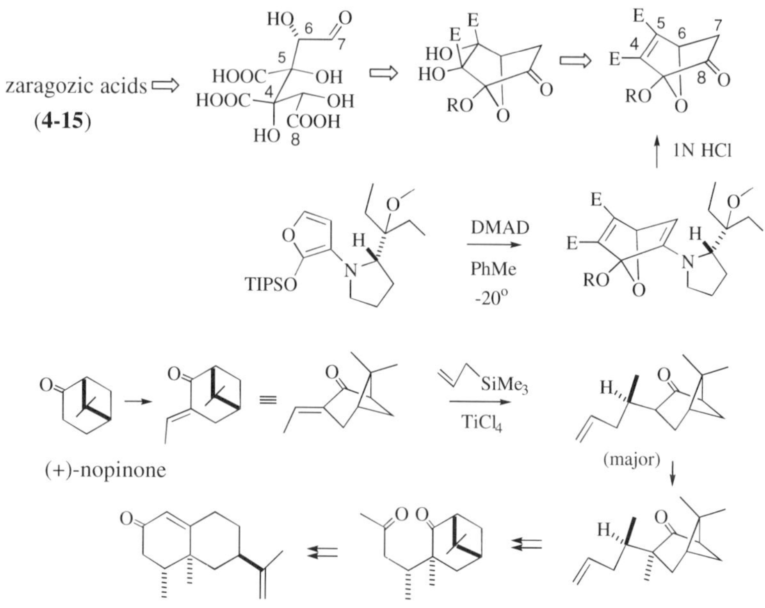

(+)-nootkatone
(**4-16**)

D-Amino acids (**4-17**) have been found as structural components of some natural products, and therefore their access is desirable. Can certain relatively inexpensive L-amino acids be used to synthesize these rare enantiomers? Starting from a glycyl dipeptide is indicated, and it requires chiral information to be transmitted from the α-carbon atom of the L-amino acid to the glycyl moiety linked to the nitrogen atom or to the original carboxyl function. While simple derivatives cannot be considered because of their conformational mobility, cyclic dipeptides composed of a glycine and another bulky amino acid subunit offer the best opportunity. For reasons of availability and effectiveness, cyclo(L-Val-Gly) is an excellent compromise. A derived bislactim ether is subject to 1,4-asymmetric induction during alkylation, and the products can be hydrolyzed to L-valine and the D-amino acids.[18–20]

Note that reaction of an exocyclic methylene derivative (**4-18**) with an organomercurial reagent followed by borohydride reduction can lead to the opposite enantiomer.[21]

A Heck reaction cascade was the central theme for the construction of the BCD-ring system of scopadulcic acid-A (**4-19**).[22] Besides meeting the structural requirements, three stereocenters at the ring junctions must also be affixed in the

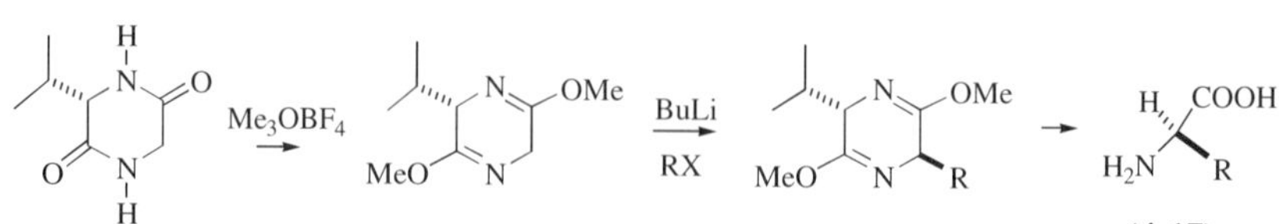

(**4-17**)

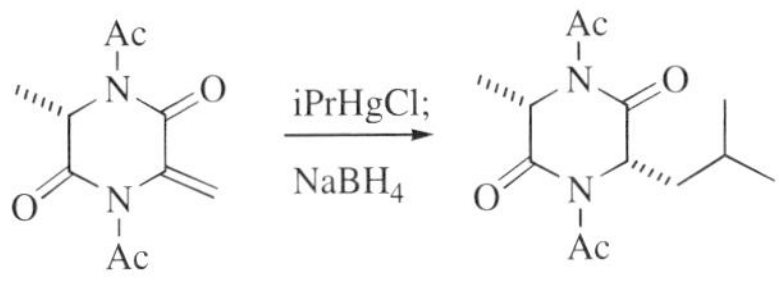

(**4-18**) 100 % ds

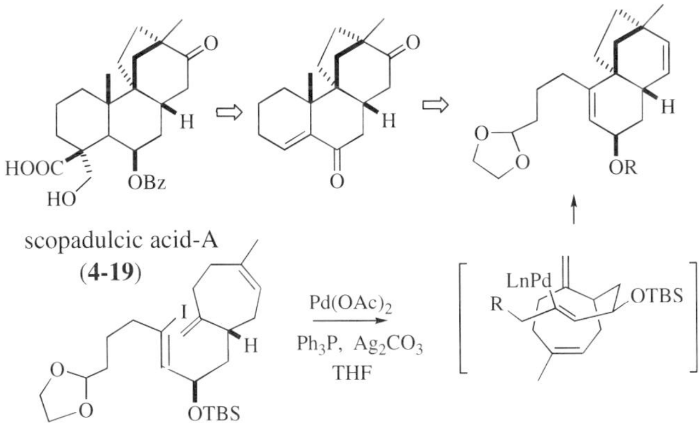

scopadulcic acid-A (**4-19**)

process. Placing a bulky siloxy functionality (at $C_{(6)}$) to occupy an equatorial position determined the transition state of the first cyclization and thus the mission was accomplished.

For a projected approach to taxusin (**4-20**),[23] effective transmission of steric information from the A-ring of a B-seco compound was of great importance to the establishment of the quaternary center which is common to B- and C-rings. The relative population of the two conformations, and therefore the steric course of reaction, can be influenced by bulky substituents.

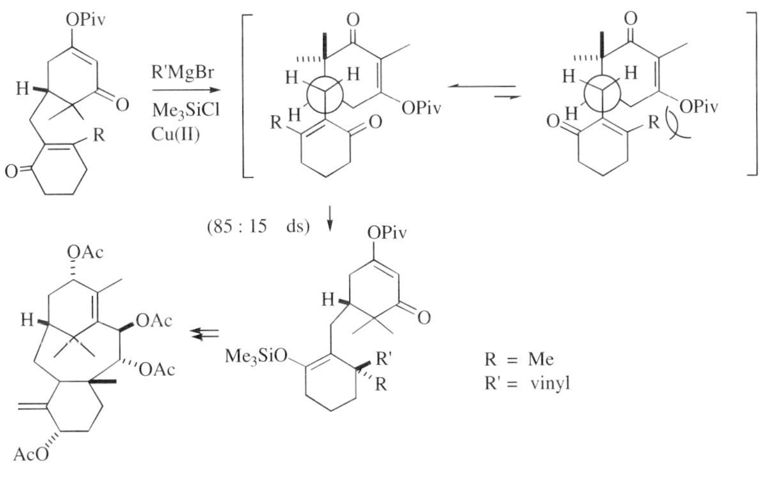

taxusin (**4-20**)

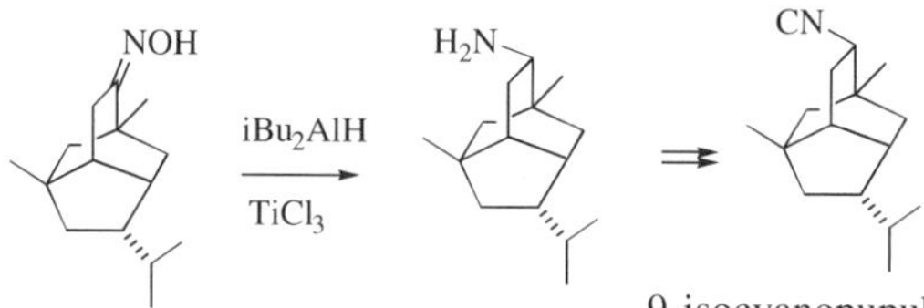

9-isocyanopupukeanane
(**4-21**)

At the late stage of a synthetic study toward 9-isocyanopupukeanane (**4-21**),[24] the stereoselective conversion of a ketone to the isocyano group was required. Reduction of the oxime was dictated by the distal methyl substituent at a bridgehead position, and therefore a rather easy solution to the problem presented itself.

In an elegant synthesis of (+)-miroestrol (**4-22**),[25] the possibly biomimetic double annulation was preceded by the preparation of a polysubstituted cyclohexenone. Analysis suggested a regio- and stereoselective epoxidation of a cross-conjugated dienone precursor which can be rendered in a *p*-benzoquinone monoacetal. When the acetal group was derived from a chiral 2,4-pentanediol, sufficiently different steric shieldings on the two faces of the more electron-rich double bond were manifested. An 85:15 mixture of diastereomers favoring the desired epoxide was obtained. This was the result of 1,4-asymmetric induction by virtue of an axial methyl group which blocked the approach of the bulky trityl hydroperoxide reagent from its side.

There are several problems pertaining to synthesis of ancistrocladine and related alkaloids. The difficulty (failure) in effecting intermolecular biaryl coupling was noted, and issues of regio- and stereoselectivity have yet to be settled. However, preassociation in the form of an ester to change the coupling process to an intramolecular reaction cleared all of the hurdles.[26] The atropisomeric lactones were separated and degraded accordingly. Only (–)-ancistrocladine (**4-23**) was obtained if the amine was protected as a trifluoroacetyl derivative.

The discovery of many bioactive natural products which contain tetrahydrofuran subunits has prompted intensive studies of their synthesis. Induced cyclization of δ,ε-unsaturated oxygenated compounds (**4-24, 4-25**) represents one approach and asymmetric induction is very important.[27,28]

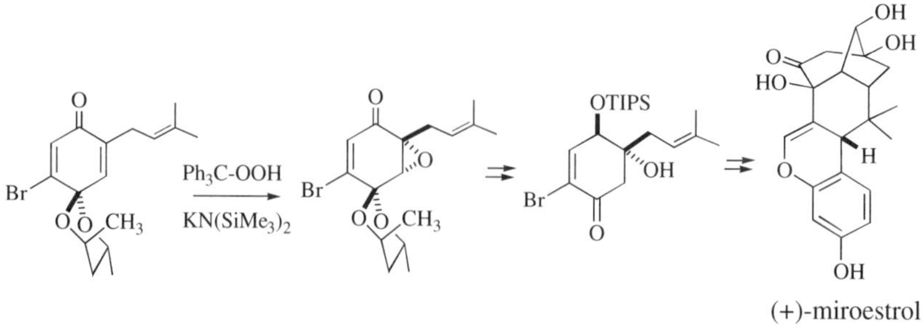

(+)-miroestrol
(**4-22**)

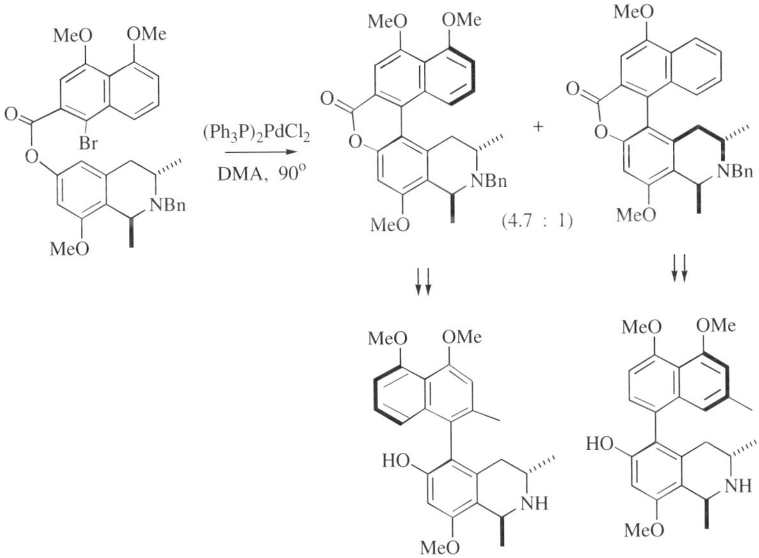

(-)-ancistrocladine (**4-23**) (+)-hamatine

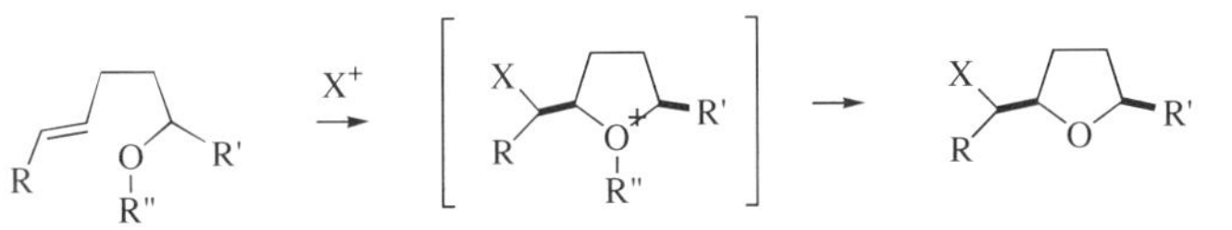

(**4-24**)

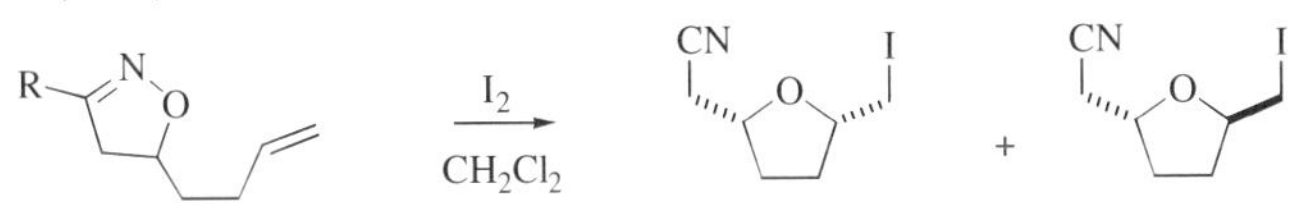

(**4-25**)

R = Ph_3C	1	:	4
R = Me_3C	1	:	1
R = Me_3Si	6.6	:	1

4.2. 1,5-STEREOINDUCTION

Imides in which the nitrogen atom is bonded to a menthane nucleus show excellent stereoselectivity. A method for the preparation of chiral β-lactam building blocks (**4-26**) for penems and carbapenems, based on aldol condensation of the proper derivatives, has been developed.[29]

The Birch reduction product of a benzoic ester (**4-27**) in which an *o*-benzylic carbon is linked to a prolinol ether was found to undergo alkylation stereoselectively.[30] However, the loss of stereoselectivity in reactions with benzylic and allylic halides is presently unexplainable.

(*S*)-(+)- 4-Methyl-3-heptanone (**4-28**), the principal alarm pheromone of the leaf-cutting ant *Atta texana*, should be available from propylation of 3-pentanone. The

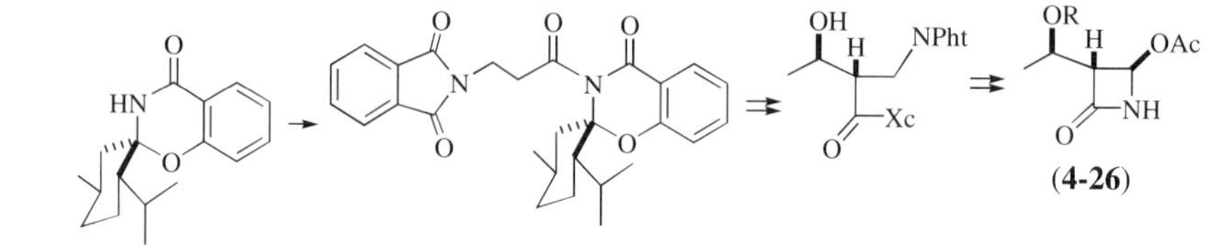

(4-26)

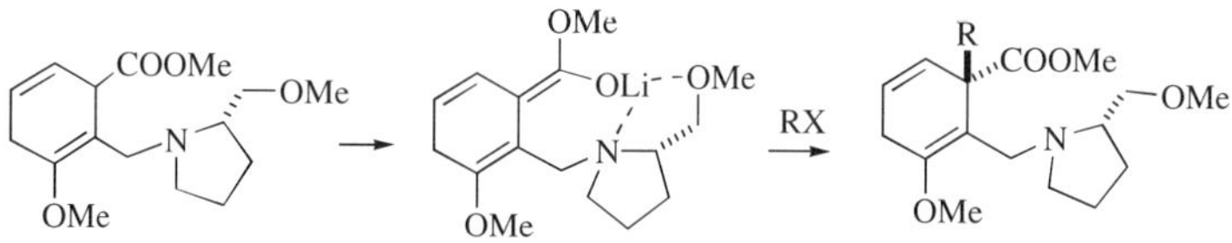

(4-27)

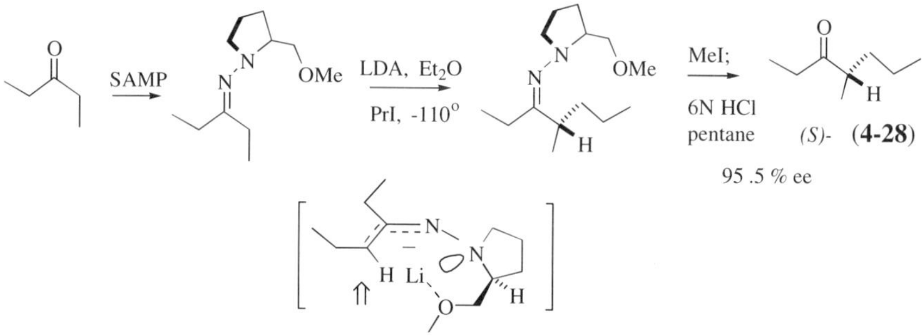

(S)- (4-28)

95 .5 % ee

synthetic challenge is to find a reliable method to achieve asymmetric induction. In this regard the alkylation of (*S*)-1-amino-2-methoxymethylpyrrolidine (SAMP) hydrazone[31] stands out and the accessibility of the chirally complementary RAMP auxiliary enhances the general utility.[32]

The key to a two-step approach to chiral 5-alkyl-2(5*H*)-furanones (**4-29**)[33] concerns reaction of an α-sulfonyl carbanion with an aldehyde. The chirality of the α-phenylamide four bonds away is felt.

Effective stereoinduction of a γ-trityloxymethyl group on reaction at the α- and β-positions of the saturated and unsaturated butyrolactones has been demonstrated on many occasions. A rarer variation was involved in the 1,2-difunctionalization of the (*Z*)-ethylidene derivative for the preparation of an intermediate of (–)-megaphone (**4-30**).[34]

1,4-Addition of Grignard reagents to conjugated imines in which the α′-position of the chiral substituent has a bulky group can lead to products of excellent enantiopurity. Exploitation of this 1,5-stereoinduction to a synthesis of (+)-ivalin

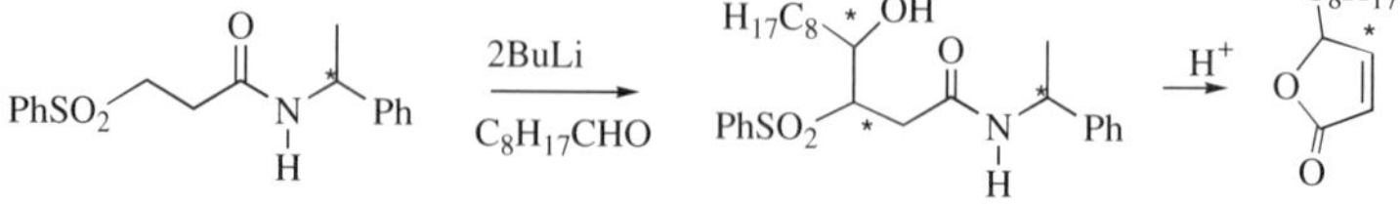

(4-29)

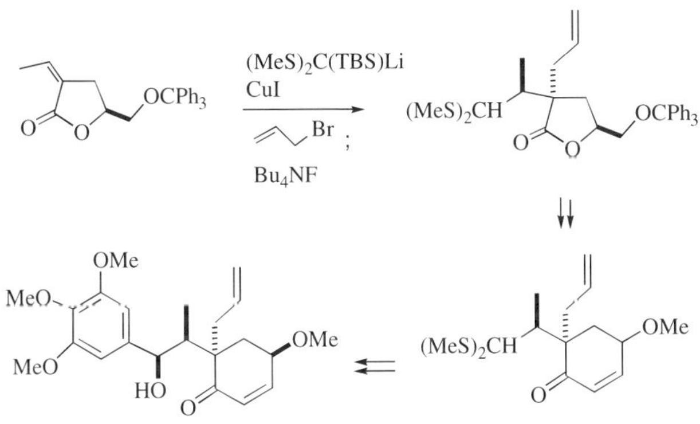

(-)-megaphone (**4-30**)

(**4-31**)[35] was quite successful, except that it was marred by the much lower stereoselectivity for the trapping step, which may have been caused by an unfavorable interaction with one arm of the ethyleneacetal.

Matched and mismatched transition states resulting from directed delivery of the nucleophile were invoked for the formation of two diastereomers from a Grignard reaction during an investigation of an ikarugamycin (**4-32**) synthesis.[36] Fortunately, the major product was the desired isomer.

Acetals of conjugated aldehydes derived from C_2-symmetrical diols undergo reactions with nucleophiles in such a way that variation of conditions can give rise to products of defined stereochemistry. By following these directions, an intermediate (**4-33**) of vitamin K and E[37] and of the California red scale pheromone (**4-34**)[38,39] were synthesized.

A monoacetal of 2,2-dimethyl-3-oxo-5-hexenal was identified as starting material for the synthesis of pedamide (**4-35**),[40] an intermediate of the insect poison pederin. When a C_2-symmetric 2,3-butanediol was used to form the acetal, the reduction of the ketone group was amenable to asymmetric induction. This 1,5-induction arose from internal hydride transfer from a metal chelate in which the conformation was influenced by the chirality of the dioxolane ring.

A weak 1,5-stereoinduction (10% optical yield) was observed in the borohydride reduction of a menthyl dehydroquinolizidinecarboxylate (**4-36**).[41] Apparently, the system was too flexible and no definite conformation was favored.

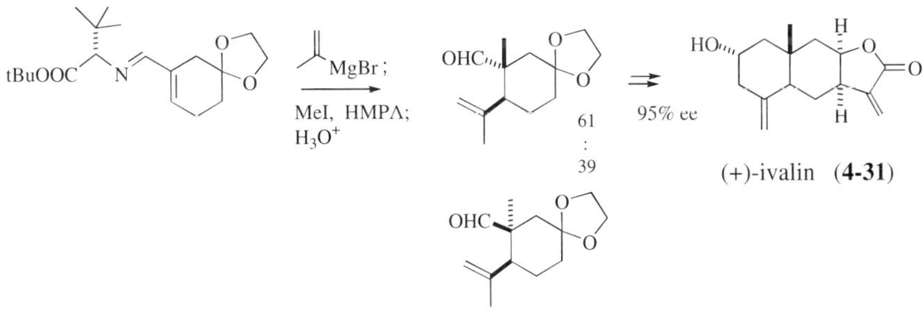

(+)-ivalin (**4-31**)

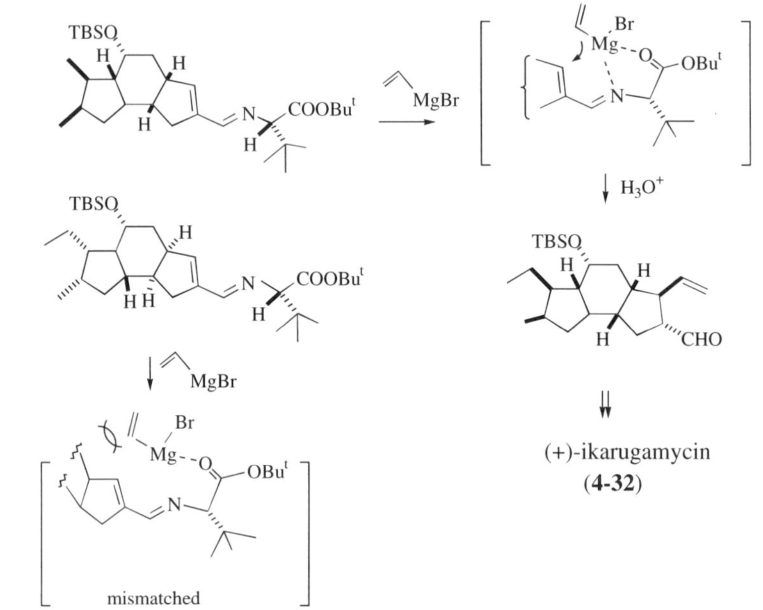

(+)-ikarugamycin
(**4-32**)

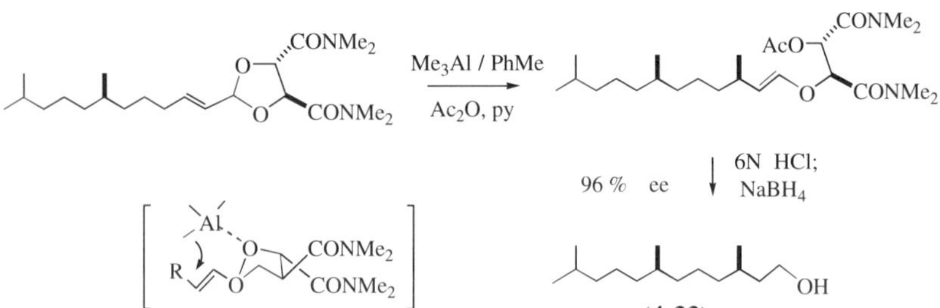

(**4-33**)

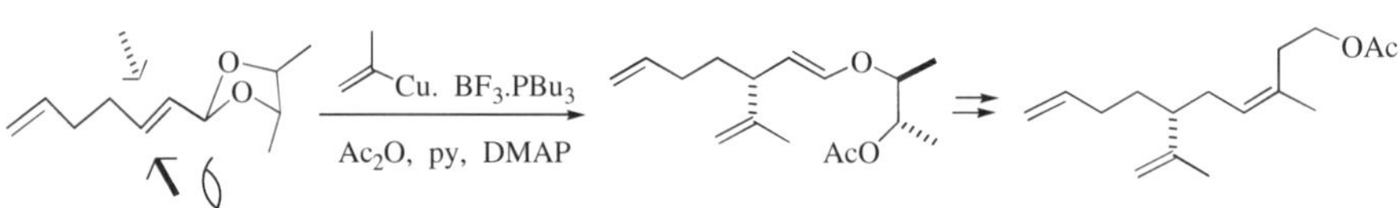

California red scale
pheromone (**4-34**)

The mixture of two diastereomers obtained from esterification of *p*-toluenesulfonic acid with (–)-menthol could be equilibrated to the pure (–)-menthyl (*S*)-*p*-toluenesulfonate. Grignard reaction gave (*R*)-alkyl tolyl sulfones (**4-37**).[42]

Unusually high degrees of 1,5-*anti* stereoinduction in boron-mediated aldol reactions of β-oxygenated methyl ketones have been observed.[43] Since the ketone group of such ketols (**4-38**) can be reduced to give *syn*- or *anti*-1,3-diols, the method contributes to solving problems pertaining to synthesis of polyketide natural products.

Ring closure by a Hiyama reaction to give only 11,0(3)-dihydropseudopterolide (**4-39**),[44] although in low yield, was surprising. It is not known why the corresponding transition state was favored to such an extent.

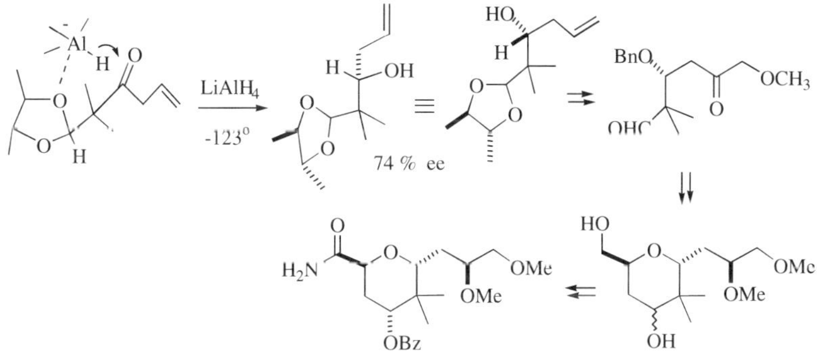

(+)-pedamide
(**4-35**)

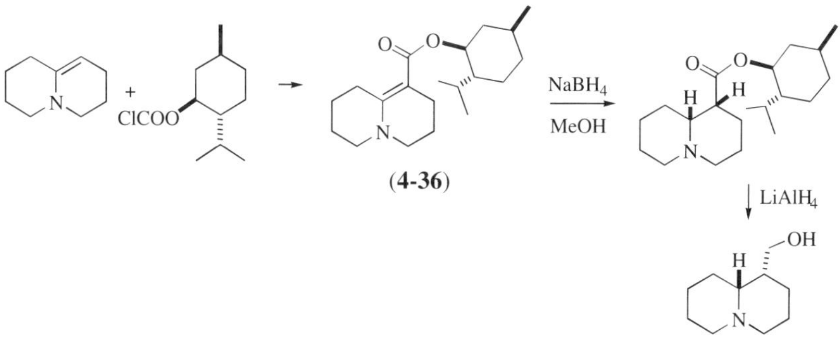

(**4-36**)

(-)-lupinine

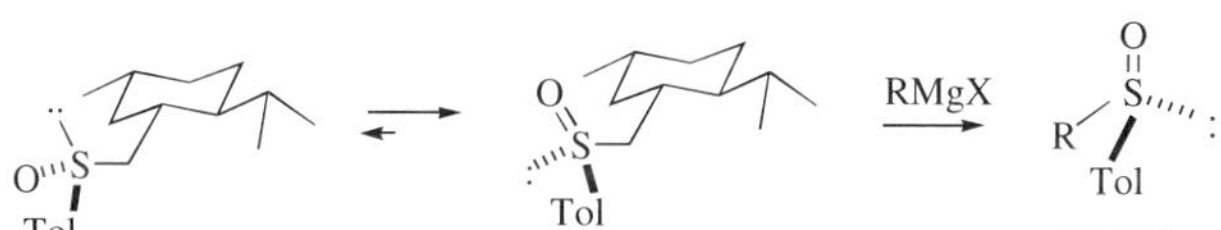

(**4-37**)

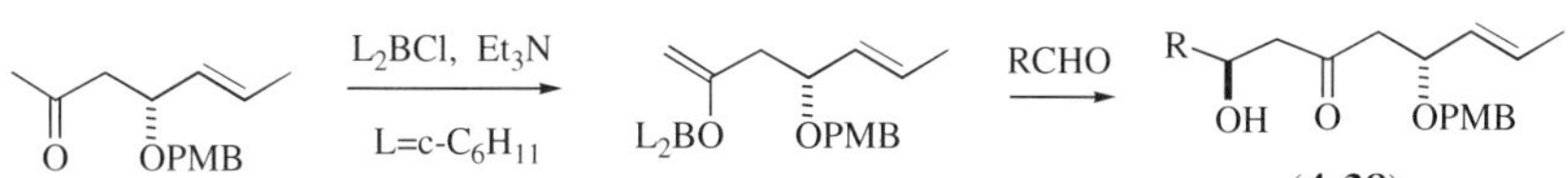

(**4-38**)

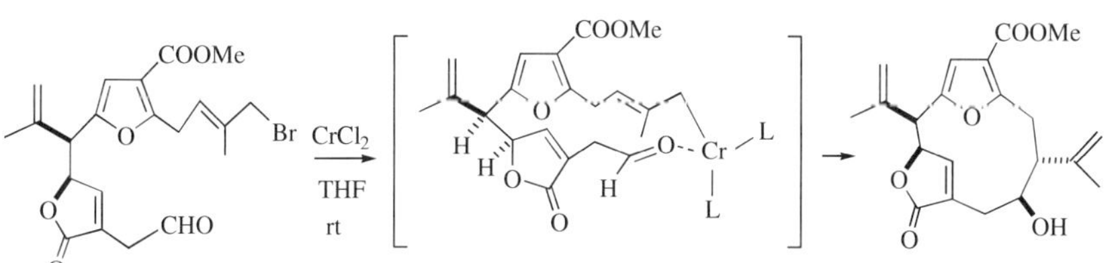

11, 0(3)-dihydropseudopterolide
(**4-39**)

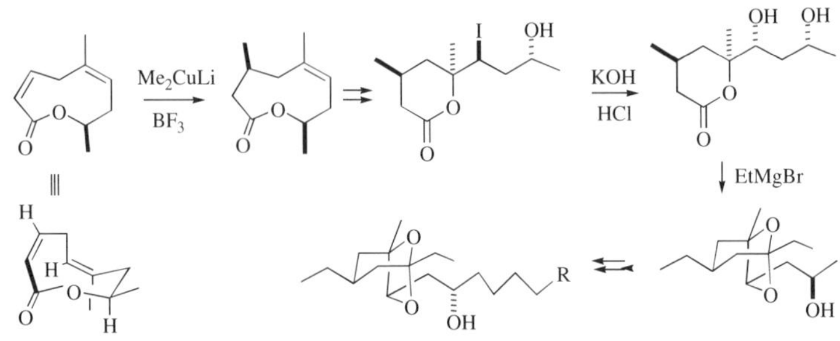

(4-40)

A single substituent in a macrocycle is sufficient to determine its conformation and to prescribe reactions at its periphery. Accordingly, remote stereoinduction of this type may be used as an indirect means to attain acyclic stereoselectivity, with the proviso of a subsequent ring opening. This principle has been demonstrated in a synthesis of the lyophilic segment (**4-40**) of palytoxin[45] in which the secondary methyl group of the cyclononadienolide intermediate ensured the steric course of the conjugate addition to afford the *cis* product.

Stereoselective Diels–Alder reactions of an axially chiral anilide in the presence of iodine have been demonstrated.[45a] Such reactions seem to proceed via the *s-trans* form of the 2-vinyloxazolinium ion (**4-40a**) in which the diastereoselectivity arises from the presence of the *t*-butyl group on the aromatic ring, which is five bonds away.

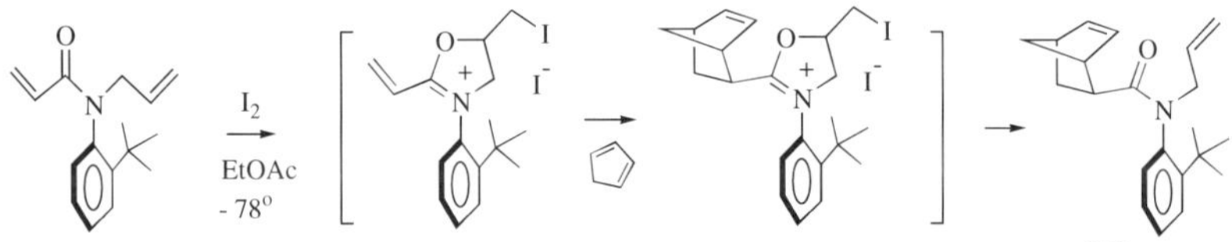

(4-40a) *endo* : *exo* = 30 92%

4.3. REMOTE STEREOINDUCTION

1,6-Stereoinduction and those at an even longer distance are rarer, although they can be quite effective. Two reports have concerned the use of chiral boronate complexes (**4-41, 4-42**) of 1,4-naphthoquinones in intermolecular Diels–Alder reactions.[46,47] Excellent diastereofacial and enantioselection were observed.

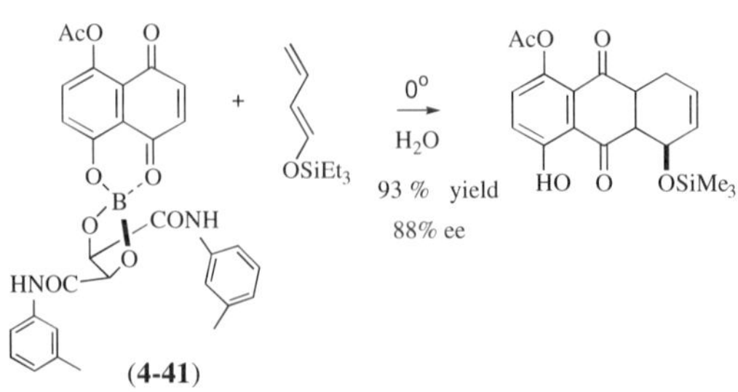

(4-41)

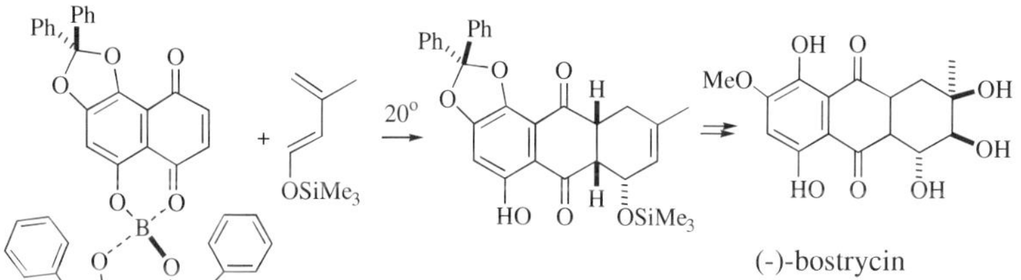

(**4-42**)

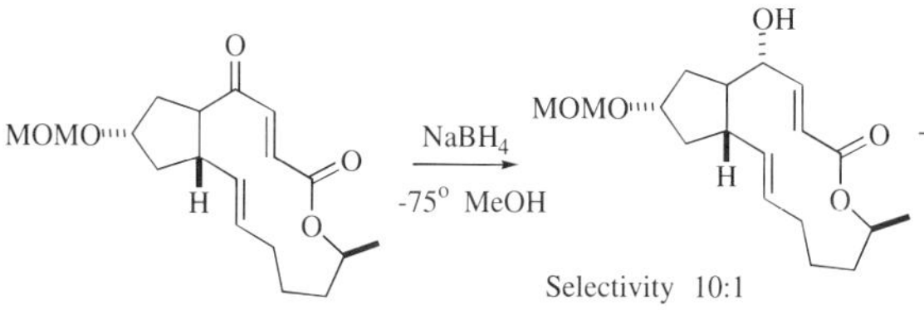

brefeldin-A
(**4-43**)

Perhaps quite unexpectedly, in a synthesis of brefeldin-A (**4-43**)[48] the reduction of the subangular ketone seemed to be influenced by the methyl group at the ethereal end of the macrolactone. The proper isomer induced the generation of the secondary alcohol with the correct relative configuration.

1,7-Induction by substituents in cyclic boronates (**4-44**) on reduction of a ketone group[49] was due to internal coordination and the attack of the reducing agent from a less hindered path.

Different conformations adopted by a δ-sulfinyl γ,δ-unsaturated ketone (**4-45**), as a result of specific coordination with the metal species present, were responsible for the reduction courses.[50] The behavior was recognized and used as a basis for generating various isomers of an insect pheromone.

1,6-Stereoinduction to produce a specific geometrical isomer of an alkylidenetetrahydropyran (**4-46**) is a significant finding.[51] A trityloxy group apparently blocked off its side (selectivity > 9:1) when the ketone underwent an Emmons–Wadsworth reaction. This result simplified the assembly of the $C_{(10)}$–$C_{(16)}$ segment of the bryostatins.

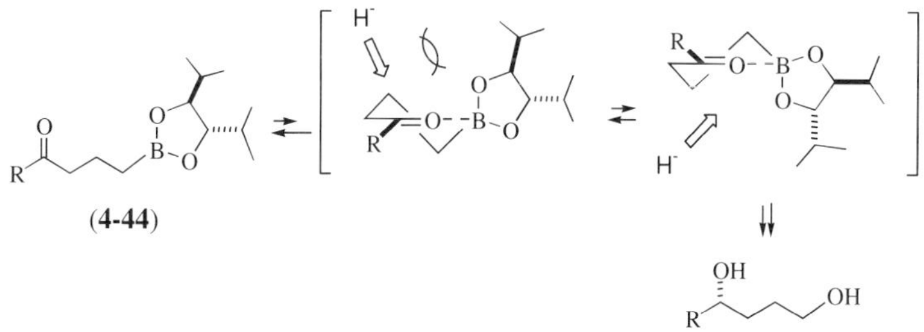

(**4-44**)

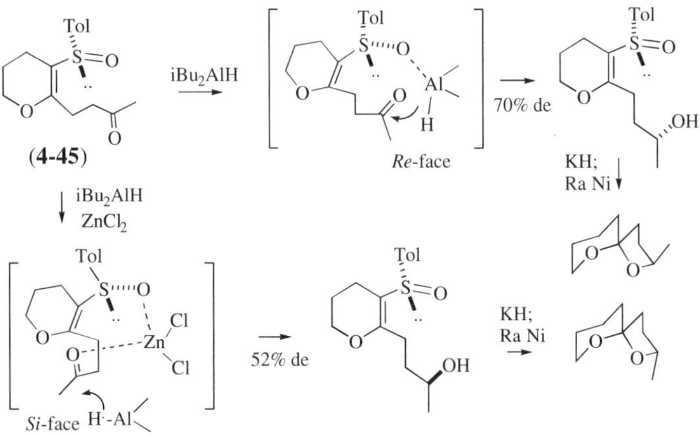

(4-45)

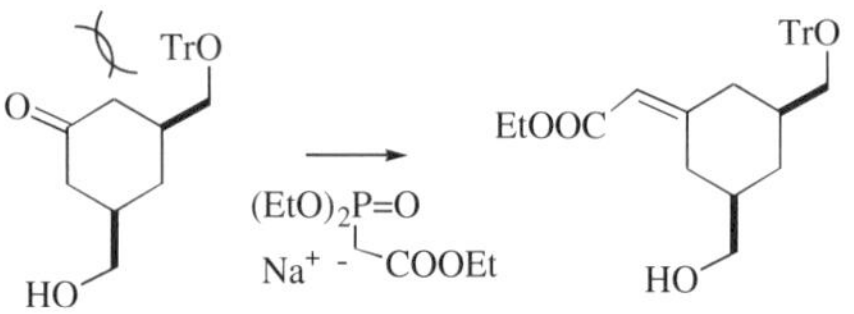

(4-46)

The local symmetry of a 2,2-disubstituted 1,3-cyclopentanedione (**4-47**) was broken by a remote stereogenic element when the molecule underwent aldol-type cyclization.[52,53] (+)-Estr-4-ene-3,17-dione was readily prepared from the product.

The cleavage of an equatorial C–O bond of C_2-symmetrical acetals is greatly preferred in the Lewis acid catalyzed cyclization by participation of a remote double bond.[54] This reaction constitutes a powerful method for the synthesis of polycyclic compounds. Thus, calcitriol can be prepared from (**4-48**).

A 1,7-induction has been witnessed in the Mukaiyama aldol reactions of π-allyltricarbonyliron lactone and lactam complexes bearing trimethylsilyl enol ether

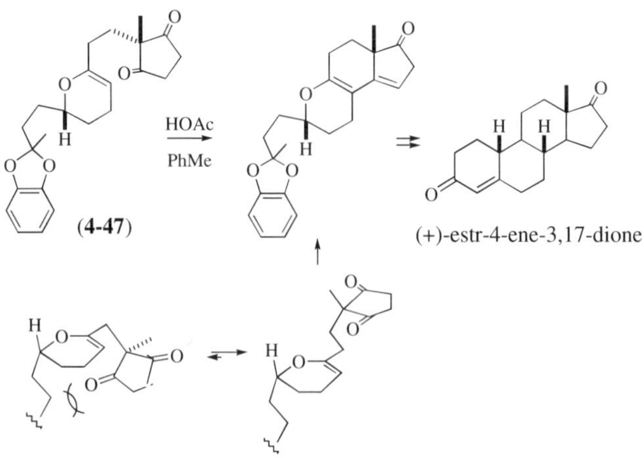

(4-47)

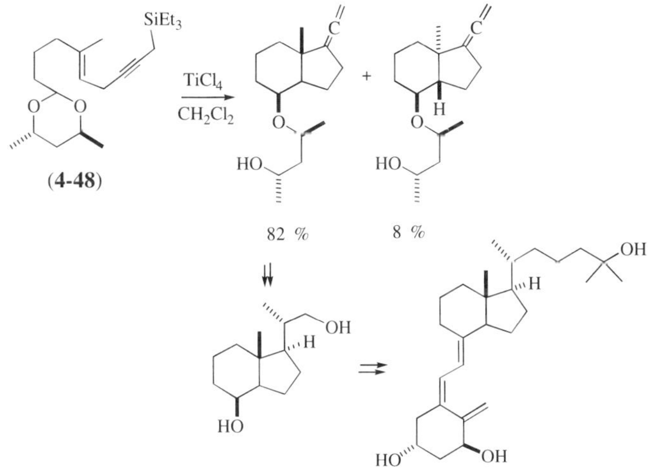

(4-48)

calcitriol

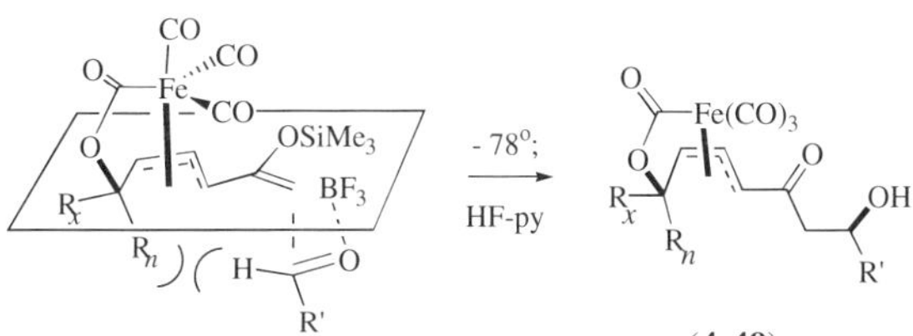

(4-49)

side-chains. A bulky endo substituent at the carbinyl center acts as a discriminator for the orientation of the aldehyde.[55]

The highly ordered transition state for the catalyzed Diels–Alder reaction involving an oxazaborolidine (**4-50**) derived from (α*S*, β*R*)-β-methyltryptophan[56] has enabled the asymmetric synthesis of intermediates of many complex natural products.[57] Clearly it is a case of 1,7-stereoinduction.

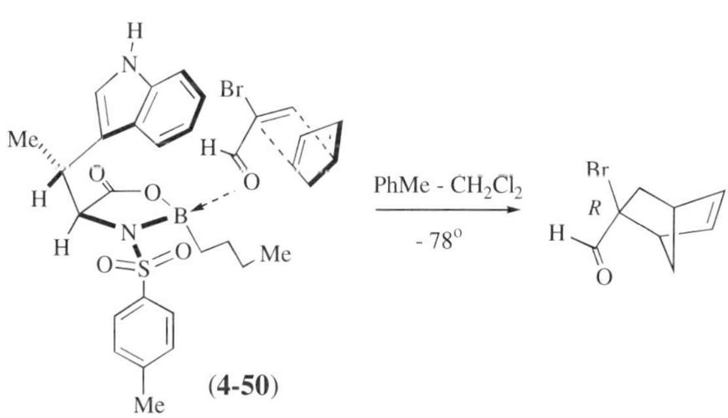

(4-50)

REFERENCES

1. Danishefsky, S.; Tsuzuki, K. *J. Am. Chem. Soc.* **1980**, *102*, 6891.
2. Ireland, R.E.; Dow, W.C.; Godfrey, J.D.; Thaisrivongs, S. *J. Org. Chem.* **1984**, *49*, 1001.
3. Guingant, A. *Tetrahedron: Asymmetry* **1991**, *2*, 411, 415.
4. Mekouar, K.; Ambroise, L.; Desmaële, D.; d'Angelo, J. *Synlett* **1995**, 529.
5. Kim, D.; Choi, W.J.; Hong, J.Y.; Park, I.Y.; Kim, Y.B. *Tetrahedron Lett.* **1996**, *37*, 1433.
6. Evans, D.A. In *Asymmetric Synthesis*; Morrison, J.D., Ed.; Academic Press: Orlando, 1984; Vol. *3*, Part B, pp 1–110.
7. Sibi, M.P.; Deshpande, P.K.; La Loggia, A.J. *Synlett* **1996**, 343.
8. Wulff, W.D.; Powers, T.S.; Shi, Y.; Wilson, K.J. *J. Org. Chem.* **1994**, *59*, 6882.
9. Schreiber, S.L.; Meyers, H.V. *J. Am. Chem. Soc.* **1988**, *110*, 5198.
10. Porter, N.A.; Lacher, B.; Chang, V.H.-T.; Magnin, D.R. *J. Am. Chem. Soc.* **1989**, *111*, 8309.
11. Suzuki, H.; Aoyagi, S.; Kibayashi, C. *Tetrahedron Lett.* **1995**, *36*, 6709.
12. Corey, E.J.; McCaully, R.J.; Sachdev, H.S. *J. Am. Chem. Soc.* **1970**, *92*, 2476.
13. Corey, E.J.; Sachdev, H.S.; Gougoutas, J.Z. *J. Am. Chem. Soc.* **1970**, *92*, 2488.
14. Schultz, A.G.; McCloskey, P.J.; Court, J.J. *J. Am. Chem. Soc.* **1987**, *109*, 6493.
15. McCloskey, P.J.; Schultz, A.G. *Heterocycles* **1987**, *25*, 437.
16. Schlessinger, R.H.; Wu, X.-H.; Pettus, T.R.R. *Synlett* **1995**, 536.
17. Yanami, T.; Miyashita, M.; Yoshikoshi, A. *J. Org. Chem.* **1980**, *45*, 607.
18. Schöllkopf, U.; Groth, U.; Deng, C. *Angew. Chem. Int. Ed. Engl.* **1981**, *20*, 798.
19. Schöllkopf, U. *Pure Appl. Chem.* **1983**, *55*, 1799.
20. Schöllkopf, U.; Pettig, D.; Busse, U.; Egert, E.; Dyrbusch, M. *Synthesis* **1986**, 737.
21. Chai, C.L.L.; King, A.R. *Tetrahedron Lett.* **1995**, *36*, 4295.
22. Kucera, D.J.; O'Connor, S.J.; Overman, L.E. *J. Org. Chem.* **1993**, *58*, 5304.
23. Furukawa, T.; Seto, M.; Horiguchi, Y.; Kuwajima, I. *Chem. Lett.* **1993**, 1279.
24. Yamamoto, H.; Sham, H.L. *J. Am. Chem. Soc.* **1979**, *110*, 1609.
25. Corey, E.J.; Wu, L.I. *J. Am. Chem. Soc.* **1993**, *115*, 9327.
26. Bringmann, G.; Jansen, J.R.; Rink, H.-P. *Angew. Chem. Int. Ed. Engl.* **1986**, *25*, 913.
27. Rychnovsky, S.D.; Bartlett, P.A. *J. Am. Chem. Soc.* **1981**, *103*, 3963.
28. Kurth, M.J.; Rodriguez, M.J. *J. Am. Chem. Soc.* **1987**, *109*, 7577.
29. Seki, M.; Miyake, T.; Yamanaka, T.; Ohmizu, H. *Synlett* **1996**, 455.
30. Schultz, A.G.; Taylor, R.E. *J. Am. Chem. Soc.* **1992**, *114*, 3937.
31. Enders, D.; Eichenauer, H.; Baus, U.; Schubert, H.; Kremer, K.A.M. *Tetrahedron* **1984**, *40*, 1345.
32. Enders, D. In Morrison, J.D., ed. *Asymmetric Synthesis* **1984**, *3*, 275.
33. Tanaka, K.; Wakita, H.; Yoda, H.; Kaji, A. *Chem Lett.* **1984**, 1359.
34. Tomioka, K.; Kawasaki, H.; Iitaka, Y.; Koga, K. *Tetrahedron Lett.* **1985**, *26*, 903.
35. Tomioka, K.; Masumi, F.; Yamashita, T.; Koga, K. *Tetrahedron Lett.* **1984**, *25*, 333.
36. Paquette, L.A.; Macdonald, D.; Anderson, L.G. *J. Am. Chem. Soc.* **1990**, *112*, 9292.

37. Fujiwara, J.; Fukutani, Y.; Hasegawa, M.; Maruoka, K.; Yamamoto, H. *J. Am. Chem. Soc.* **1984**, *106*, 5004.
38. Mangeney, P.; Alexakis, A.; Normant, J.-F. *Tetrahedron Lett.* **1987**, *28*, 2363.
39. Alexakis, A.; Mangeney, P.; Ghribi, A.; Marek, I.; Sedrani, R.; Guir, C.; Normant, J.-F. *Pure Appl. Chem.* **1988**, *60*, 49.
40. Yanagiya, M.; Matsuda, F.; Hasegawa, K.; Matsumoto, T. *Tetrahedron Lett.* **1982**, *23*, 4039.
41. Goldberg, S.I.; Ragade, I. *J. Org. Chem.* **1967**, *32*, 1046.
42. Solladie, G.; Hutt, J.; Girardin, A. *Synthesis* **1987**, 173.
43. Paterson, I.; Gibson, K.R.; Oballa, R.M. *Tetrahedron Lett.* **1996**, *37*, 8585.
44. Paquette, L.A.; Rayner, C.M.; Doherty, A.M. *J. Am. Chem. Soc.* **1990**, *112*, 4078.
45. Still, W.C.; Galynker, I. *J. Am. Chem. Soc.* **1982**, *104*, 1774.
45a. Kitagawa, O.; Izawa, H.; Sato, K.; Dobashi, A.; Taguchi, T.; Shiro, M. *J. Org. Chem.* **1998**, *63*, 2634.
46. Kelly, T.R.; Whiting, A.; Chandrakumar, N.S. *J. Am. Chem. Soc.* **1986**, *108*, 3510.
47. Maruoka, K.; Sakurai, M.; Fujiwara, J.; Yamamoto, H. *Tetrahedron Lett.* **1986**, *27*, 4895.
48. Corey, E.J.; Wollenberg, R.H. *Tetrahedron Lett.* **1976**, 4701.
49. Molander, G.A.; Bobbitt, K.L. *J. Am. Chem. Soc.* **1993**, *115*, 7517.
50. Iwata, C.; Moritani, Y.; Sugiyama, K.; Fujita, M.; Imanishi, T. *Tetrahedron Lett.* **1987**, *28*, 2255.
51. Lampe, T.F.J.; Hoffmann, H.M.R. *Tetrahedron Lett.* **1996**, *37*, 7695.
52. Rosenberger, M.; Fraher, T.P.; Saucy, G. *Helv. Chim. Acta* **1971**, *54*, 2857.
53. Rosenberger, M.; Duggan, A.J.; Borer, R.; Müller, R.; Saucy, G. *Helv. Chim. Acta* **1972**, *55*, 2663.
54. Johnson, W.S.; Elliott, J.D.; Hanson, G.J. *J. Am. Chem. Soc.* **1984**, *106*, 1138.
55. Ley, S.V.; Cox, L.R.; Middleton, B.; Worrall, J.M. *Chem. Commun.* **1998**, 1339.
56. Corey, E.J.; Loh, T.-P. *Tetrahedron Lett.* **1993**, *34*, 3979.
57. Corey, E.J.; Guzman-Perez, A.; Loh, T.-P. *J. Am. Chem. Soc.* **1994**, *116*, 3611.

5

GROUP DIRECTED REACTIONS AND CHELATION EFFECTS

In recent years many synthetic methods have been developed which depend on the characteristics of the reagents to guide the steric courses. It is also popular to introduce auxiliaries and ligands to substrates to create or enhance asymmetric reactions.[1] Regarding the latter practice, they can be categorized in various 1,*n*-stereoinductions. This chapter will concentrate on polar functionalities of molecules and show how these intrinsic functionalities may help various synthetic endeavors in the stereoselective sense.[2]

5.1. DIRECTED OXIDATIONS

5.1.1. Epoxidation of Alkenes with Peracids

The pioneering work of Henbest on allylic alcohols[3] revealed the ability of a proximal hydroxyl group to direct epoxidation with peracids. Hydrogen bonding between the hydroxyl functionality and one of the peracid oxygen atoms guides the electrophile to one particular π-face of the double bond. Allylic carbamates show the same propensity. Support for the contention that the stereoselectivity is due to hydrogen bonding and not the polarity of the substituent is the observation of random epoxidation or opposite steric course with the corresponding ethers as substrates.

2-Cycloalkenols usually afford the *cis*-epoxides and a pseudoequatorial hydroxyl group appears to be more effective in the direction. The hydrogen bonding facilitation of epoxidation is due to the fact that the leaving group is a carboxylic acid instead of the carboxylate ion. For good results epoxidation should not be conducted in solvents which can compete with the substrate in providing a hydrogen bond donor.

Homoallylic alcohols also exhibit the same effect, but pseudoaxial alcohols are much better directors. Numerous examples are found in the chemical literature describing the service of directed epoxidation, generally with *m*-chloroperbenzoic acid, but only a few cases can be mentioned here.

The glycidol system of paniculide-A (**5-1**) is most invitatory of the technique and indeed received due attention. Unfortunately, the stereoselectivity was rather modest (3:1).[4] In a synthesis of vernolepin (**5-2**),[5] the allylic alcohol segment which underwent epoxidation was created by way of iodolactonization, dehydroiodination, and hydrolysis. The hydroxyl group was important as epoxidation of the lactone itself gave poor results. A methylenecyclohexanol derived from quebrachitol was transformed into (–)-ovalicin (**5-3**)[6] the first operation being epoxidation with *m*-chloroperbenzoic acid. Also in the penultimate step of the synthesis the treatment with VO(acac)$_2$-*t*BuOOH gave two diastereomers in a 65:35 ratio, favoring the desired product.

The unique structure of aranorosin (**5-4**) contains two epoxide units in a *cis* relationship. It was fortunate that the organometallic reaction of a cyclohexadienone intermediate was stereoselective and the resulting alcohol could be used to direct the epoxidation.[7]

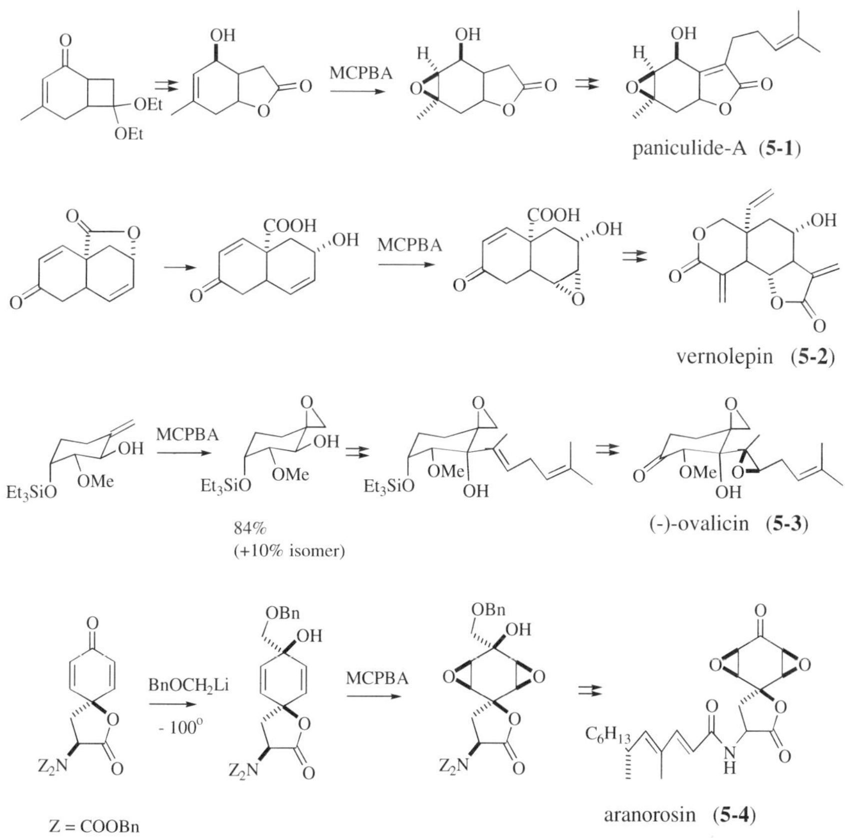

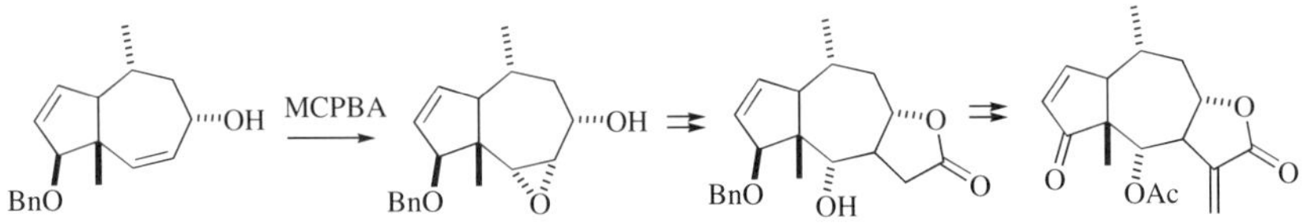

bigelovin (**5-5**)

Bigelovin (**5-5**) is a pseudoguaianolide sesquiterpene in which the two oxygenated functionalities in the seven-membered ring are *cis* to one another. Since they are *trans* to the $C_{(7)}$ branch, the pattern can be created from a *cis*-glycidol unit and ultimately the cycloheptenol.[8]

An allylic tosylamido group in cycloheptene derivatives was successfully engaged to construct epoxy intermediates for the synthesis of scopine (**5-6**) and pseudoscopine (**5-7**).[9]

Homoallylic hydroxyl groups serving an important role in synthesis are found in precursors of alliacol-A,[10] trichodermin,[11] verrucarol,[12] and monensin.[13] Thus, it can be located in a fused ring, a bridged ring, or an aliphatic chain.

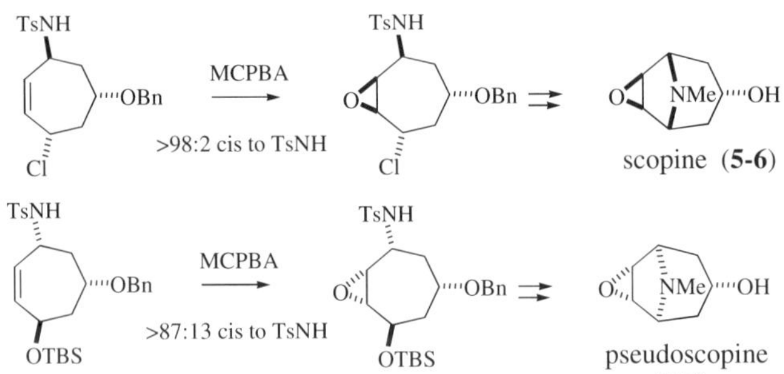

scopine (**5-6**)

pseudoscopine (**5-7**)

5.1.2. Metal-Catalyzed Epoxidations

The discovery of alkene epoxidation[14] by *t*-butyl hydroperoxide in the presence of $VO(acac)_2$ and $Mo(CO)_6$ and the subsequent clarification of regio- and stereoselective aspects for allylic/homoallylic alcohols in the vanadium-catalyzed reactions[15] has had great impact on organic synthesis which also has implications in the production of certain fine chemicals. Generally these reactions show rate enhancements as contrasted to deactivation observed for peracid epoxidation of the hydroxylated compounds. It is now recognized that vanadium gathers the substrate and the oxidizing agent together to change the reaction into an intramolecular version. Accordingly the ease of complex formation is of prime importance, as shown by the relatively low diastereoselection (3:1, same as in peracid epoxidation) in the reaction at the hindered endo side of the bicyclo[4.2.0]octenol intermediate for a synthesis of paniculide-A.[4] Much more successful cases include the preparation of the precursors (**5-8**) of anthracycline antibiotics,[16] and of trichodermol (**5-9**)[17] and taxol (**5-10**).[18,19]

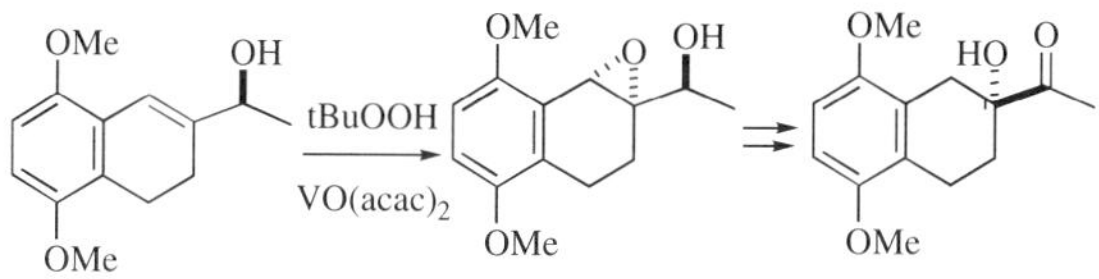

(**5-8**)

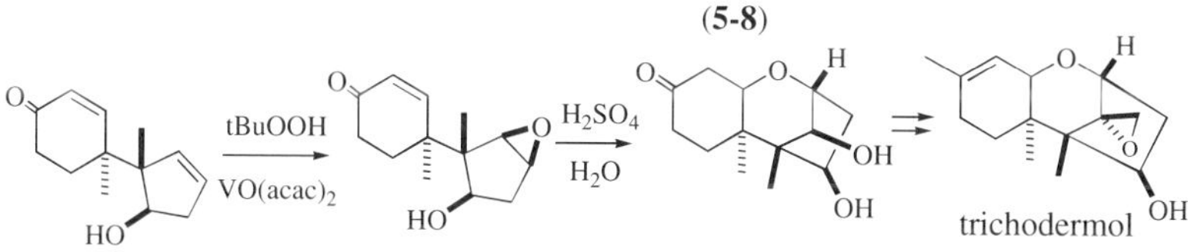

trichodermol
(**5-9**)

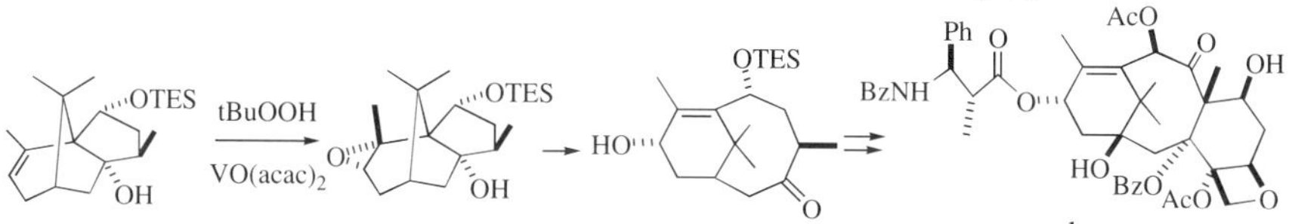

taxol
(**5-10**)

Topographical factors in the assisted epoxidation produced favorable outcomes during the syntheses of 3α,4β-dihydroxyclavulara-1(15),17-diene (**5-11**),[20] hybridalactone (**5-12**),[21] and cytochalasin-B (**5-13**).[22]

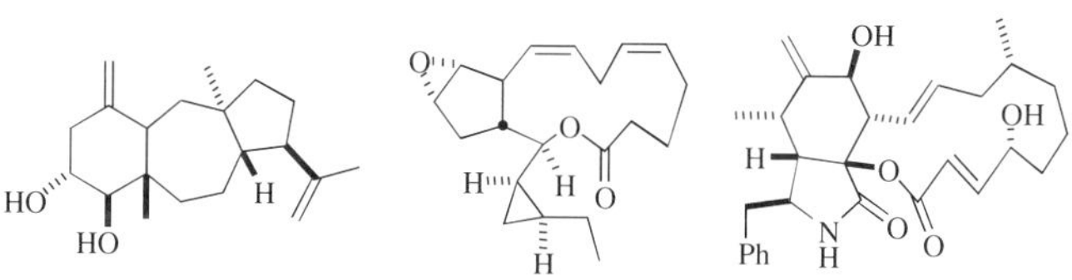

3α,4β-dihydroxyclavulara-1(15),17-diene (**5-11**) hybridalactone (**5-12**) cytochalasin-B (**5-13**)

Although used less frequently, the molybdenum-catalyzed reaction is of special advantage in reaching a more distant double bond. Thus, the $\Delta^{17(20)}$ double bond of a pregn-4-en-3α-ol ester (**5-14**) was selectively epoxidized, whereas without the hydroxyl-containing acyl unit the A-ring was attacked.[23] A bridged ring homoallylic alcohol was epoxidized en route to breynolide (**5-15**).[24]

Acyclic systems (**5-16, 5-17**) have more variables to affect epoxidation but extensive studies have led to structural correlations[25] as shown below. By introducing a trimethylsilyl group to the double bond to increase steric interactions during epoxidation, it is possible to obtain *erythro* products (**5-18**) after desilylation with retention of configuration by fluoride ion.[26]

Simultaneous epoxidation of two allylic alcohol units was the prelude to the establishment of the two trisubstituted double bonds in a precursor of cecropia juvenile hormone JH-I (**5-19**).[27] Stereochemically homogeneous glycol systems were required.

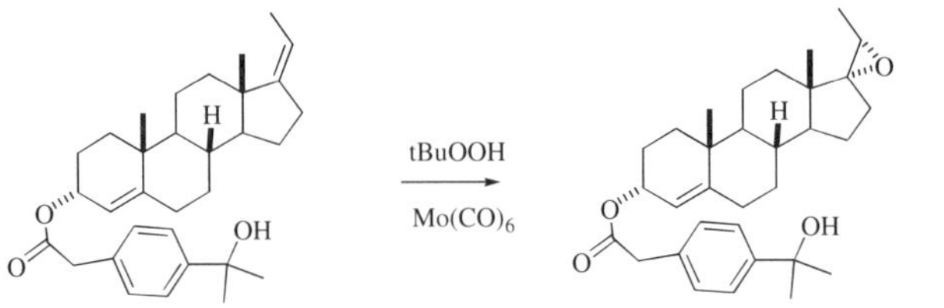

(**5-14**)

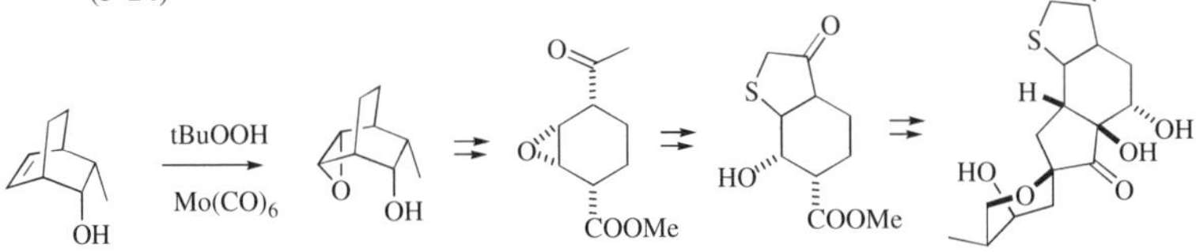

breynolide
(**5-15**)

Generally, tertiary allylic alcohols do not exhibit, good diastereoselectivity in epoxidation. However, an outstanding exception was discovered during a synthesis of ovalicin (**5-3**).[28] Only one product was obtained in the last step.

Titanium(IV) isopropoxide can also serve as catalyst. Thus, in a synthesis of maysine (**5-20**)[29] the allylic alcohol was primary and the stereoinducer was on the other extension of the double bond. Nowadays the majority of cases involving Ti catalysts concern establishment of absolute configuration.[30,31] This is made possible by adding a chiral tartrate ester as cocatalyst.

The behavior of homoallylic alcohols embedded in more flexible systems such as (**5-21**) has also been studied.[32] The most influential steric factor seems to concern the severity of $A^{1,3}$- strain in the transition state. Also it is advantageous to incorporate a removable silyl substituent to bring about diastereoselection, as effectively realized in the creation of a building block for 6-deoxyerythronolide-B, lankanolide, and oleandonolide.[33]

When a *trans*-alkene contains oxygen functionalities in both allylic and homoallylic sites, the homoallylic hydroxyl is more important for determining

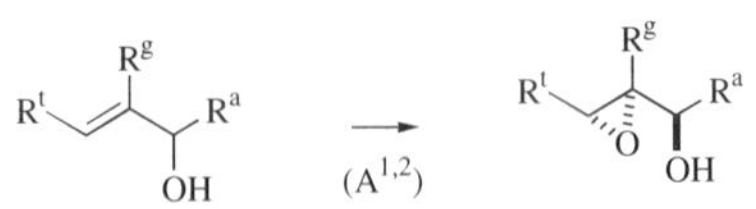

erythro

(**5-16**)

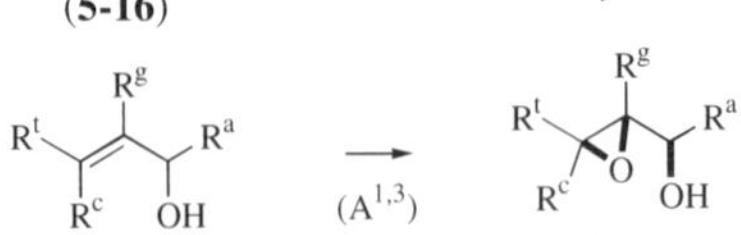

threo

(**5-17**)

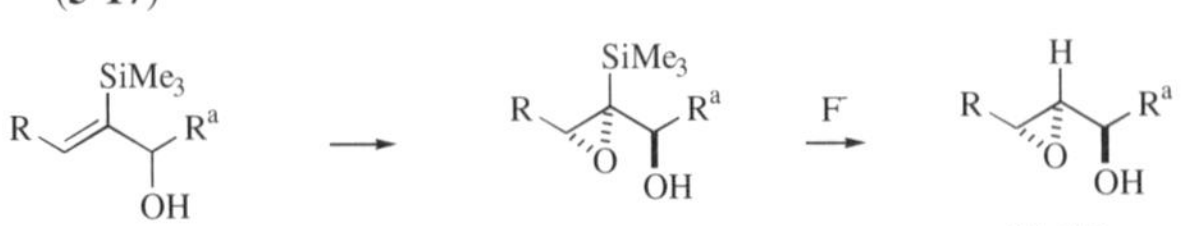

(**5-18**)

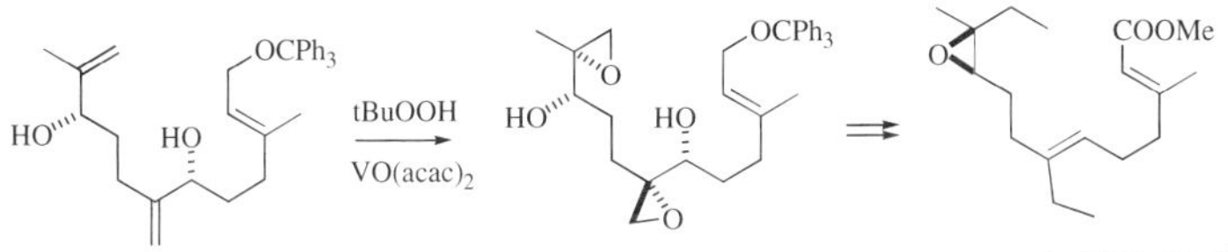

cecropia JH-I (**5-19**)

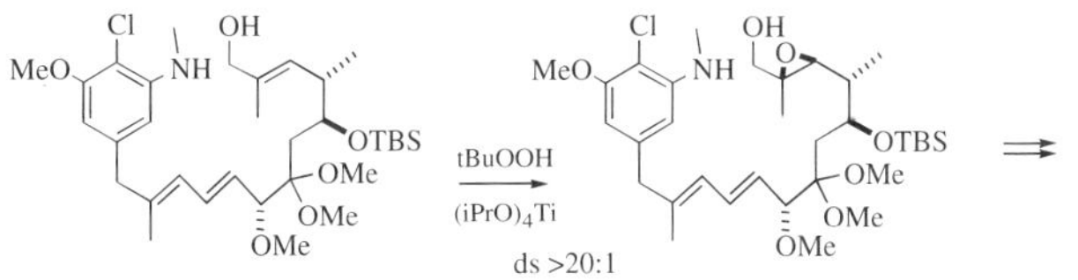

maysine (**5-20**)

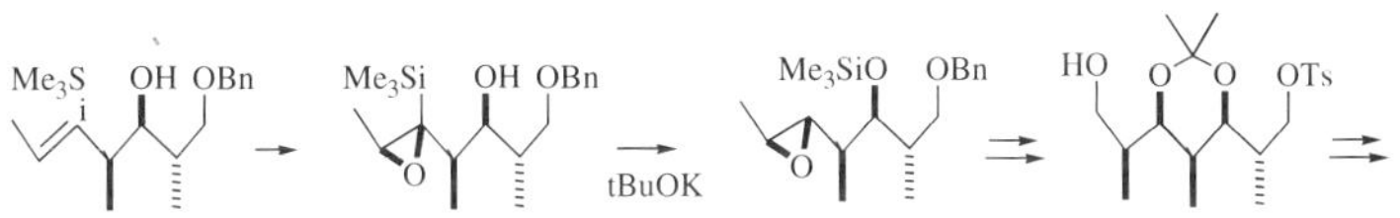

6-deoxy-erythronolide-B

(**5-21**)

diastereoselection.[34,35] Diastereoselectivity is lost when it is masked as an ether while leaving the allylic alcohol free. The situation might be different for the *cis*-alkenes and rigid systems.[36,37]

A partially rigid homoallylic alcohol underwent the desired directed epoxidation to give a synthetic intermediate of euonyminol (**5-22**).[37] It was proposed that the cyclohexanone ring to which the hydroxyl and isopropenyl groups were attached adopted a boat conformation in the transition state. Significantly, the reaction with peracid led to the epoxide of opposite configuration.

There is evidence for direction extending to bishomoallylic and trishomoallylic alcohols. The relative configuration of substituents in the α and β positions to the hydroxyl group, which influences the population of various conformations, is crucial to the stereochemical outcome. Such was the case in the construction of tetrahydrofuran and tetrahydropyran rings of lasalocid-A,[38] ionomycin (**5-23**),[39] and teurilene.[40]

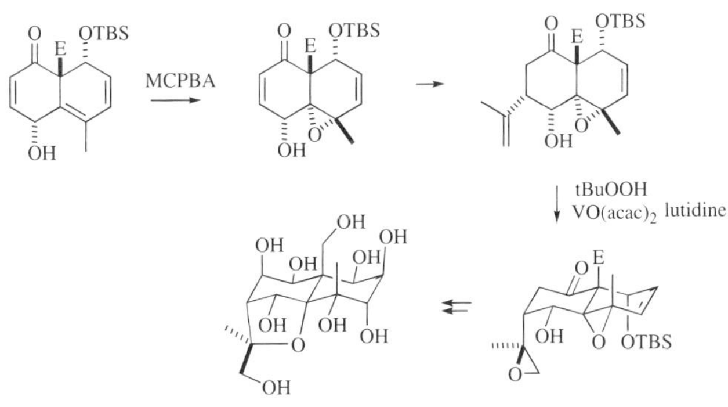

euonyminol (**5-22**)

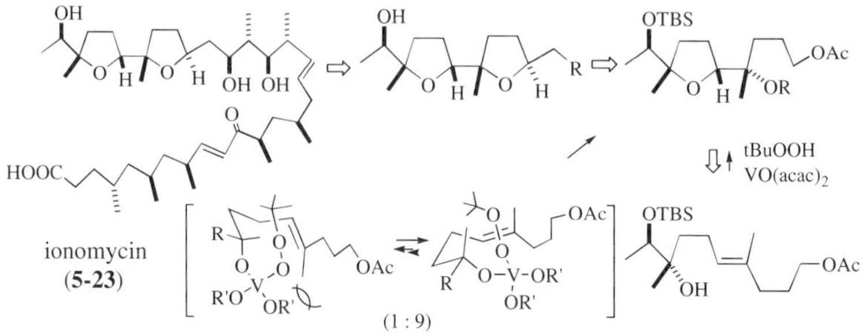

The effect of other metal species such as aluminum, tin, or lanthanides has been evaluated. In one report[41] dibutyltin oxide was described as a superior catalyst to discriminate between allylic and homoallylic alcohols (by internal competition). *erythro*-Glycidols were the predominant products.

5.1.3. Dihydroxylations

Reports of directed dihydroxylation of alkenes with osmium tetroxide are sporadic. Sulfinyl,[42] sulfoximinyl,[43] and nitro[44] groups have been implicated as possessing such ability. Interestingly, the directing effect of the nitro group in osmylation of **5-24** rescinded such a protocol for the synthesis of aristeromycin which has an α-oriented diol system. The problem was solved by changing the reagent to potassium permanganate.

A tertiary amine was shown to mediate the delivery of OsO_4 to the indolic double bond.[45] The resulting diol (**5-25**) underwent pinacolic rearrangement to afford a spirocyclic compound which can be correlated with alstonisine.

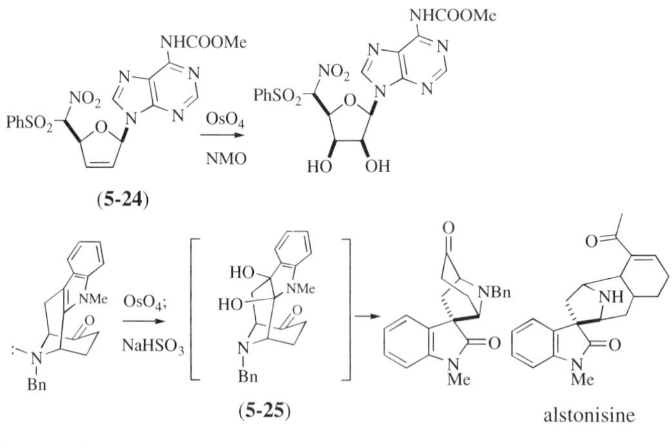

The hydroxyl group of a bishomoallylic alcohol (**5-26**) can deliver rhenium(VII) oxide to the double bond,[46] thereby creating an α-hydroxyalkyltetrahydrofuran of defined stereochemistry (correlated with the double bond configuration). This methodology has great potential in the elaboration of the annonaceous polyketides.[47]

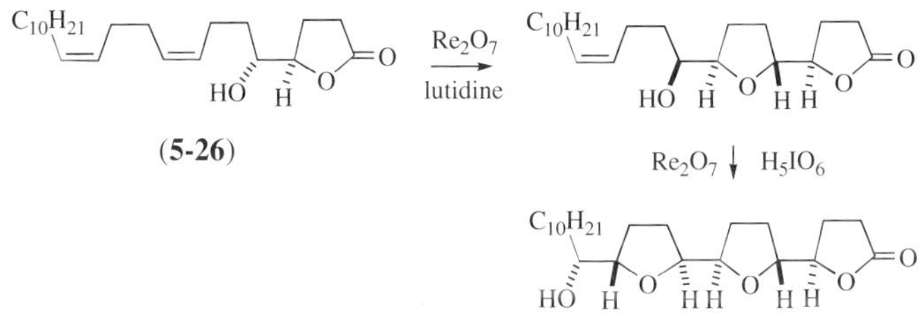

(**5-26**)

Pyridinium dichromate in the presence of acetic acid[48] is another reagent of similar capability, but the yields of such reactions (e.g., of **5-27**) are lower.

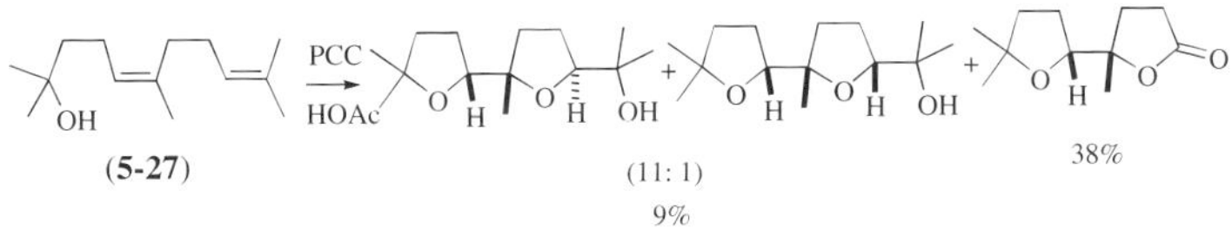

(**5-27**)

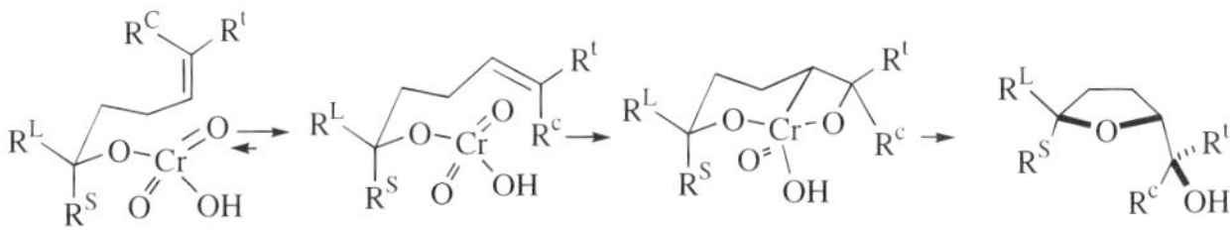

5.1.4. Hydration via Hydroboration

A simple question for mental exercise: Is there a good way to synthesize rishitinol (**5-28**)? The question really asks for a method to set up two *cis* substituents in the tetralin core, and a reasonable answer to the positive invokes the preparation and hydroboration of a homoallylic alcohol. The directed reaction generated regio- and stereoselectively the required diol as a major product.[49]

Because phosphorus is a soft base, unsaturated phosphinites such as **5-29** show directive effects in rhodium-mediated hydroboration of the alkene moieties.[50]

A study on the directed hydroboration of benzyl 3-cyclohexen-1-yl ether in the presence of rhodium complexes shows the importance of transient $\eta^5 \rightarrow \eta^3$ ring slippage of a cyclopentadienyl ligand on Rh to provide the necessary coordinative unsaturation for access of the double bond.[50a] The relatively high propensity of an indenyl group correlates with its effectiveness.

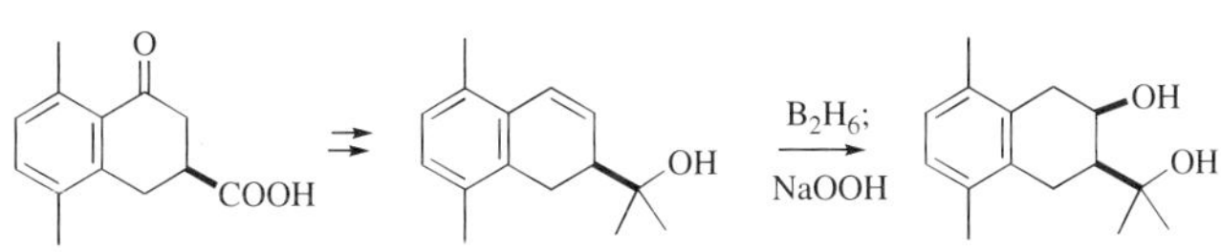

rishitinol (**5-28**)

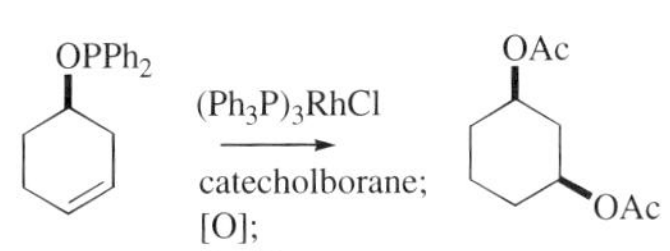

(**5-29**)

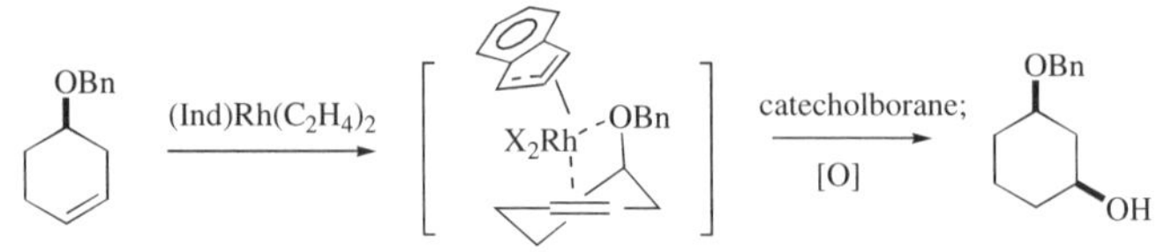

(**5-29a**)

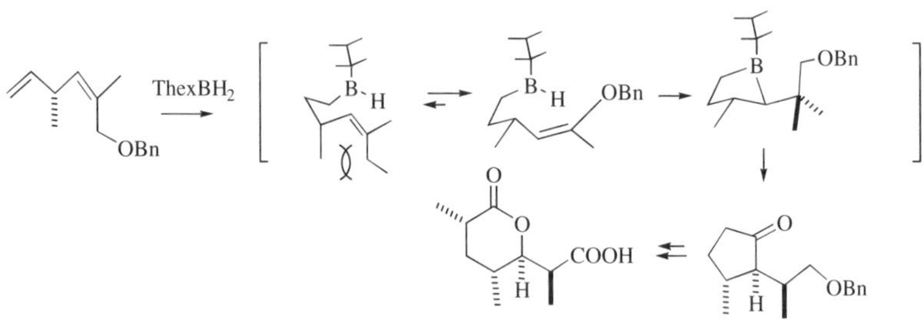

Prelog-Djerassi
lactone (**5-30**)

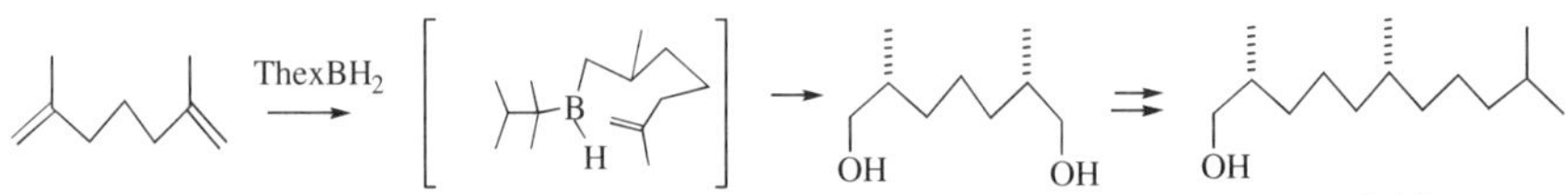

(**5-31**)

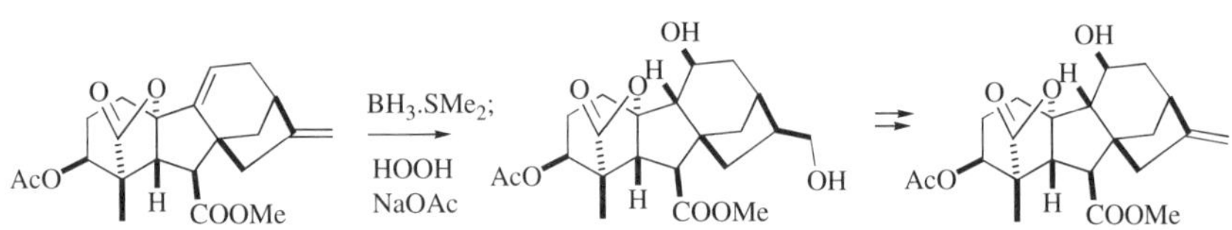

gibberellin-A$_{35}$ methyl ester
(**5-32**)

It can be argued that functionalization of 1,n-dienes by means of hydroboration and manipulation of the cyclic boranes is a directed reaction. It is a synthetically useful method for substrates with a much more reactive terminal alkene. According to this reasoning and the expected conformational selection for the intramolecular step, novel synthetic routes to the Prelog-Djerassi lactone (**5-30**)[51–53] and the side chain (**5-31**) of vitamin E[54] have been designed. An even more interesting maneuver is that involving in a synthesis of gibberellin-A$_{35}$ methyl ester (**5-32**)[55] in which the β-alcohol of the C-ring was introduced by relay, using the exocyclic methylene group as the pivot.

5.2. DIRECTED REDUCTIONS

5.2.1. Reduction of Double Bonds

Stereoselective hydrogenation due to certain functional groups was observed long ago, and for illustration, the hydrogenation of a tricyclic enone for a synthesis of

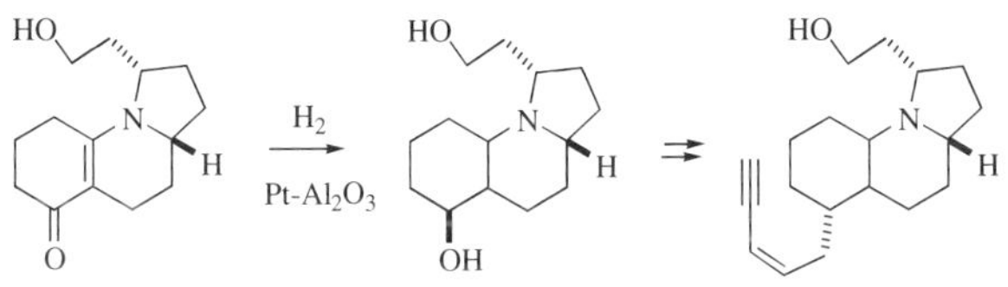

diastereosel. 12:1 gephyrotoxin (**5-33**)

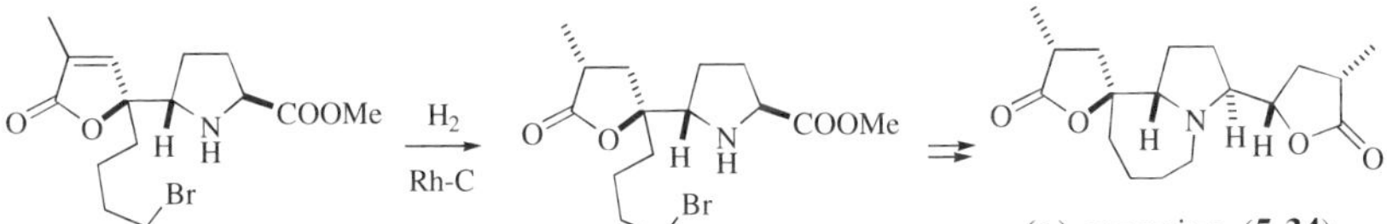

(+)-croomine (**5-34**)

gephyrotoxin (**5-33**)[56] and a butenolide in an intermediate of (+)-croomine (**5-34**)[57] are mentioned here. In the latter process there was a proline ester in the same molecule and the nitrogen atom played a crucial role in the hydrogenation.

In recent years the advent of homogeneous hydrogenation brought the aspect into a sharper focus. A homogeneous catalyst gathers the substrate and hydrogen molecule into the ligand sphere, and splitting of the hydrogen allows successive migratory insertion into the complexed alkene and reductive elimination to occur. When the catalytic cycle is complete, the saturated product is released. The confinement of the substrate within the metal complex subjects the former to strict regulation with respect to molecular alignment and hence a high degree of stereoselectivity usually accompanies the hydrogenation. The most popular homogeneous catalysts for hydrogenation are based on rhodium (e.g., Wilkinson's catalyst) and iridium.

Polar groups can direct homogeneous hydrogenation of a proximal double bond because those groups as well as the π-bond would simultaneously be ligated to the metal ion. Thus, when a *trans*-perhydroindanone was required for synthesis of retigeranic acid (**5-35**),[58] it was obtained from the allylic alcohol by hydrogenation with a rhodium(I) complex, enlisting the hydroxyl group to guide the delivery of hydrogen.

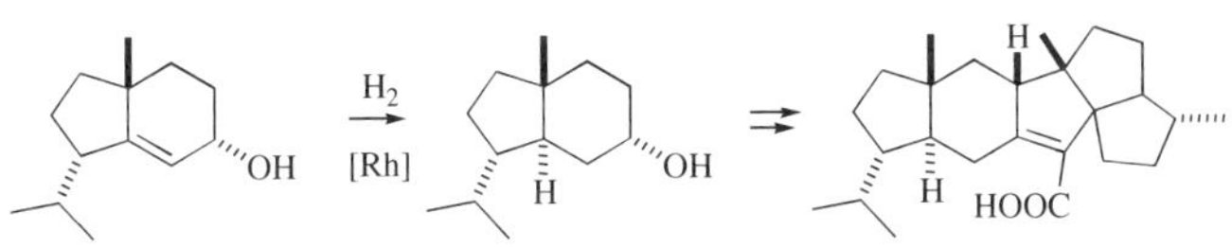

retigeranic acid-A
(**5-35**)

Examples of iridium-catalyzed hydrogenation which provided crucial aids to synthesis include those of modhephene (**5-36**)[59] and cuanzine (**5-37**).[60] The carbonyl group is an excellent director.

A classical stereoselective semireduction of the pyridone appeared in the late stage of a synthesis of strychnine (**5-38**).[61] With the tertiary alcohol scavenging the reagent, it was possible to deliver the hydride ion inside the bowl-shaped molecule.

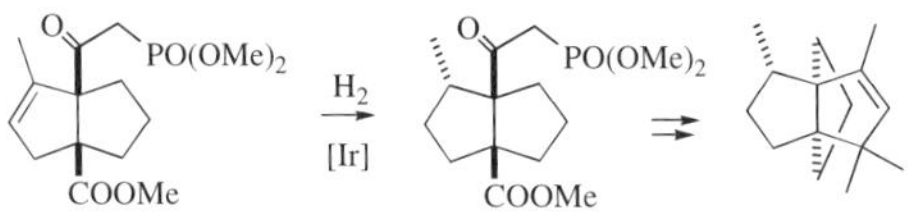

modhephene
(**5-36**)

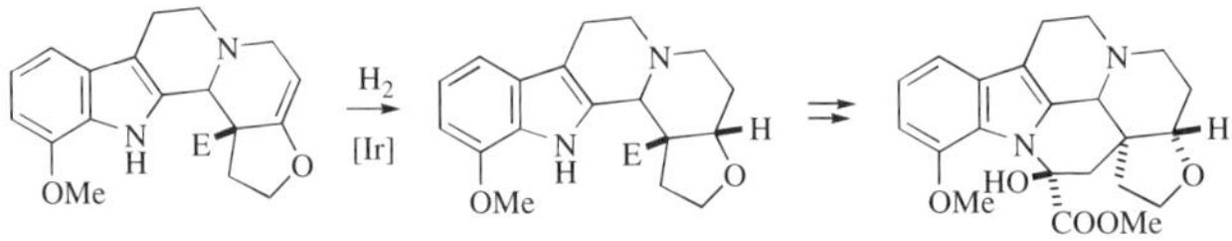

cuanzine (**5-37**)

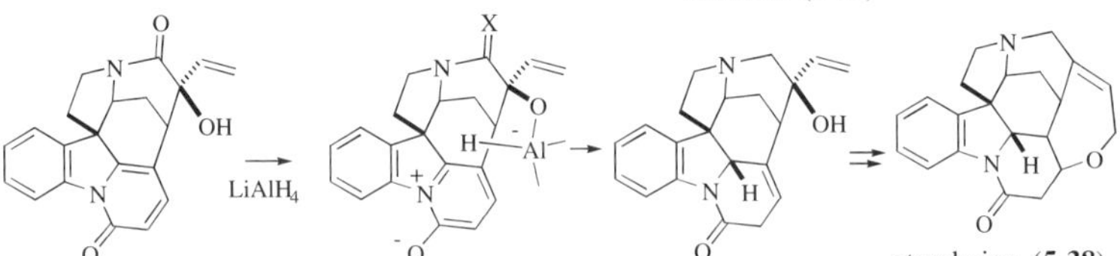

strychnine (**5-38**)

Ketones with neighboring groups such as hydroxyl, alkoxy, keto, alkylthiocarbonyl, and dialkylamino functions undergo stereoselective hydrogenation in the presence of RuX_2(binap) catalysts[62] to give predominantly the *anti* products (**5-39**). A significant application of the method is for the synthesis of a carbapenem intermediate (**5-40**).[63]

Allylic and homoallylic alcohols are liable to directed reduction of the double bond by complex metal hydrides, particularly when the latter is part of a conjugate system. Such processes contributed significantly to synthetic efforts toward vernolepin (**5-2**),[64] stoechospermol (**5-41**),[65] and a pseudoguaianolide intermediate (**5-42**).[66]

A hydroxyl group may supply proton to a prostereogenic center during reduction of an enone with lithium in liquid ammonia. This phenomenon was exploited in the synthesis of solavetivone (**5-43**)[67] and stemodinone.[68]

The differential acidity of similarly located hydroxyl and carboxyl groups is the cause for divergent results of reduction. Unless water is added to the reducing milieu, a completely ionized carboxyl group (e.g., in **5-44**) is incapable of direction.[69]

The inherent instability of a complex bicyclo[2.2.2]octadiene toward various reagents almost aborted a route to patchouli alcohol (**5-45**).[70] Thus, the ultimate success depended on a hydroxyl-directed reduction of the *syn* double bond.

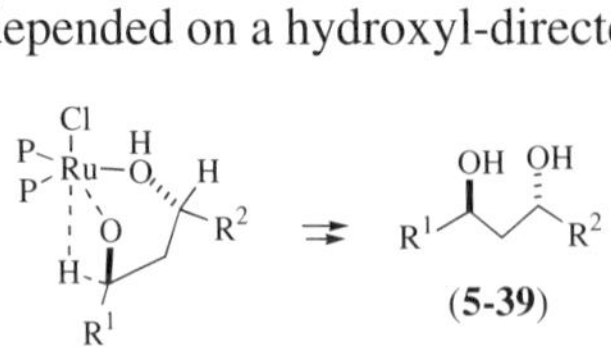

(**5-39**)

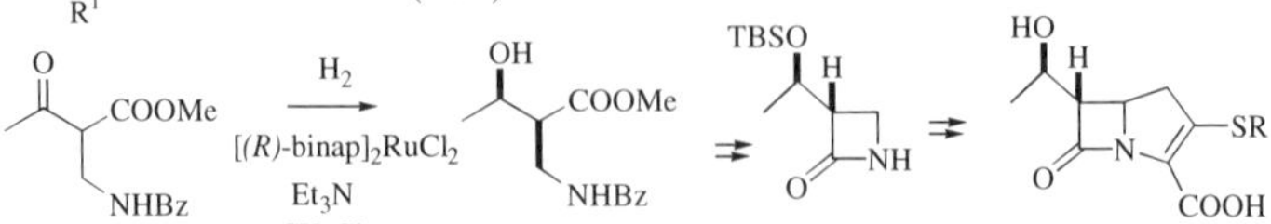

(**5-40**)

carbapenems

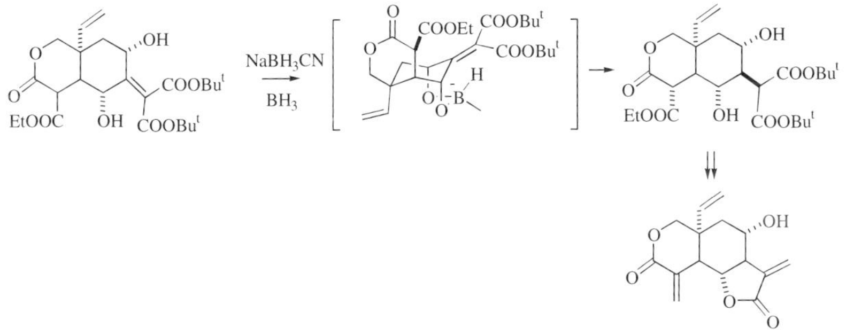

vernolepin (**5-2**)

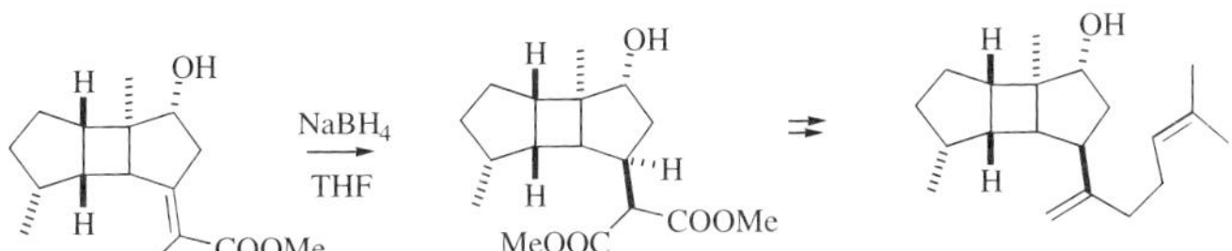

stoechospermol
(**5-41**)

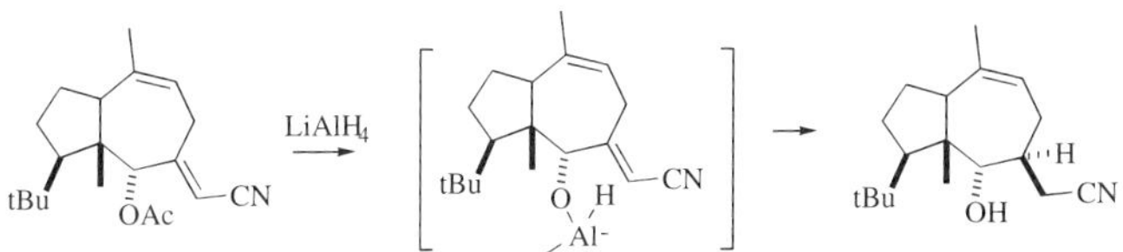

(**5-42**)

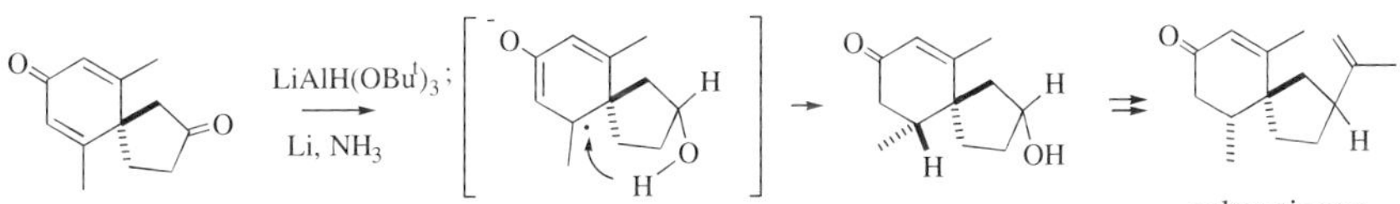

solavetivone
(**5-43**)

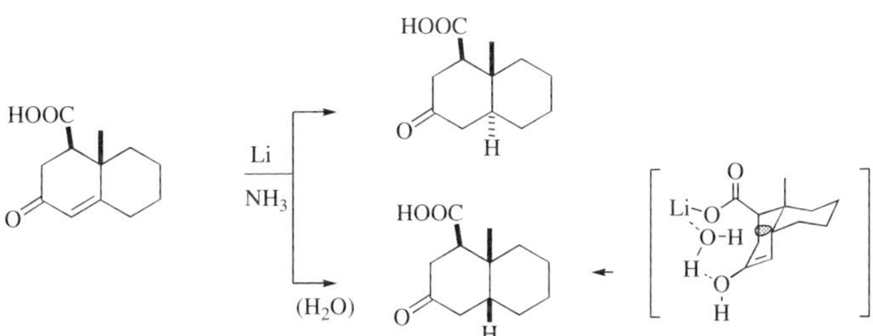

(**5-44**)

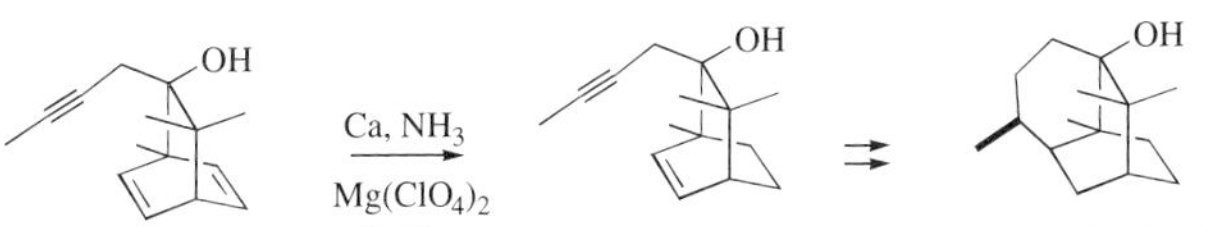

patchouli alcohol
(**5-45**)

5.2.2. Metal Hydride Reduction of C=X Bonds

Hydroxyl, amino, and related functionality react with metal hydride reagents very rapidly. The reduction of a carbonyl group is subject to involvement of such a functionality nearby and the hydride is delivered from a bound state.

The final steps in a synthesis of illudin-M (**5-46**)[71] involved reduction of the hydroxy diketone with excess lithium aluminum hydride and reoxidation to the α-ketol system. Interestingly, the tertiary alcohol which is farther away took control in the reduction of the cyclopentenone. The reason may be that the axial branch of the cyclopropane ring inhibited the inward orientation of the secondary alkoxyaluminum hydride unit to reach across the ring.

The reduction of the 3-acylindole intermediate in a synthesis of sporidesmin-A (**5-47**) profited from the participation of the amido nitrogen.[72] The result also implied a conformation of the complex in which the indolyl moiety oriented away from the dithioacetal bridge, and the oxygen atom of the carbonyl group pointing outside of the dilactam ring.

Preassociation of diisobutylaluminum hydride with the tertiary amine in quininone (**5-48**) and quinidinone (**5-49**) was likely the cause for their stereoselective reduction.[73] α-(Arenesulfinyl) ketones give (*RS,SR*)-alcohols (**5-50**) in which the new stereocenter has the absolute configuration opposite to the sulfinyl group, but in the presence of zinc chloride, the reduction products (**5-51**) are mainly (*RS,RS*).[74] Hydride attack comes from the side of the sulfur lone electron-pair in both circumstances, but the open conformation is different from the chelate. The open conformation disposes the two oxygen atoms as far apart as possible to minimize dipole–dipole interactions.

The synthetic potential of this chemistry is nicely demonstrated by the elaboration of both enantiomers of disparlure (**5-52, 5-52′**)[75] from the two isomeric sulfinyl alcohols derived from the same ketone. Because of the chelated dianions, the direction of alkylation was governed by the existing substituent.

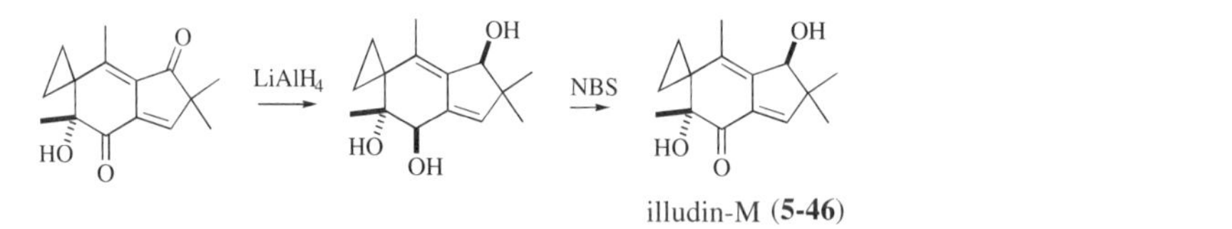

illudin-M (**5-46**)

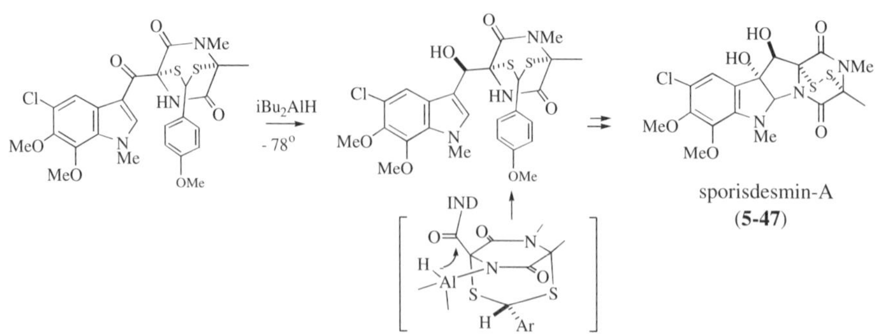

sporisdesmin-A
(**5-47**)

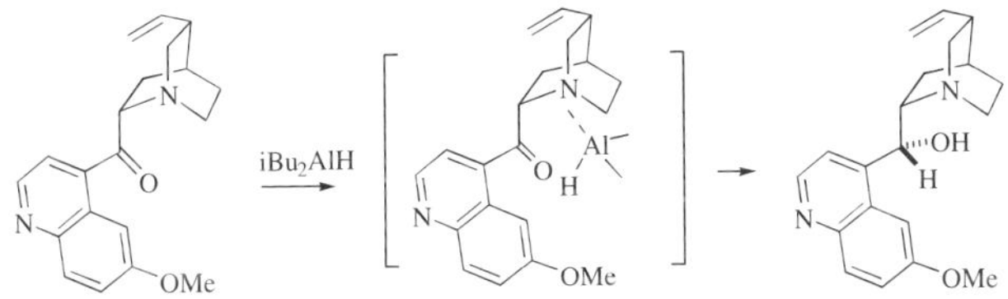

(5-48) quinine

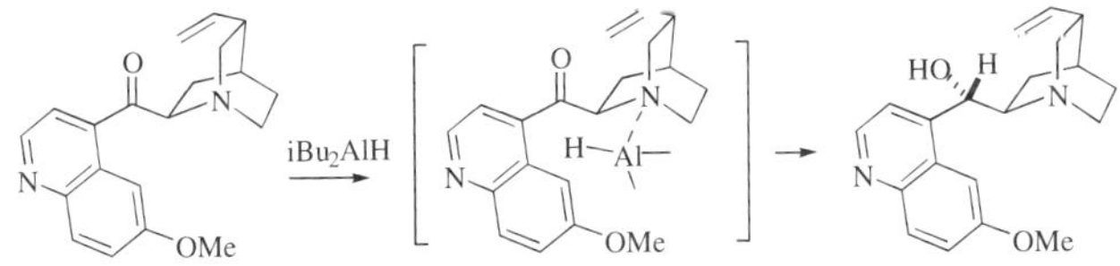

(5-49) quinidine

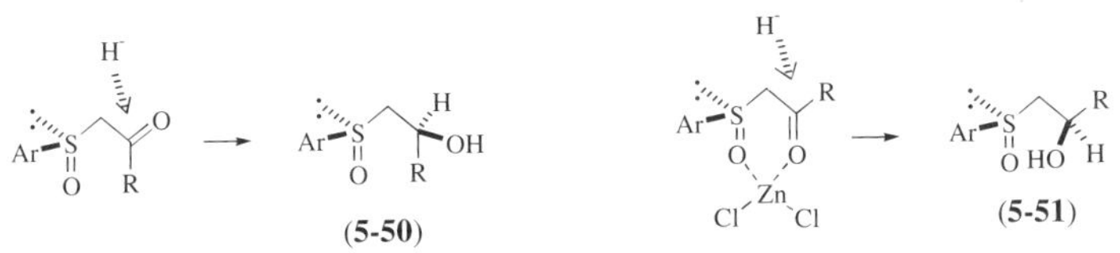

(5-50) (5-51)

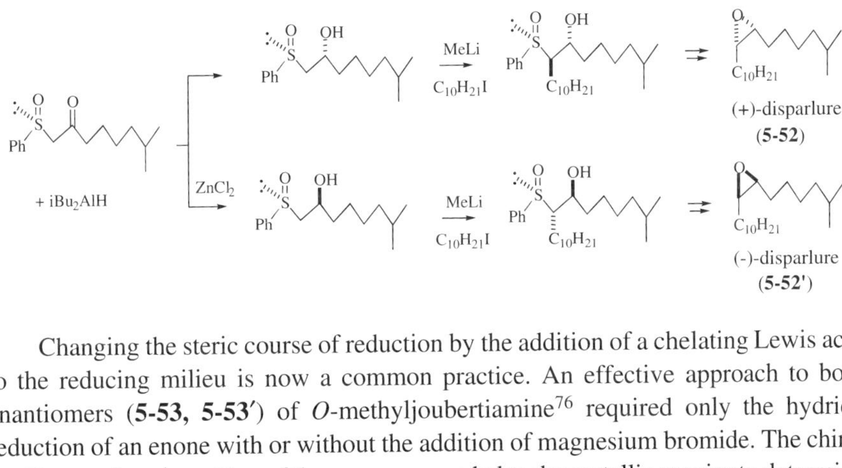

(+)-disparlure (5-52)

(-)-disparlure (5-52')

Changing the steric course of reduction by the addition of a chelating Lewis acid to the reducing milieu is now a common practice. An effective approach to both enantiomers (**5-53, 5-53′**) of *O*-methyljoubertiamine[76] required only the hydride reduction of an enone with or without the addition of magnesium bromide. The chiral auxiliary at the α′-position of the enone responded to the metallic species to determine the manner by which hydride was received.

Zinc borohydride alone can play the dual role of chelator and reductant for the reduction of functionalized ketones. For example, in a synthesis of indanomycin (**5-54**),[77] the dioxanone intermediate was prepared from an α-alkoxy ketone.

While sodium triacetoxyborohydride discriminates between aldehydic and ketonic carbonyl groups,[78] it attacks ketols[79] through ligand exchange to evolve an intramolecular reaction. Contrary to external hydride addition to a chelated β-ketol which yields mainly the *syn*-1,3-diol (**5-55**),[80] internal delivery of hydride leads to the *anti* isomer (**5-56**). The countercation of the hydride reagent may be replaced by the tetramethylammonium ion.[81]

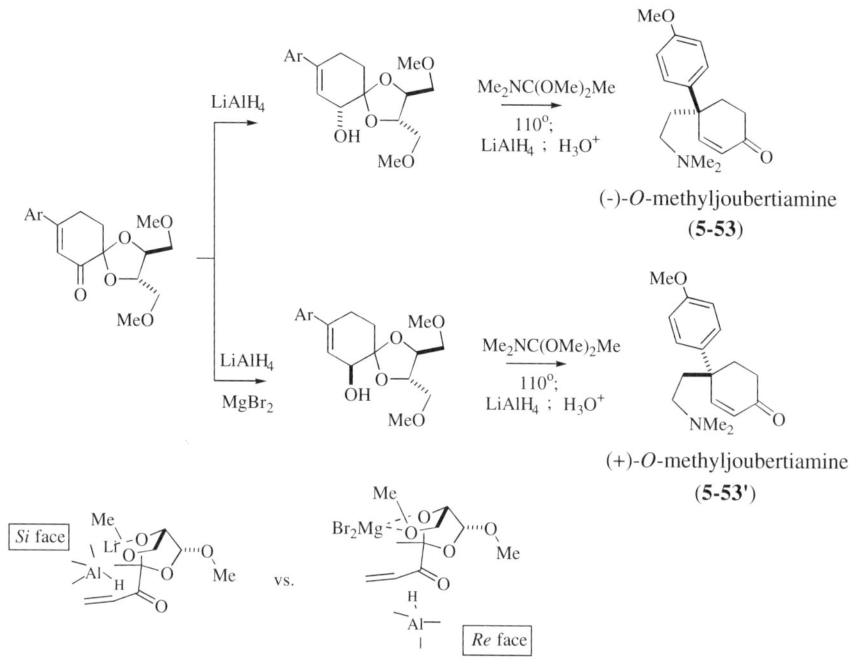

(-)-*O*-methyljoubertiamine
(**5-53**)

(+)-*O*-methyljoubertiamine
(**5-53'**)

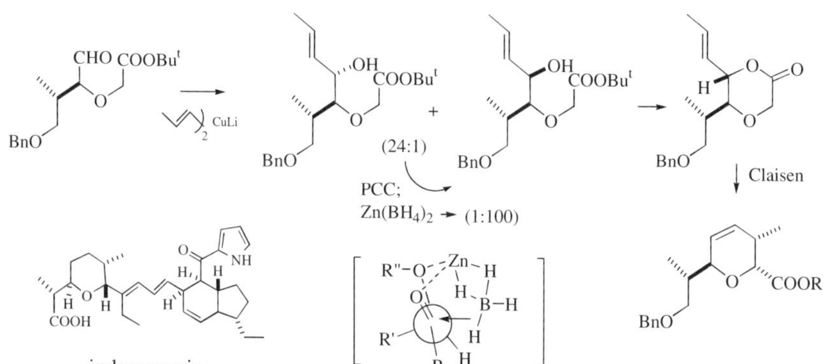

indanomycin
(**5-54**)

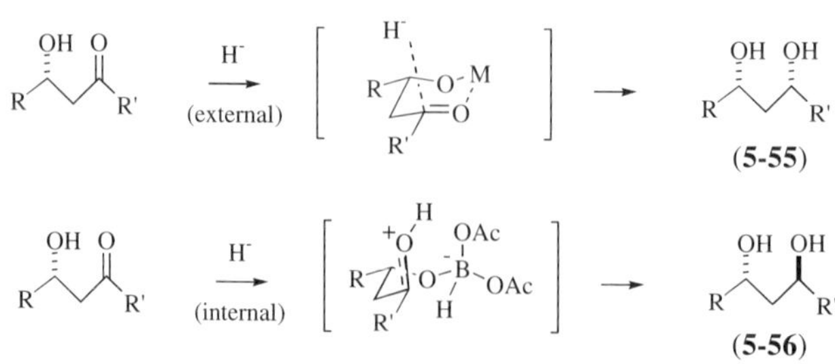

(**5-55**)

(**5-56**)

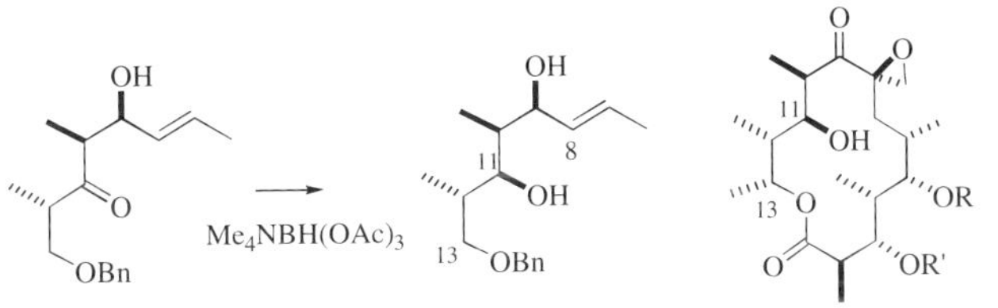

oleandomycin
(**5-57**)

Interestingly, the configuration of $C_{(11)}$ of oleandomycin (**5-57**) was established in a synthesis[82] employing triacetoxyborohydride reduction. Since the $C_{(8)}$–$C_{(13)}$ segment was initially formed by an aldol reaction, there was actually a net exchange of redox state at these two sites.

In connection with the synthesis of phorbol (**5-58**),[83] the genesis of the secondary alcohol in the six-membered ring relied on the intervention of the hydroxyl group at the B/C ring junction. This is one of those cases which require a more circuitous maneuver due to undesired topological response of a precursor lacking such a hydroxyl group to reducing reagents.

Establishment of the *trans*-diol system of (+)-allopumiliotoxin-267A (**5-59**) by reduction of the α-ketol precursor[84] is particularly attractive, since its success would make available many other analogues based on alkylidenation of the indolizidine. And as the final step to reach rocaglamide (**5-60**),[85] the angular hydroxyl group played a crucial role in directing the reduction of the cyclopentanone.

L-Selectride reduction of an acetonyl side chain completed a synthesis of ramulosin (**5-61**).[86] A chelated complex that underwent reduction was implied. In fact, this conformation is quite reasonable because the pseudoaxial chain should be favored ($A^{1,2}$-strain).

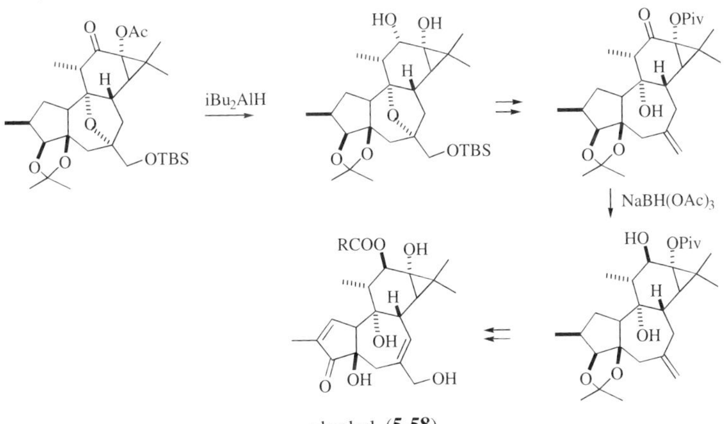

phorbol (**5-58**)

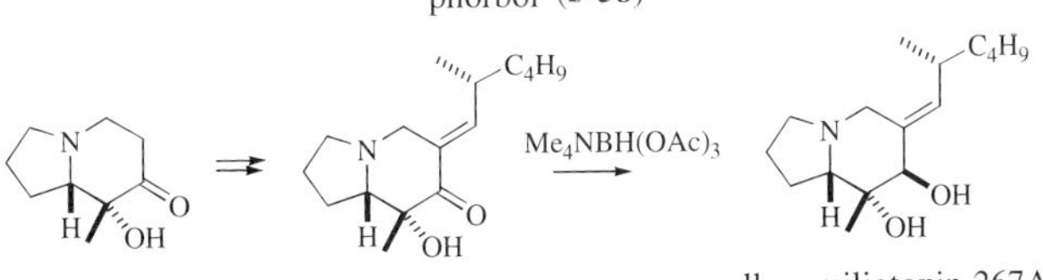

allopumiliotoxin-267A
(**5-59**)

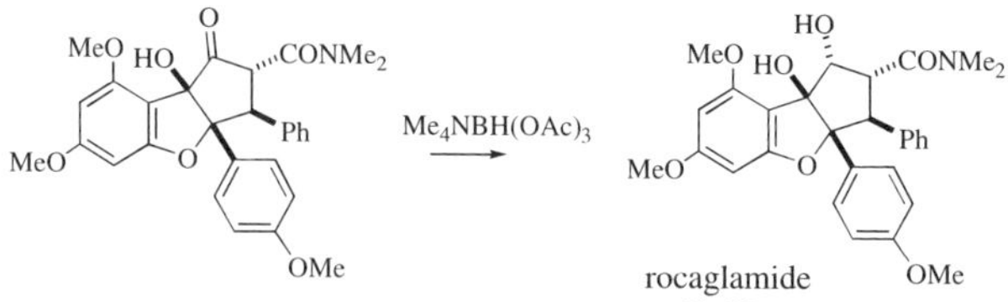

rocaglamide
(**5-60**)

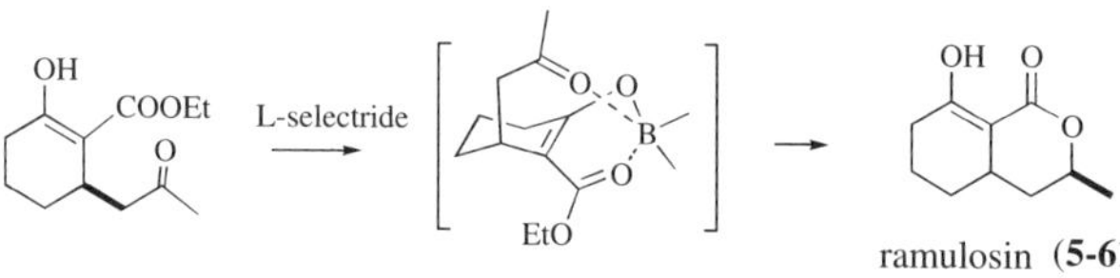

ramulosin (**5-61**)

Introduction of the last stereogenic center of perhydrohistrionicotoxin (**5-62**)[87] in the imine reduction step was fortunate in acquiring the assistance of the secondary alcohol. Hydride transfer occurred via a chelate. An interesting observation is that a β-hydroxyl group located in an oxime ether[88] can influence the reduction of the latter functionality in a manner according to the *syn/anti* isomerism of the oxime ether.

Cyclic acetals derived from C_2-symmetric 1,3-propanediols can be cleaved with a variety of reagents in the presence of Lewis acids which prefer coordination with the oxygen atom geminal to the axial susbtituent. With external reducing agents the reductive cleavage occurs with inversion of configuration. However, neutral aluminum hydrides effect cleavage by frontside transfer of the hydride ion to the incipient carbenium ion.[89] Reduction of spiroacetals (**5-63**) also shows dependence on structural features and reagents.[90]

β-Ketols (**5-64**) undergo stereoselective Tishchenko reaction in the presence of an aldehyde and samarium diiodide.[91] A cyclic transition state accounts for the generation of the *anti*-diol monoesters (selectivity > 99:1). An elegant application of this method in a synthesis of (–)-rapamycin (**5-65**)[92] served to introduce the pipecolinic ester concomitant with the stereoselective reduction.

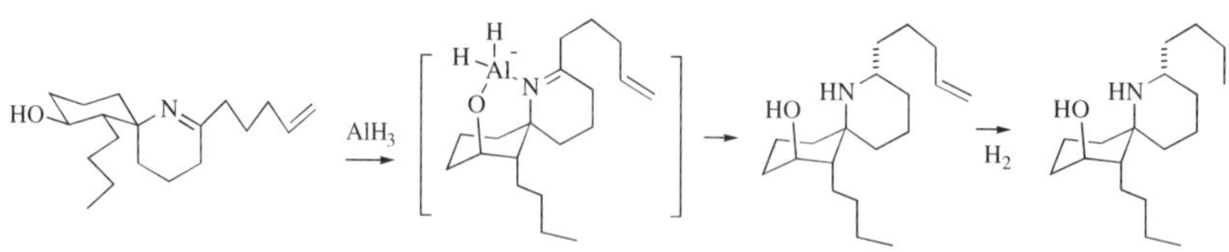

perhydro-
histrionicotoxin
(**5-62**)

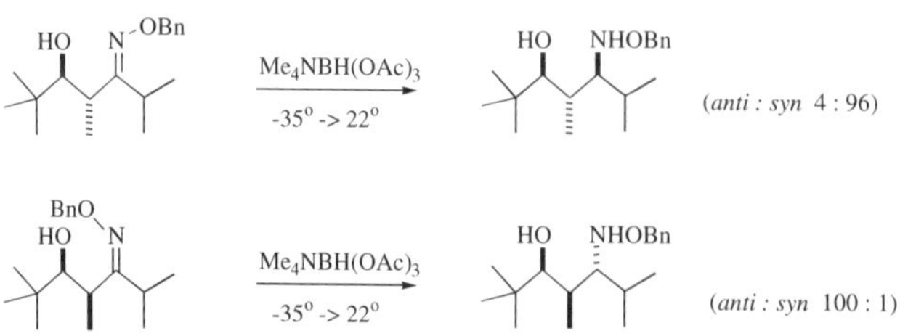

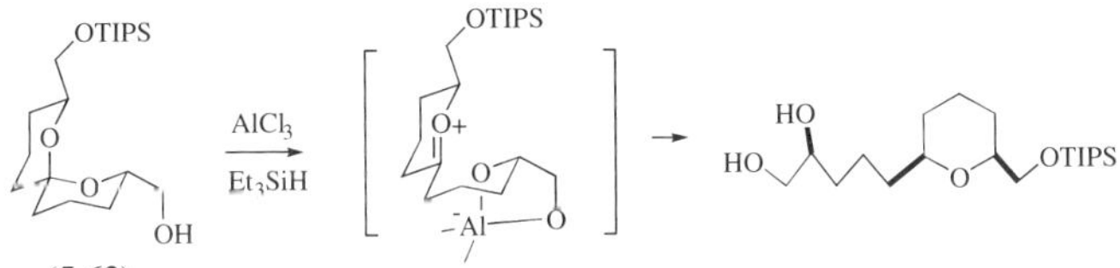

(5-63)

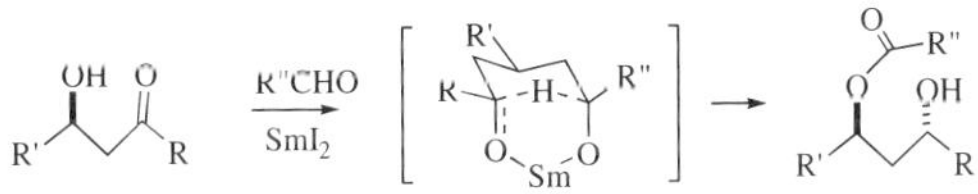

(5-64)

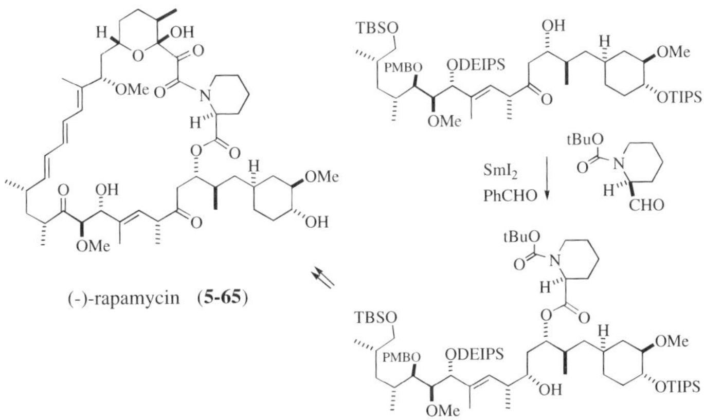

(-)-rapamycin (5-65)

The selective reduction of α,β-epoxy ketones with lithium in liquid ammonia in the presence of ammonium chloride as proton source has been exploited in a synthesis of compressanolide (**5-66**).[93] Protonation of the intermediate enolate probably occurred after the protonation of the alkoxide ion.

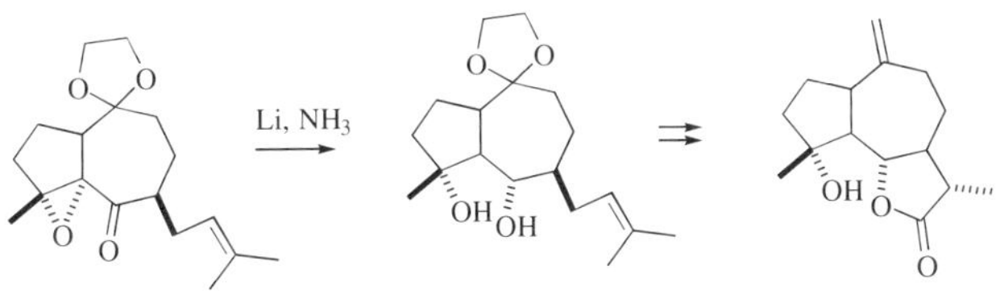

compressanolide
(**5-66**)

5.3. DIRECTED C–C BOND FORMATION

5.3.1. Simmons–Smith Reaction

The facilitation of the Simmons–Smith cyclopropanation[94] by allylic and homoallylic hydroxyl and alkoxy groups is well known. More importantly, reactions in their presence are *syn* selective due to interception of the organozinc reagent by the polar functionality. In addition to the standard zinc–copper couple, there are several modifications such as use of a zinc–silver couple[95] and diethylzinc.[96] In the last variant there is a rapid exchange of ligands to form ICH_2ZnEt, a homogeneous cyclopropanation reagent. Samarium is also effective.[97]

It is still not totally clear as to the nature (covalent or dative) of the intermediate from reaction of an allylic alcohol with the (iodomethyl)zinc iodide species. However, conformer with pseudoequatorial hydroxyl group of the cyclic allylic alcohol seems to be involved during the reaction.

The first prominent application of stereochemical correlation was in a synthesis of thujopsene (**5-67**).[98] If the chiral terpene is needed, the β-hydroxysulfoximine derivatives may be used as substrates.[99] It should be noted that the enone reacted to give an 8:2 mixture in favor of the unsuitable isomer.[100]

The Simmons–Smith has met its greatest challenge in the elaboration of the antifungal agent FR-900848 (**5-68**) which contains five cyclopropane units. These components were formed in three stages.[101]

With regard to utility, the reaction is not restricted to synthetic targets containing a cyclopropane ring. Valeranone (**5-69**) was acquired after further modification of the Simmons–Smith adduct and acid cleavage.[102] Similarly, in the synthesis of azadiradione (**5-70**),[103] alnusenone (**5-71**),[104] and taxusin (**5-72**),[105] the cyclopropanation step served to introduce the angular methyl group indirectly but stereoselectively. In an approach to 11β-hydroxytestosterone derivative (**5-73**)[106] the reductive cleavage step was not necessary.

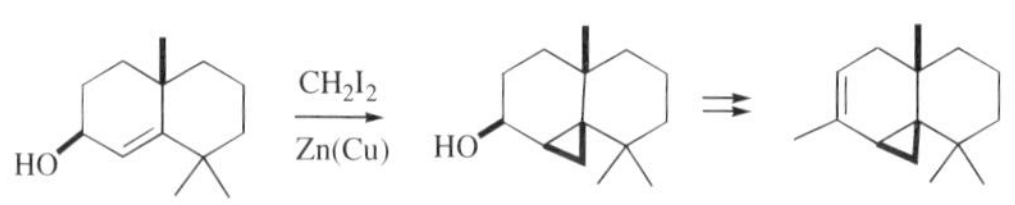

thujopsene (**5-67**)

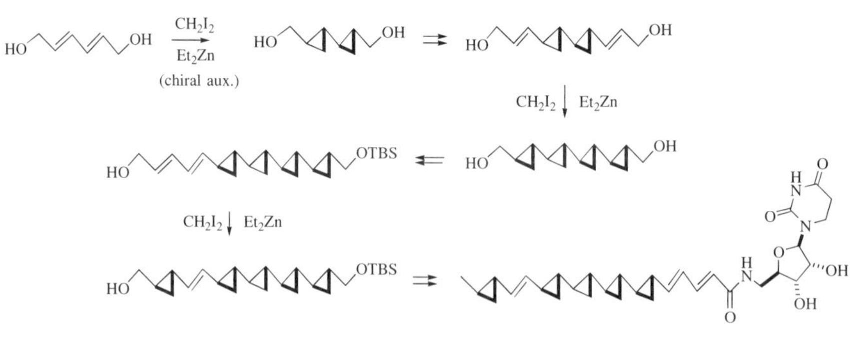

antifungal agent FR-900848 (**5-68**)

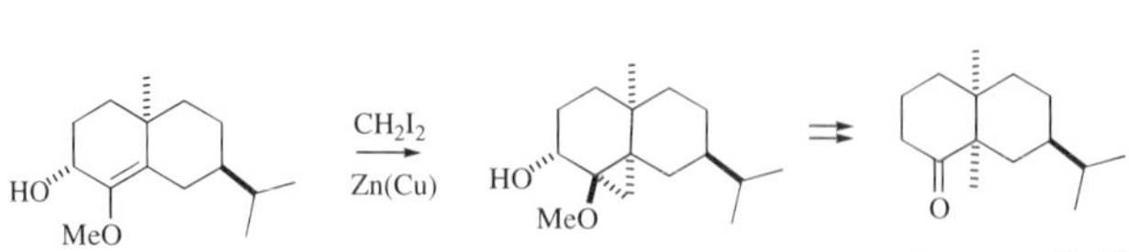

valeranone (**5-69**)

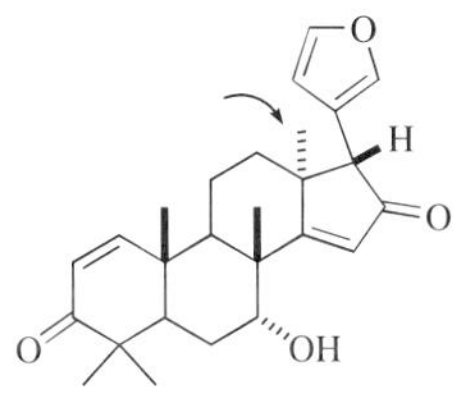

azadiradione (**5-70**)

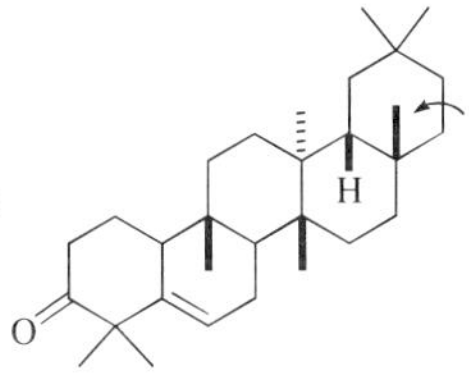

alnusenone (**5-71**)

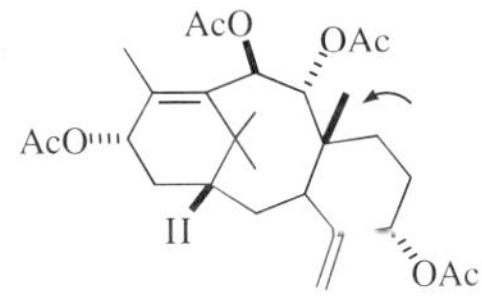

taxusin (**5-72**)

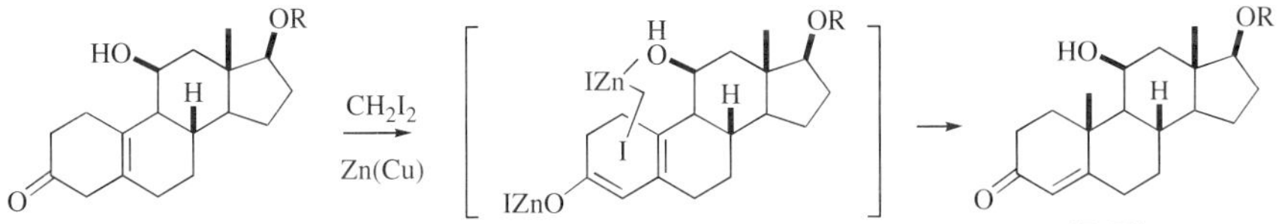

(**5-73**)

Complete facial selectivity was observed in the cyclopropanation of an exocyclic double bond connected to a bicyclo[3.3.1]nonane skeleton. Thus, it lent great support toward a synthesis of trifarienol-A (**5-74**) and trifarienol-B (**5-75**).[107] It is also interesting that organozinc reagents were involved in both the cyclization and cyclopropanation steps which led to a chiral precursor of (–)-muscone (**5-76**).[108]

Two stereocenters were generated in the cyclopropanation step during a synthesis of isocomene (**5-77**).[109] The *exo*-methyl selectivity was also important although there was opportunity for epimerization of the adduct after its rearrangement to a cyclobutanone.

With a chiral acetal to direct cyclopropanation of a diquinene, and subsequent transformation of the adduct into an iodo compound and then to a known precursor[110] of modhephene (**5-36**), the absolute configuration of the sesquiterpene was clarified.

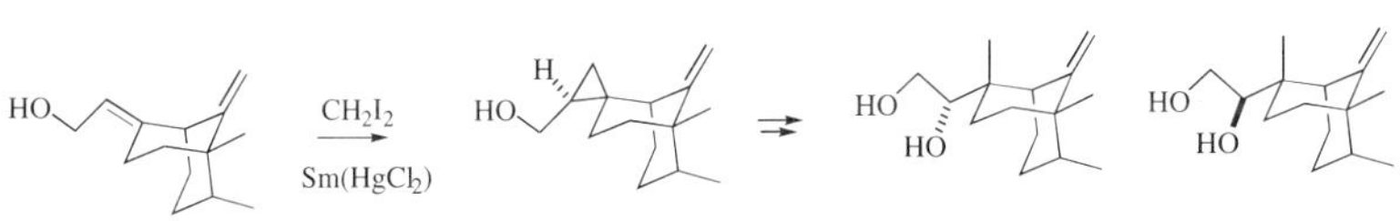

trifareinol-A (**5-74**)

trifareinol-B (**5-75**)

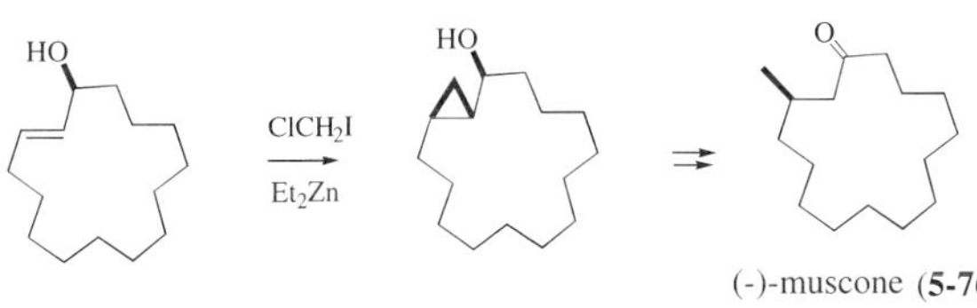

(-)-muscone (**5-76**)

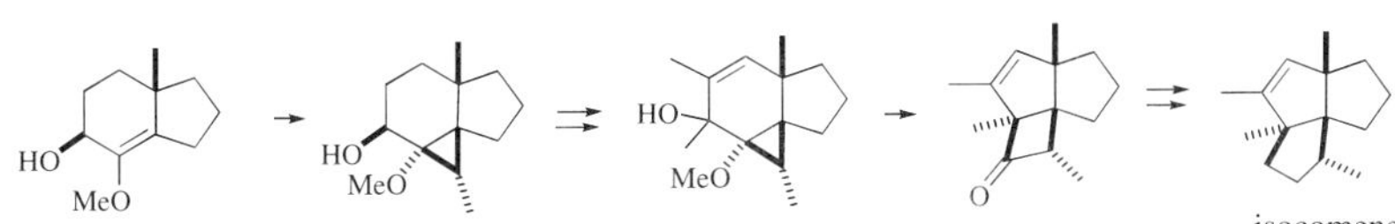

isocomene (**5-77**)

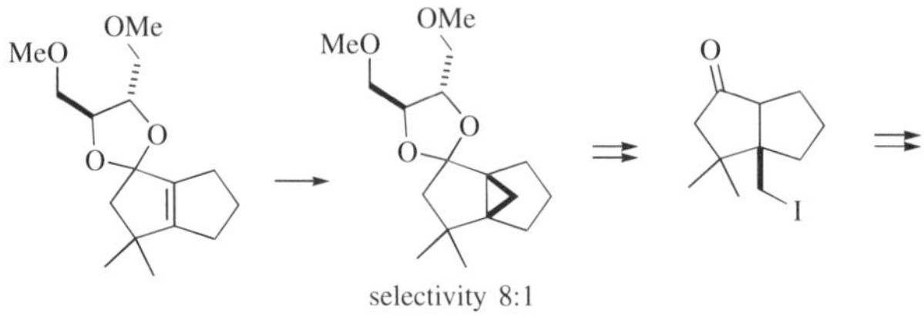

modhephene
(**5-36**)

Members of the vitamin D series such as calcitriol (**5-78**) can be formed by rearrangement of the allylic alcohol which is further conjugated to a fused cyclopropane.[111] Since the configuration of the cyclopropane correlates with that of the emerging secondary alcohol, it is essential to obtain the proper intermediate. A Simmons–Smith reaction of a cyclopentanol paved the way.[112]

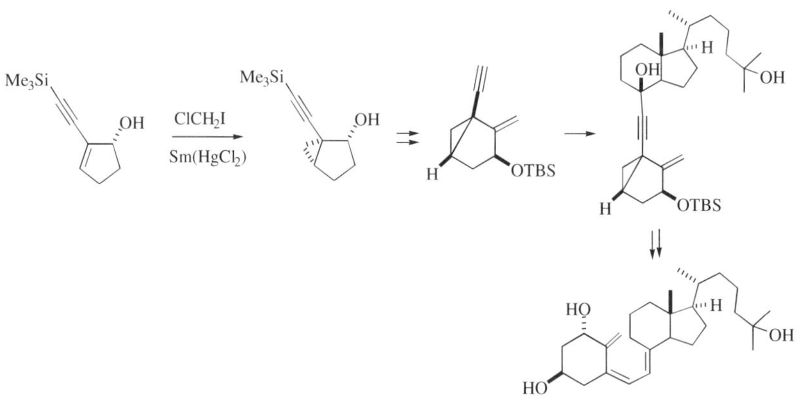

calcitriol (**5-78**)

5.3.2. Alkylations via Chelates

Alkylation of enolates is subject to stereoinduction by bulky groups in proximity. However, problems similar to that concerning establishment of the quaternary center α to the carboxyl group of hirsutic acid (**5-79**) are not easy to solve. Nonetheless, one solution emerged via the introduction of a polyether chain to an *endo*-hydroxyl substituent in a diquinane intermediate.[113] The chelated lithium enolate also attracted the methylating agent to the metal to achieve a subsequent alkylation.

Carbonyl compounds possessing an n-donor at the α-position usually react with organometallic reagents via a chelate. Bond formation is from the less hindered side. As examples are innumerable, very few will be mentioned. In a synthesis of zaragozic

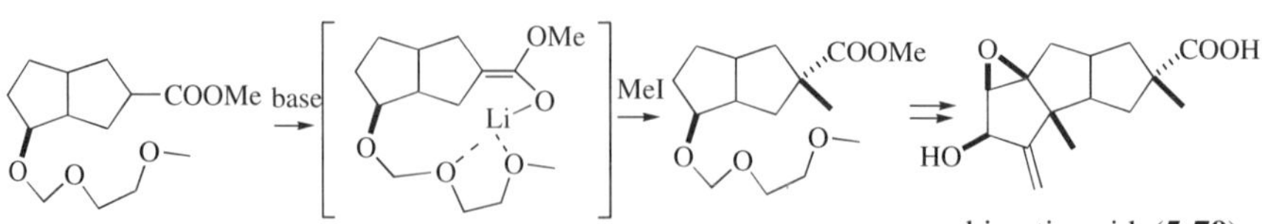

hirsutic acid (**5-79**)

acid-C (**5-80**),[114] the carboxylic acid attaching to the quaternary bridgehead was introduced in the latent form of a vinyl substituent by a Grignard reaction with a ketone precursor.

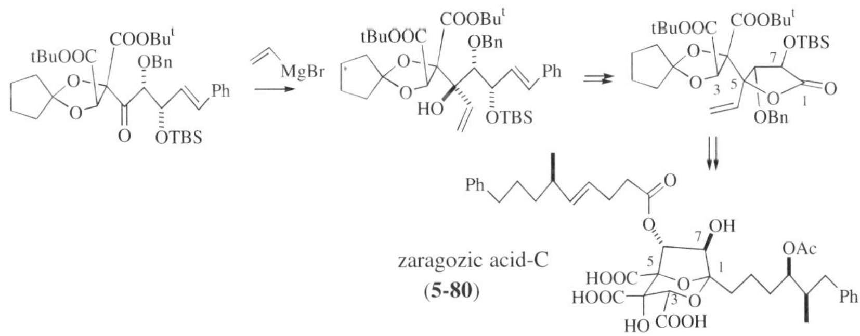

zaragozic acid-C (**5-80**)

A method for stereoselective synthesis of acyclic 1,2-dioxy compounds is through Grignard reaction of 2-acyl-1,3-oxathianes. By using a chiral 1,3-hydroxythiol to obtain the heterocycle, it is possible to prepare tertiary alcohols of predetermined absolute configuration. The reaction takes place after coordination of the reagent to both the carbonyl group and the oxygen atom of the oxathiane because the magnesium ion is a hard acid. From pulegone, routes to several natural products have been developed, including (+)-mevalonolactone (**5-81**) and (–)-mevalonolactone (**5-81′**)[115] and (*R*)-frontalin (**5-82**).[116] On the other hand, the organometallic reaction in the presence of a lanthanide salt exhibits opposite diastereoselectivity.[117] For an explanation of the changed stereochemical course, one may consider the participation of all three heteroatoms in the complexation with the lanthanide ion, which can share a ligand (e.g., Cl) with Grignard reagent or organolithium, thereby directing the addition to the carbonyl group. The side of the sulfur is more open because the C–S bond is longer than the C–O bond.

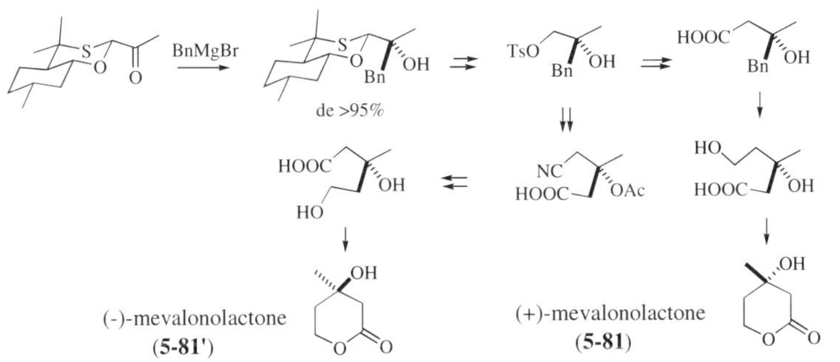

(-)-mevalonolactone (**5-81'**)

(+)-mevalonolactone (**5-81**)

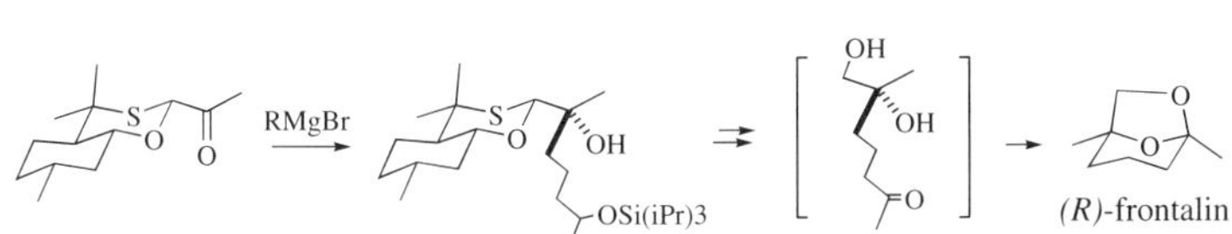

(*R*)-frontalin (**5-82**)

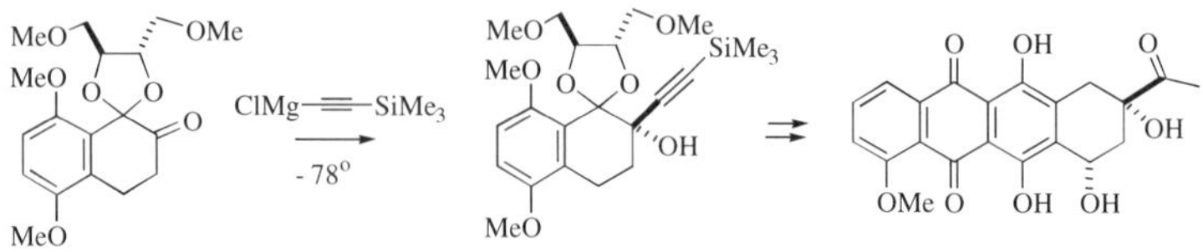

(+)-daunomycinone (**5-83**)

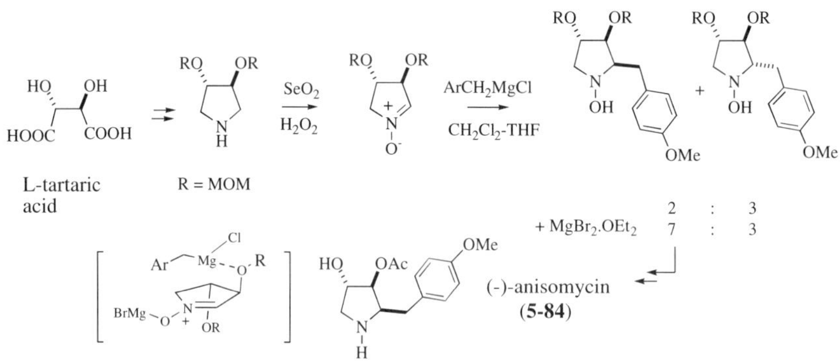

Delivery of the reagent within a chelate is characterized by the exclusive formation of a propargylic alcohol[118] which was used in the synthesis of (+)-daunomycinone (**5-83**). In a preparation of the (–)-anisomycin (**5-84**) precursor[119] from a cyclic nitrone, the desired stereoselectivity manifested when $MgBr_2$ was added. Apparently the ethereal oxygen could direct the Grignard reaction only after the nitrone was harnessed.

An expedient access to the Prelog–Djerassi lactone (**5-30**) is by reaction of a *syn*-2,4-dimethylglutaraldehydic ester with crotyltributyltin in the presence of boron trifluoride etherate.[120] A 7-centered crown-shaped complex derived from the substrate was the reactive species.

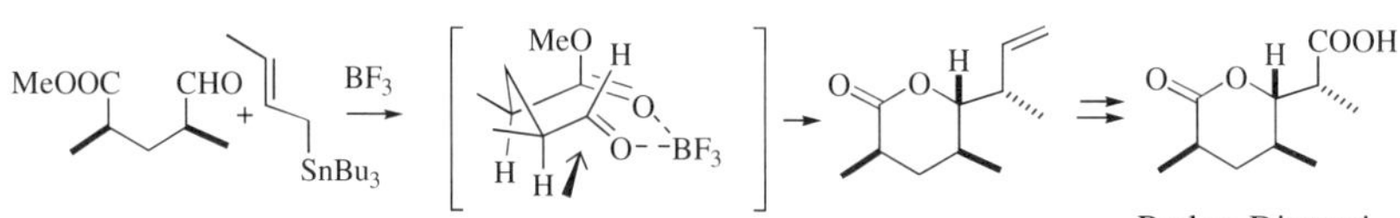

Prelog-Djerassi lactone (**5-30**)

The chemistry of allylboronates is highly developed. It has been shown that these reagents transfer the allyl group to aldehydes with excellent stereoselectivity. Presently, double asymmetric induction is practiced in earnest for solving synthetic problems, and analysis[121] helps identify matched and mismatched pairings to allow the choice of appropriate substrates and reagents. For simple diastereoselection one may refer to a convenient synthesis of *exo*-brevicomin (**5-85**),[122] and a double asymmetric coupling which was presented in a route to invictolide (**5-86**).[123]

There are many allylmetal reagents and each possesses certain characteristic properties. Metallated allyl carbamates, for example, **5-87** and **5-88**,[124] are internally chelated and their reaction with aldehydes is interesting.

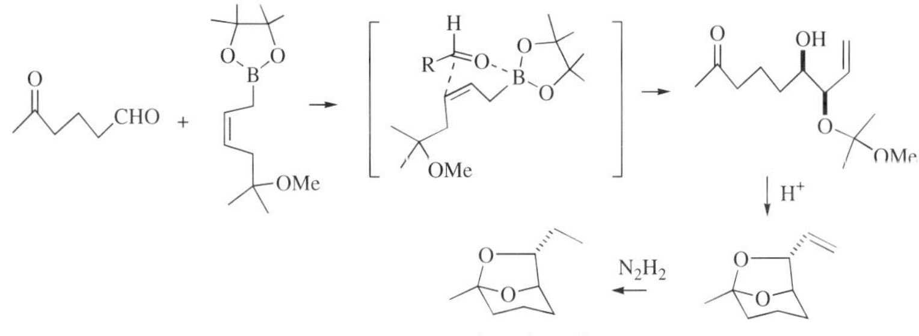

exo-brevicomin (**5-85**)

88% de

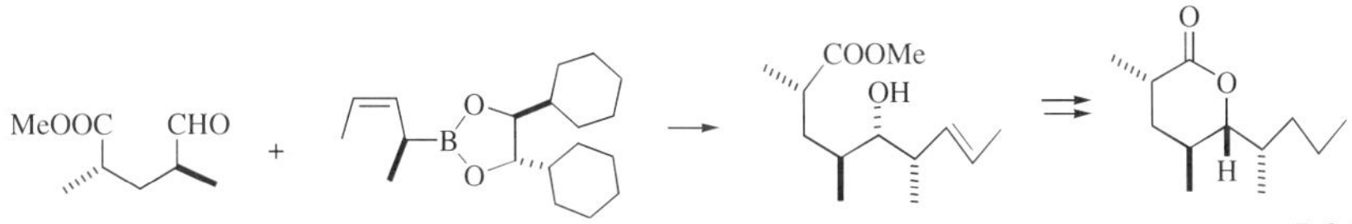

invictolide (**5-86**)

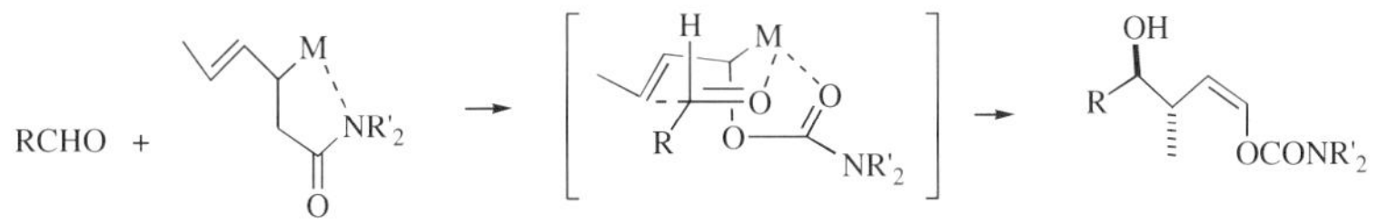

(**5-87**) *(Z)-anti-*

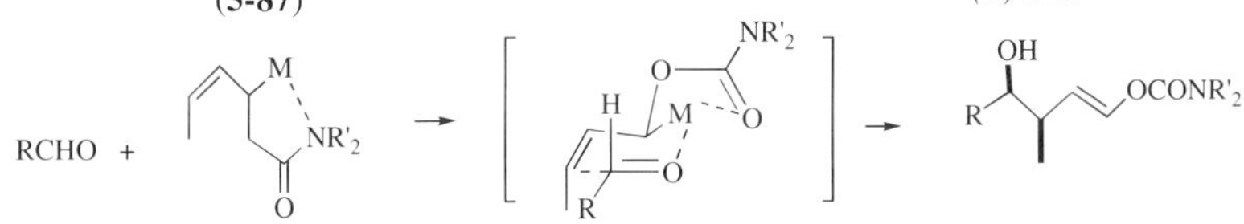

(**5-88**) *(E)-syn-*

The revival of the aldol reaction[125,126] as a modern synthetic method is due to the development in metal enolate generation whose geometry can be correlated with the kinetic products. The α- and β-substituents of a metal enolate (note: not the ketone!) contribute a major steric influence to the outcome, particularly for the Group I and II metals. With a bulky α-substituent the predominant aldol has the *syn* (or *erythro*) configuration, but a demanding β-substituent gives preferably the *anti* (or *threo*) product. The results agree with the Zimmerman–Traxler cyclic transition state model quite well.

The amide enolate geometry is easiest to control, and many applications have thus been realized using amide auxiliaries. The (*Z*)-isomer (ratio > 97:3) of enolates derived from *N*-acyl-1,3-oxazolidin-2-ones (**5-89**)[127] favored by the allylic strain factor is reinforced by chelation effect. Boryl enolates generally afford better aldol diastereoselection.

Chelated ester enolates are also well behaved. An exploitation appeared in a synthesis of maytansine (**5-90**).[128] The nucleophilic reagent approached the aldehyde with the sulfur's lone electron-pair which incurred least interference from the

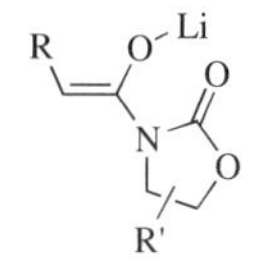

(**5-89**)

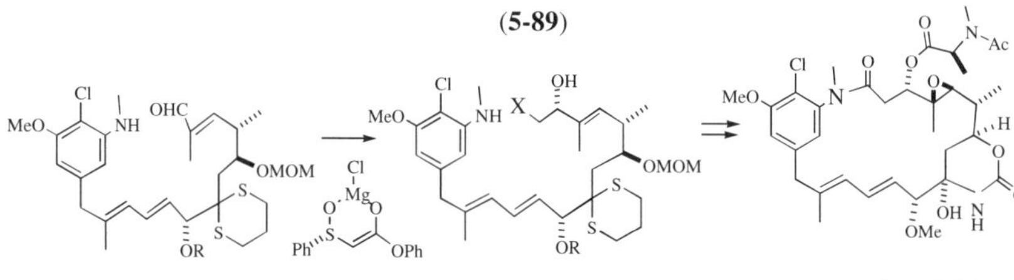

X = COOPh

maytansine (**5-90**)

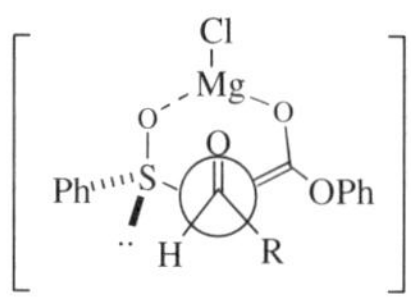

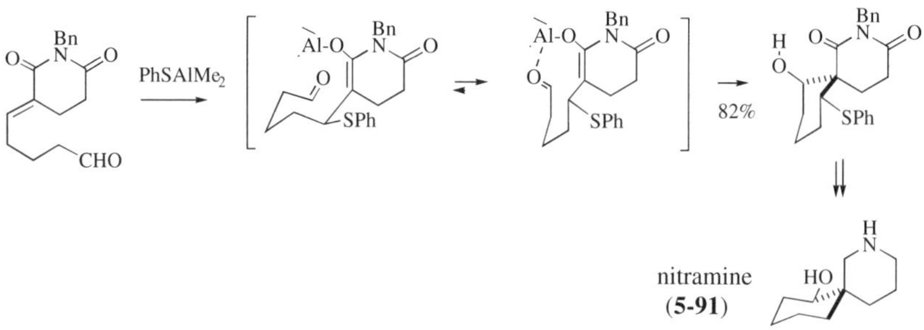

nitramine (**5-91**)

hydrogen of the aldehyde, while the flattened enolate and the larger group of the aldehyde were juxtaposed.

Intramolecular aldol reaction in tandem with Michael addition was the key step in a synthesis of nitramine (**5-91**).[129] Chelation ensured the emerging stereocenters corresponding to the target molecule.

The Mukaiyama modification of the aldol condensation involves reaction of enol ethers (silyl ethers) with acetals in the presence of a Lewis acid. An intramolecular version is the contractive rearrangement of 1,3-dioxepins to tetrahydrofuranaldehydes. Due to the organization of the two oxygen ligands around the catalytic metal ion after the ring cleavage, the stereochemical outcome can be predicted. Examination of a synthesis of (+)-sesamin (**5-92**)[130] serves to reveal the details.

2-Cyclohexenones and 2,5-cyclohexadienones carrying an oxy substituent at $C_{(4)}$ have been shown to undergo stereoselective conjugate addition with hard organometallic reagents[131–133] from the side of the oxygen substituent.

The *threo*-selective Michael reaction between methyl crotonate and the 7-phenylmenthyl ester of a heptenoic acid was used to start construction of an intramolecular Diels–Alder reaction substrate for eventual elaboration of 7,20-diisocyanoadociane (**5-93**).[134]

Reflexive Michael reactions initiated by kinetic enolates of conjugated ketones are very valuable to synthesis. These reactions are also frequently characterized by good stereoselectivity as the transition states are highly ordered. The application to a synthesis of atisine (**5-94**)[135] is exemplary.

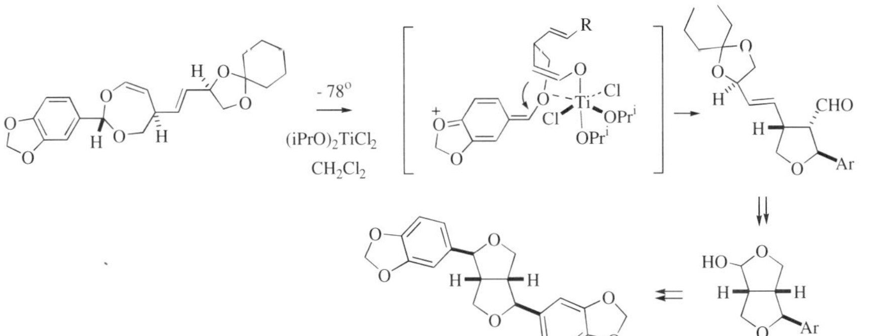

(+)-sesamin
(**5-92**)

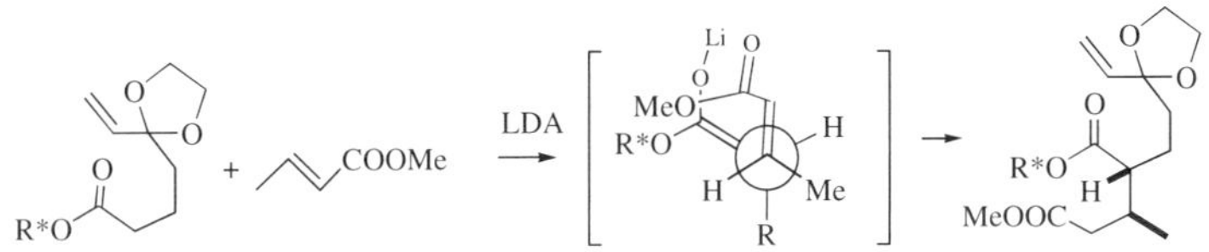

7,20-diisocyanoadociane
(**5-93**)

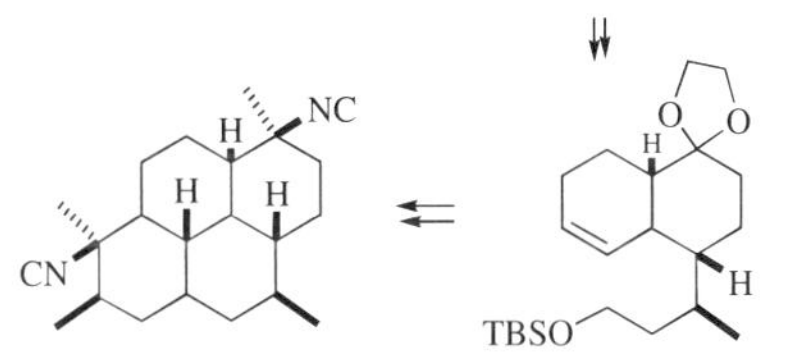

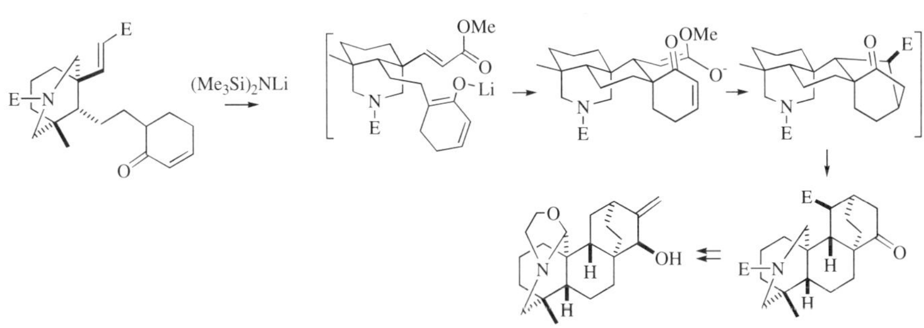

atisine (**5-94**)

Chiral acetals of conjugated aldehydes (**5-95, 5-96**) undergo S_N2' displacement with organoaluminum reagents in nonpolar solvents.[136] The observed stereoselectivities are consistent with chelated transition states. The process is useful for assembly of vitamins E and K. On the other hand, the reaction of such acetals with organocopper reagents in the presence of boron trifluoride showed opposite diastereoselectivity.[137]

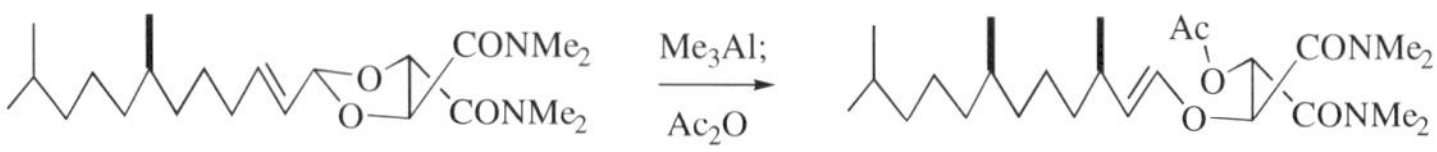

(5-95)

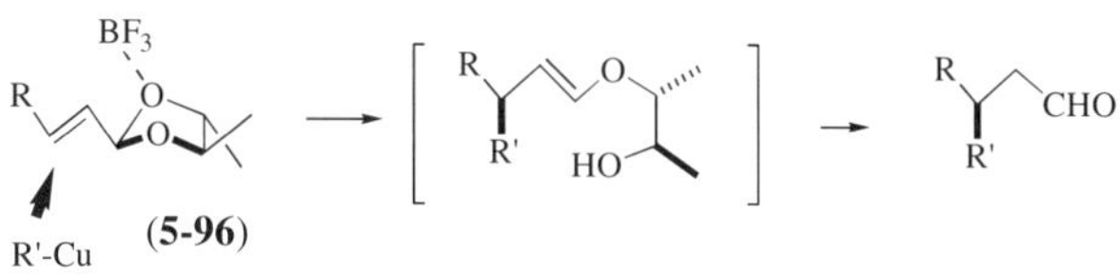

(5-96)

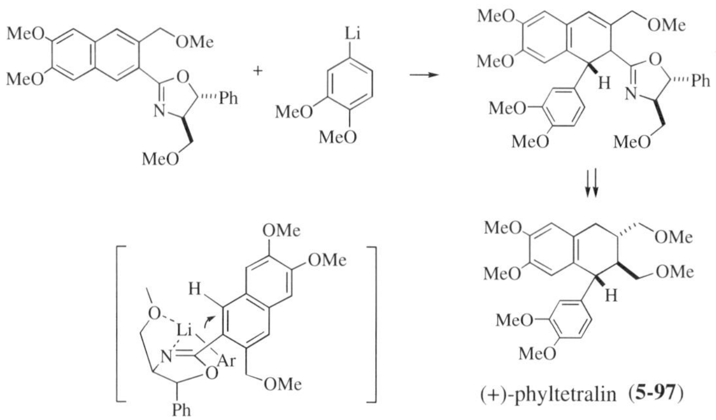

(+)-phyltetralin (5-97)

A dearomatization approach to (+)-phyltetralin (**5-97**)[138] was based on the chelation power of 4-methoxymethyloxazoline which affixed the aryllithium reagent prior to its attack. A [1,5]-sigmatropic rearrangement of the aryl group was invoked to explain the stereochemical course.

α-Arenesulfinyl 2-cycloalkenones (**5-98**) and unsaturated lactones are bidentate ligands, and on complexation with metal ions they present two diastereofaces to react with Michael donors. With an organometallic nucleophile, the steric course depends

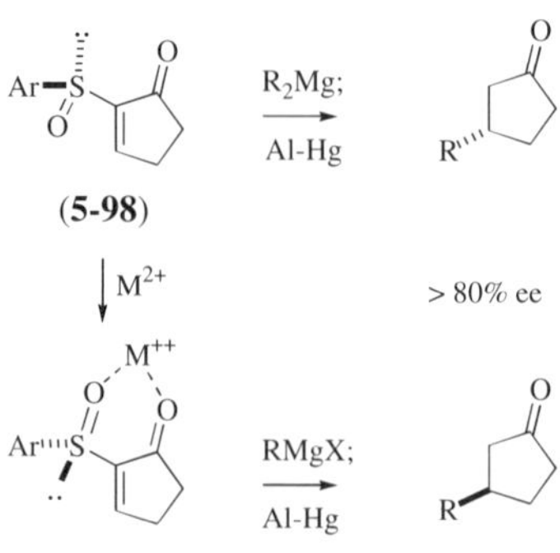

(5-98)

on whether a precoordinated state exists. Thus, adjustment of reaction conditions can affect the stereochemical results.[139]

A key step in a synthesis of aphidicolin (**5-99**) is the union of the kinetic enolate of a cyclohexenone with a sulfinyl butenolide.[140] The highly ordered transition state was arranged by the lithium ion.

(+)-Talaromycin-A (**5-100**) and (–)-talaromycin-B (**5-101**) have been constructed from a common intermediate which was obtained from internal addition of a side chain alcohol to the sulfinyldihydropyran.[141] Potassium ion directed the process through simultaneous chelation to three oxygen atoms.

A powerful tactic for stereoselective functionalization of unsaturated sulfones in the β-carbon is to attach it to a pyran ring. Depending on whether an additional hydroxyl group is present in the carbon adjacent to the Michael acceptor chain or not,

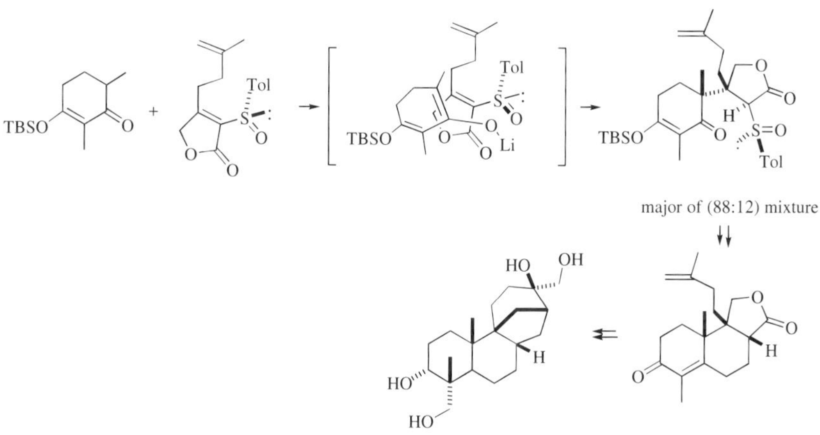

aphidicolin (**5-99**)

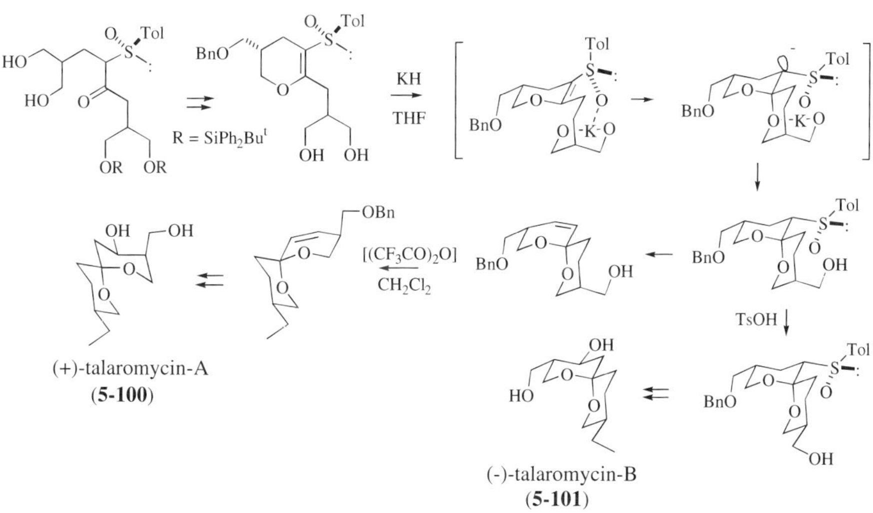

(+)-talaromycin-A
(**5-100**)

(-)-talaromycin-B
(**5-101**)

an organometallic reaction occurs preferentially from one of the two faces. The effectiveness of the technique is substantiated in the synthesis of (–)-maytansinol (**5-102**)[142] and okadaic acid (**5-103**).[143] A similar concept was based on the conjugate addition of a formyl anion equivalent to an unsaturated ester when the $C_{(1)}$–$C_{(10)}$ segment of carbomycin (**5-104**) was assembled.[144]

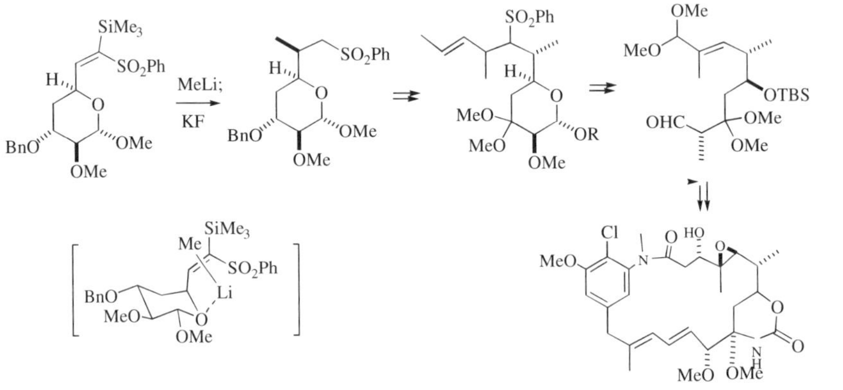

(-)-maytansinol (**5-102**)

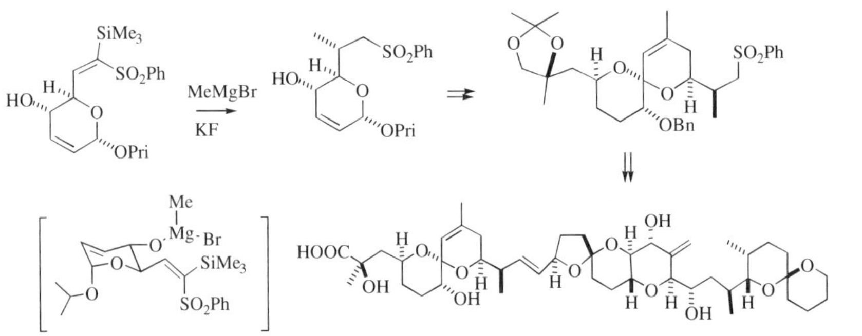

okadaic acid (**5-103**)

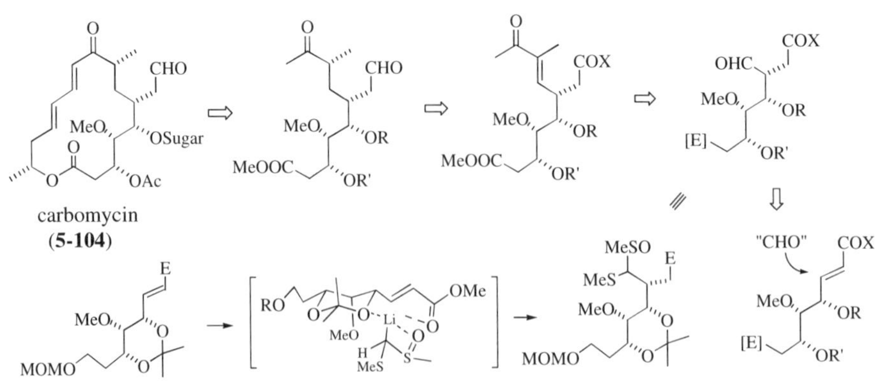

carbomycin (**5-104**)

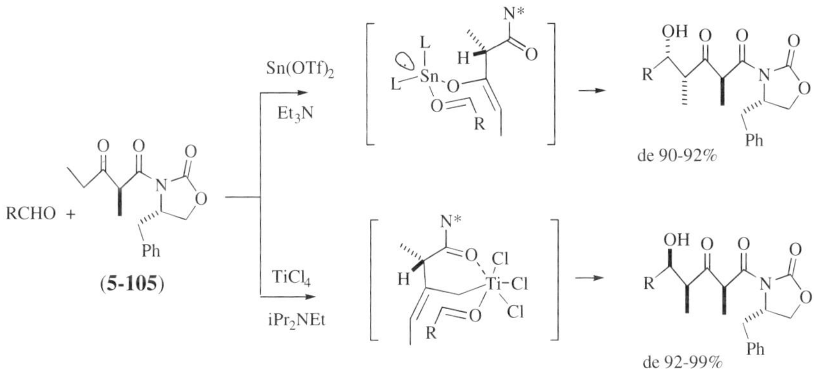

An unexpected *cis*-selective 1,4-addition of lower order cyanocuprates RCu(CN)M to 5-alkoxy-2-cyclohexenones[144a] has been ascribed to coordination of the copper atom in the d,π*-complexes[144b] to the alkoxy substituent. Even the bulky *t*-butyldimethylsiloxy group is effective despite the requirement for occupying an axial orientation. The corresponding reactions with the higher order cuprate reagents $R_2Cu(CN)Li_2$ afforded the *trans*-adducts.

Valuable methodologies for the assembly of *syn*- and *anti*-1,3-dimethyl segments employ titanium and stannous reagents, respectively.[145] It is remarkable that the β-ketoamide system (**5-105**) was not disturbed in the apparently kinetic enolization.

The stereoselectivity of cycloaddition reactions can be improved by chelation. Consider the assembly of dihydropyranones which are useful building blocks for many structural units. The Lewis acid catalyzed condensation between activated dienes and aldehydes is an important contribution to methodology which permits fine tuning of stereochemistry by modification of the dieneophiles (e.g., chelatable versus nonchelatable) and the catalyst. With BF_3 and $TiCl_4$, the reaction proceeds by an aldol condensation pathway, differences appearing when the aldehyde is alkoxylated at the α- or β-position. A chelate can be formed with the titanium catalyst but not with BF_3. Pericyclic transition states are adopted when reactions are conducted in the presence of $ZnCl_2$ and $MgBr_2$ in THF, but again there are differences as $ZnCl_2$ is not effective

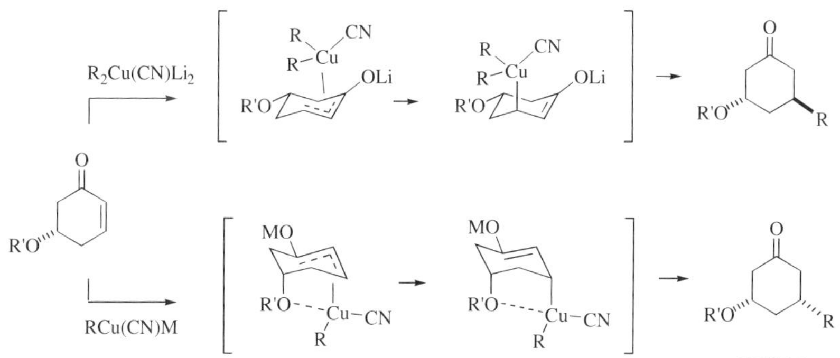

(**5-104a**)

for chelation with α-alkoxyaldehydes. This method of dihydropyranone assembly has been used repeatedly for extension of the skeletal chain with alternate hydroxylated and methylated carbon atoms in a synthesis of zincophorin (**5-106**).[146]

2-Furyl carbinols were shown to undergo [4+3]-cycloadditions with 2-oxyallyl cations generated from α,α′-dibromoketones with a Zn–Ag couple, with results consistent with chelated transition states (**5-107**).[147] In extreme circumstances a temporary link can be used to impose regiochemical as well as stereochemical control, like fulfilling the critical need in a synthesis of a taxol intermediate (**5-108**).[148]

A ketyl species generated by samarium(II) reduction of carbonyl compounds can add to a proximal double bond. A juxtaposed hydroxyl group stabilizes the intermediate by chelation to the metal ion and through such chelation the configuration of the alcohol to be formed becomes controllable. An excellent opportunity for completing the CD-ring component of grayanotoxin-III (**5-109**) was presented.[149]

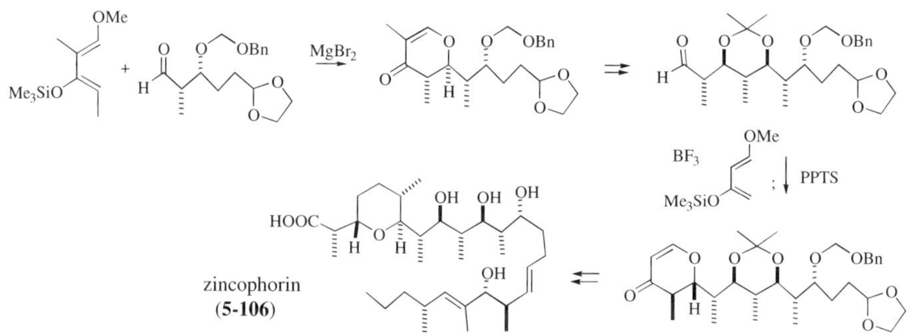

zincophorin (**5-106**)

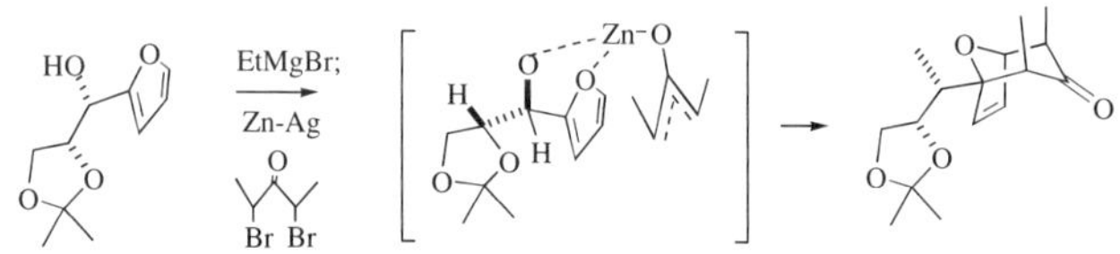

(**5-107**)

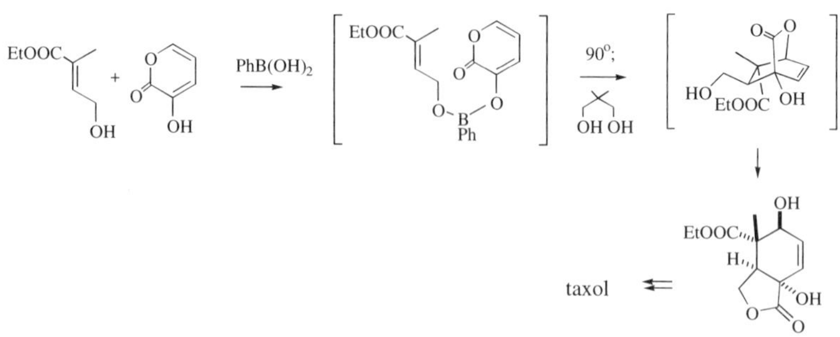

(**5-108**)

Activation of a double bond toward attack of carbon nucleophiles by coordination with low-valent metal species accomplishes carbometallation. The utility of this stereoselective process is enhanced in the presence of a functional group capable of sequesting the metal ion to direct the C–Nu bond formation that follows. An intramolecular carbozincation process followed by in situ alkylation was designed to construct a precursor of (–)-methyl cucurbate (**5-110**),[150] taking advantage of a chelated transition state in which all substituents are pseudoequatorial.

An elegant albeit uneconomical synthesis of prostaglandin-$F_{2\alpha}$ intermediate (**5-111**)[151] was based on such a maneuver using a palladium salt to introduce three stereocenters in the cyclopentane nucleus.

The Heck reaction enjoys current popularity as a C–C bond forming method. An internal coordination by a nonreacting vinyl group of the substrate (**5-112**) to reverse the steric course for a spiroannulation process has been noted.[152] This finding was related to a synthetic investigation on the complex indole alkaloid gelsemine.

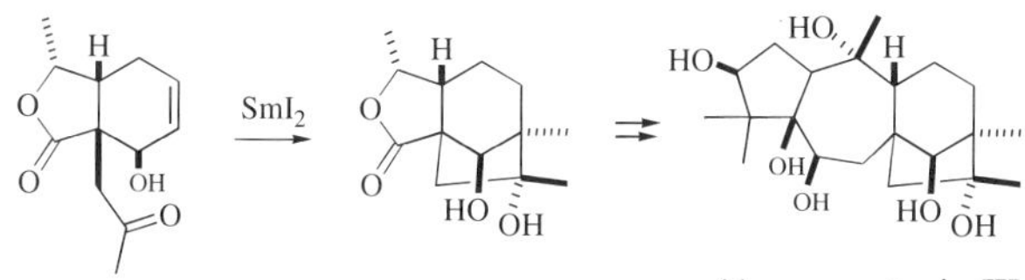

(-)-grayanotoxin-III
(**5-109**)

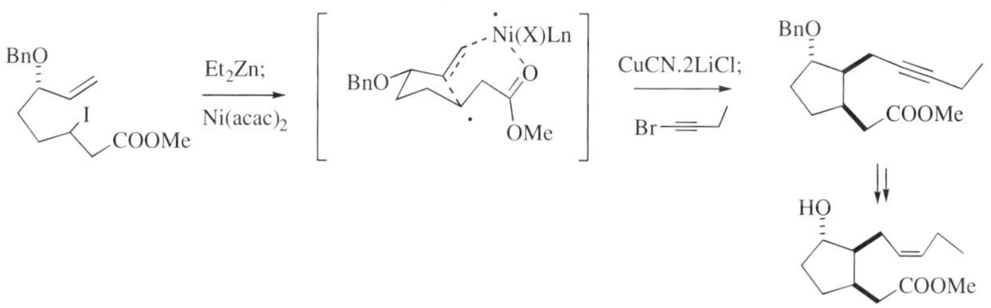

(-)-methyl cucurbate
(**5-110**)

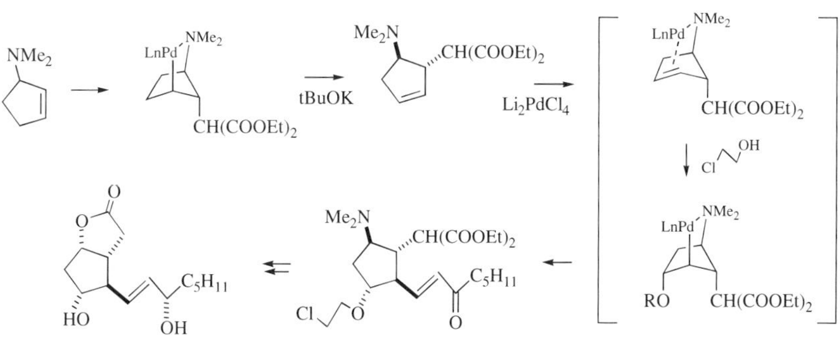

(**5-111**)

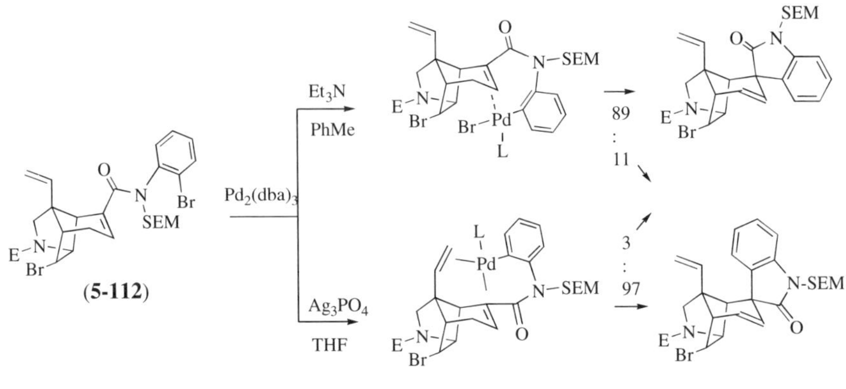

(**5-112**)

5.4. INTRAMOLECULAR REACTIONS

In synthesis it is always advantageous to use an existing functionality to form a covalent intermediate in the course of group transformation. Difficult stereochemical problems are often resolved by this tactic.

Halolactonization served to establish the relative configuration of the two stereocenters of sesbanine (**5-113**).[153] Similarly, in a synthesis of erythronolide-B (**5-114**)[154] the $C_{(1)}$–$C_{(9)}$ segment was elaborated on a six-membered ring template by a bromolactonization to initiate a stereogenic reaction sequence.

A convenient way to construct the B-ring of forskolin (**5-115**) while creating three stereocenters involved an intramolecular Michael reaction followed by aldol

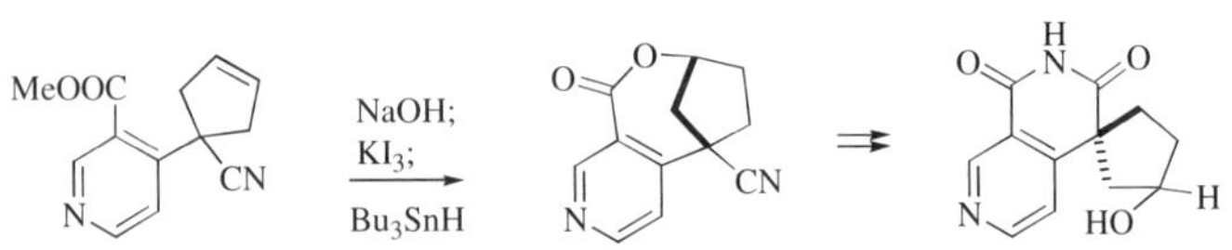

sesbanine (**5-113**)

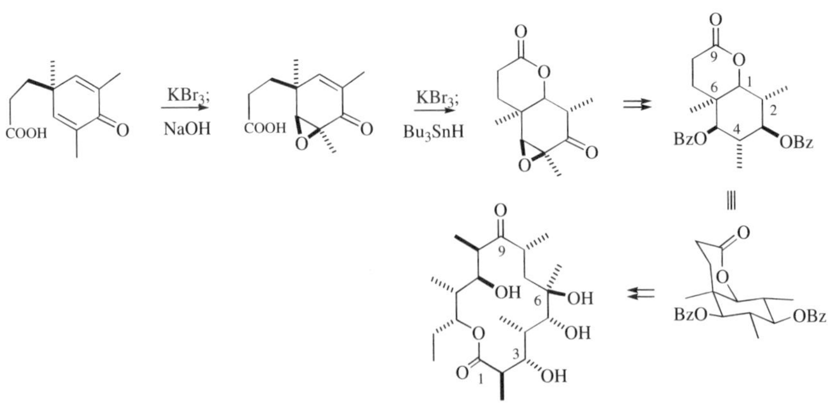

erythronolide-B (**5-114**)

condensation.[155] Completion of the tricyclic system of isoclovene (**5-116**) by bridging a *cis*-hydrindenone using an intramolecular Michael reaction[156] was also very effective.

A critical issue in the synthesis of talaromycin-A (**5-100**) is the placement of an axial hydroxymethyl chain. A solution to this problem is to apply Stork's method of intramolecular addition of a silicon-tethered radical to a double bond.[157] The short chain ensured formation of a *cis*-fused ring.

Intramolecular cycloaddition reactions including reaction of metal carbenoids and [2+2]-cycloaddition to meet stereochemical demands are illustrated in the

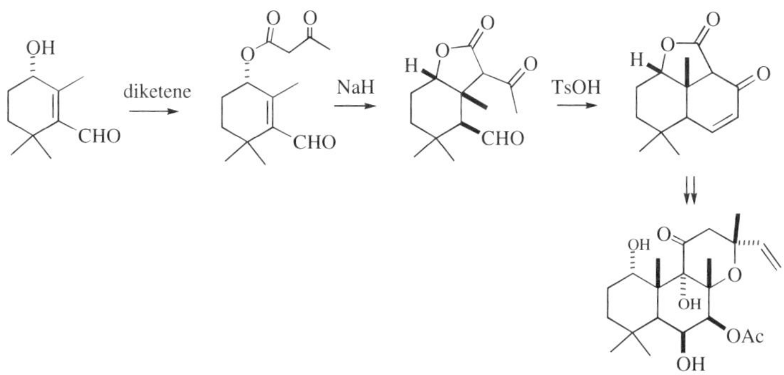

forskolin (**5-115**)

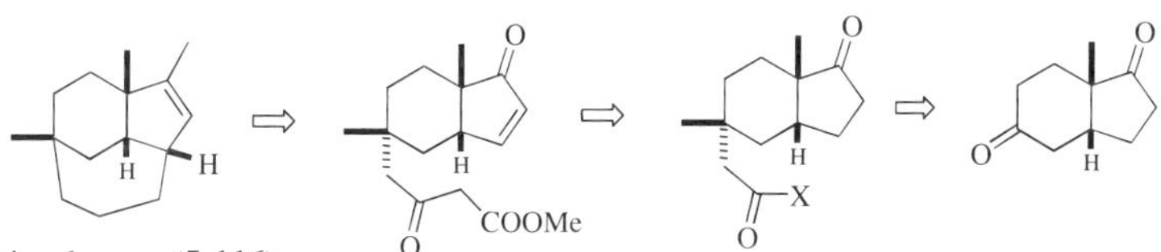

isoclovene (**5-116**)

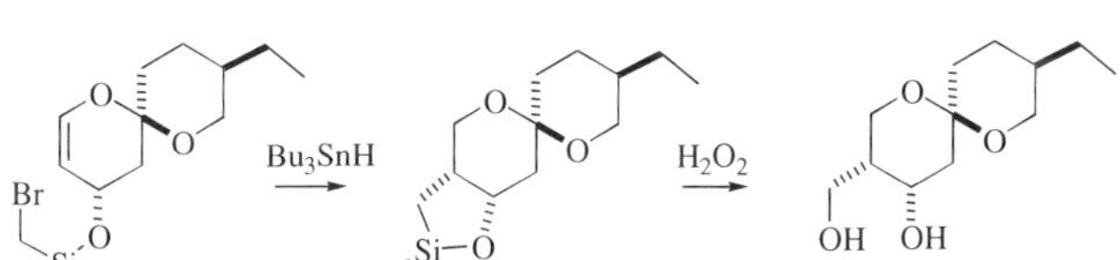

talaromycin-A (**5-100**)

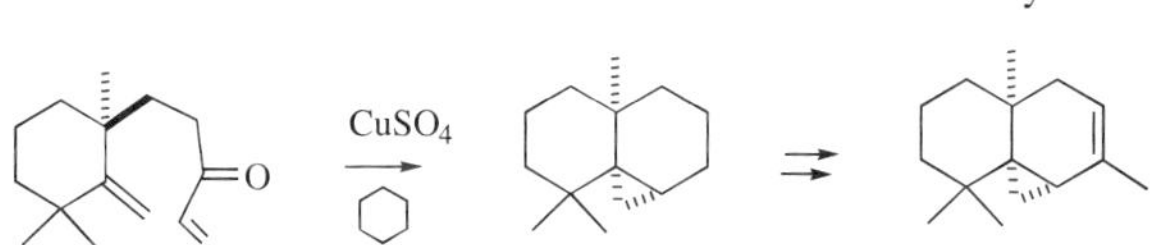

thujopsene (**5-67**)

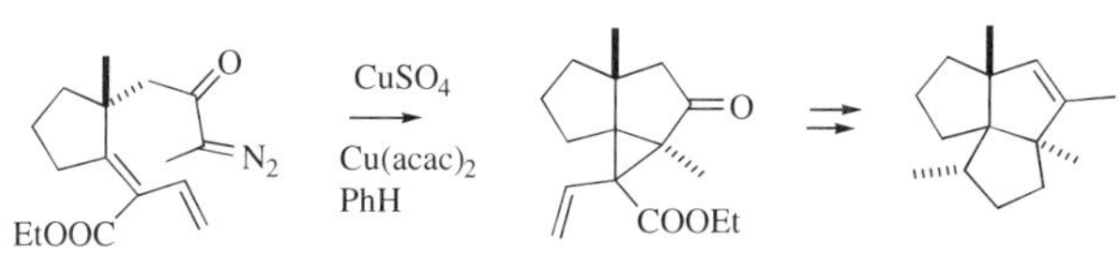

isocomene (**5-77**)

following syntheses: thujopsene (**5-67**),[158] isocomene (**5-77**),[159] α-/β-panasinsene (**5-117, 5-118**),[160] silphinene (**5-119**),[161] reserpine (**5-120**),[162] and trihydroxydecipiadiene (**5-121**).[163]

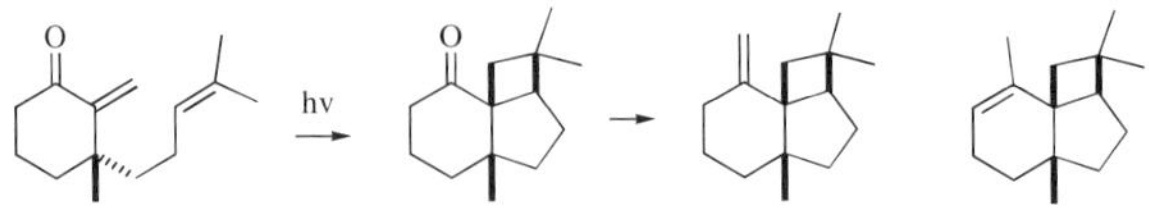

β-panasinsene (**5-118**) α-panasinsene (**5-117**)

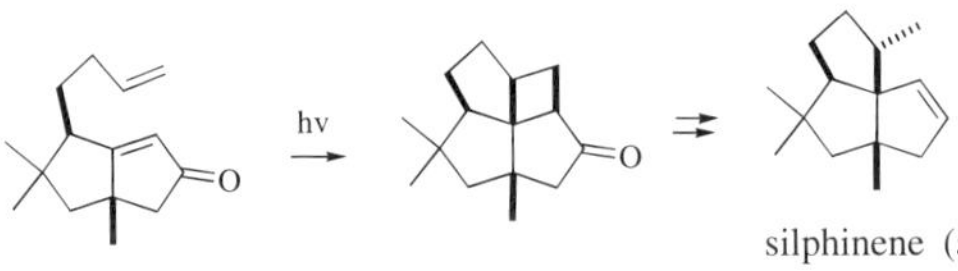

silphinene (**5-119**)

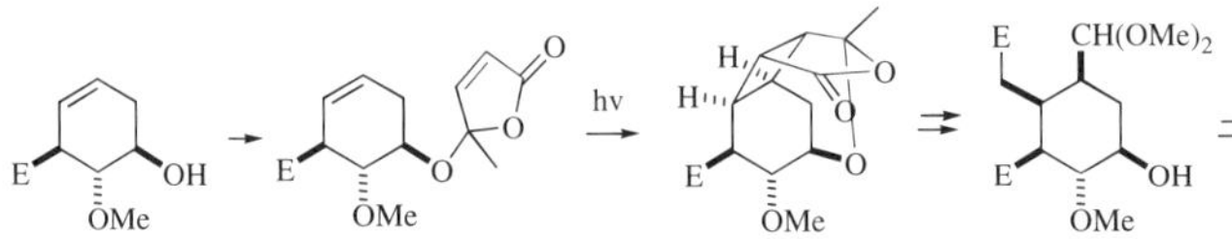

reserpine (**5-120**)

E = COOMe

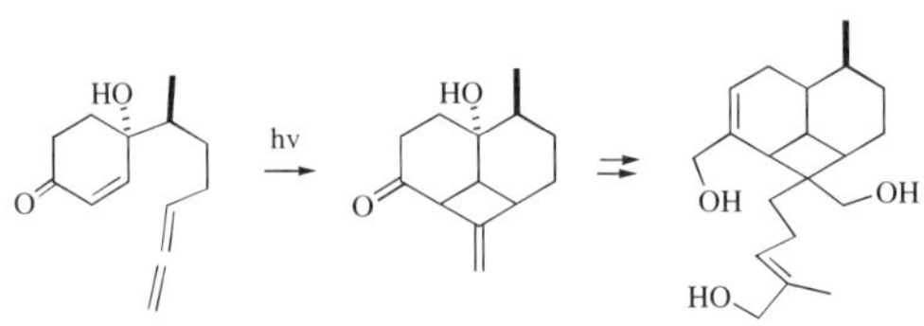

trihydroxydecipiadiene (**5-121**)

The intramolecular ene reaction has been exploited to form a stereocenter in a new ring. Examples include syntheses of β-acorenol (**5-122**),[164] modhephene (**5-36**),[165] and isocomene (**5-76**).[166]

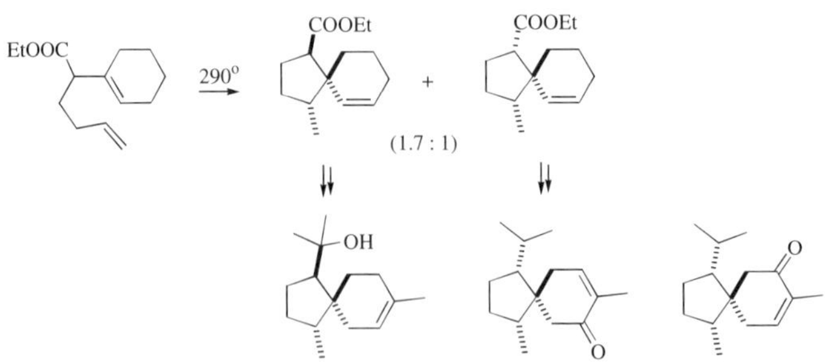

β-acorenol (**5-122**) acorenone acorenone-B

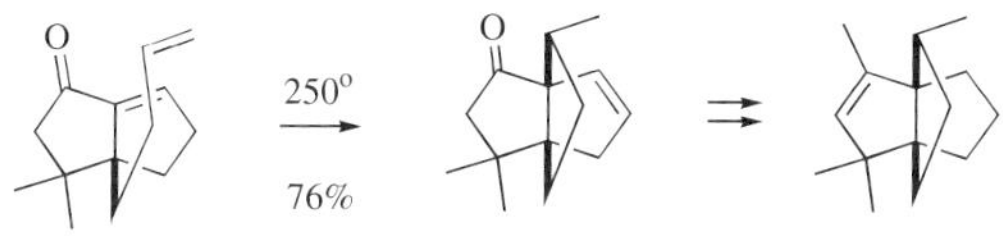

modhephene (**5-36**)

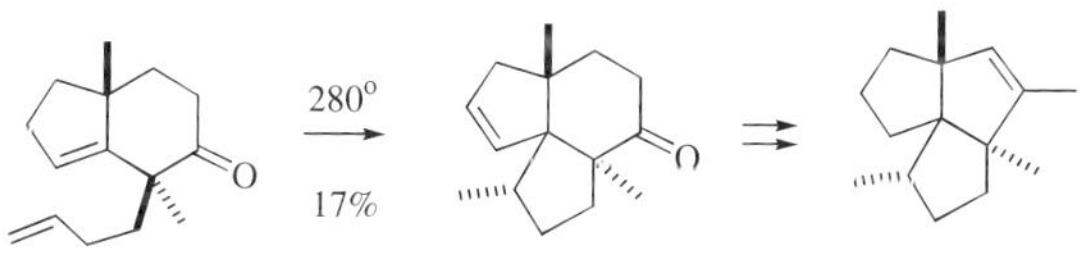

isocomene (**5-76**)

Intramolecular metallo-ene reactions, particularly the magnesium version, can render even greater value to synthesis, as an active site is transferred to a new location in an orderly manner. Some applications have been delineated: $\Delta^{9(12)}$-capnellene (**5-123**),[167] sinularene (**5-124**),[168] chokol-A (**5-125**),[169] and khusimone (**5-126**).[170]

Stereocontrolled reduction of 2-acetylphenol to afford (**5-127R**) and (**5-127S**) has been achieved via an optically active 3-hydroxy-1-methylbutyl ether.[171] Treatment with aluminum isopropoxide induced an intramolecular hydride transfer which involved a chelate as the intermediate.

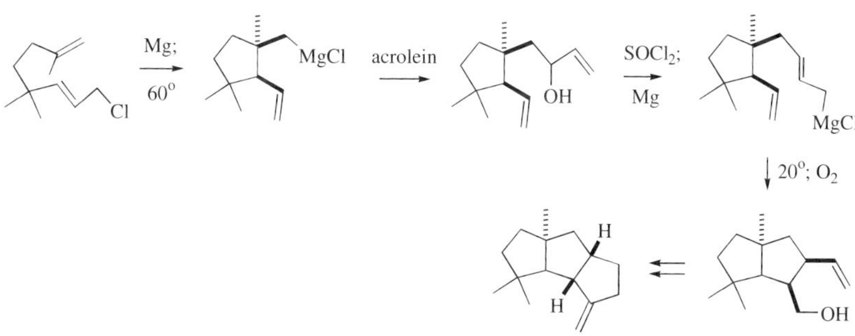

$\Delta^{9(12)}$-capnellene
(**5-123**)

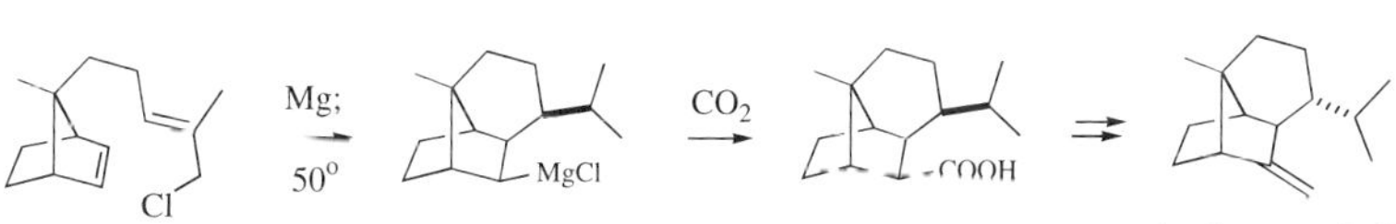

sinularene (**5-124**)

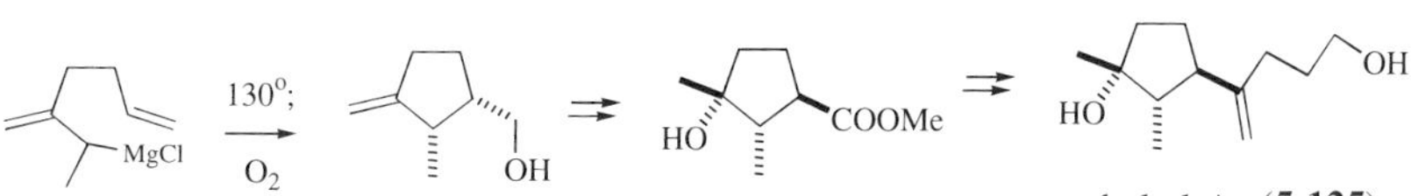

chokol-A (**5-125**)

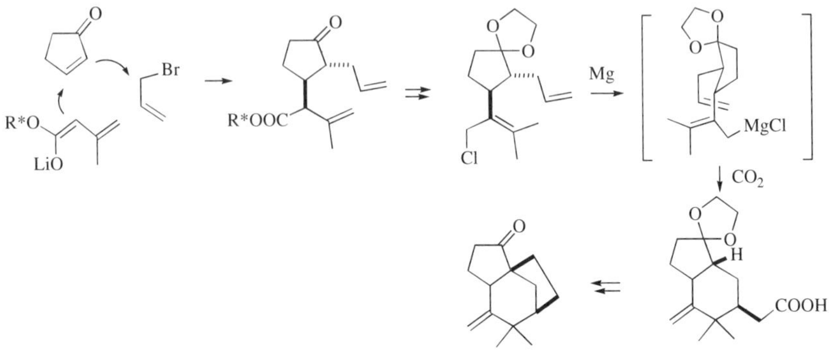

(+)-khusimone
(**5-126**)

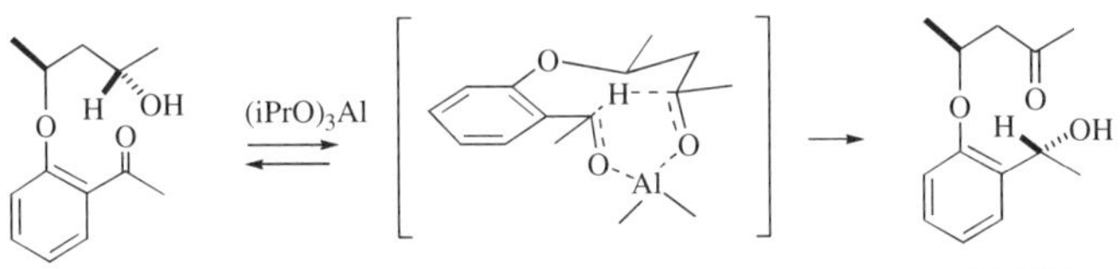

(**5-127R**)

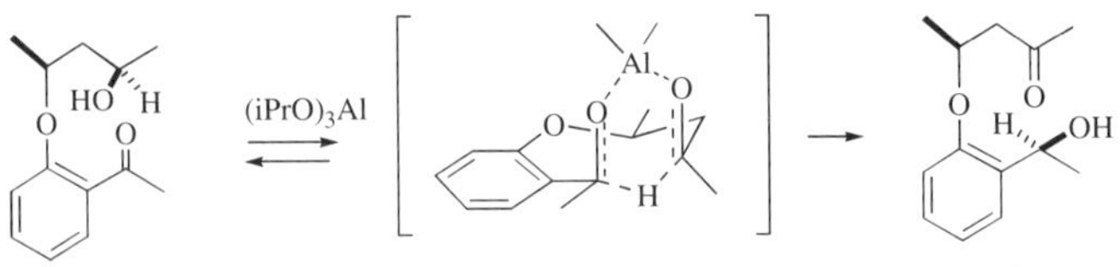

(**5-127S**)

REFERENCES

1. Hoyveda, A.H.; Evans, D.A.; Fu, G.C. *Chem. Rev.* **1993**, *93*, 1307.
2. Seyden-Penne, J. *Chiral Auxiliaries and Ligands in Asymmetric Synthesis*; Wiley: New York, 1995.
3. Henbest, H.B.; Wilson, R.A.L. *J. Chem. Soc.* **1957**, 1958.
4. Smith, A.B., III; Richmond, R.E. *J. Org. Chem.* **1981**, *46*, 4814.
5. Danishefsky, S.; Schuda, P.F.; Kitahara, T.; Etheredge, S.J. *J. Am. Chem. Soc.* **1977**, *99*, 6066.
6. Bath, S.; Billington, D.C.; Gero, S.D.; Quiclet-Sire, B.; Samadi, M. *Chem. Commun.* **1994**, 1495.
7. Wipf, P.; Kim, Y.; Fritch, P.C. *J. Org. Chem.* **1993**, *58*, 7195.
8. Grieco, P.A.; Ohfune, Y.; Majetich, G. *J. Org. Chem.* **1979**, *44*, 3092.
9. Schink, H.E.; Pattersson, H.; Bäckvall, J.E. *J. Org. Chem.* **1991**, *56*, 2769.
10. Lansbury, P.T.; La Clair, J.J. *Tetrahedron Lett.* **1993**, *34*, 4431.

11. Colvin, E.W.; Malchenko, S.; Raphael, R.A.; Roberts, J.S. *J. Chem. Soc. Perkin Trans. 1* **1973**, 1989.
12. Roush, W.R.; D'Ambra, T.E. *J. Org. Chem.* **1981**, *46*, 5045.
13. Fukuyama, T.; Wang, C.-L.J.; Kishi, Y. *J. Am. Chem. Soc.* **1979**, *101*, 260.
14. Sheng, M.N.; Zajacek, J.G. *J. Org. Chem.* **1970**, *35*, 1839.
15. Sharpless, K.B.; Michaelson, R.C. *J. Am. Chem. Soc.* **1973**, *95*, 6136.
16. Tanno, N.; Terashima, S. *Chem. Pharm. Bull.* **1983**, *31*, 821.
17. Still, W.C.; Tsai, M.-Y. *J. Am. Chem. Soc.* **1980**, *102*, 3654.
18. Holton, R.A.; Somoza, C.; Kim, H.-B.; Liang, F.; Biediger, R.J.; Boatman, P.D.; Shindo, M.; Smith, C.C.; Kim, S.; Nadizadeh, H.; Suzuki, Y.; Tao, C.; Vu, P.; Tang, S.; Zhang, P.; Murthi, K.K.; Gentile, L.N.; Liu, J.H. *J. Am. Chem. Soc.* **1994**, *116*, 1597.
19. Holton, R.A.; Kim, H.-B.; Somoza, C.; Liang, F.; Biediger, R.J.; Boatman, P.D.; Shindo, M.; Smith, C.C.; Kim, S.; Nadizadeh, H.; Suzuki, Y.; Tao, C.; Vu, P.; Tang, S.; Zhang, P.; Murthi, K.K.; Gentile, L.N.; Liu, J.H. *J. Am. Chem. Soc.* **1994**, *116*, 1599.
20. Williams, D.R.; Coleman, P.J.; Henry, S.S. *J. Am. Chem. Soc.* **1993**, *115*, 11654.
21. Corey, E.J.; De, B. *J. Am. Chem. Soc.* **1984**, *106*, 2735.
22. Stork, G.; Nakahara, Y.; Greenlee, W.G. *J. Am. Chem. Soc.* **1978**, *100*, 7775.
23. Breslow, R.; Maresca, L.M. *Tetrahedron Lett.* **1977**, 623.
24. Smith, A.B., III; Empfield, J.R.; Rivero, R.A.; Vaccaro, H.A. *J. Am. Chem. Soc.* **1991**, *113*, 4037.
25. Sato, F.; Kobayashi, Y.; Takahashi, O.; Chiba, T.; Takeda, Y.; Kusakabe, M. *Chem. Commun.* **1985**, 1636.
26. Chan, T.H.; Lau, P.W.K.; Li, M.P. *Tetrahedron Lett.* **1976**, 2667.
27. Yasuda, A.; Tanaka, S.; Yamamoto, H.; Nozaki, H. *Bull. Chem. Soc. Jpn.* **1979**, *52*, 1701.
28. Corey, E.J.; Dittami, J.P. *J. Am. Chem. Soc.* **1985**, *107*, 256.
29. Kitamura, M.; Isobe, M.; Ichikawa, Y.; Goto, T. *J. Am. Chem. Soc.* **1984**, *106*, 3252.
30. Sharpless, K.B. *Chemtech* **1985**, 692.
31. Finn, M.G.; Sharpless, K.B. In *Asymmetric Synthesis*; Morrison, J.D., Ed.; Academic Press: Orlando, FL, 1985; Vol. 5, p 247.
32. Mihelich, E.D.; Daniels, K.; Eickhoff, D.J. *J. Am. Chem. Soc.* **1981**, *103*, 7690.
33. Kobayashi, Y.; Uchiyama, H.; Kanbara, H.; Sato, F. *J. Am. Chem. Soc.* **1985**, *107*, 5541.
34. Johnson, M.R.; Kishi, Y. *Tetrahedron Lett.* **1979**, 4347.
35. Oikawa, Y.; Nishi, T.; Yonemitsu, O. *J. Chem. Soc. Perkin Trans. 1* **1985**, 7.
36. Stevens, R.V.; Chang, J.H.; Lapalme, R.; Schow, S.; Schlageter, M.G.; Shapiro, R.; Weller, H.N. *J. Am. Chem. Soc.* **1983**, *105*, 7719.
37. White, J.D.; Cutshall, N.S.; Kim, T.-S.; Shin, H. *J. Am. Chem. Soc.* **1995**, *117*, 9780.
38. Nakata, T.; Schmid, G.; Vranesic, B.; Okigawa, M.; Smith-Palmer, T.; Kishi, Y. *J. Am. Chem. Soc.* **1978**, *100*, 2933.
39. Hanessian, S.; Cooke, N.G.; De Hoff, B.; Sakito, Y. *J. Am. Chem. Soc.* **1990**, *112*, 5276.
40. Hashimoto, M.; Harigaya, H.; Yanagiya, M.; Shirahama, H. *J. Org. Chem.* **1991**, *56*, 2299.
41. Kanemoto, S.; Nonaka, T.; Oshima, K.; Utimoto, K.; Nozaki, H. *Tetrahedron Lett.* **1986**, *27*, 3387.

42. Hauser, F.M.; Ellenberger, S.R.; Clardy, J.C.; Bass, L.S. *J. Am. Chem. Soc.* **1984**, *106*, 2458.
43. Johnson, C.R. *Pure Appl. Chem.* **1987**, *59*, 969.
44. Trost, B.M.; Kuo, G.-H.; Benneche, T. *J. Am. Chem. Soc.* **1988**, *110*, 621.
45. Peterson, A.C.; Cook, J.M. *J. Org. Chem.* **1995**, *60*, 120.
46. Tang, S.; Kennedy, R.M. *Tetrahedron Lett.* **1992**, *33*, 3729, 5299, 5303.
47. Sinha, S.C.; Sinha-Bagchi, A.; Keinan, E. *J. Am. Chem. Soc.* **1995**, *117*, 1447.
48. McDonald, F.E.; Towne, T.B. *J. Am. Chem. Soc.* **1994**, *116*, 7921.
49. Katsui, N.; Matsunaga, A.; Imaizumi, K.; Masamune, T.; Tomiyama, K. *Bull. Chem. Soc. Jpn.* **1972**, *45*, 2871.
50. Evans, D.A.; Fu, G.C.; Hoveyda, A.H. *J. Am. Chem. Soc.* **1992**, *114*, 6671.
50a. Garrett, C.E.; Fu, G.C. *J. Org. Chem.* **1998**, *63*, 1370.
51. Morgans, D.J. *Tetrahedron Lett.* **1981**, *22*, 3721.
52. Still, W.C.; Shaw, K.R. *Tetrahedron Lett.* **1981**, *22*, 3725.
53. Yokoyama, Y.; Kawashima, H.; Kohno, M.; Ogawa, Y.; Uchida, S. *Tetrahedron Lett.* **1991**, *32*, 1479.
54. Still, W.C.; Darst, K.P. *J. Am. Chem. Soc.* **1980**, *102*, 7385.
55. Mander, L.N.; Patrick, G.L. *Tetrahedron Lett.* **1990**, *31*, 423.
56. Fujimoto, R.; Kishi, Y.; Blount, J.F. *J. Am. Chem. Soc.* **1980**, *102*, 7154.
57. Martin, S.F.; Barr, K.J. *J. Am. Chem. Soc.* **1996**, *118*, 3299.
58. Corey, E.J.; Desai, M.C.; Engler, T.A. *J. Am. Chem. Soc.* **1985**, *107*, 4339.
59. Kraus, G.A.; Shi, J. *J. Org. Chem.* **1991**, *56*, 4147.
60. Palmisano, G.; Danieli, B.; Lesma, G.; Passarella, D.; Toma, L. *J. Org. Chem.* **1991**, *56*, 2380.
61. Woodward, R.B.; Cava, M.P.; Ollis, W.D.; Hunger, A.; Daeniker, H.U.; Schenker, K. *Tetrahedron* **1963**, *19*, 247.
62. Kitamura, M.; Ohkuma, T.; Inoue, S.; Sayo, N.; Kumobayashi, H.; Akutagawa, S.; Ohta, S.; Takaya, H.; Noyori, R. *J. Am. Chem. Soc.* **1988**, *110*, 629.
63. Murahashi, S.I.; Naota, T.; Kuwabara, T.; Saito, T.; Kumobayashi, H.; Akutagawa, S. *J. Am. Chem. Soc.* **1990**, *112*, 7820.
64. Isobe, M.; Iio, H.; Kawai, T.; Goto, T. *J. Am. Chem. Soc.* **1978**, *100*, 1940.
65. Salomon, R.G.; Sachinvala, N.D.; Raychaudhuri, S.R.; Miller, D.B. *J. Am. Chem. Soc.* **1984**, *106*, 2211.
66. Lansbury, P.T.; Vacca, J.P. *Tetrahedron Lett.* **1982**, *23*, 2623.
67. Iwata, C.; Fusaka, T.; Fujiwara, T.; Tomita, K.; Yamada, M. *Chem. Commun.* **1981**, 463.
68. Iwata, C.; Morie, T.; Maezaki, N.; Shimamura, H.; Tanaka, T.; Imanishi, T. *Chem. Commun.* **1984**, 930.
69. McMurry, J.E.; Blaszczak, L.C.; Johnson, M.A. *Tetrahedron Lett.* **1978**, 1633.
70. Magee, T.V.; Stork, G.; Fludzinski, P. *Tetrahedron Lett.* **1995**, *36*, 7607.
71. Kinder, F.R.; Bair, K.W. *J. Org. Chem.* **1994**, *59*, 6965.
72. Kishi, Y.; Fukuyama, T.; Nakatsuka, S. *J. Am. Chem. Soc.* **1973**, *95*, 6490.
73. Gutzwiller, J.; Uskokovic, M.R. *Helv. Chim. Acta* **1973**, *56*, 1494.
74. Solladie, G. *Pure Appl. Chem.* **1988**, *60*, 1699.

75. Sato, T.; Itoh, T.; Fujisawa, T. *Tetrahedron Lett.* **1987**, *28*, 5677.
76. Fujioka, H.; Annoura, H.; Murano, K.; Kita, Y.; Tamura, Y. *Chem. Pharm. Bull.* **1989,** *37*, 2047.
77. Burke, S.D.; Armistead, D.M.; Fevig, J.M. *Tetrahedron Lett.* **1985**, *26*, 1163.
78. Gribble, G.W.; Ferguson, D.C. *Chem. Commun.* **1975**, 535.
79. Saksena, A.K.; Mangiaracina, P. *Tetrahedron Lett.* **1983**, *24*, 273.
80. Evans, D.A.; Hoveyda, A.H. *J. Org. Chem.* **1990**, *55*, 5190.
81. Evans, D.A.; Chapman, K.T.; Carreira, E.M. *J. Am. Chem. Soc.* **1988**, *110*, 3560.
82. Paterson, I.; Ward, R.A.; Romea, P.; Norcross, R.D. *J. Am. Chem. Soc.* **1994**, *116*, 3623.
83. Wender, P.A.; Kogen, H.; Lee, H.Y.; Munger, J.D.; Wilhelm, R.S.; Williams, P.D. *J. Am. Chem. Soc.* **1989**, *111*, 8957.
84. Goldstein, S.W.; Overman, L.E.; Rabinowitz, M.H. *J. Org. Chem.* **1992**, *57*, 1179.
85. Trost, B.M.; Greenspan, P.D.; Yang, B.V.; Saulnier, M.G. *J. Am. Chem. Soc.* **1990**, *112*, 9022.
86. Enders, D.; Kaiser, A. *Synthesis* **1996**, 209.
87. Evans, D.A.; Thomas, E.W.; Cherpeck, R.E. *J. Am. Chem. Soc.* **1982**, *104*, 3695.
88. Williams, D.R.; Osterhout, M.H. *J. Am. Chem. Soc.* **1992**, *114*, 8750.
89. Maruoka, K.; Yamamoto, H. *Angew. Chem. Int. Ed. Engl.* **1985**, *24*, 668.
90. Crimmins, M.T.; Rafferty, S.W. *Tetrahedron Lett.* **1996**, *37*, 5649.
91. Evans, D.A.; Hoveyda, A.H. *J. Am. Chem. Soc.* **1990**, *112*, 6447.
92. Romo, D.; Meyer, S.D.; Johnson, D.D.; Schreiber, S.L. *J. Am. Chem. Soc.* **1993**, *115*, 7906.
93. Devreese, A.A.; De Clercq, P.J.; Vandewalle, M. *Tetrahedron Lett.* **1980**, *21*, 4767.
94. Simmons, H.E.; Cairns, T.L.; Vladuchick, S.A.; Hoiness, C.M. *Org. React.* **1973**, *20*, 1.
95. Denis, J.M.; Conia, J.M. *Tetrahedron Lett.* **1972**, 4593.
96. Furukawa, J.; Kawabata, N.; Nishimura, J. *Tetrahedron* **1968**, *24*, 53; Charette, A.B.; Lebel, H. *J. Org. Chem.* **1995**, *60*, 2966.
97. Molander, G.A.; Etter, J.B. *J. Org. Chem.* **1987**, *52*, 3942.
98. Dauben, W.G.; Ashcraft, A.C. *J. Am. Chem. Soc.* **1963**, *85*, 3673.
99. Johnson, C.R.; Barbachyn, M.R. *J. Am. Chem. Soc.* **1982**, *104*, 4290.
100. Amice, P.; Conia, J.M. *C.R. Acad. Sci.* **1970**, *271C*, 948.
101. Barrett, A.G.M.; Kasdorf, K. *Chem. Commun.* **1996**, 325.
102. Wenkert, E.; Berges, D.E.; Golob, N.F. *J. Am. Chem. Soc.* **1978**, *100*, 1263.
103. Corey, E.J.; Hahl, R.W. *Tetrahedron Lett.* **1989**, *30*, 3023.
104. Ireland, R.E.; Baldwin, S.W.; Dawson, D.J.; Dawson, M.I.; Dolfini, J.E.; Newbould, J.; Johnson, W.S.; Brown, M.; Crawford, R.J.; Hudrlik, P.F.; Rasmussen, G.H.; Schmiegel, K.K. *J. Am. Chem. Soc.* **1970**, *92*, 5743.
105. Hara, R.; Furukawa, T.; Horiguchi, Y.; Kuwajima, I. *J. Am. Chem. Soc.* **1996**, *118*, 9186.
106. Turnbull, P.; Syhora, K.; Fried, J.H. *J. Am. Chem. Soc.* **1966**, *88*, 4764.
107. Huang, H.; Forsyth, C.J. *J. Org. Chem.* **1995**, *60*, 5746.
108. Oppolzer, W.; Radinov, R.N. *J. Am. Chem. Soc.* **1993**, *115*, 1593. For an approach involving cyclopropanation of an unsaturated chiral acetal, see Nelson, K.A.; Mash, E.A. *J. Org. Chem.* **1986**, *51*, 2721.

109. Wenkert, E.; Arrhenius, T.S. *J. Am. Chem. Soc.* **1983**, *105*, 2030.

110. Mash, E.A.; Math, S.K.; Flann, C.J. *Tetrahedron Lett.* **1988**, *29*, 2147.

111. Wilson, S.R.; Venkatesan, A.M.; Augelli-Szafran, C.E.; Yasmin, A. *Tetrahedron Lett.* **1991**, *32*, 2339.

112. Kabat, M.; Kiegiel, J.; Cohen, N.; Toth, K.; Wovkulich, P.M.; Uskokovic, M.R. *Tetrahedron Lett.* **1991**, *32*, 2343.

113. Yamazaki, M.; Shibasaki, M.; Ikegami, S. *J. Org. Chem.* **1983**, *48*, 4402.

114. Evans, D.A.; Barrow, J.C.; Leighton, J.L.; Robichaud, A.J.; Sefkow, M. *J. Am. Chem. Soc.* **1994**, *116*, 12111.

115. Frye, S.V.; Eliel, E.L. *J. Org. Chem.* **1985**, *50*, 3402.

116. Ohwa, M.; Eliel, E.L. *Chem. Lett.* **1987**, 41.

117. Utimoto, K.; Nakamura, A.; Matsubara, S. *J. Am. Chem. Soc.* **1990**, *112*, 8189.

118. Fujioka, H.; Yamamoto, H.; Annoura, H.; Miyazaki, M.; Kita, Y. *Chem. Pharm. Bull.* **1990**, *38*, 1872.

119. Ballini, R.; Marcantoni, E.; Petrini, M. *J. Org. Chem.* **1992**, *57*, 1316.

120. Maruyama, K.; Ishihara, Y.; Yamamoto, Y. *Tetrahedron Lett.* **1981**, *22*, 4235.

121. Masamune, S.; Choy, W.; Petersen, J.S.; Sita, L.R. *Angew. Chem. Int. Ed. Engl.* **1985**, *24*, 1.

122. Hoffmann, R.W.; Kemper, B.; Metternich, R.; Lehmeier, T. *Liebigs Ann. Chem.* **1985**, 2246.

123. Hoffmann, R.W.; Ditrich, K.; Köster, G.; Stürmer, R. *Chem. Ber.* **1989**, *122*, 1783.

124. Hoppe, D. *Angew. Chem. Int. Ed.* **1984**, *23*, 932; Hoppe, D.; Lichtenberg, F. *Angew. Chem. Int. Ed.* **1984**, *23*, 239.

125. Evans, D.A.; Nelson, J.V.; Taber, T.R. *Top. Stereochem.* **1982**, *13*, 1.

126. Heathcock, C.H. *Aldrichimica Acta* **1990**, *23*, 99.

127. Evans, D.A.; Ennis, M.D.; Mathre, D.J. *J. Am. Chem. Soc.* **1982**, *104*, 1737.

128. Corey, E.J.; Weigel. L.O.; Chamberlin, A.R.; Cho, H.; Hua, D.H. *J. Am. Chem. Soc.* **1980**, *102*, 6613.

129. Wanner, M.J.; Koomen, G.-J. *Tetrahedron* **1992**, *48*, 3935.

130. Samizu, K.; Ogasawara, K. *Chem. Lett.* **1995**, 543.

131. Maruoka, K.; Nonoshita, K.; Yamamoto, H. *Tetrahedron Lett.* **1987**, *28*, 5723.

132. Swiss, K.A.; Hinkley, W.; Maryanoff, C.A.; Liotta, D.C. *Synthesis* **1992**, 127.

133. Sonoda, S.; Houchigai, H.; Asaoka, M.; Takei, H. *Tetrahedron Lett.* **1992**, *33*, 3145.

134. Corey, E.J.; Magriotis, P.A. *J. Am. Chem. Soc.* **1987**, *109*, 287.

135. Ihara, M.; Suzuki, M.; Fukumoto, K.; Kabuto, C. *J. Am. Chem. Soc.* **1990**, *112*, 1164.

136. Fujiwara, J.; Fukutani, Y.; Hasegawa, M.; Maruoka, K.; Yamamoto, H. *J. Am. Chem. Soc.* **1984**, *106*, 5004.

137. Alexakis, A.; Mangeney, P.; Ghribi, A.; Marek, I.; Sedrani, R.; Guir, C.; Normant, J.F. *Pure Appl. Chem.* **1988**, *60*, 49.

138. Meyers, A.I.; Roth, G.P.; Hoyer, D.; Barner, B.A.; Laucher, D. *J. Am. Chem. Soc.* **1988**, *110*, 4611.

139. Posner, G.H.; Mallamo, J.P.; Hulce, M.; Frye, L.L. *J. Am. Chem. Soc.* **1982**, *104*, 4180.

140. Holton, R.A.; Kennedy, R.M.; Kim, H.B.; Krafft, M.E. *J. Am. Chem. Soc.* **1987**, *109*, 1597.

141. Iwata, C.; Fujita, M.; Moritani, Y.; Hattori, K.; Imanishi, T. *Tetrahedron Lett.* **1987**, *28*, 3135.

142. Kitamura, M.; Isobe, M.; Ichikawa, Y.; Goto, T. *J. Am. Chem. Soc.* **1984**, *106*, 3252.

143. Isobe, M.; Ichikawa, Y.; Funabashi, Y.; Mio, S.; Goto, T. *Tetrahedron* **1986**, *42*, 2863.

144. Tatsuta, K.; Yamauchi, T.; Kinoshita, M. *Bull. Chem. Soc. Jpn.* **1978**, *51*, 3035.

144. a. Hareau-Vittini, G.; Hikichi, S.; Sato, F. *Angew. Chem. Int. Ed.* **1998**, *37*, 2099. b. Corey, E.J.; Hannon, F.J.; Boaz, N.W. *Tetrahedron* **1989**, *45*, 545; Corey, E.J.; Hannon, F.J. *Tetrahedron Lett.* **1990**, *31*, 1393.

145. Evans, D.A.; Clark, J.S.; Metternich, R.; Novack, V.J.; Sheppard, G.S. *J. Am. Chem. Soc.* **1990**, *112*, 866.

146. Danishefsky, S.J.; Selnick, H.G.; Zelle, R.E.; DeNinno, M.P. *J. Am. Chem. Soc.* **1988**, *110*, 4368.

147. Lautens, M.; Aspiotis, R.; Colucci, J. *J. Am. Chem. Soc.* **1996**, *118*, 10930.

148. Nicolaou, K.C.; Yang, Z.; Liu, J.J.; Ueno, H.; Nantermet, P.G.; Guy, R.K.; Claiborne, C.F.; Renaud, J.; Couladouros, E.A.; Paulvannan, K.; Sorensen, E.J. *Nature* **1994**, *367*, 630.

149. Kan, T.; Hosokawa, S.; Nara, S.; Oikawa, M.; Ito, S.; Matsuda, F.; Shirahama, H. *J. Org. Chem.* **1994**, *59*, 5532.

150. Stadtmüller, H.; Knochel, P. *Synlett* **1995**, 463.

151. Holton, R.A. *J. Am. Chem. Soc.* **1977**, *99*, 8083.

152. Madin, A.; Overman, L.E. *Tetrahedron Lett.* **1992**, *33*, 4859.

153. Bottaro, J.C.; Berchtold, G.A. *J. Org. Chem.* **1980**, *45*, 1176.

154. Corey, E.J.; Trybulski, E.J.; Melvin, L.S.; Nicolaou, K.C.; Secrist, J.A.; Lett, R.; Sheldrake, P.W.; Falck, J.R.; Brunelle, D.J.; Haslanger, M.F.; Kim, S.; Yoo, S.-e. *J. Am. Chem. Soc.* **1978**, *100*, 4618; Corey, E.J.; Kim, S.; Yoo, S.-e.; Nicolaou, K.C.; Melvin, L.S.; Brunelle, D.J.; Falck, J.R.; Trybulski, E.J.; Lett, R.; Sheldrake, P.W. *J. Am. Chem. Soc.* **1978**, *100*, 4620.

155. Li, T.-T.; Wu, Y.-L. *Tetrahedron Lett.* **1988**, *29*, 4039.

156. Baraldi, P.G.; Barco, A.; Benetti, S.; Pollini, G.P.; Simoni, D. *Tetrahedron Lett.* **1983**, *24*, 5669.

157. Crimmins, M.T.; O'Mahony, R. *J. Org. Chem.* **1989**, *54*, 1157.

158. McMurry, J.E.; Blaszczak, L.C. *J. Org. Chem.* **1974**, *39*, 2217.

159. Short, R.P.; Revol, J.M.; Ranu, B.C.; Hudlicky, T. *J. Org. Chem.* **1983**, *48*, 4453.

160. McMurry, J.E.; Choy, W. *Tetrahedron Lett.* **1980**, *21*, 2477.

161. Crimmins, M.T.; Mascarella, S.W. *J. Am. Chem. Soc.* **1986**, *108*, 3435.

162. Pearlman, B.A. *J. Am. Chem. Soc.* **1979**, *101*, 6404.

163. Dauben, W.G.; Shapiro, G. *Tetrahedron Lett.* **1985**, *26*, 989; Dauben, W.G.; Rocco, V.P.; Shapiro, G. *J. Org. Chem.* **1985**, *50*, 3155.

164. Oppolzer, W.; Mahalanabis, K.K.; Bättig, K. *Helv. Chim. Acta* **1977**, *60*, 2388.

165. Oppolzer, W.; Bättig, K. *Helv. Chim. Acta* **1981**, *64*, 2489.

166. Oppolzer, W.; Bättig, K.; Hudlicky, T. *Tetrahedron* **1981**, *37*, 4359.

167. Oppolzer, W.; Bättig, K. *Tetrahedron Lett.* **1982**, *23*, 4669.
168. Oppolzer, W.; Strauss, H.F.; Simmons, D.P. *Tetrahedron Lett.* **1982**, *23*, 4673.
169. Oppolzer, W.; Cunningham, A.F. *Tetrahedron Lett.* **1986**, *27*, 5467.
170. Oppolzer, W.; Pitteloud, R. *J. Am. Chem. Soc.* **1982**, *104*, 6478.
171. Fujita, M.; Takarada, Y.; Sugimura, T.; Tai, A. *Chem. Commun.* **1997**, 1631.

6

CONFORMATIONAL EFFECTS

Sachse and Mohr's recognition that cyclic compounds such as cyclohexane have nonplanar structures, and the revival of this concept with further evaluation (conformational analysis) of its consequences by Barton in the 1950s turned a new leaf in organic chemistry. Nowadays organic chemists can hardly describe their work without some reference to stereochemistry and conformational issues. Even the very specialized arena of aromatic/heteroaromatic chemistry cannot escape from this discussion as fullerenes and novel compounds such as corannulene are brought into light.

Conformational selection of transition states is regarded as critically important for many reactions. Accordingly, chemists can intervene by structural modification of reactants to eliminate altogether adverse situations or just to accentuate the stereochemical bias in the decided direction.

6.1. GROUP ANCHORING

Installation of an even moderately bulky group to a strategic position can confer stability to a certain conformation, and direct it along the reaction pathway. The group may lie inside or outside the circuit defined by the reactive sites. When the circuit involves five- or six-carbon or heteroatoms, the largest group prefers to occupy a pseudoequatorial position.

Oxabicyclo[3.2.1]octenones are available from intramolecular cycloaddition of pentadienyl cations derived from a γ-pyrone with a tethered alkene. When applied to the synthesis of phorbol (**6-1**),[1] it led to a useful intermediate containing four stereocenters. The well-placed allylic methyl group, insisting on being equatorial, provided the necessary stereocontrol in claiming the proper transition state. A very

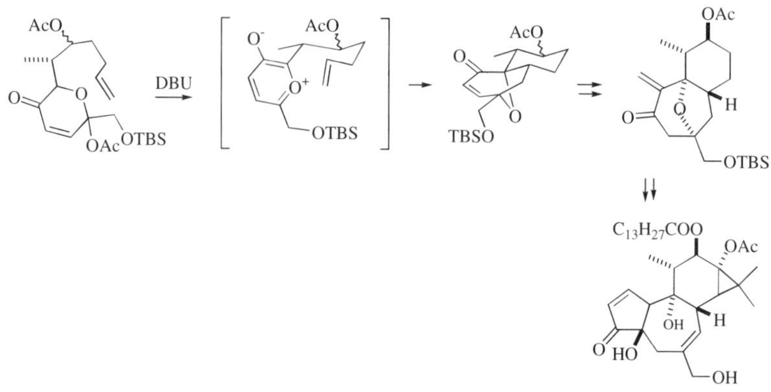

phorbol (**6-1**)

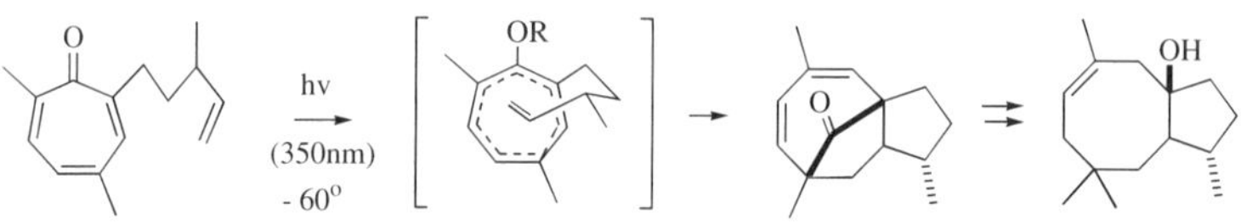

dactylol (**6-2**)

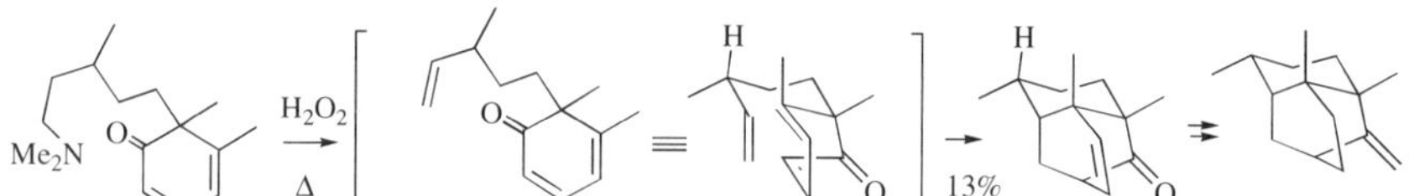

seychellene
(**6-3**)

similarly emplaced methyl group in a photocycloaddend enabled the generation of a dactylol (**6-2**) precursor.[2]

The intramolecular Diels–Alder approach to seychellene (**6-3**)[3] and patchouli alcohol (**6-4**)[4] selected transition states which disposed the secondary methyl equatorially. Although the yields were low, the reactions did not cause problems such as giving rise to diastereomers which would have required effort in separation. Note that this preference was reinforced by the presence of the vinylic methyl group.

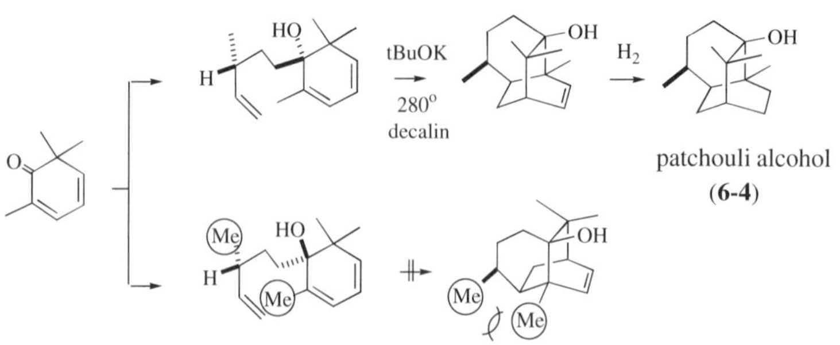

patchouli alcohol
(**6-4**)

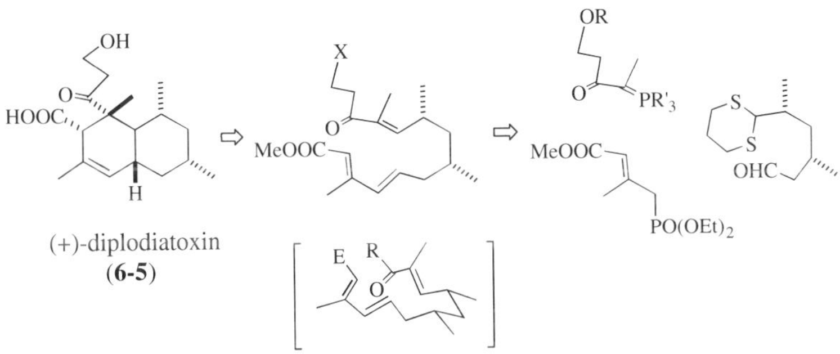

(+)-diplodiatoxin
(**6-5**)

favored TS

Generation of the observed products represented an avoidance action of incipient 1,3-diaxial interaction.

Four transition states are available for the acyclic triene to undergo an intramolecular Diels–Alder reaction leading to an advanced intermediate of (+)-diplodiatoxin (**6-5**).[5] Evidently the two transition states featuring 1,3-related pseudoaxial methyl groups were strongly disfavored. The desired product was derived from an *exo* stacking of the diene and dienophile which did not suffer excessive steric interactions. However, at higher temperatures other diastereomers began to appear.

Acromelic acid-A (**6-6**) has been prepared[6] by way of an intramolecular 1,3-dipolar cycloaddition in which the 1,3-dipole was generated by thermolysis. The stereocontrol element was a homoallylic substituent. Shortening of the chain between the two reactive components affects their folding mode and consequently the absolute configuration of new stereocenters as induced by the stereocontrol element. Thus, platynecine (**6-7**) can be synthesized accordingly.[7]

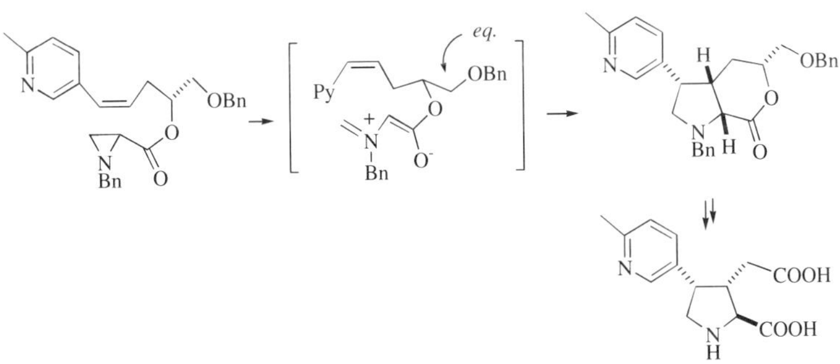

acromelic acid-A (**6-6**)

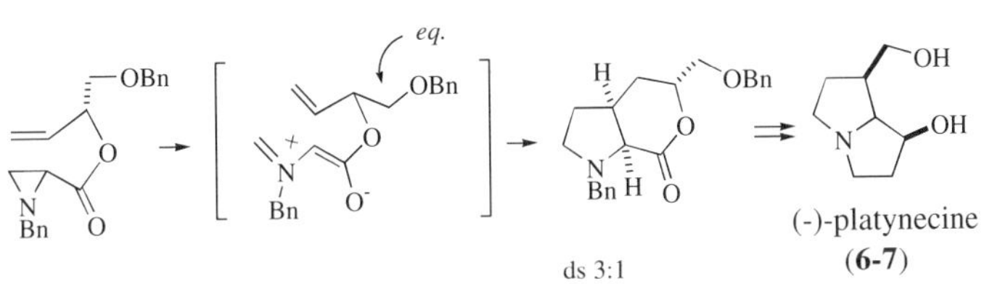

ds 3:1

(-)-platynecine
(**6-7**)

Another [3+2]-cycloaddition was employed in a synthesis of (+)-cassiol (**6-8**)[8] and the quaternary center was induced by the (*S,S*)-configuration of the substrate and the equatorial occupancy of the two substituents in a chairlike transition state. These two stereocenters disappeared subsequently.

The key steps for the elaboration of perhydrohistrionicotoxin (**6-9**)[9] were intramolecular [2+2]-photocycloaddition and subsequent fragmentation to generate the spirocyclic system. In the former reaction the substituent at the α-position of the

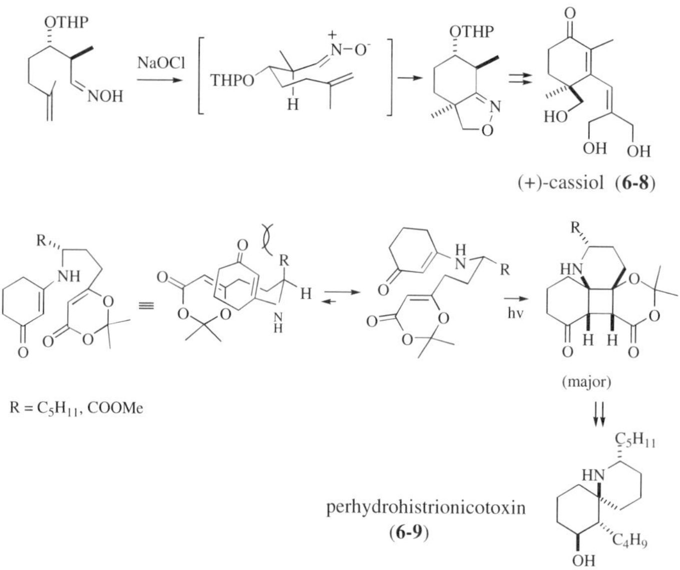

(+)-cassiol (**6-8**)

perhydrohistrionicotoxin (**6-9**)

amino group became the determining factor of the folding which led to one or the other diastereomeric products. This group's preferred occupancy was in the pseudoequatorial orientation.

Stereoinduction at two centers by a secondary methyl group during a magnesium-ene cyclization leading to the iridane skeleton has been observed.[10] Thus, it constitutes a powerful method for stereoselective construction of compounds including (+)-α-iridodiol, (+)-α-skytanthine (**6-10**), and (+)-δ-skytanthine (**6-11**).

The overwhelming effect of two large *cis*-1,3 substituents was exploited in the total synthesis of bilobalide (**6-12**)[11] and 8-epiprostaglandin-$F_{2\alpha}$ (**6-13**).[12] It should be noted that if the substrate for cyclization to give the epiprostaglandin precursor was the unprotected diol, the reaction pursued another pathway as favored by internal hydrogen bondings.

Copper-catalyzed decomposition of a diazoketone which was tethered to the nitrogen atom of a proline ester led to quinolizidine derivatives.[13] As the major product was converted to (–)-epilupinine (**6-14**), the rearrangement process can be interpreted as a stereoselective 1,2-migration with retention of configuration. Formation of spirocyclic intermediates was likely and the predisposition of the ester group and the side chain bearing the ketocarbenoid in a *trans*-diequatorial relationship

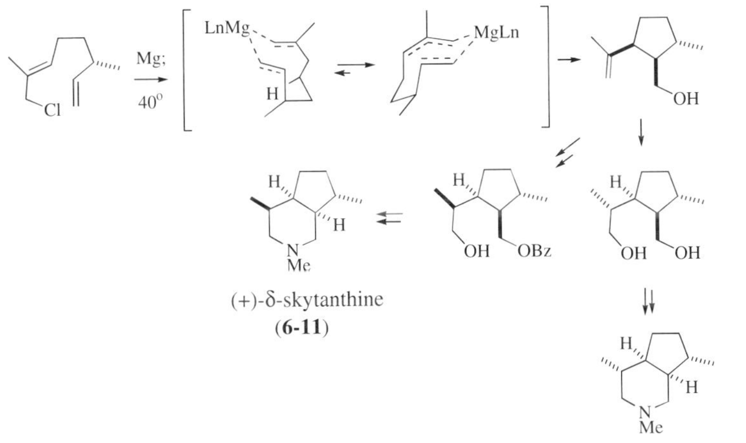

(+)-δ-skytanthine (**6-11**)

(+)-α-skytanthine (**6-10**)

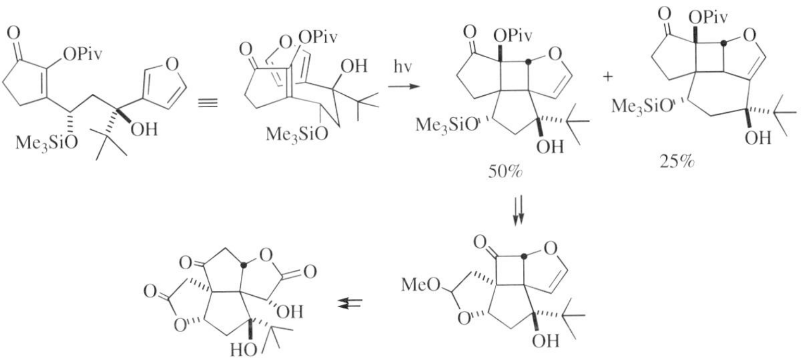

bilobalide (**6-12**)

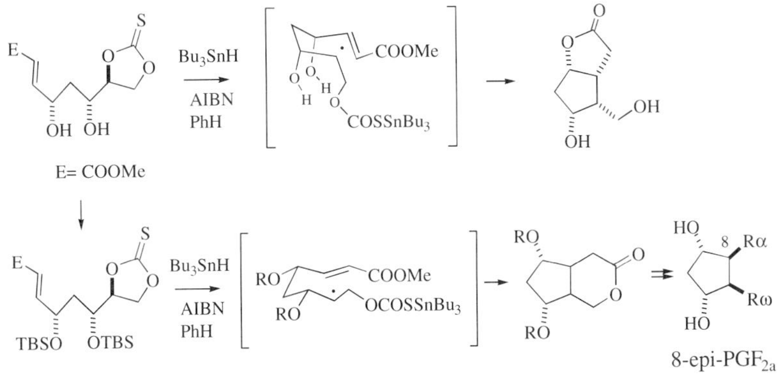

8-epi-PGF_{2a} (**6-13**)

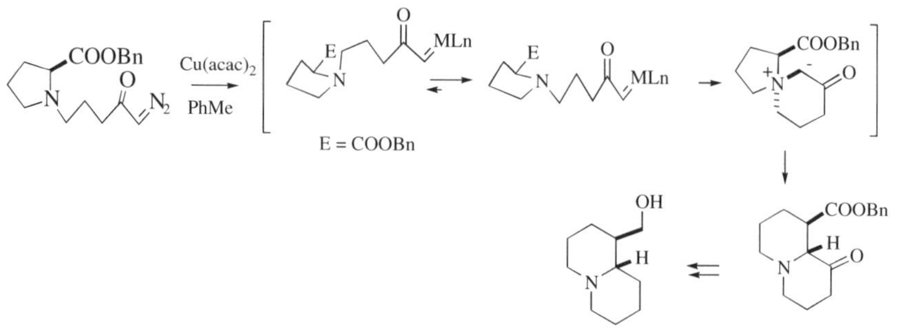

(-)-epilupinine

(**6-14**)

must be energetically more favorable. Pyrimidalization of the nitrogen atom is implied.

The stereoselective dihydroxylation of an alkene in a stepwise fashion by intramolecular participation of an alkoxydisilane unit is a useful synthetic method.[14] Palladium insertion into the Si–Si bond and coordination with the double bond determined the stereochemistry of the product. As demonstrated in a synthetic route to avenaciolide (**6-15**), the stereoselectivity (9:1) reflects the preference for more pseudoequatorial substituents in the chairlike transition states.

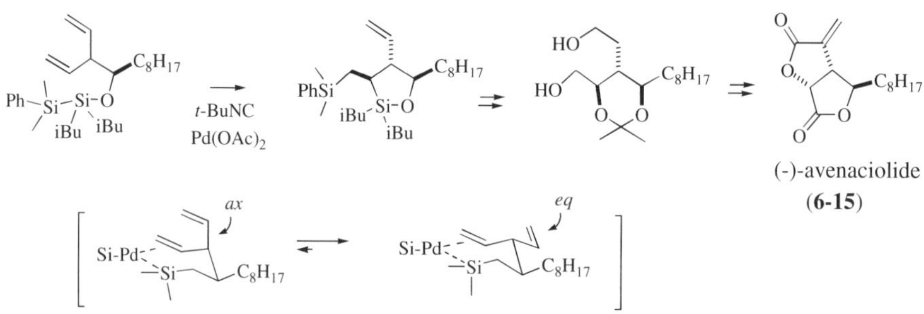

(-)-avenaciolide

(**6-15**)

A spiroacetal diol (**6-16**) is an apparent intermediate for synthesis of milbemycin-β_1 based on chain elaboration and Mitsunobu macrolactonization. As a result of conformational stability the dimethyltetrahydropyran provides a stereochemical reference for the establishment of the secondary alcohol.[15] The objective can be achieved by exploiting the characteristic alkylation of

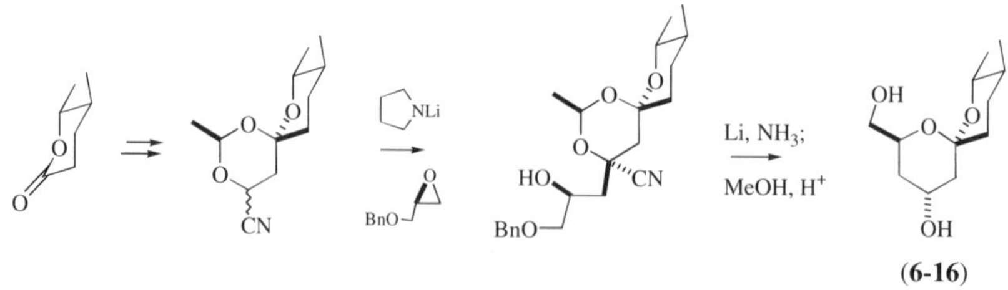

(**6-16**)

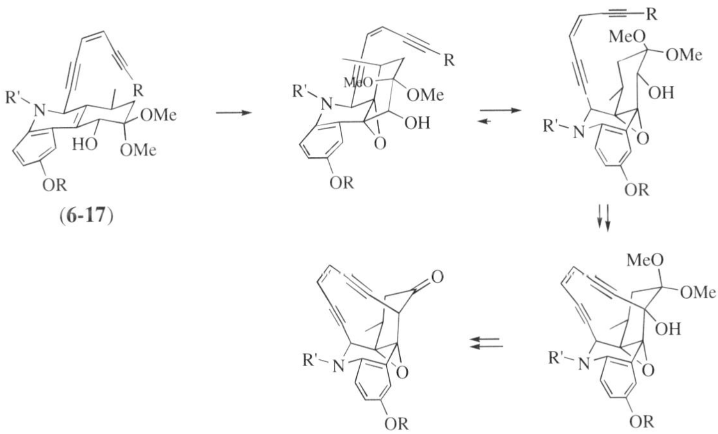

(6-17)

4-cyano-1,3-dioxanes which places the alkyl group in an equatorial position and reductive cyanation with retention of configuration, if the alkylating agent is equipped with a latent diol of proper configuration. On cleavage of the dioxane the spiroacetal will assemble by itself.

All of the synthetic hurdles of dynemicin-A are concentrated on the enediyne unit and the astride rings. With a strategy focusing on bridging the enediyne by an intramolecular attack on the heteronuclear ketone,[16] this nucleophile must maintain an axial conformation. It is easier to achieve the required form when the epoxide is already in place because conformational differences are more conspicuous than the corresponding cycloalkene. In fact, multiple equatorial substituents in a precursor would become axial and cause conformational change in each component ring. Thus, it is possible to render the enediyne chain of such a precursor (**6-17**) into an attacking position, for example, by exploiting the mutual repulsion of a secondary methyl and an axial methoxy group in a 1,3-diaxial relationship.

Desymmetrization of 2-substituted 1,3-propanediols is synthetically significant. In cyclic acetals derived from such diols with (–)- or (+)-menthone, the larger group would occupy the equatorial position. When coupled with stereocontrol by the isopropyl group which determines the conformation of the spiroannulated dioxane

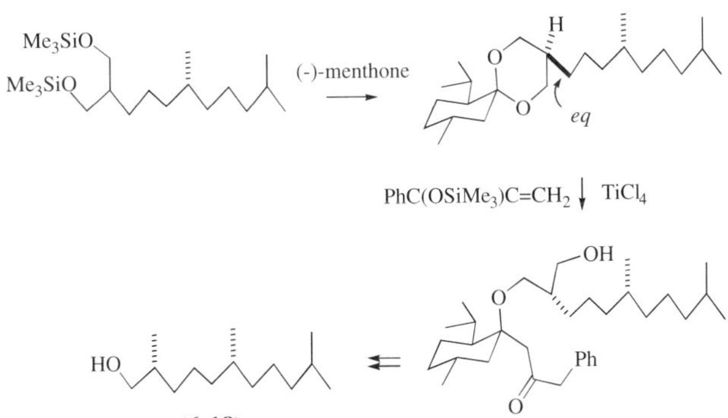

(6-18)

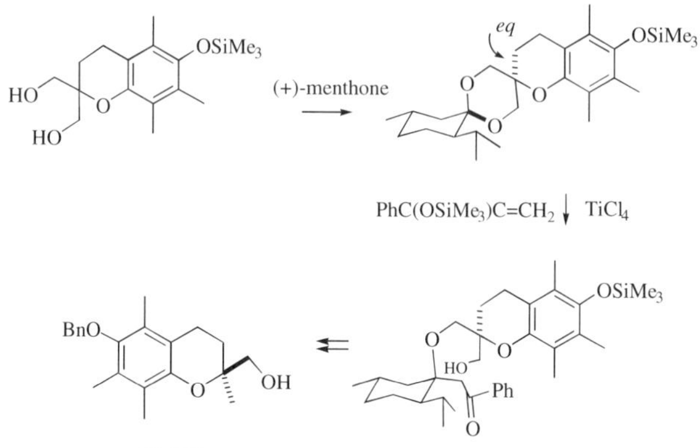

(**6-19**)

ring (flip from one chair form to another), the selective cleavage of the acetals (equatorial C–O bond)[17] has enormous value. Most remarkably, both the chiral side chain (**6-18**) and the benzopyran unit (**6-19**) of α-tocopherol can be obtained by the maneuver.

Completion of the tetracyclic array of ishwarane (**6-20**)[18] by a tandem process of carbene addition and insertion depended on the conformation of the intermediate. Because the steroidal conformation was disfavored by methyl:methyl 1,3-diaxial interaction, the result was not complicated by reactions at positions other than the designated angular methyl group. Of course there was a prerequisite that the dibromocarbene added from the convex face of the *cis*-octalin precursor.

Absence of stereoselectivity was evident in the hydrogenation of dehydro intermediates of valeranone, even if such compounds are *cis*-octalins. Conformational mobility is probably the culprit and other structural elements must be introduced to bias one conformation. In this regard the ethyleneacetal gave improved results (selectivity 5:1), and a reasonable synthesis of valeranone (**6-21**) could be developed.[19]

Introduction of an acetic acid chain by epoxide opening from the concave side of a *cis*-oxadecalin was instrumental to the success of a vernolepin (**6-22**) synthesis.[20]

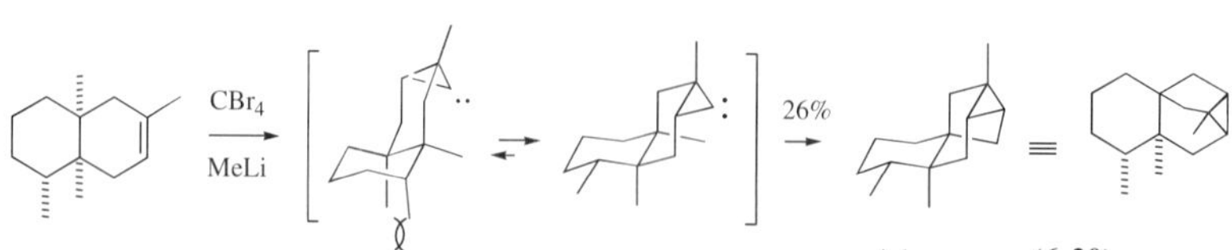

ishwarane (**6-20**)

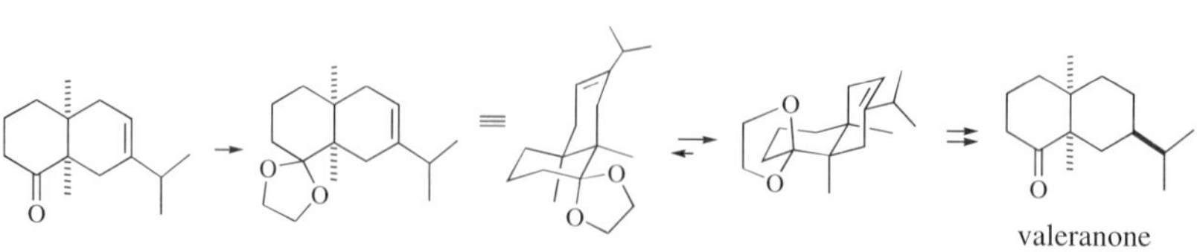

valeranone
(**6-21**)

This task turned out to be rather straightforward when the A-ring lactone was protected as an orthoester. The ethylenedioxy group demanded a conformation which made the attack on $C_{(7)}$ from the axial angle. This conformation was also favored by the possibility of hydrogen bond formation of the epoxide with the neighboring hydroxyl group.

Linderalactone (**6-23**) undergoes thermal equilibration with isolinderalactone (**6-24**) at 160°C by a reversible Cope rearrangement and thereby a synthesis of the latter provided an access route to linderalactone.[21] Interestingly, the diol (**6-25**) gave the elemane isomer in which the new stereocenters are opposite to those of isolinderalactone.[22] The effect of conformational locking by a lactone moiety in the former case must be a crititcal issue.

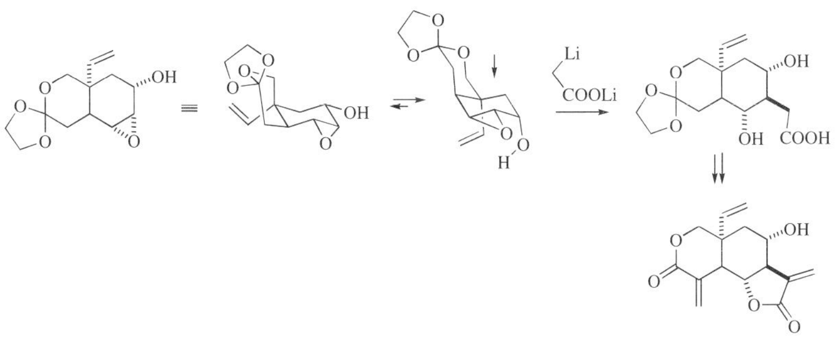

vernolepin (**6-22**)

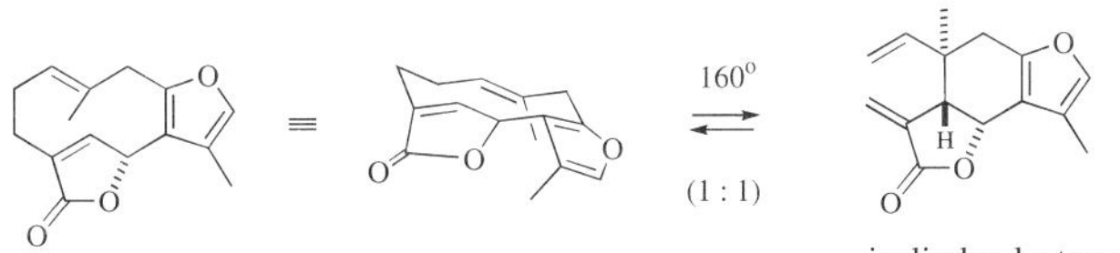

linderalactone
(**6-23**)

isolinderalactone
(**6-24**)

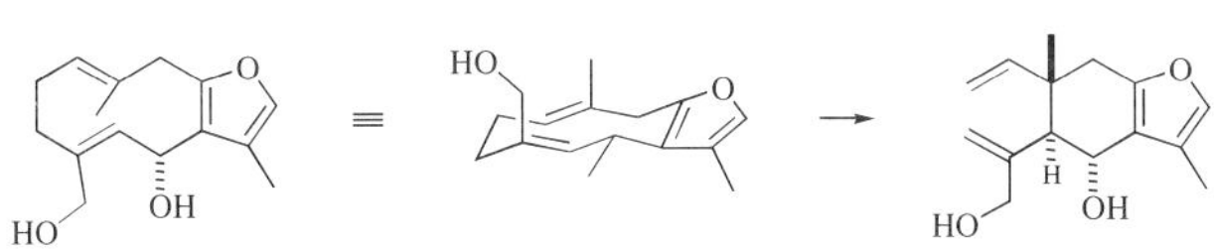

(**6-25**)

Chelation is a very effective determinant of conformation. For example, it has been shown that, starting from ethyl *O*-benzyl (*S*)-lactate,[23] two remote stereocenters in (+)-brefeldin-A (**6-26**) can be affixed by the Ireland–Claisen rearrangement with the aid of chelated enolates and then the Johnson–Claisen rearrangement.

Currently gaining popularity is the use of silicon-containing tethers for intramolecular group delivery to achieve regio- and stereoselective bond formation. Furthermore, the 5-*exo* mode of cyclization can be shifted to the 6-*endo* mode due to the longer C–Si bond which renders the molecular system easily accommodating to the latter transition state. In the assembly of a 21-hydroxylated steroid side chain (**6-27**)[24] by a free radical reaction, the preference for pseudoequatorial orientation of a methyl and an isobutyl group determined the final product.

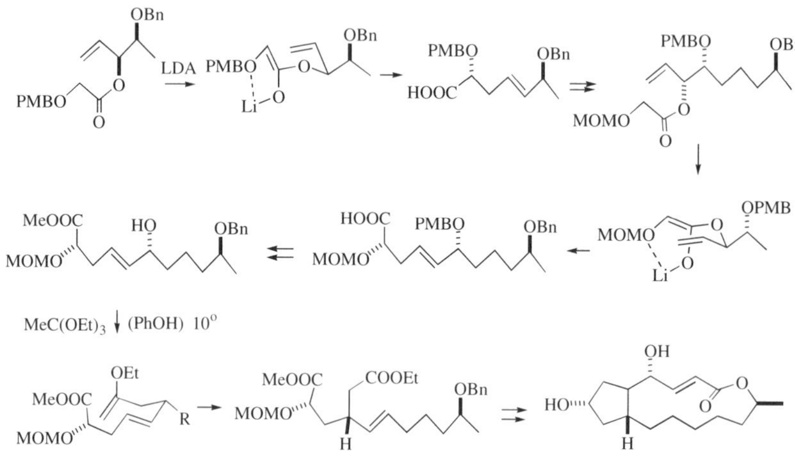

(+)-brefeldin-A (**6-26**)

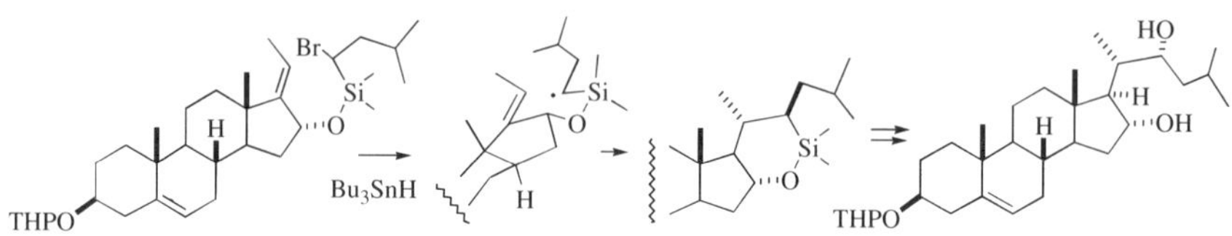

(**6-27**)

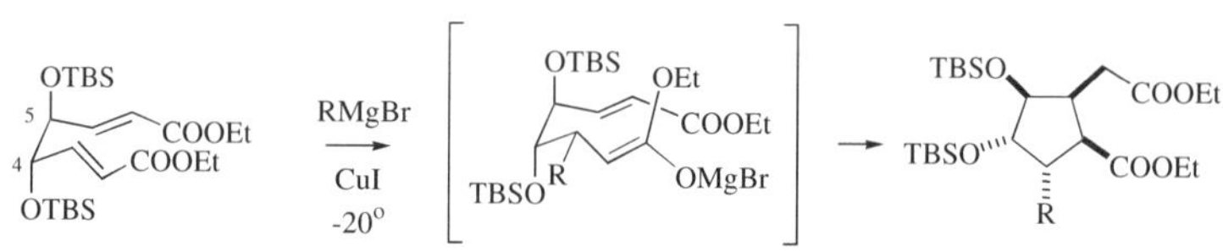

(**6-28**)

It is unusual to find a conformational preference such as in (*E,E*)(*4SR,5SR*)-4,5-bis(*t*-butyldimethylsiloxy)-2,6-octadienoic esters (**6-28**) that place the two bulky substituents in axial positions.[25] However, the largest separation of the groups is understandable, and rotation of the $C_{(4)}$–$C_{(5)}$ bond must be energetically much more costly than most commonly encountered cases. Each double bond is exposed on one face to reagents and the two faces are homotopic.

The conformational fluidity of compounds with a *cis*-5/6 ring fusion can cause problems in achieving stereoselectivity. However, it is sometimes sufficient to manipulate with the aid of even a methyl group, as shown in the preparation of a lactone[26] en route to a building block for tetronomycin (**6-29**).

Unfavorable conformational bias in mobile systems is not always uncompromising. If the reaction can only proceed from the minor conformer,

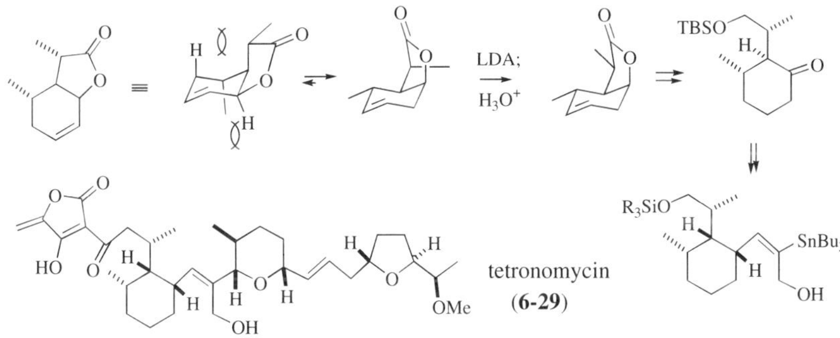

tetronomycin (6-29)

probably only the rate would be affected. The nickel(0)-mediated cyclization involving an alkenyl iodide and an enone to give an intermediate of akuammicine (**6-30**)[27] formed an axial C–C bond. Thus, the *cis*-hexahydroindol-4-one with an equatorial aryl substituent must undergo conformational change before the ring closure. Coordination of the alkenylnickel species with the double bond maintained the active transition state. (Compare Heck reaction on a synthesis of dehydrotubifoline.[28])

(*Z*)-Alkylidene derivatives of an oxazepinedione prepared from ephedrine are obtained by condensation with aldehydes. The unusual reactivity of these compounds in that they tend to be attacked by reagents from the same side as the phenyl and methyl groups may reflect the rather flat surfaces and perhaps influences by pyramidalization at the nitrogen atom. In a use for synthesis of (–)-rocaglamide (**6-31**)[29] based on a [3+2]-cycloaddition, inversion of the phenyl group originating from the benzylidene derivative was required.

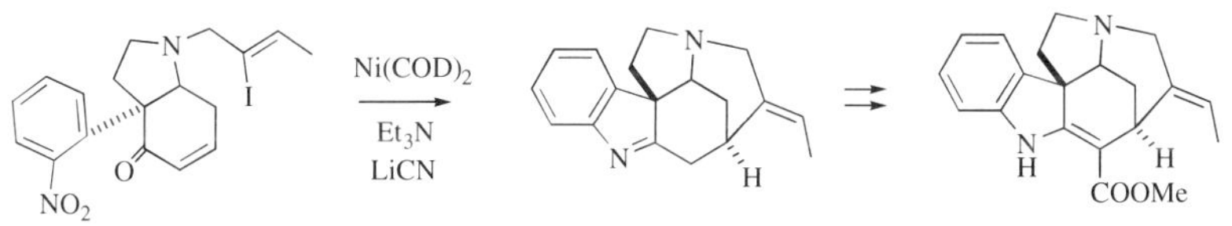

akuammicine (**6-30**)

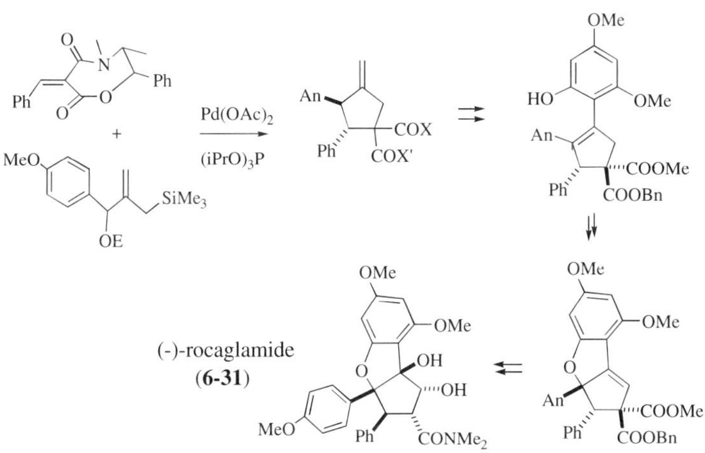

(-)-rocaglamide (**6-31**)

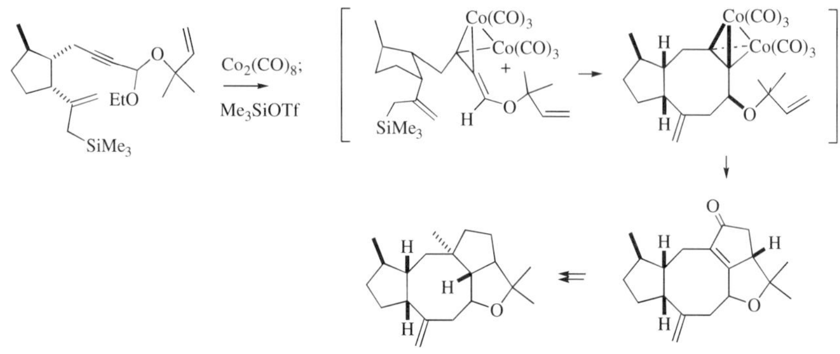

(+)-epoxydictymene
(**6-32**)

A Nicholas reaction for closing an eight-membered ring of epoxydictymene (**6-32**) from a mixed acetal was found to be chemo- and stereoselective.[30] Ionization of the small ethoxy group upon silylation was favored sterically. The complete stereoselectivity reflected a least encumbered transition state with the bulky ether group pointing away from potential interference.

Hydrogen bonding and metal chelation prescribe conformation even more effectively. Thus, an intermediate of lupanine (**6-33**) containing five stereocenters was obtained by hydrogenation of a pyridylquinolizinone,[31] although only four such centers are present in lupanine.

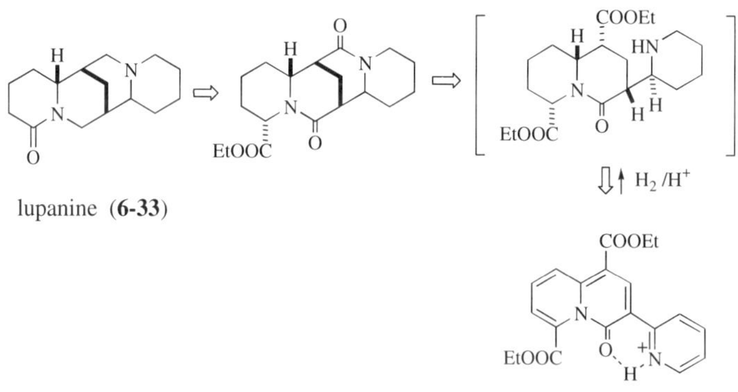

lupanine (**6-33**)

Maintenance of the proper conformation by intramolecular hydrogen bonding ensured the participation of the alkene instead of the alkyne chain in a Cope rearrangement which eventually led to gnididione (**6-34**).[32]

The Diels–Alder reaction generally shows predictable stereoselectivity when the dienophile is internally hydrogen-bonded[33] or there is association between the reacting components by hydrogen bonding.[34] Control of the ring junction stereochemistry by metal chelation is shown in a synthesis of β-eudesmol (**6-35**).[35]

Perhaps the most effective technique to consolidate a conformation is by tethering the reactive groups with a short chain. Thus, a synthesis of quadrone (**6-36**)[36]

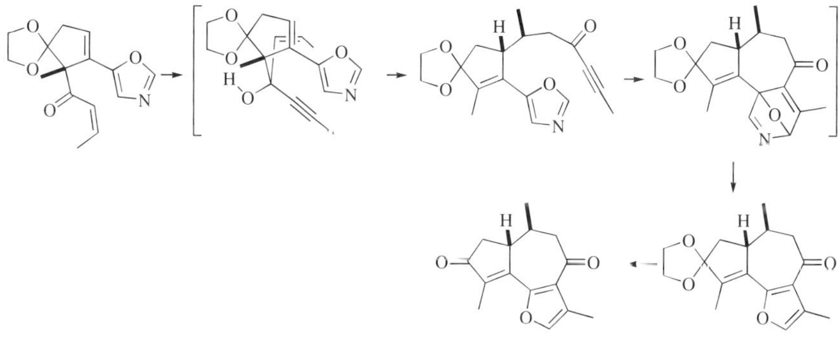

gnididione (**6-34**)

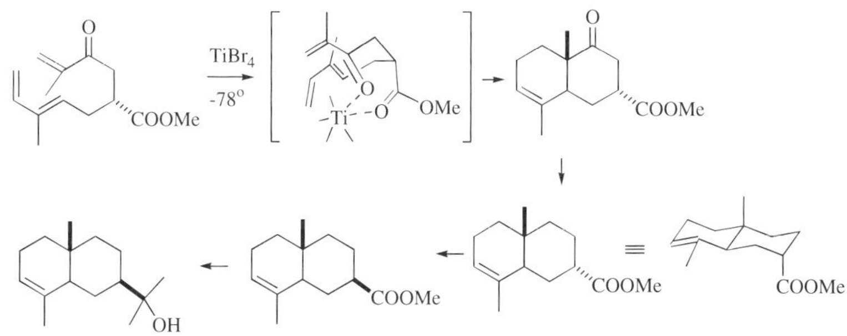

β-eudesmol (**6-35**)

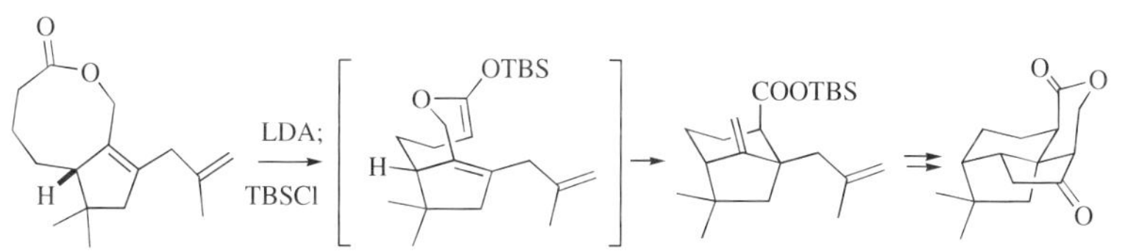

quadrone (**6-36**)

was based on this premise to establish not only the bridged ring system but also an axial carboxylic acid pendant for lactone formation.

6.2. NONBONDED INTERACTIONS

Steric repulsion that inhibits the attainment of low-energy transition states usually continues to plague a product unless it is relieved by some other reactions or conformational changes. Generally these situations are avoided in transition state selection.

Allylic 1,2-strain ($A^{1,2}$-strain) is now widely recognized. Thus, the enamine formation from *cis*-2,4-dimethylcyclohexanone was accompanied by epimerization of the 2-methyl group to an axial orientation to minimize its interference with the N-CH_2

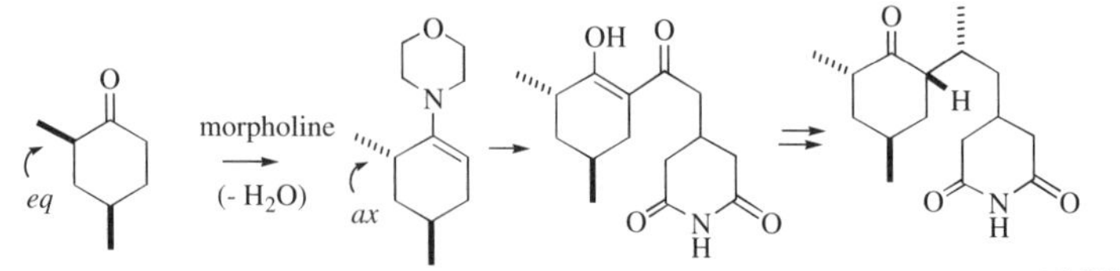

cycloheximide (**6-37**)

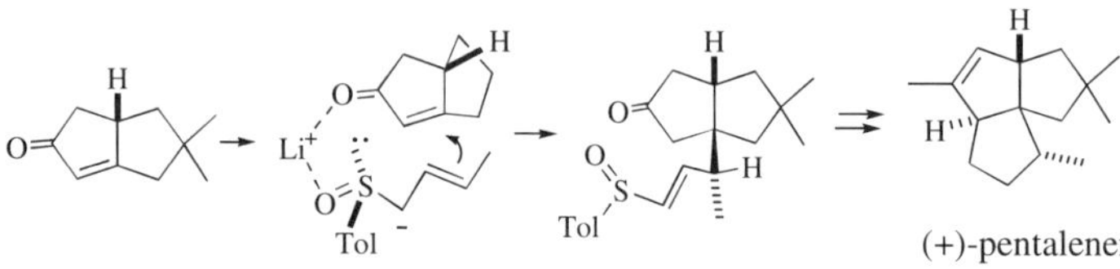

(+)-pentalenene
(**6-38**)

segment.[37] This phenomenon can be exploited, for example, in a synthesis of cycloheximide (**6-37**). In other words, it is not necessary to start with the pure *trans*-2,4-dimethylcyclohexanone, even a diastereomeric mixture can be used.

Angular annexation of the cyclopentane containing a secondary methyl group to a functionalized diquinane to complete the synthesis of pentalenene (**6-38**)[38] or pentalenic acid[39] started with conjugate addition with an anion of crotyl sulfoxide and crotylphosphonate, respectively. Thermodynamic reasons ensured approach of the reagent from the same side as the angular hydrogen atom, but the vinylic methyl must be on the outside to avoid unwanted compression. Thus, two stereocenters were established during the Michael reaction.

Spiroannulation of cyclic acyliminium ions that forms a cyclohexane ring is stereoselective in that the protonated nitrogen atom prefers an equatorial position.[40] The reason is that 1,3-diaxial interactions involving the methylene group are less severe. This observation had implications in applying the method to the synthesis of perhydrohistrionicotoxin (**6-39**). A side chain with (*E*)-double bond fulfilled the requirement.

The higher steric demand of an oxygen atom than a hydrogen seemed to direct the solvolytic cyclization of an acyliminium ion[41] to give a tricyclic intermediate of

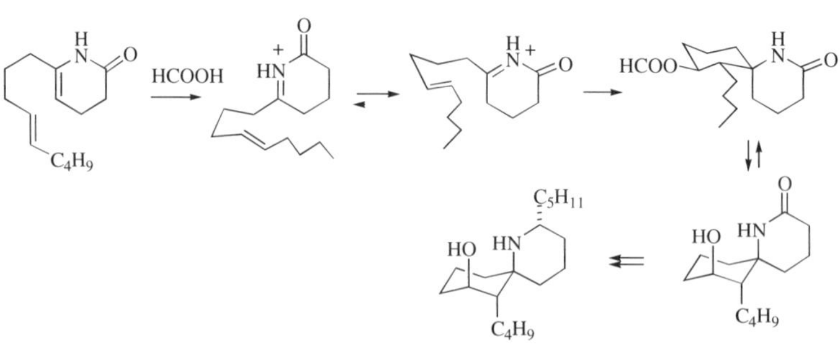

perhydrohistrionicotoxin
(**6-39**)

gephyrotoxin (**6-40**). The new stereocenter featured an angular hydrogen *trans* to those between the six-membered rings.

Intramolecular conjugate addition of a conformationally rigid piperidine to the unsaturated sulfone side chain can lead to the quinolizidine system characteristic of clavepictine-A (**6-41**) and clavepictine-B (**6-42**).[42] Long-range nonbonding interactions seem to be the cause of the specific alignment of the acceptor unit during the cyclization.

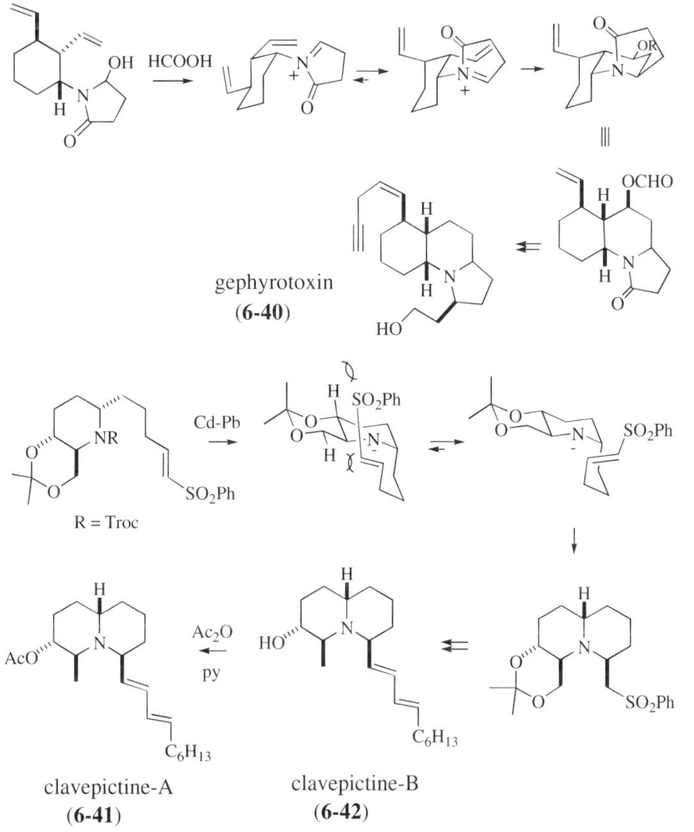

gephyrotoxin (**6-40**)

clavepictine-A (**6-41**)

clavepictine-B (**6-42**)

Folding of the conjugated ester encountered less opposition to reach one conformation and favored its participation in ring closure to give a proper intermediate of sanadaol (**6-43**).[43] It is not known to what degree the dipole–dipole interaction contributed to this selectivity.

Palladium-mediated reductive cyclization of enynes is an excellent method for carbocycle construction. As shown by a synthesis of β-necrodol (**6-44**),[44] nonbonded interactions of the ligated palladium atom with a remote substituent (in this case an ester four bonds away) determined the structure of the major product.

A model study[45] has pointed the way to a potential solution concerning construction of the trisubstituted cyclopropane unit of ambruticin (**6-45**). Solvolysis

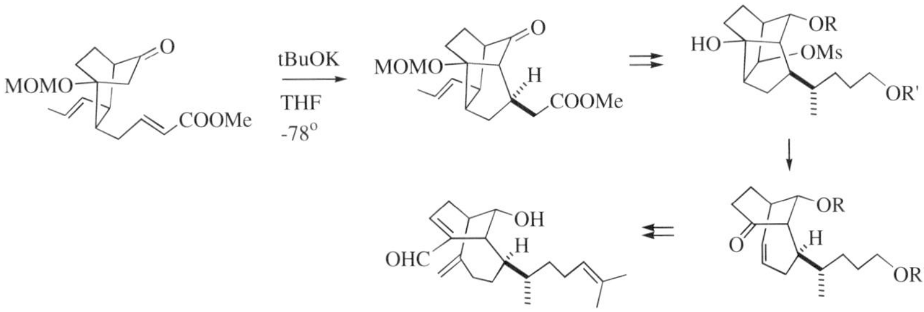

(+)-sanadaol (**6-43**)

of a homoallylic triflate followed a pathway traversing the least encumbered transition state.

Condensation of a (γ-alkoxyallyl)boronate with a chiral aldehyde assembled a functionalized carbon chain which was to be incorporated into (+)-olivin (**6-46**).[46] The reaction seemed to involve a Felkin–Ahn transition state, although a Cornforth transition state is more likely as a serious nonbonded interaction exists in the former.

A more recent synthetic approach to aphidicolin (**6-47**)[47] was based on an intramolecular Diels–Alder reaction to form the AB-ring segment. One stereochemical issue is that an *exo* cycloadduct is much preferred. This situation was realized (selectivity 3:1) because the bridged ring system interfered with the diene in the *endo* transition state.

By incorporating a trimethylsilyl group to a diene unit it has been possible to direct the intramolecular Diels–Alder reaction to form a *trans*-octalin which totally conformed to the requirement of parlaying into (–)-chlorothricolide (**6-48**).[48] This extra group provided nonbonded interactions of $A^{1,3}$ and more distant types to suppress the other transition states.

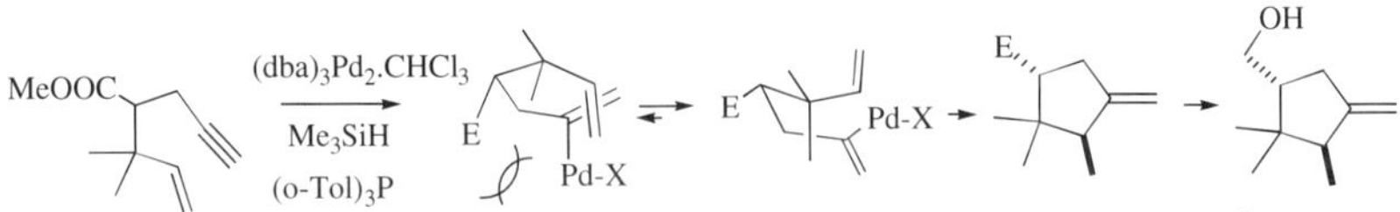

β-necrodol
(**6-44**)

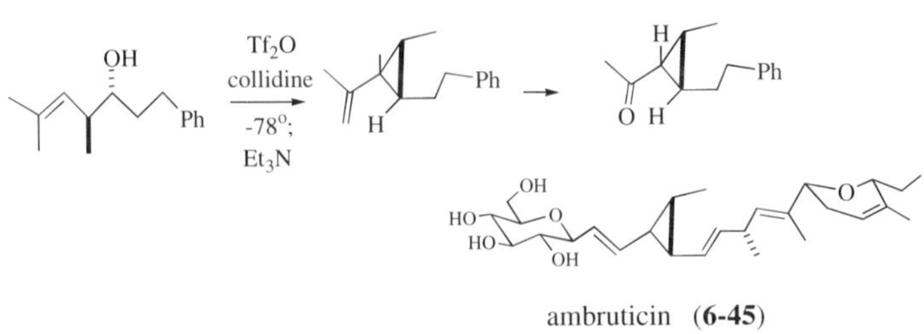

ambruticin (**6-45**)

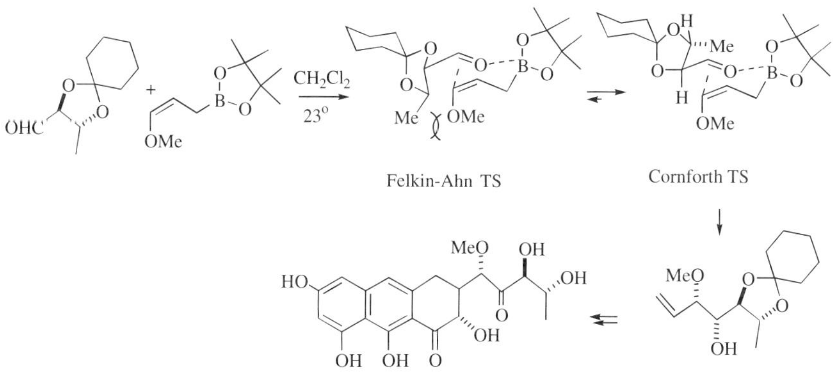

(+)-olivin (**6-46**)

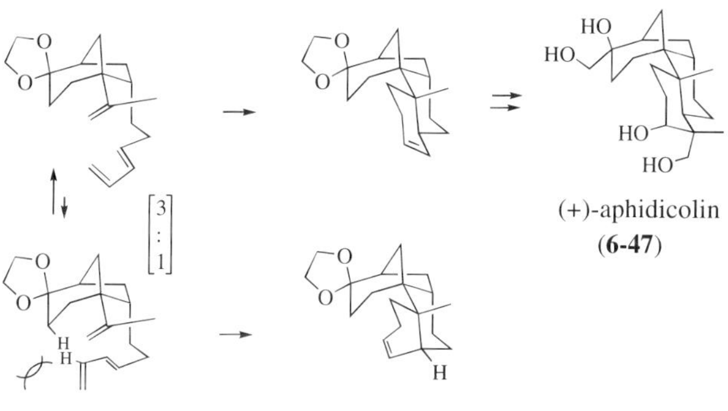

(+)-aphidicolin
(**6-47**)

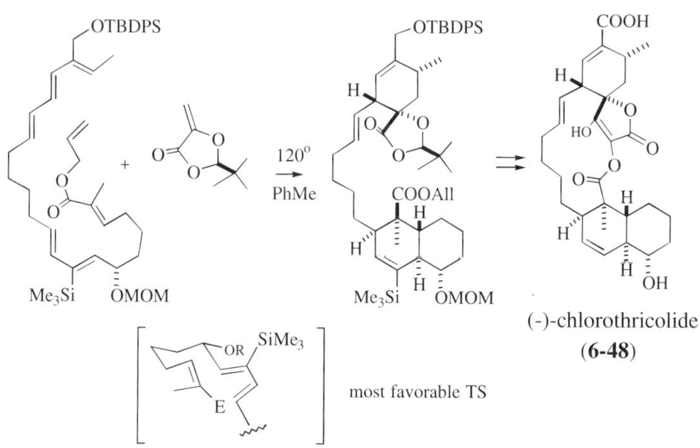

(-)-chlorothricolide
(**6-48**)

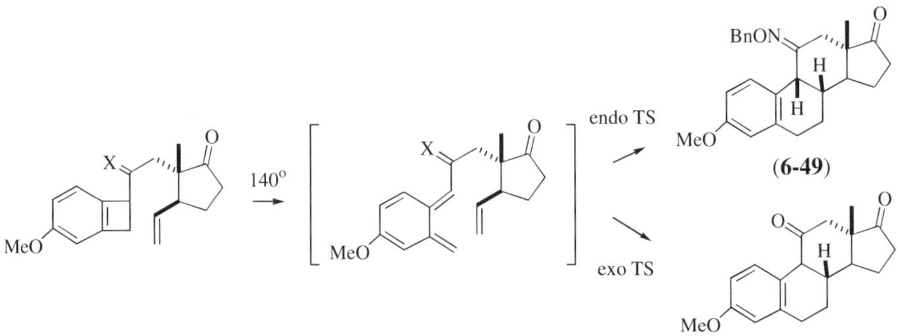

(**6-49**)

(**6-50**)

For synthesis of estrone derivatives from nascent *o*-quinodimethanes, a change in the preference for the *endo* and *exo* transition states, respectively for an oxime ether and the ketone, led to *cis*- or *trans*-B/C ring fusion products (**6-49, 6-50**).[49]

The three cases mentioned above are among thousands of applications of the intramolecular Diels–Alder reaction. Those involving diene and dienophile moieties separated by a short tethering chain (three or four atoms) give rise to fused ring products. If activating groups are present in the dienophile, diastereoselectivity differs when the activating group is located terminally or internally. Most such reactions are concerted but asynchronous.

The tendency to form *cis*-fused products from substrates containing an internal ketone is favorable for application to synthesis of compounds such as α-himachalene (**6-51**), β-himachalene (**6-52**),[50] and 11-oxosteroids (**6-53**),[51] although in the latter instance there was subsequently a stereochemical amendment at $C_{(9)}$ by epimerization.

Nonatrienes with connecting amide functionality generally show a preference for delivery of the *cis*-fused products,[52] unless a terminal activating group is present in the dienophilic moiety. The low reactivity of the corresponding esters may be due to poor overlap of the nonbonding electrons of the ethereal oxygen and the carbonyl group in

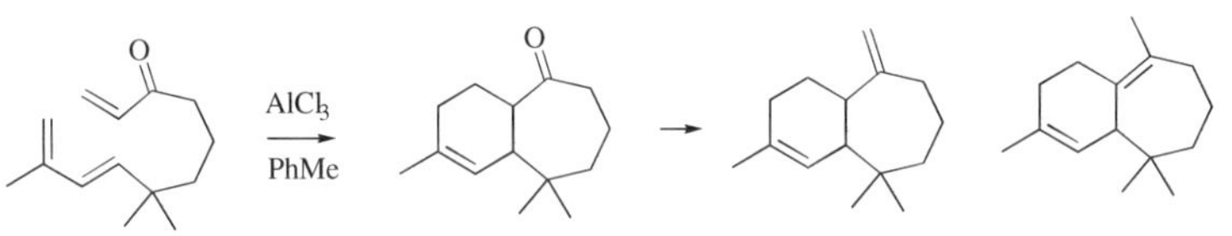

α-himachalene (**6-51**)   β-himachalene (**6-52**)

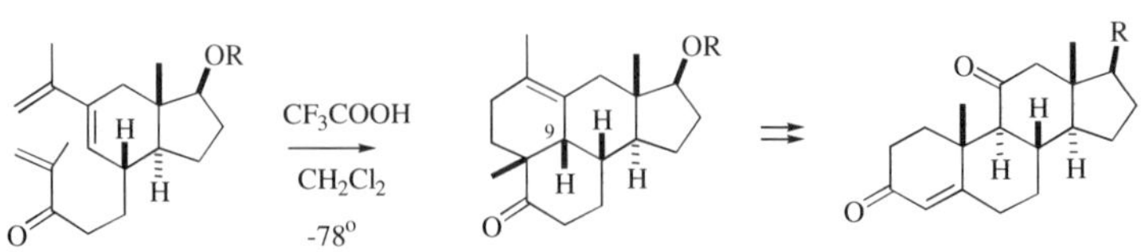

sel. 94 β-Me : 6 αMe

11-oxosteroids (**6-53**)

the transition state. Such compounds are reactivated by an additional terminal substituent on the diene. However, *trans* isomers (**6-54**) prevail, except when the terminal group is a carboxylic acid (**6-55**).[53]

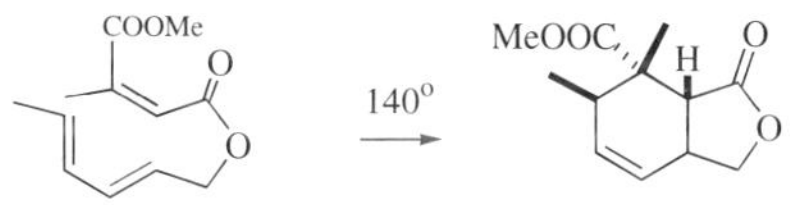

(**6-54**)

(**6-55**)

Construction of the dicarbocyclic core (**6-56**) of compactin by an intramolecular Diels–Alder reaction[54] was most appropriate as four contiguous stereocenters emerged naturally. The allylic ether prefers to line up pseudoaxially so as to minimize the $A^{1,3}$-strain. Note an alternative tactic is by axial locking the oxygen functionality in a γ-lactone (**6-57**).[55]

Diastereoselection may be determined by destabilization of certain transition states due to nonbonded interactions, leaving the substrate with only one choice. For example, an *exo* mode of cyclization was found for a cyclopentenone with an (*E*)-diene unit in a side chain attached to $C_{(3)}$ for a synthesis of 3-oxosilphinene (**6-58**).[56] In the corresponding *endo* transition state, the side chain would overlay the ring, and the two other modes of cycloaddition were inhibited by strong $A^{1,3}$-strain (Me:Me repulsion). Interestingly, the isomer with a (*Z*)-diene in the side chain failed to undergo cycloaddition.

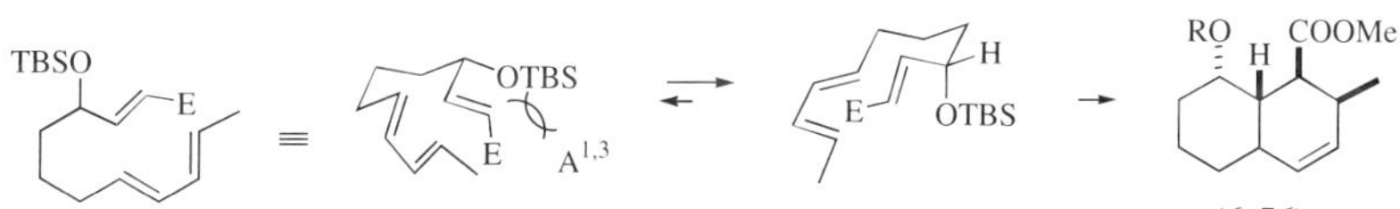

(**6-56**)

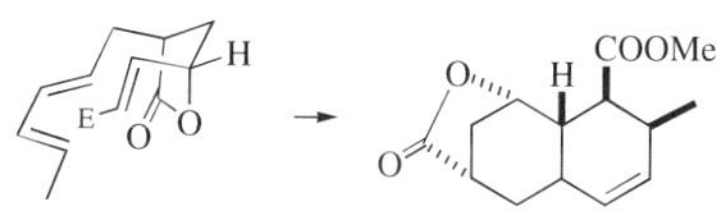

(**6-57**)

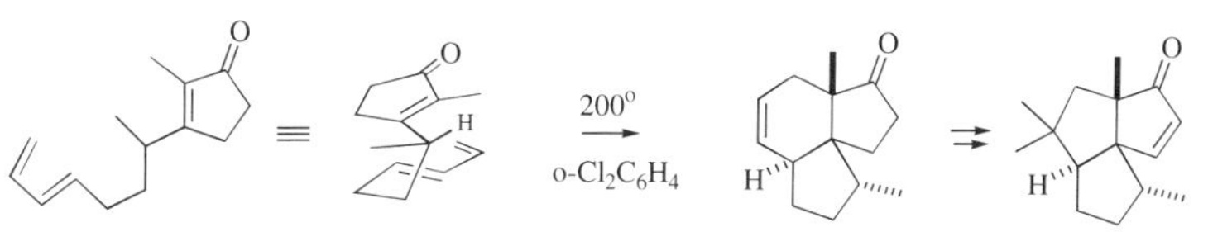

3-oxosilphinene
(**6-58**)

Stereofacial discrimination of the dioxenone ring by installation of a spirocyclic acetal auxiliary (from menthone) can be very successful, as demonstrated by synthesis of both enantiomers of grandisol (**6-59, 6-59′**).[57] To avoid steric interaction with the isopropyl group in the carbocycle, the dioxenone ring would bend away, thereby making the proximal (to the isopropyl group) face more exposed to external reagents.

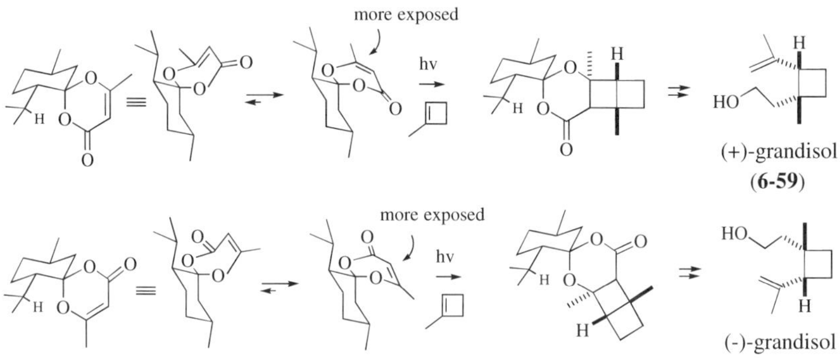

(+)-grandisol (**6-59**)

(-)-grandisol (**6-59'**)

It was gratifying to observe the 4:1 selectivity for the formation of the desired stereoisomer (**6-60**) for synthesis of ikarugamycin[58] in a photoinduced electrocyclization of a triene. The outcome was due to the preference for coiling in a direction to minimize nonbonded interactions between the allylic ethyl group and a vinylic hydrogen.

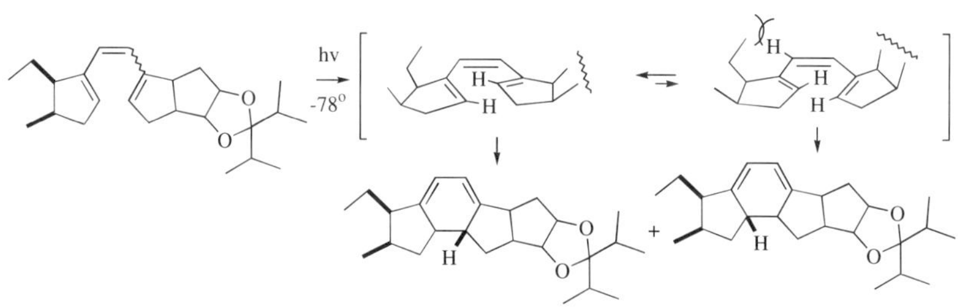

(**6-60**) (4 : 1)

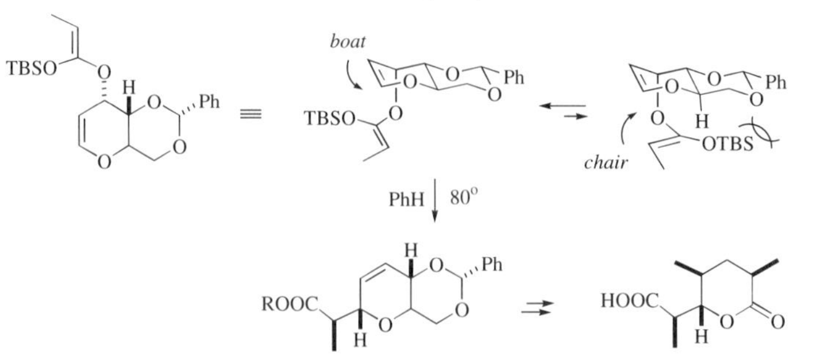

Prelog-Djerassi lactone (**6-61**)

Increase of nonbonded interactions such as *t*-butyldimethylsilylation of enolate anions to force the adoption of boatlike transition states in Claisen rearrangements was actually exploited in the synthesis of the Prelog–Djerassi lactone (**6-61**)[59] and nonactic acid (**6-62**).[60] It should be emphasized that the energy difference between chair and boat transition states for [3,3]-sigmatropic rearrangements is not substantial, steric factors can be significant.

Early examples of the chair/boat dichotomy emerged from work related to vitamin D synthesis.[61,62] The choice was dictated by overcrowded transition states which were to be avoided.

A pleasant surprise was the exclusive formation of phoracantholide-J (**6-63**) from rearrangement of a cyclic keteneacetal.[63] Of the two available conformations, one that was selected was less strained because of lesser transannular interactions. Transannular nonbonded interactions were also the major factors in the ring contraction rearrangement which led to a potential intermediate of ingenol (**6-64**).[64]

The ring contraction of a 13-membered allyl propargyl ether by [2,3]-Wittig rearrangement led to a *trans*-alcohol (**6-65**) only.[65] This stereoselcctivity arose from the substantial difference in steric compression between two conformations, one of

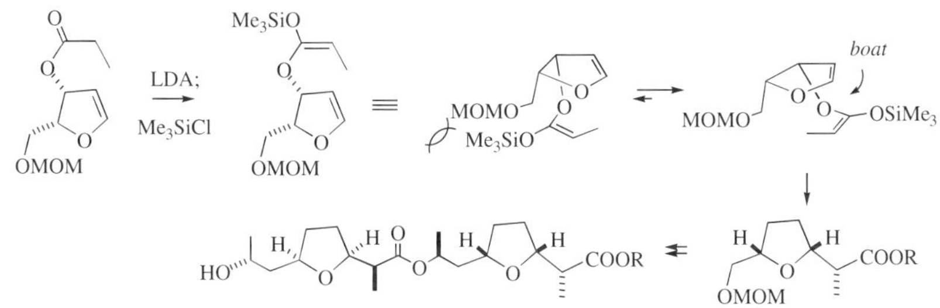

nonactic acid (**6-62**)

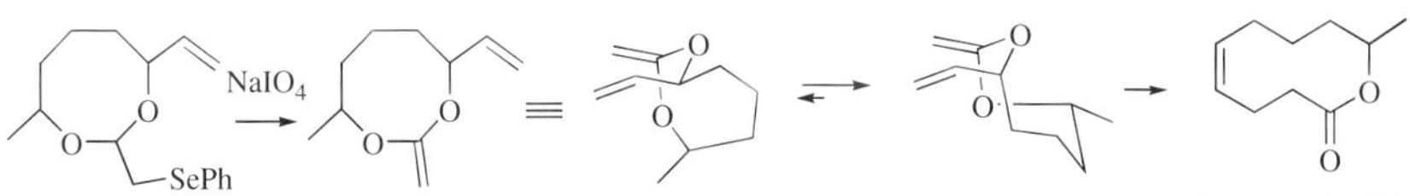

phoracantholide-J
(**6-63**)

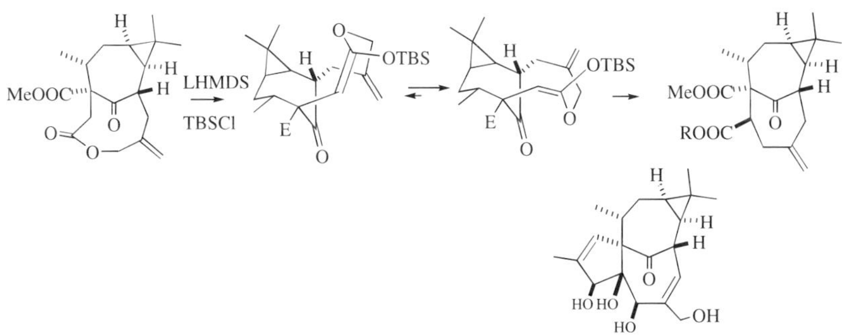

ingenol (**6-64**)

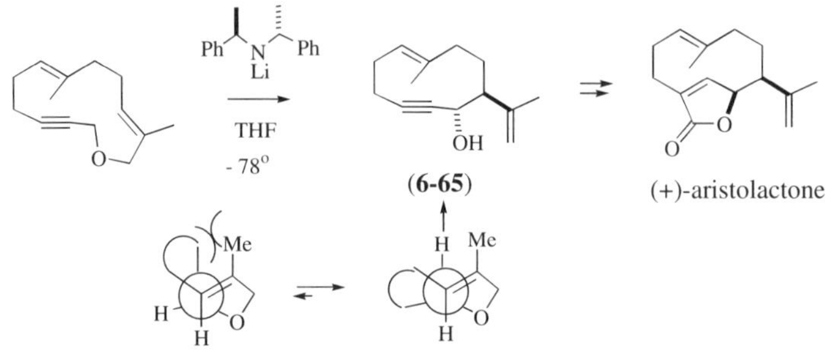

(6-65)

(+)-aristolactone

which featured the vinylic methyl group and a methylene four bonds away in the ring in much too close proximity.

6.3. CHAIR VERSUS BOAT CONFORMATION

The three most well-known forms of a six-membered ring are the chair, the boat, and the twist-boat. Generally, the boat form is less stable because of torsional strain and transannular interactions of the inward atoms attaching to the bow and stern. However, exceptions are not infrequent as other structural factors can upset the balance of forces. I shall examine a few normal examples and then discuss some intriguing cases in which the boatlike transition states predominate.

A simple synthesis of laurene (**6-66**)[66] involved 5-*exo* ring closure of a substituted 5-hexenyllithium in which the metal was originally attached to a double bond. The important feature is that the reactive termini circumscribed a chairlike transition state with the geminal tolyl and methyl groups inside the chain slanting toward pseudoaxial and pseudoequatorial directions, respectively. Note that in a cyclohexane a phenyl group is slightly better accommodated in an axial position than a methyl group.

The aldol-type condensation of an enamino ester with a chiral aldehyde was found to proceed via a chairlike Cram transition state in which all substituents occupied equatorial positions. It gave mainly the *threo* product,[67] thus establishing three contiguous stereocenters corresponding to those of the Prelog–Djerassi lactone (**6-61**). With the proper reactants a concise synthesis of the lactone was accomplished.

Acyliron complexes (**6-67**) in which the metal is also coordinated to carbon monoxide, triphenylphosphine, and η^5-cyclopentadienyl ligands are excellent precursors of enolate surrogates, with regard to stereoselection. The bulky phosphine ligand defines a forbidden zone to reactants which can select an orientation toward

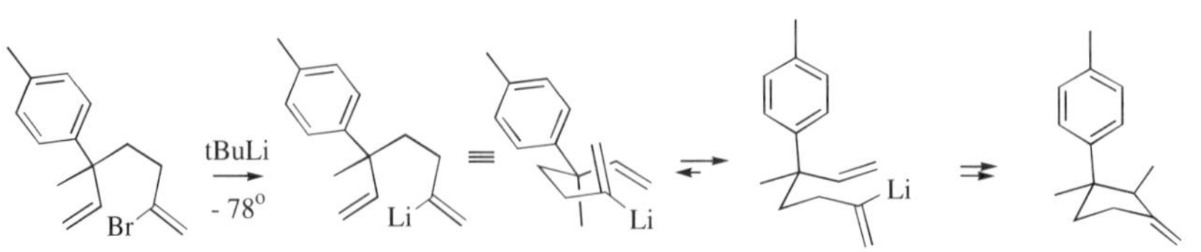

laurene (**6-66**)

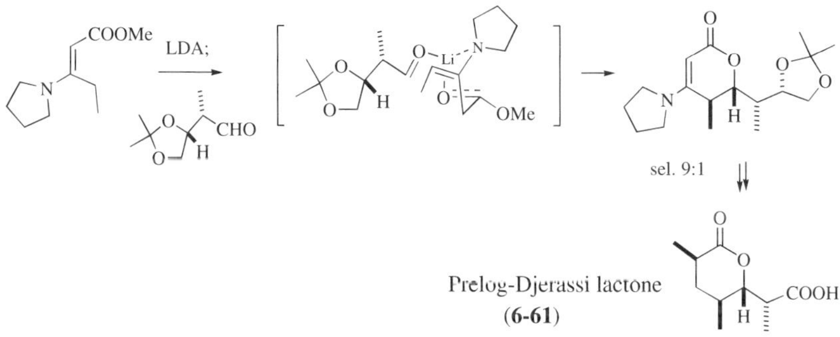

Prelog-Djerassi lactone
(**6-61**)

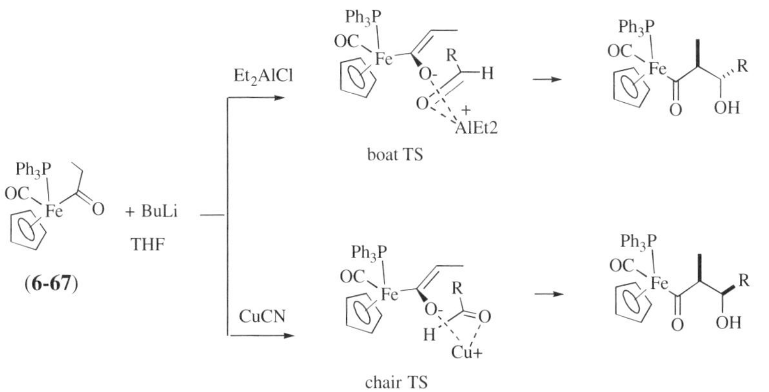

(**6-67**)

either the CO or Cp group. For aldol-type reactions[68] it has been demonstrated that promotion by diethylaluminum chloride and cuprous cyanide led to different diastereomers apparently derived from boat and chair transition states, respectively. The substituents on aluminum while keeping themselves away from the Cp ligand determine a boat transition state (with the Lewis acid complexed aldehyde in an *anti* conformation[69]). On the other hand, small ligands around the copper atom would allow its proximate location to the Cp group to assume a chair transition state.

It has been argued that complexation of aldehydes with boron Lewis acids also amasses the formyl C–H and an electronegative ligand of the boron atom into a five-centered coplanar hydrogen-bonded substructure. This model explains many enantioselective reactions involving chiral catalysts which organize the substrates.[70]

The utility of a Claisen rearrangement for a synthesis of neosporol (**6-68**)[71] depended on the high stereoselectivity (16:1) favoring the chairlike transition state. Previous efforts[72,73] were much less successful due to inability to control the configuration of the vinyl ether or ester enolate,[74] even though the chair transition state was also preferred.

Cyclooctenones are now readily accessible by employing the Claisen rearrangement to achieve ring expansion. The selection of a chairlike transition state by a substrate was crucial to the elaboration of the kalmanol (**6-69**) skeleton.[75]

Intramolecular alkylation of an anisole with an acyliminium ion was the key step in a synthesis of lycopodine (**6-70**).[76] Generation of the electrophilic species by protonation of the acylenamine can give rise to two stereoisomeric structures but obviously the one with a chairlike cyclohexane ring was heavily favored.

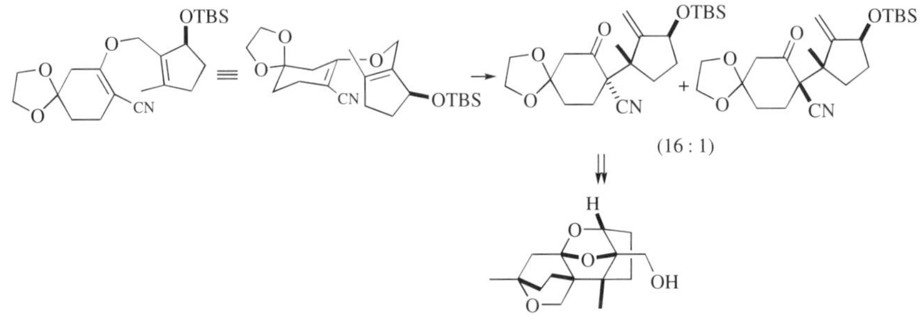

neosporol (**6-68**)

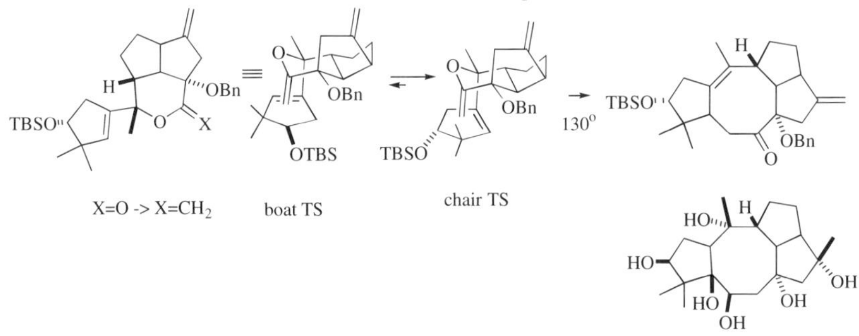

kalmanol (**6-69**)

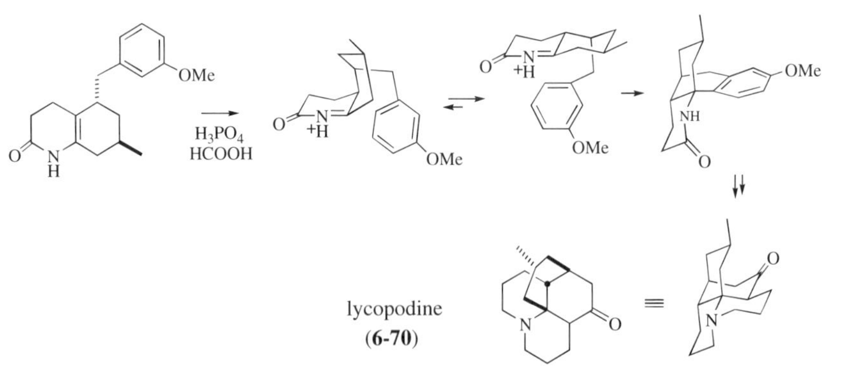

lycopodine
(**6-70**)

The Lewis acid catalyzed intramolecular ene reaction has found applications in many syntheses, e.g., methyl corynantheate (**6-71**).[77] The six-centered transition state usually has a chairlike alignment.

The most thoroughly investigated metallo-ene reactions are those involving magnesium.[78] One of the intriguing synthetic applications is for khusimone[79] in which cyclization was accompanied by a transfer of the magnesium atom to a position originally in the other carbon chain. The transition state was a double chair and its preference is apparent. The alternative transition state featured a boat in the emerging cyclohexane which also suffered a severe bow–stern impingement.

Very impressive results have been obtained from cationic polyene cyclization.[80] Necessarily the substrates must present themselves in such a way that termini of each double bond are in proper contact with those of the next unsaturated component. When an array of chairlike units can be grouped together, the cyclization would be most efficient. An allylic hydroxyl group (destined to become 11α) is sufficient to induce five new stereocenters and thereby constitutes a very elegant method for gaining access to 11-oxygenated steroids (**6-72**).[81] Interestingly, interruption of the cyclization by inserting an internal alkyne led to surprises. The modification did enable the synthesis of euphol (**6-73**) and tirucallol (**6-74**).[82]

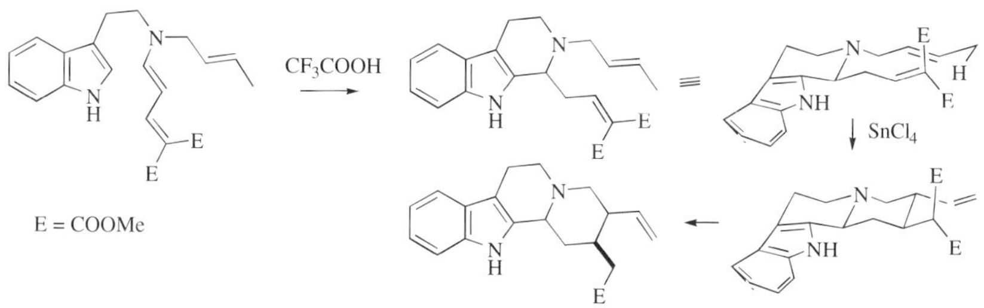

methyl corynantheate
(**6-71**)

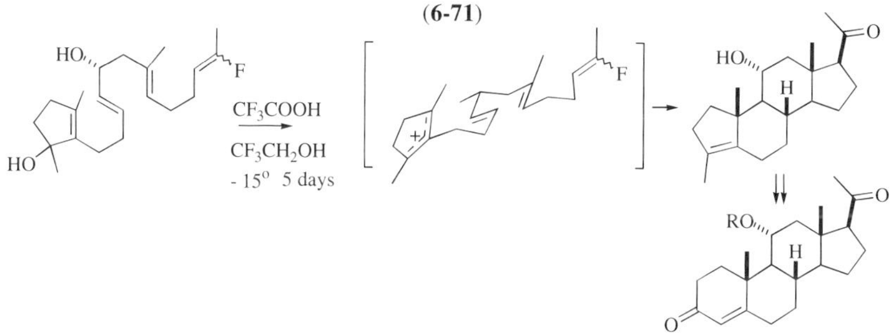

(**6-72**)

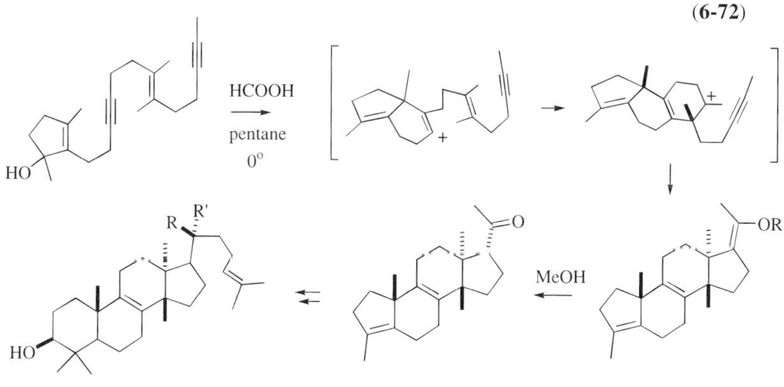

R=H, R'=Me euphol (**6-73**)
R=Me, R'=H tirucallol (**6-74**)

Another significant observation[83] is that a β-oriented silyl group at $C_{(8)}$ (terpene numbering) promoted cyclization of an epoxide through a boatlike transition state, leading to a precursor (**6-75**) of lanosterol. It is known that the biosynthesis of lanosterol involves cyclization of squalene epoxide in the chair–boat–chair–boat conformation.

A fruitful extension of the cationic aza-Cope rearrangement/Mannich reaction tandem is a method for the preparation of 3-acyltetrahydrofurans (**6-76**),[84] although the pathway proved to be different. It is characterized by a high stereoselectivity (>20:1) for giving *cis*-2,5-disubstituted tetrahydrofurans (**6-76**). The same product is generated from a mixture of four diastereoisomeric acetals. *trans*-Kumausyne (**6-77**) affords a prototypical structure for a test of this methodology. Actually, starting from a cyclopentanediol the process effected a ring enlarging annulation and the synthesis was smoothed by the assured stereochemistry at the ring junction.[85] Baeyer–Villiger oxidation and follow-up reactions could be brought to bear on completing the synthesis.

Because of conformational constraints, certain Claisen rearrangements are known to adopt boatlike transition states.[86] This behavior is manifested in the

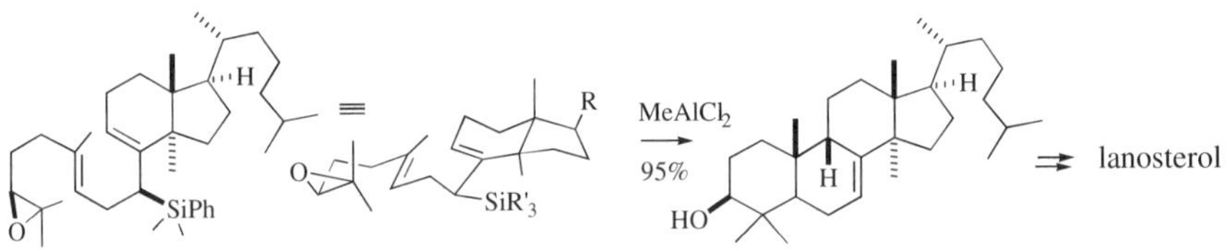

(**6-75**)

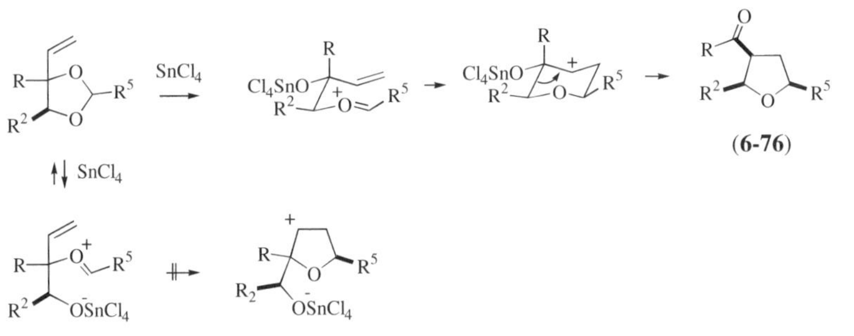

(**6-76**)

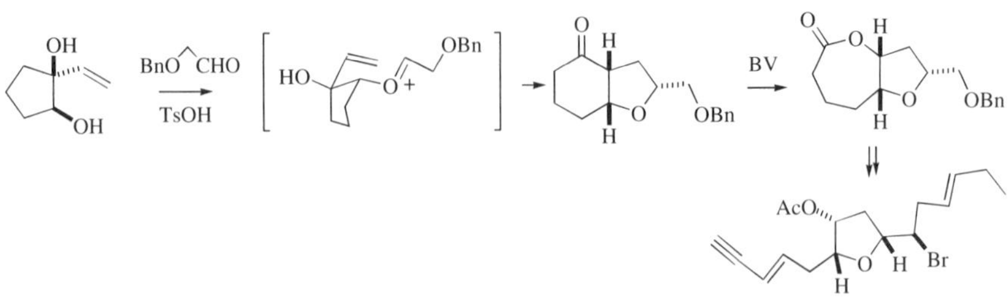

trans-kumausyne

(**6-77**)

rearrangement of homonuclear unsaturated lactones and 6-alkenyl-1,4-dioxan-2-ones, from which the synthesis of many compounds can profit: iridomyrmecin (**6-78**),[87] *cis*-chrysanthemic ester,[88] *N*-benzoylmeroquinene,[89] α-kainic acid,[90] and the $C_{(1)}/C_{(5)}$ fragment of erythronolide-B.[91]

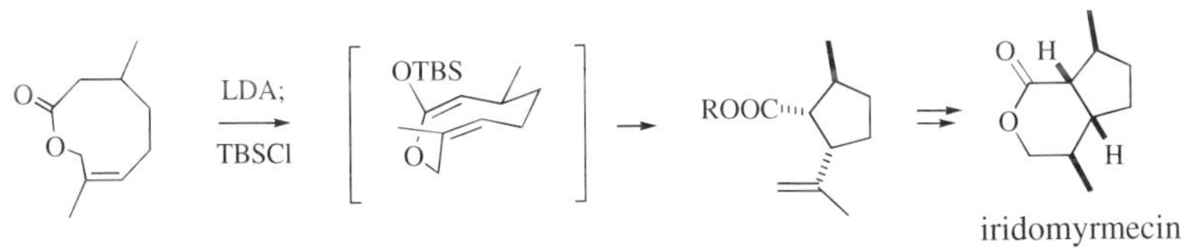

(**6-78**)

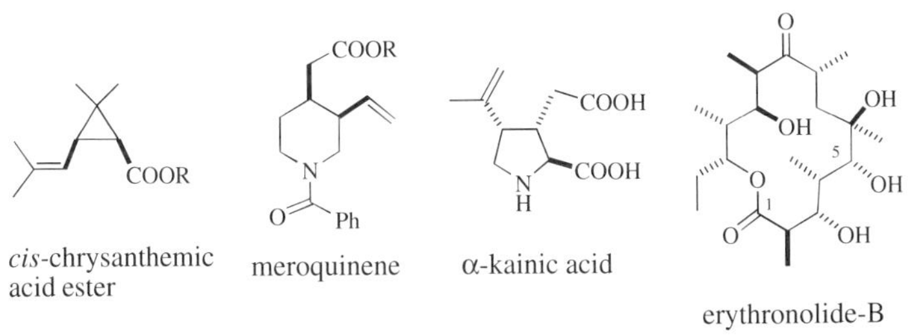

A synthesis of (–)-cycloaraneosene (**6-79**)[92] from two iridane building blocks proceeded from an intermediate which was submitted to a siloxy-Cope rearrangement via a chair transition state. While dictymal (**6-80**) is a B-seco aldehyde with the same C–C connectivity as cycloaraneosene between the two iridanes, it cannot be prepared in exactly the same manner because there are several different nonepimerizable centers of the product. In other words, only a boat transition state can lead to the required intermediate of dictymal. Fortunately, when the silyl ether was replaced with

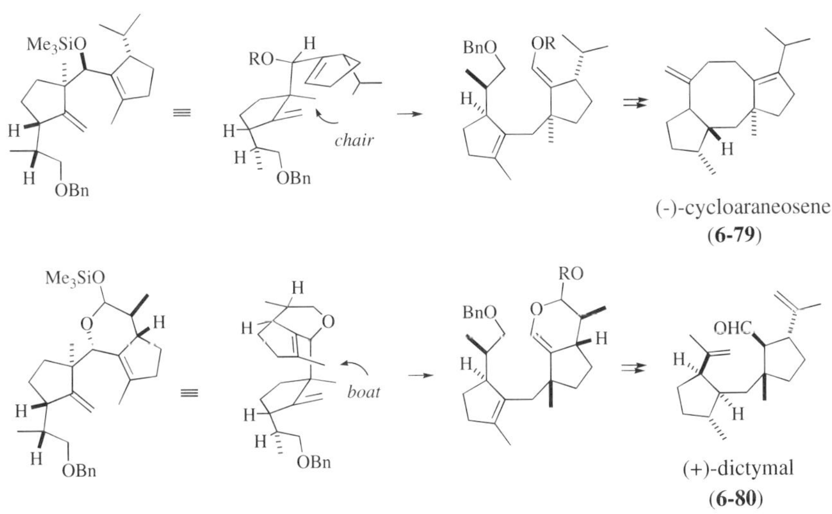

(**6-79**)

(**6-80**)

a δ-lactol derivative, the rearrangement was forced to adopt the desired conformation.[93]

An interesting discovery is that a 9-azabicyclo[3.3.1]nonenemethanol underwent Claisen rearrangement by choosing either a chairlike (**6-81**)[94] or boatlike (**6-82**) transition state.[95] Furthermore, the study indicates an opposite diastereofacial preference for orthoester Claisen (boat transition state) and the more conventional Claisen rearrangement (chair transition state).[96]

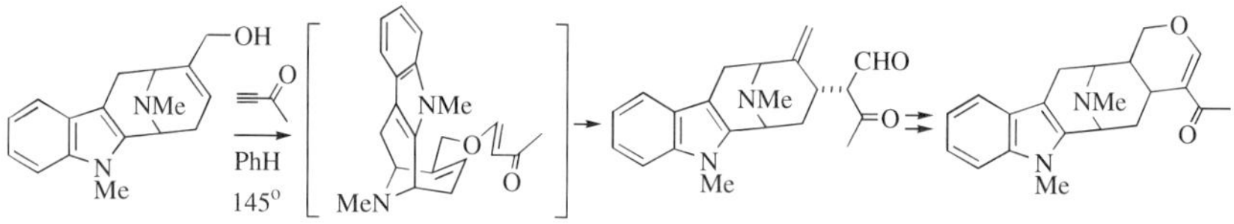

chair TS (**6-81**) (-)-alstonerine

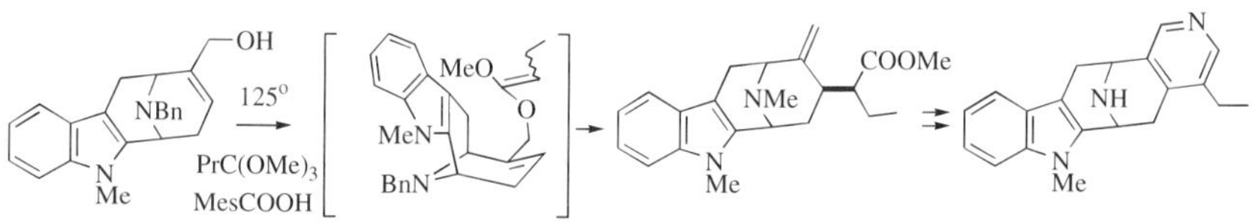

boat TS (**6-82**) suaveoline

It is remarkable that both (*E*)- and (*Z*) -allyl α-(trimethylsilyl)vinyl ethers can be converted to the same acylsilanes (**6-83**) by using a bulky alane catalyst.[97] In the coordinated form the (*Z*) -isomers must adopt a boatlike conformation.

$A^{1,3}$-Strain in the transition state for a homo-Cope rearrangement of a dialkenylcyclopropane (**6-84**) was inherited from the (*Z*)-double bond of a diene. It was manifested in the high reaction temperature which was not required by its double bond isomer.[98]

Decatrienones frequently undergo intramolecular Diels–Alder reaction via boatlike transition states[99] because most of the destabilizing eclipsing and nonbonded interactions are absent, whereas the chairlike counterparts have *gauche* instead of *skew* conformations. It is interesting to note that the trichothecane skeleton (**6-85**) was assembled[100] on such a premise.

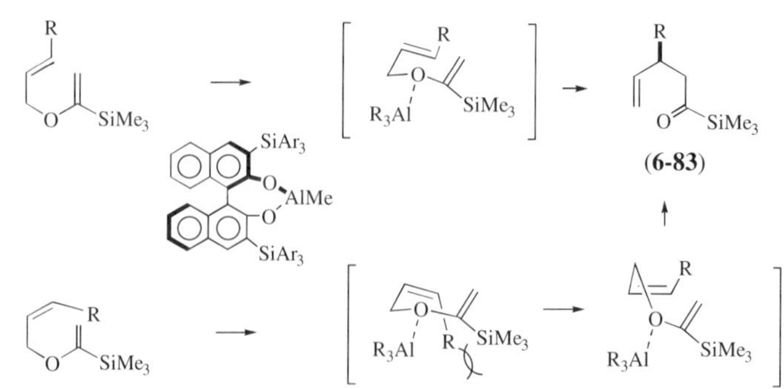

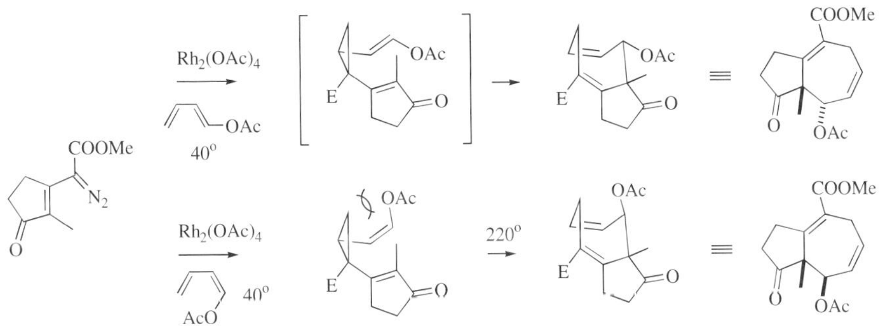

(6-84)

Intramolecular hetero-Diels–Alder reaction involving an acylnitroso group as dienophile is a useful method for synthesis of 2-substituted azacycles, as the adducts undergo N–O bond cleavage very readily. There are many applications of the reaction sequence and in a route to (–)-swainsonine (**6-86**)[101] the cycloaddition was conducted in water to maximize stereoselectivity. The desired product was formed from a boatlike *endo* transition state.

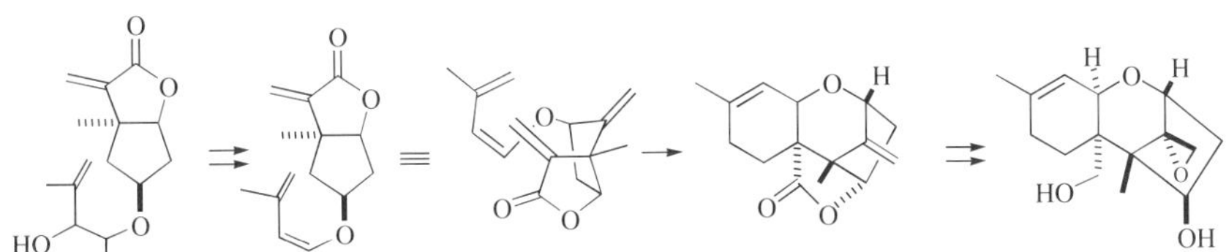

(**6-85**) verrucarol

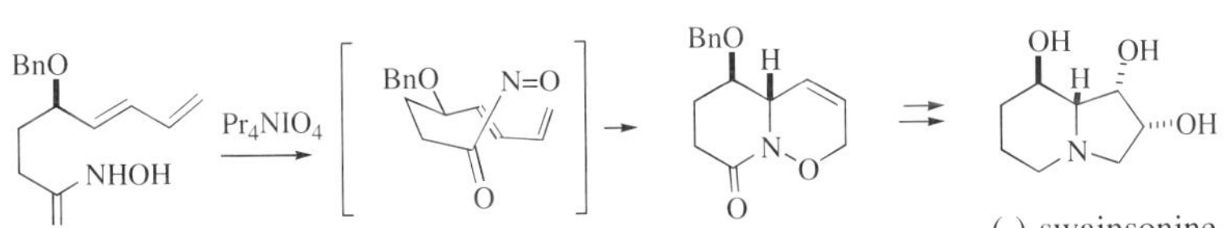

major

(-)-swainsonine (**6-86**)

The remarkable stereoselectivity (9:1) leading to the heteroyohimboid skeleton (**6-87**)[102] by an intramolecular Diels–Alder reaction is a testament to the advances in synthesis design and execution. The result indicates a boat/boat transition state wherein the amide resonance could be maintained. The alternative half chair/boat transition state that would have led to other isomeric cycloadducts did not enjoy the same electronic stabilization.

Useful results have been obtained from alkylation studies of 1,4-dihydrobenzamides.[103] When an (*S*)-prolinol methoxymethyl ether was part of the amide (**6-88**), reaction occurred predominantly on the *Si*-face; but with the oxygen atom of prolinol linked to the aromatic ring to form an oxazepinone (**6-89**), which is conformationally restricted, the *Re*-face attack by electrophiles was witnessed. More intriguingly, the stereochemical course for the same open-chain substrate can be

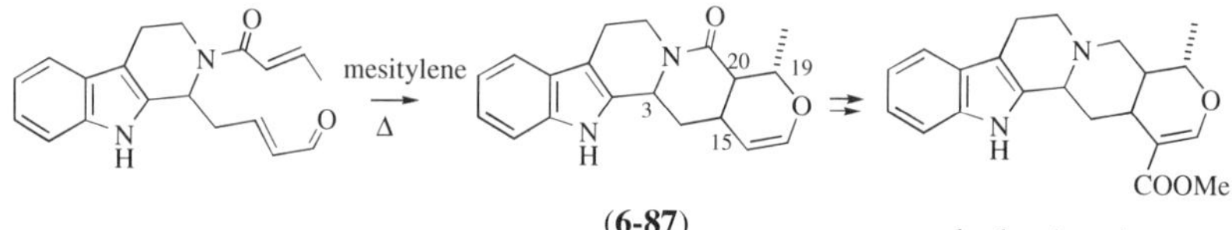

(6-87)

tetrahydroalstonine

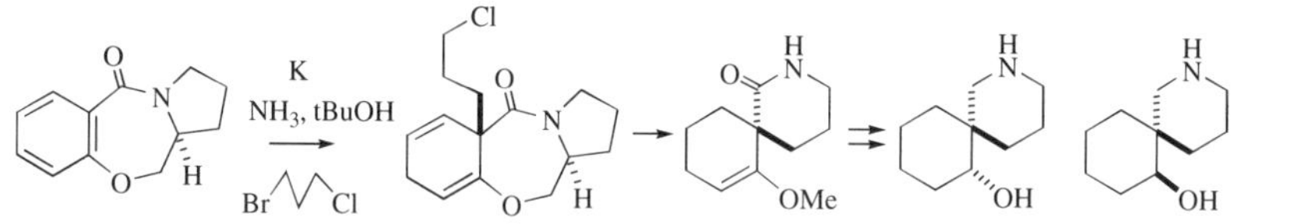

(6-89)

(+)-nitramine (-)-isonitramine

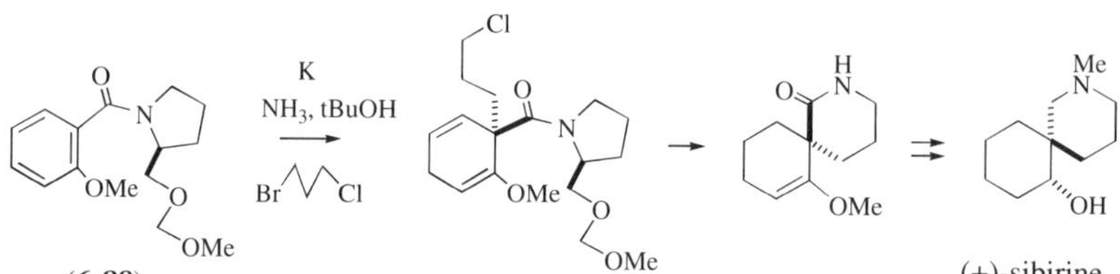

(6-88)

(+)-sibirine

changed completely when ammonia (from the Birch reduction) is retained or totally removed prior to methylation.[104]

6.4. MACROCYCLIC SYSTEMS

Even if there is a higher degree of uncertainty in conformational preferences of macrocyclic compounds, the presence of restricting elements such as multiple bonds can make analysis feasible in many cases.

Conformational effects are responsible for the greatly diverse behavior toward cyclization of two seco acid derivatives (**6-90, 6-91**)[105,106] of erythronolide-A. The differences in NMR coupling constants of proton pairs along a large portion of the skeleton in the acid and the lactone already indicated the degree of conformational change required during the cyclization.

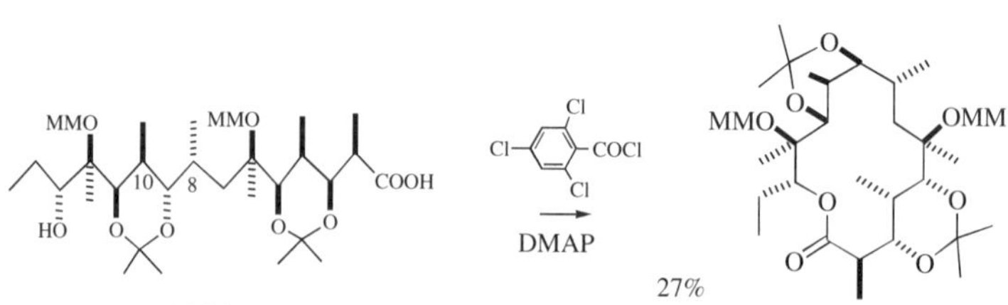

(6-90)

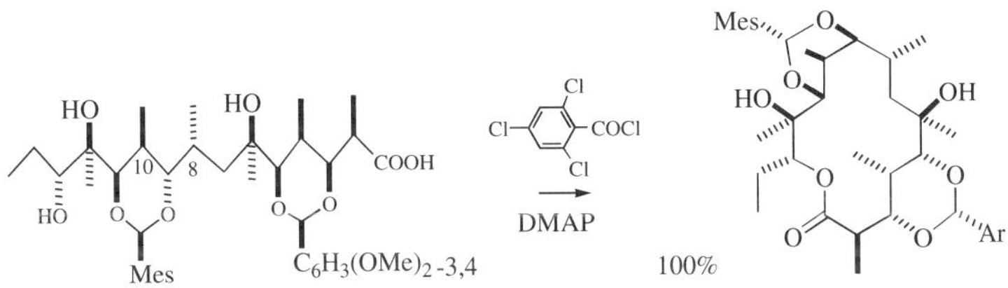

(6-91)

A cardinal rule related to reactions of macrocyclic compounds is peripheral attack by external reagents. Such is a manifestation of kinetic diastereoselection owing to inherent conformational and topographical aspects of medium and large rings which are dominated by transannular interactions. A trigonal center is essentially blocked on one face by the remote constituent parts of the molecule, leaving the only uncertainty about the local conformation. Such analysis was of enormous importance to the elucidation of the periplanone-B (**6-92**) structure[107] while various stereoisomers were being constructed by functionalization of keto polyene intermediates. A similar tactic was also used in a synthesis of (+)-dihydrocostunolide (**6-93**).[108]

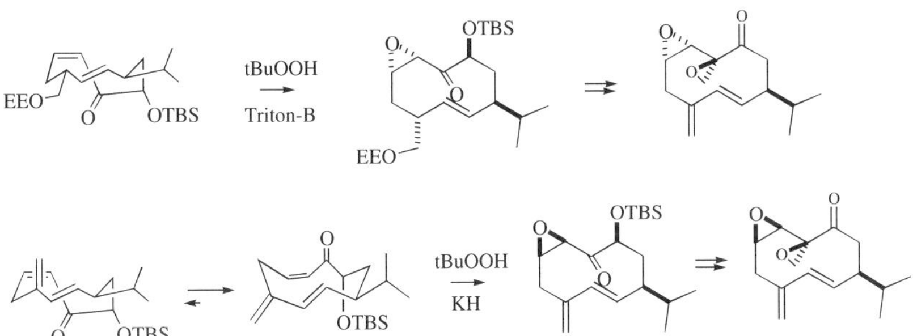

periplanone-B
(**6-92**)

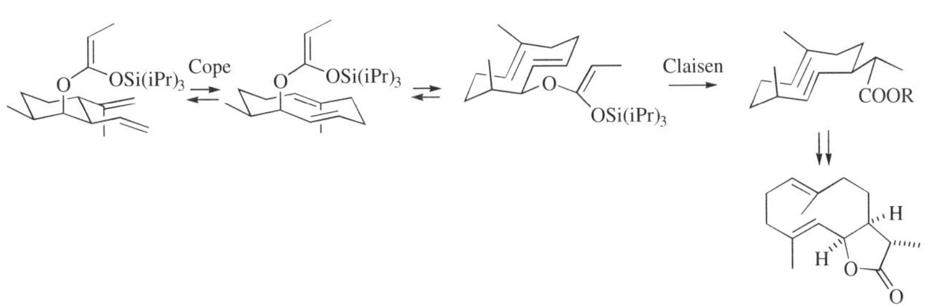

(+)-dihydrocostunolide
(**6-93**)

Both alkylation of macrocyclic lactones and conjugate addition to the unsaturated members show excellent stereoselectivity.[109] However, the conjugate addition proved unsuitable when applied to a more complex (highly substituted) lactone to link up with other structural units for a synthesis of (+)-lepicidin-A (**6-94**).[110]

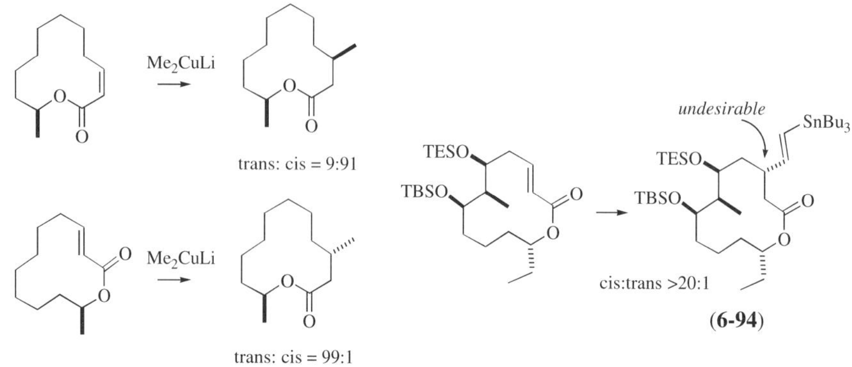

(**6-94**)

In pleuromutilin (**6-95**) the eight-membered ring has the most stable boat–chair conformation which is adorned with all substituents except the vinyl group projecting equatorially. Advantage can be taken in synthesis by elaborating the skeleton with only key functionality in that ring.[111] Thus, replacing the *gem*-disubstituted carbon with a ketone the two hydroxyl groups can be introduced by electrophilic substitution and, after inserting a double bond on the other side, conjugate hydration.

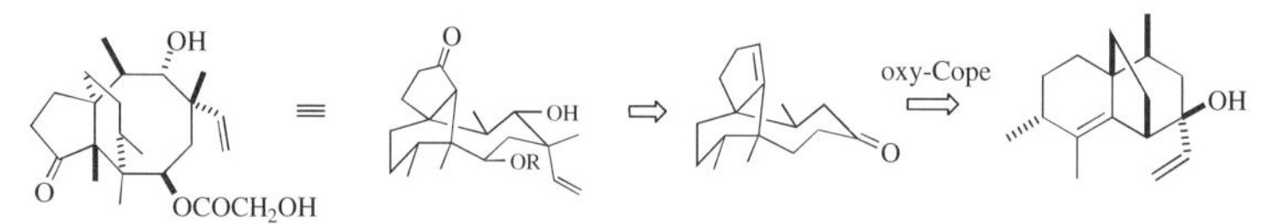

pleuromutilin
(**6-95**)

Ionic reduction of oxocenes was found to occur from the unencumbered (peripheral) face of a conformation with minimal A^{1,3}-strain leading to *cis*-2,8-disubstituted derivatives.[112] (This work was related to a synthetic study of gloeosporone which was originally considered to possess an oxocane skeleton.)

A ring expansion strategy for methynolide (**6-96**) synthesis[113] was pivoted by a [2,3]-sigmatropic rearrangement of a thiacyclic compound to increase the ring size from 5 to 8 to 11. For example, the pseudoaxial methyl group generated from diimide reduction of a thiacyclooctene intermediate became pseudoequatorial by a

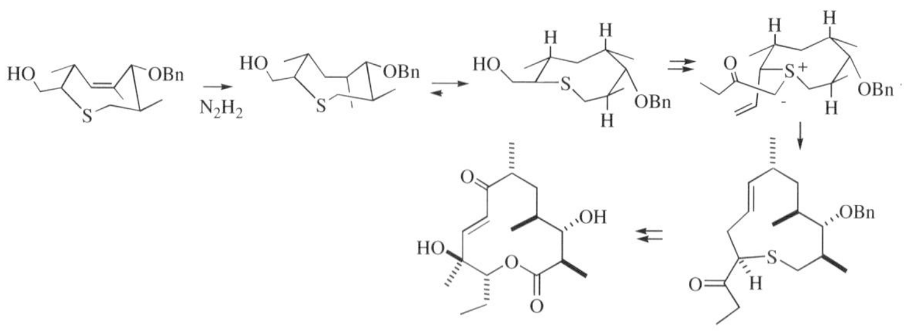

methynolide (**6-96**)

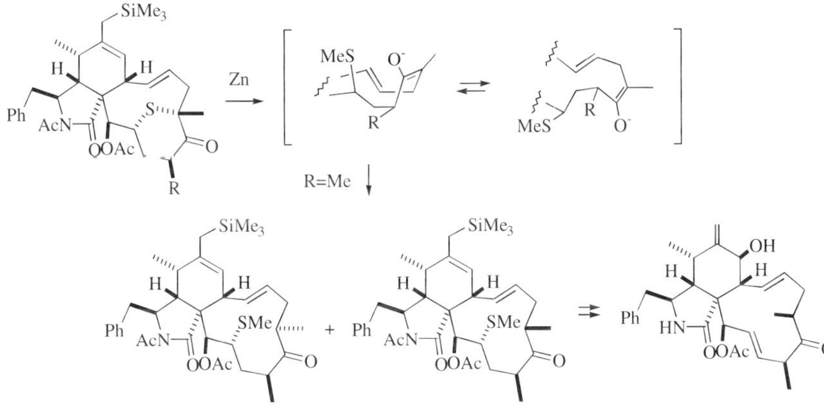

zygosporin-E (**6-97**)

conformational change. Peripheral *S*-alkylation prior to each ring expansion was also instrumental to the stereoselective process. A theoretical study[114] indicates that the *exo* preference of the sulfur lone pair (away from the allyl fragment) in combination with ring strain factor accounts for the stereoselectivity of the ring expansion.

The unraveling of macrocycles by cleavage of bridged systems is expedient. However, conformational aspects are very important and in untoward circumstances they border on being ruinous to synthetic efforts. Thus, the reductive ring opening of a thiabridged intermediate for synthesis of zygosporin-E (**6-97**)[115] led to an epimeric ketone in larger amounts (ratio 1:2.6) due to protonation from a more favorable conformation forcing the methyl group in the unnatural configuration.

A crucial step in a synthesis of taxusin (**6-98**)[116] is closure of the B-ring by an intramolecular Mukaiyama reaction. The important issue is that two contiguous stereocenters were established correctly, probably favored by the pseudoequatorial configuration of the benzyloxy groups.

An interesting synthesis of (–)-muscone (**6-99**)[117] was based on peripheral addition of lithium dimethylcuprate to a 20-membered chiral dilactone which embodied one conjugated double bond. The nascent enolate anion was in an excellent position to undergo a Dieckmann cyclization.

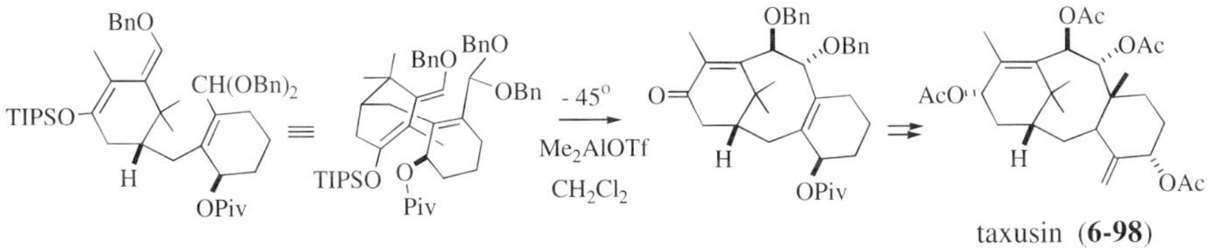

taxusin (**6-98**)

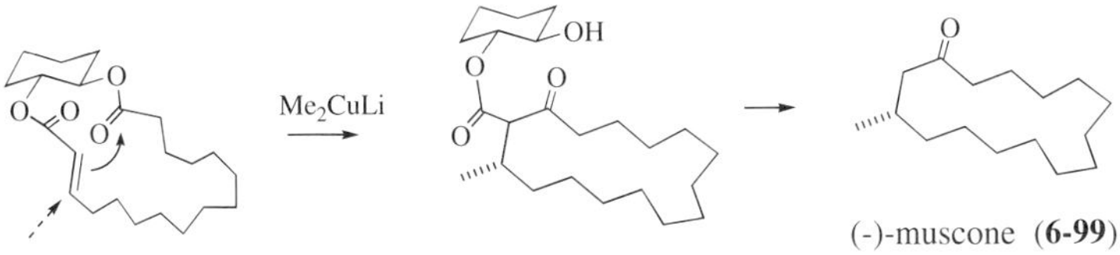

(-)-muscone (**6-99**)

Remarkably, two synthetic approaches to biotin (**6-100**)[118,119] which initially encountered some difficulties were extricated on modifying the intermediates to a macrocyclic state for directing intramolecular 1,3-dipolar cycloadditions. One route formed the thiophane with properly oriented substituents and the other route maintained the unfavorable rotamer in a (*Z*)-cycloalkene to engage in the stereoselective redistribution of an azido group into an amide and a diazonium ion for further transformation.

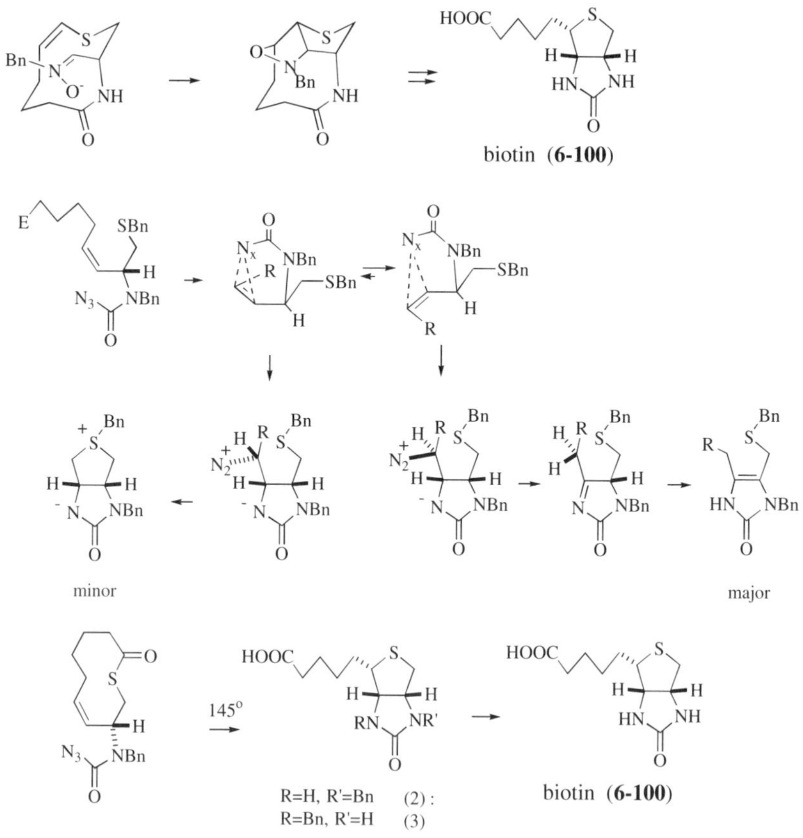

Some macrocyclic trienes which are properly segmented into conjugate diene and dienophile subunits can be coaxed to undergo the Diels–Alder reaction. Intriguing conformational effects have been evaluated and several compounds have been identified as useful precursors of steroids (**6-101, 6-102**)[120,121] and macrolides (**6-103**).[122]

Of interest also is the temperature dependence in the recyclization of a tricyclic iminium species when a secocleavamine derivative was allowed to couple with vindoline.[123] Only at low temperature could the vinblastine precursor (**6-104**) be obtained.

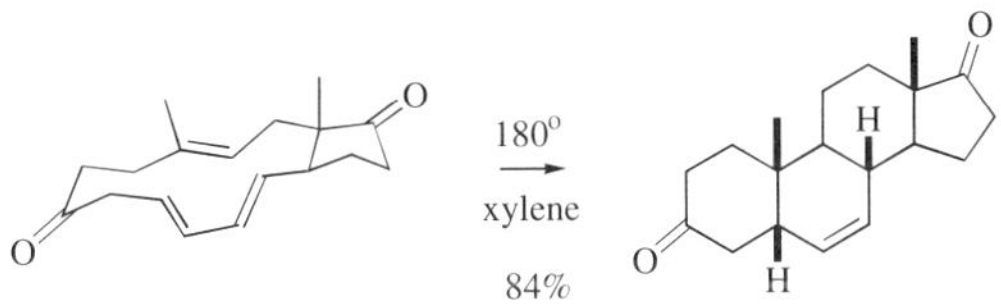

(**6-101**)

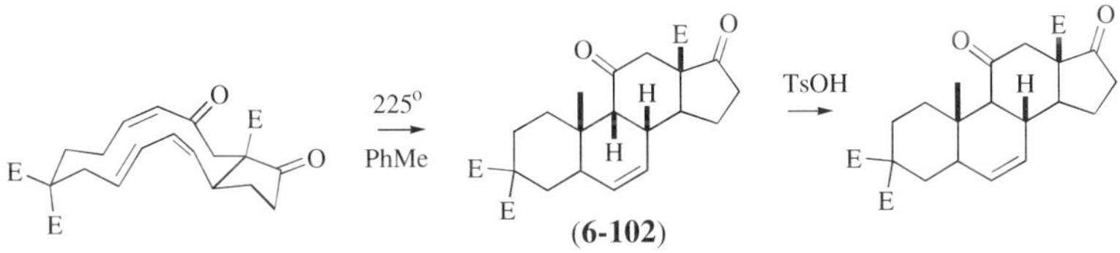

(**6-102**)

E = COOMe

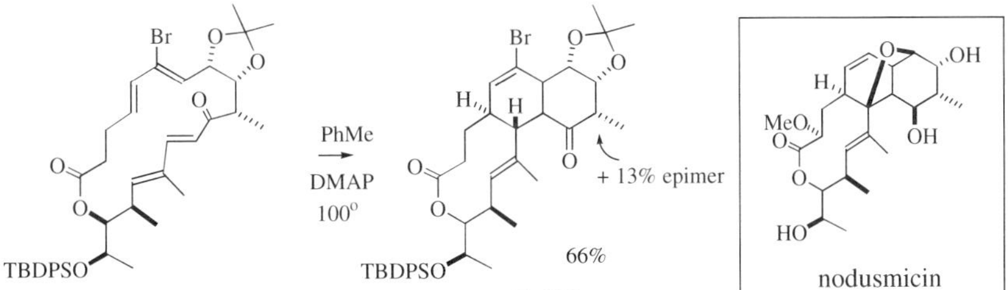

(**6-103**)

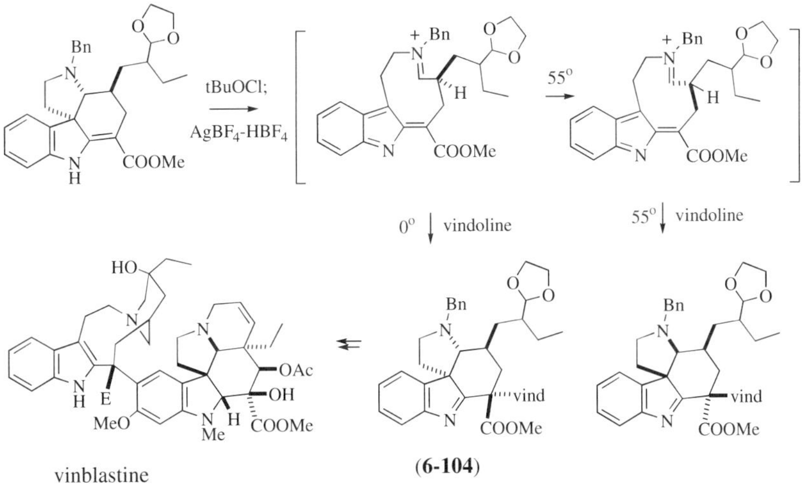

(**6-104**)

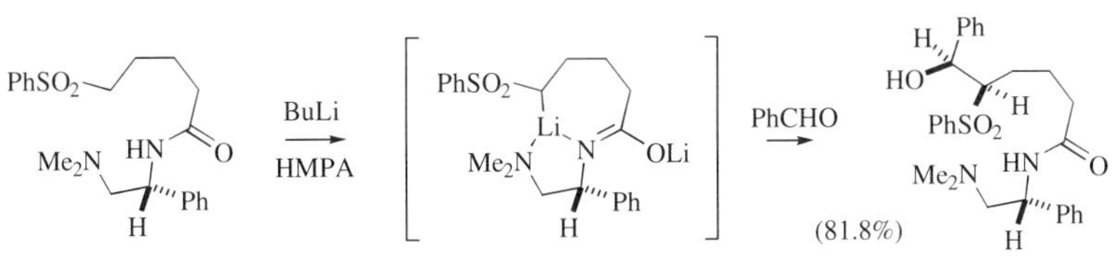

(**6-104a**)

Conformational effects must be responsible for the remote stereoinduction in the condensation of amide/sulfone dianions with benzaldehyde.[124] Formation of one particular diastereomer (**6-104a**) (81.8% yield) in a favorable case was observed. The dianion is thought to adopt a macrobicyclic structure in which the lithium ion is ligated to two nitrogen atoms.

6.5. *s-cis* VERSUS *s-trans* CONFORMATIONS

The facial selectivity of conjugate addition is intimately related to the conformation of the acceptor. Reaction from the same face gives different diastereomers by changing the *cisoid* to *transoid* conformation, and vice versa. The natural predilection of one or the other conformation can sometimes be reversed by complexation, as shown by enone (**6-105**).[125]

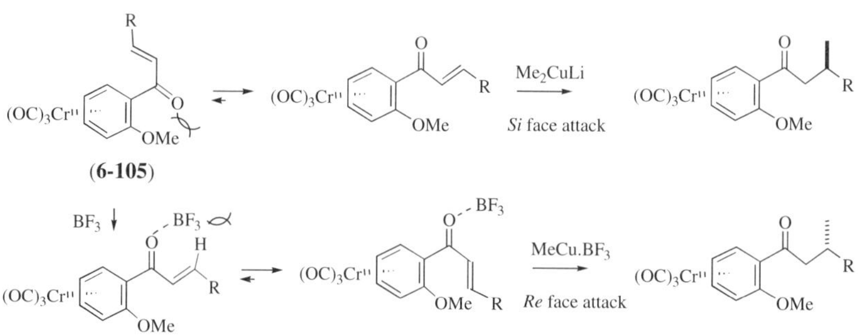

N-Alkenoylbornane-10,2-sultams exist in either the *s-cis* or *s-trans* form on chelation with metal ions. This characteristic can be exploited in the synthesis of enantiomeric series of β-substituted carboxylic acid derivatives.[126,127] The lithium chelate (*s-trans*) was applied to a synthesis of β-necrodol (**6-44**).[128]

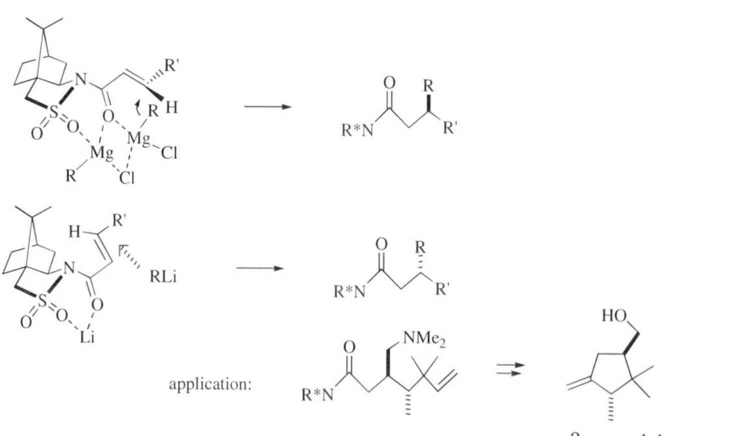

β-necrodol
(**6-44**)

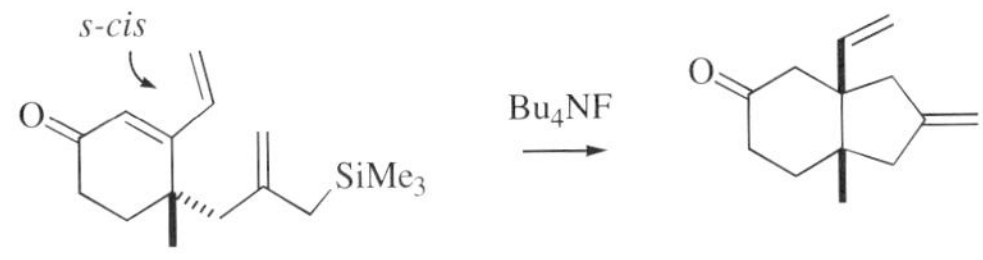

(**6-106**)

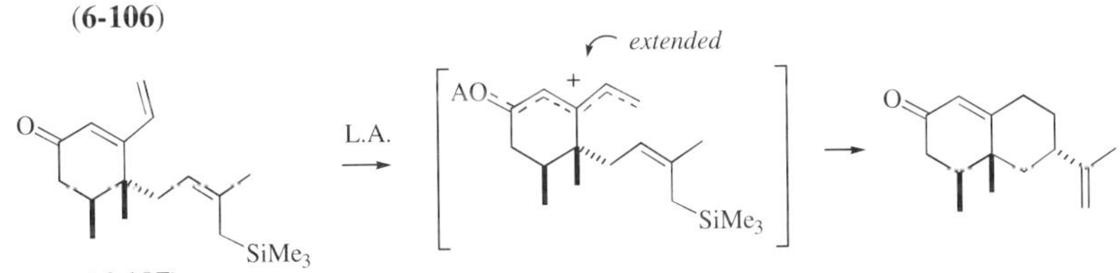

(**6-107**)

Very different modes of cyclization occur when 3-vinyl-2-cyclohexenones (e.g., **6-106**, **6-107**) which carry an allylsilane unit in a substituent at $C_{(4)}$ are exposed to fluoride ion[129] or a Lewis acid.[130] Probably in the ground state of such molecules the diene moiety is in an *s-cis* form, but the formation of a cationic species by coordination with a Lewis acid changes it into a zigzag arrangement.

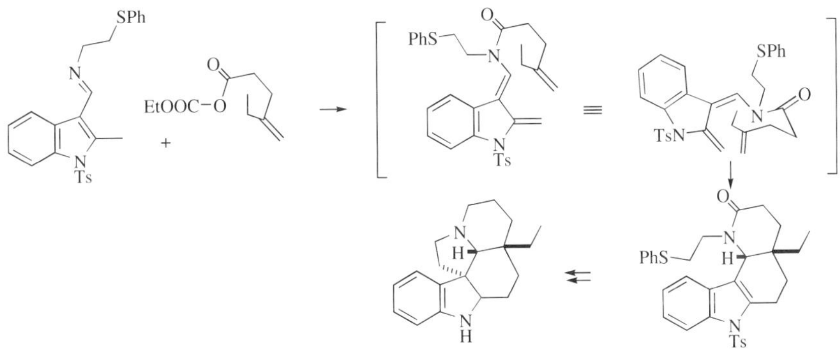

aspidospermidine
(**6-108**)

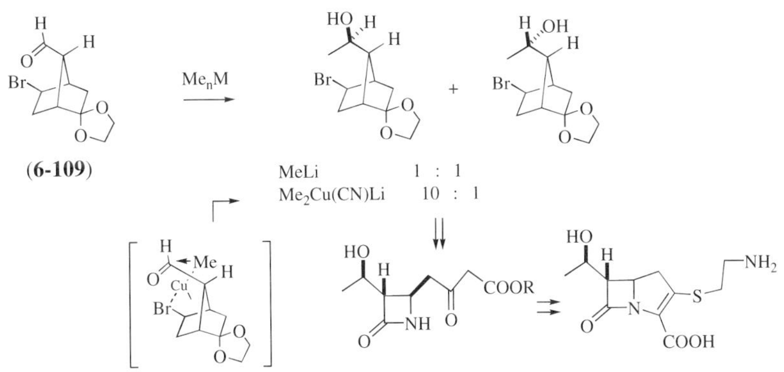

(**6-109**)

thienamycin

In situ intramolecular Diels–Alder reaction of a 2-methylene-3-acylaminomethyleneindoline with a double bond in the acyl group leads to very valuable intermediates of indole alkaloids. Aspidospermidine (**6-108**) has been acquired from the properly designed substrate.[131] Although the formation of any isomer would be inconsequential, the synthesis was facilitated on account of the observed stereoselectivity. *N*-Acylation gave only the (*Z*)-acylenamine as repulsion of the nitrogen atom and its substituent with the benzene ring should prevent the formation of the (*E*)-isomer.

Usually only reagent-controlled reactions can differentiate the two diastereofaces of a carbonyl group. However, selectivity may manifest from reactions of compounds containing special molecular features. For example, the formation of a secondary alcohol from a norbornane-7-aldehyde (**6-109**) can be made quite stereoselective[132] by changing the reagent from methyllithium to lithium dimethylcyanocuprate. It is speculated that coordination of the copper reagent with the *exo*-bromine atom restricted the rotation of the aldehyde group before reaction occurred.

REFERENCES

1. Wender, P.A.; Lee, H.Y.; Wilhelm, R.S.; Williams, P.D. *J. Am. Chem. Soc.* **1989**, *111*, 8954; Wender, P.A.; Kogen, H.; Lee, H.Y.; Munger, J.D.; Wilhelm, R.S.; Williams, P.D. *J. Am. Chem. Soc.* **1989**, *111*, 8957.
2. Feldman, K.S.; Wu, M.-J.; Rotella, D.P. *J. Am. Chem. Soc.* **1990**, *112*, 8490.
3. Fukamiya, N.; Kato, M.; Yoshikoshi, A. *J. Chem. Soc. Perkin Trans. 1* **1973**, 1843.
4. Näf, F.; Ohloff, G. *Helv. Chim. Acta* **1974**, *57*, 1868.
5. Ichihara, A.; Kawagishi, H.; Tokugawa, N.; Sakamura, S. *Tetrahedron Lett.* **1986**, *27*, 1347.
6. Takano, S.; Iwabuchi, Y.; Ogasawara, K. *J. Am. Chem. Soc.* **1987**, *109*, 5523.
7. Hashimura, K.; Tomita, S.; Hiroya, K.; Ogasawara, K. *Chem. Commun.* **1995**, 2291.
8. Irie, O.; Fujiwara, Y.; Nemoto, H.; Shishido, K. *Tetrahedron Lett.* **1996**, *37*, 9229.
9. Winkler, J.D.; Hershberger, P.M. *J. Am. Chem. Soc.* **1989**, *111*, 4852.
10. Oppolzer, W.; Jacobsen, E.J. *Tetrahedron Lett.* **1986**, *27*, 1141.
11. Crimmins, M.T.; Jung, D.K.; Gray, J.L. *J. Am. Chem. Soc.* **1992**, *114*, 5445.
12. Hwang, S.W.; Adiyaman, M.; Khanapure, S.; Schio, L.; Rokach, J. *J. Am. Chem. Soc.* **1994**, *116*, 10829.
13. West, F.G.; Naidu, B.N. *J. Am. Chem. Soc.* **1994**, *116*, 8420.
14. Suginome, M.; Yamamoto, Y.; Fujii, K.; Ito, Y. *J. Am. Chem. Soc.* **1995**, *115*, 9608.
15. Rychnovsky, S.D.; Griesgraber, G. *Chem. Commun.* **1993**, 291.
16. Myers, A.G.; Fraley, M.E.; Tom, N.J. *J. Am. Chem. Soc.* **1994**, *116*, 11556.
17. Harada, T.; Hayashiya, T.; Wada, I.; Iwa-ake, N.; Oku, A. *J. Am. Chem. Soc.* **1987**, *109*, 527.
18. Cory, R.M.; McLaren, F.R. *Chem. Commun.* **1977**, 587.
19. Takeshita, H.; Cui, Y.-S.; Kato, N.; Mori, A. *Bull. Chem. Soc. Jpn.* **1993**, *66*, 2694.

20. Danishefsky, S.; Schuda, P.F.; Kitahara, T.; Etheredge, S.J. *J. Am. Chem. Soc.* **1977**, *99*, 6066.
21. Gopalan, A.; Magnus, P. *J. Am. Chem. Soc.* **1980**, *102*, 1756.
22. Takeda, K.; Horibe, I.; Minato, H. *J. Chem. Soc. C* **1970**, 1142.
23. Kim, D.; Lim, J.I. *Tetrahedron Lett.* **1995**, *36*, 5035.
24. Koreeda, M.; George, I.A. *Chem. Lett.* **1990**, 83.
25. Saito, S.; Hirohara, Y.; Narahara, O.; Moriwake, T. *J. Am. Chem. Soc.* **1989**, *111*, 4533.
26. Semmelhack, M.F.; Epa, W.R.; Cheung,A.W.-H.; Gu, Y.; Kim, C.; Zhang, N.; Lew, W. *J. Am. Chem. Soc.* **1994**, *116*, 7455.
27. Sole, D.; Bonjoch, J.; Bosch, J. *J. Org. Chem.* **1996**, *61*, 4194.
28. Rawal, V.H.; Michoud, C.; Monestel, R.F. *J. Am. Chem. Soc.* **1993**, *115*, 3030.
29. Trost, B.M.; Greenspan, P.D.; Yang, B.V.; Saulnier, M.G. *J. Am. Chem. Soc.* **1990**, *112*, 9022.
30. Jamison, T.F.; Shambayati, S.; Crowe, W.E.; Schreiber, S.L. *J. Am. Chem. Soc.* **1994**, *116*, 5505.
31. Clemo, G.R.; Raper, R.; Seaton, J.C. *J. Chem. Soc.* **1956**, 3390.
32. Jacobi, P.A.; Selnick, H.G. *J. Org. Chem.* **1990**, *55*, 202.
33. Choy, W.; Reed, L.A., III; Masamune, S. *J. Org. Chem.* **1983**, *48*, 1137.
34. Tripathy, R.; Carroll, P.J.; Thornton, E.R. *J. Am. Chem. Soc.* **1990**, *112*, 6743.
35. Frey, B.; Hünig, S.; Koch, M.; Reissig, H.-U. *Synlett* **1991**, 854.
36. Funk, R.L.; Abelman, M.M. *J. Org. Chem.* **1986**, *51*, 3247.
37. Johnson, F.; Starkovsky, N.A.; Paton, A.C.; Carlson, A.A. *J. Am. Chem. Soc.* **1966**, *88*, 149.
38. Hua, D.H. *J. Am. Chem. Soc.* **1986**, *108*, 3835.
39. Agnel, G.; Negishi, E. *J. Am. Chem. Soc.* **1991**, *113*, 7424.
40. Evans, D.A.; Thomas, E.W.; Cherpeck, R.E. *J. Am. Chem. Soc.* **1982**, *104*, 3695.
41. Hart, D.J.; Kanai, K. *J. Am. Chem. Soc.* **1983**, *105*, 1255.
42. Toyooka, N.; Yotsui, Y.; Yoshida, Y.; Momose, T. *J. Org. Chem.* **1996**, *61*, 4882.
43. Nagaoka, H.; Kobayashi, K.; Matsui, T.; Yamada, Y. *Tetrahedron Lett.* **1987**, *28*, 2021.
44. Trost, B.M.; Braslau, R. *Tetrahedron Lett.* **1988**, *29*, 1231.
45. Nagasawa, T.; Handa, Y.; Onoguchi, Y.; Ohba, S.; Suzuki, K. *Synlett* **1995**, 739.
46. Roush, W.R.; Michaelides, M.R.; Tai, D.F.; Lesur, B.M.; Chong, W.K.M.; Harris, D.J. *J. Am. Chem. Soc.* **1989**, *111*, 2984.
47. Toyota, M.; Nishikawa, Y.; Fukumoto, K. *Tetrahedron Lett.* **1995**, *36*, 5379.
48. Roush, W.R.; Sciotti, R.J. *J. Am. Chem. Soc.* **1994**, *116*, 6457; Roush, W.R.; Kageyama, M.; Riva, R.; Brown, B.B.; Warmus, J.S.; Moriarty, K.J. *J. Org. Chem.* **1991**, *56*, 1192.
49. Oppolzer, W.; Petrzilka, M.; Bättig, K. *Helv. Chim. Acta* **1977**, *60*, 2964; Oppolzer, W.; Bättig, K.; Petrzilka, M. *Helv. Chim. Acta* **1978**, *61*, 1945.
50. Wenkert, E.; Naemura, K. *Synth. Commun.* **1973**, *3*, 45.
51. Stork, G.; Clark, G.; Shiner, C.S. *J. Am. Chem. Soc.* **1981**, *103*, 4948.
52. Martin, S.F.; Li, W. *J. Org. Chem.* **1989**, *54*, 265.
53. White, J.D.; Sheldon, B.G. *J. Org. Chem.* **1981**, *46*, 2273.

54. Funk, R.L.; Zeller, W.E. *J. Org. Chem.* **1982**, *47*, 180.
55. Davidson, A.H.; Floyd, C.D.; Jones, A.J.; Myers, P.L. *Chem. Commun.* **1985**, 1662.
56. Ihara, M.; Kawaguchi, A.; Ueda, H.; Chihiro, M.; Fukumoto, K.; Kametani, T. *J. Chem. Soc. Perkin Trans. 1* **1987**, 1331.
57. Demuth, M. *Pure Appl. Chem.* **1986**, *58*, 1233; Demuth, M.; Palomer, A.; Sluma, H.-D.; Dey, A.K.; Krüger, C.; Tsay, Y.-H. *Angew. Chem. Int. Ed. Engl.* **1987**, *25*, 1117.
58. Whitesell, J.K.; Minton, M.A.; Tran, V.D. *J. Am. Chem. Soc.* **1989**, *111*, 1473.
59. Ireland, R.E.; Daub, J.P. *J. Org. Chem.* **1981**, *46*, 479.
60. Ireland, R.E.; Vevert, J.-P. *Can. J. Chem.* **1981**, *59*, 572.
61. Cave, R.J.; Lythgoe, B.; Metcalfe, D.A.; Waterhouse, I. *J. Chem. Soc. Perkin Trans. 1* **1977**, 1218.
62. Chapleo, C.B.; Hallett, P.; Lythgoe, B.; Waterhouse, I.; Wright, P.W. *J. Chem. Soc. Perkin Trans. 1* **1977**, 1211.
63. Petrzilka, M. *Helv. Chim. Acta* **1978**, *61*, 3075.
64. Funk, R.L.; Olmstead, T.A.; Parvez, M.; Stallman, J.B. *J. Org. Chem.* **1993**, *58*, 5873.
65. Marshall, J.A.; Lebreton, J. *J. Am. Chem. Soc.* **1988**, *110*, 2925.
66. Bailey, W.F.; Jiang, X.-L.; McLeod, C.E. *J. Org. Chem.* **1995**, *60*, 7791.
67. Schlessinger, R.H.; Poss, M.A. *J. Am. Chem. Soc.* **1982**, *104*, 357.
68. Davies, S.G.; Dordor-Hedgecock, I.M.; Warner, P. *Tetrahedron Lett.* **1985**, *26*, 2125; Davies, S.G.; Dordor-Hedgecock, I.M.; Warner, P.; Jones, R.H.; Prout, K. *J. Organomet. Chem.* **1985**, *285*, 213.
69. Shambayati, S.; Schreiber, S.L. *Compr. Org. Chem.* **1991**, *1*, 283.
70. Corey, E.J.; Rohde, J.J. *Tetrahedron Lett.* **1997**, *38*, 37.
71. Ziegler, F.E.; Nangia, A. *Tetrahedron Lett.* **1988**, *29*, 1669.
72. Suda, M. *Tetrahedron Lett.* **1982**, *23*, 427.
73. Gilbert, J.C.; Wiechman, B.E. *J. Org. Chem.* **1986**, *51*, 258.
74. Kraus, G.A.; Thomas, P.J. *J. Org. Chem.* **1986**, *51*, 503.
75. Borrelly, S.; Paquette, L.A. *J. Am. Chem. Soc.* **1996**, *118*, 727.
76. Stork, G.; Kretchmer, R.A.; Schlessinger, R.H. *J. Am. Chem. Soc.* **1968**, *90*, 1647.
77. Tietze, L.F.; Wichman, J. *Angew. Chem. Int. Ed. Engl.* **1992**, *31*, 1079.
78. Oppolzer, W. In *Comprehensive Organic Synthesis*; Trost, B.M.; Fleming, I., Eds.; 1991, Vol. 5, p 29.
79. Oppolzer, W.; Pitteloud, R. *J. Am. Chem. Soc.* **1982**, *104*, 6478.
80. Johnson, W.S. *Acc. Chem. Res.* **1968**, *1*, 1; Johnson, W.S. *Biorg. Chem.* **1976**, *5*, 51.
81. Johnson, W.S.; Lyle, T.A.; Daub, G.W. *J. Org. Chem.* **1982**, *47*, 161.
82. Bartlett, W.R.; Johnson, W.S.; Plummer, M.S.; Small, V.R. *J. Org. Chem.* **1990**, *55*, 2215.
83. Corey, E.J.; Lee, J.; Liu, D.R. *Tetrahedron Lett.* **1994**, *35*, 9149.
84. Hopkins, M.H.; Overman, L.E.; Rishton, G.M. *J. Am. Chem. Soc.* **1991**, *113*, 5354.
85. Brown, M.J.; Harrison, T.; Overman, L.E. *J. Am. Chem. Soc.* **1991**, *113*, 5378.
86. Büchi, G.; Powell, J.E. *J. Am. Chem. Soc.* **1970**, *92*, 3126.
87. Abelman, M.M.; Funk, R.L.; Munger, J.D. *J. Am. Chem. Soc.* **1982**, *104*, 4030.

88. Funk, R.L.; Munger, J.D. *J. Org. Chem.* **1985**, *50*, 707.
89. Funk, R.L.; Abelman, M.M.; Munger, J.D. *Tetrahedron* **1986**, *42*, 2831.
90. Cooper, J.; Knight, D.W.; Gallagher, P.T. *J. Chem. Soc. Perkin Trans. 1* **1991**, 705.
91. Burke, S.D.; Schoenen, F.J.; Murtiashaw, C.W. *Tetrahedron Lett.* **1986**, *27*, 449.
92. Kato, N.; Tanaka, S.; Takeshita, H. *Chem. Lett.* **1986**, 1989.
93. Kato, N.; Tanaka, S.; Kataoka, H.; Takeshita, H. *Chem. Lett.* **1987**, 2295.
94. Zhang, L.H.; Cook, J.M. *J. Am. Chem. Soc.* **1990**, *112*, 4088.
95. Trudell, M.L.; Cook, J.M. *J. Am. Chem. Soc.* **1989**, *111*, 7504.
96. Zhang, L.H.; Trudell, M.L.; Hollinshead, S.P.; Cook, J.M. *J. Am. Chem. Soc.* **1989**, *111*, 8263.
97. Maruoka, K.; Yamamoto, H. *Synlett* **1991**, 793.
98. Cantrell, W.R.; Davies, H.M.L. *J. Org. Chem.* **1991**, *56*, 723.
99. Coe, J.W.; Roush, W.R. *J. Org. Chem.* **1989**, *54*, 915.
100. Koreeda, M.; Ricca, D.J.; Luengo, J.I. *J. Org. Chem.* **1988**, *53*, 5586.
101. Naruse, M.; Aoyagi, S.; Kibayashi, C. *J. Org. Chem.* **1994**, *59*, 1358.
102. Martin, S.F.; Benage, B.; Geraci, L.S.; Hunter, J.E.; Mortimore, M. *J. Am. Chem. Soc.* **1991**, *113*, 6161.
103. McCloskey, J.; Schultz, A.G. *Heterocycles* **1987**, *25*, 437.
104. Schultz, A.G.; Macielag, M.; Sundararaman, P.; Taveras, A.G.; Welch, M. *J. Am. Chem. Soc.* **1988**, *110*, 7828.
105. Tone, H.; Nishi, T.; Oikawa, Y.; Hikota, M.; Yonemitsu, O. *Chem. Pharm. Bull.* **1989**, *37*, 1167.
106. Hikota, M.; Tone, H.; Horita, K; Yonemitsu, O. *Tetrahedron* **1990**, *46*, 4613.
107. Still, W.C. *J. Am. Chem. Soc.* **1979**, *101*, 2493.
108. Raucher, S.; Chi, K.-W.; Hwang, K.-J.; Burks, J.E. *J. Org. Chem.* **1986**, *51*, 5503.
109. Evans, D.A.; Black, W.C. *J. Am. Chem. Soc.* **1993**, *115*, 4497.
110. Still, W.C.; Galynker, I. *Tetrahedron* **1981**, *37*, 3981.
111. Boeckman, R.K.; Springer, D.M.; Alessi, T.R. *J. Am. Chem. Soc.* **1989**, *111*, 8284.
112. Vedejs, E.; Buchanan, R.A.; Conrad, P.C.; Meier, G.P.; Mullins, M.J.; Schaffhausen, J.G.; Schwartz, C.E. *J. Am. Chem. Soc.* **1989**, *111*, 8421; Vedejs, E.; Buchanan, R.A.; Watanabe, Y. *J. Am. Chem. Soc.* **1989**, *111*, 8430.
113. Wu, Y.-D.; Houk, K.N. *J. Org. Chem.* **1991**, *56*, 5657.
114. Schreiber, S.L.; Kelly, S.E.; Porco, J.A.; Sammakia, T.; Suh, E.M. *J. Am. Chem. Soc.* **1988**, *110*, 6210.
115. Vedejs, E.; Rodgers, J.D.; Wittenberger, S.J. *J. Am. Chem. Soc.* **1988**, *110*, 4822.
116. Hara, R.; Furukawa, T.; Horiguchi, Y.; Kuwajima, I. *J. Am. Chem. Soc.* **1996**, *118*, 9186.
117. Ogawa, T.; Fang, C.-L.; Suemune, H.; Sakai, K. *Chem. Commun.* **1991**, 1438.
118. Baggiolini, E.G.; Lee, H.L.; Pizzolato, G.; Uskokovic, M.R. *J. Am. Chem. Soc.* **1982**, *104*, 6460.
119. Deroose, F.D.; De Clercq, P.J. *J. Org. Chem.* **1995**, *60*, 321.
120. Takahashi, T.; Shimizu, K.; Doi, T.; Tsuji, J.; Fukazawa, Y. *J. Am. Chem. Soc.* **1988**, *110*, 2674.
121. Couturier, M.; Deslongchamps, P. *Synlett* **1996**, 1140.

122. Roush, W.R.; Koyama, K.; Curtin, M.L.; Moriarty, K.J. *J. Am. Chem. Soc.* **1996**, *118*, 7502.
123. Kuehne, M.E.; Zebovitz, T.C. *J. Org. Chem.* **1987**, *52*, 4331.
124. Linnane, P.; Magnus, N.; Magnus, P. *Nature* **1997**, *385*, 799.
125. Uemura, M.; Oda, H.; Minami, T.; Hayashi, Y. *Tetrahedron Lett.* **1991**, *32*, 4565.
126. Oppolzer, W.; Poli, G.; Kingma, A.J.; Starkemann, C.; Bernardinelli, G. *Helv. Chim. Acta* **1987**, *70*, 2201.
127. Oppolzer, W.; Kingma, A.J.; Poli, G. *Tetrahedron* **1989**, *45*, 479.
128. Oppolzer, W.; Schneider, P. *Helv. Chim. Acta* **1986**, *69*, 1817.
129. Majetich, G.; Defauw, J.; Hull, K.; Shawe, T. *Tetrahedron Lett.* **1985**, *26*, 2755.
130. Majetich, G.; Behnke, M.; Hull, K. *J. Org. Chem.* **1985**, *50*, 3615.
131. Gallagher, T.; Magnus, P. *J. Am. Chem. Soc.* **1983**, *105*, 4750.
132. Grieco, P.A.; Flynn, D.L.; Zelle, R.E. *J. Am. Chem. Soc.* **1984**, *106*, 6414.

7

TOPOGRAPHICAL AND TEMPLATE EFFECTS

7.1. TOPOGRAPHICAL INFLUENCE OF STERIC COURSES

7.1.1. Condensed Ring Systems

Fusion of two small rings in the *cis* mode is obligatory or greatly preferred due to minimization of angular strain. Bicyclic arrays with constituting rings larger than 5/6 may favor the *trans* alternative. Many other factors can influence the thermodynamic stability, with special reference to ring fusion, of a particular compound. Generally, cyclization or annulation leading to small condensed ring systems is preordained.

When a polycyclic compound is generated by intramolecular cyclization onto a ring sized at five or smaller, the new junction is almost invariably *cis*. Exceptions may be found only when the existing ring is a macrocycle. This propensity is a valuable guideline for the design of synthetic routes, and there are numerous exploitations: in the synthesis of biotin,[1] dendrobine,[2] hirsutene,[3] and at the starting point of the AD-component preparation in the synthesis of vitamin B_{12}.[4]

Compounds typified by amauromine (**7-1**) invoke a synthetic challenge which focuses the attachment of the reverse prenyl group. Biogenetically, this alkaloid is a tryptophan derivative and the pursuit of its synthesis from the amino acid is intriguing. Since the use of the prenyl cation to initiate cyclization is not likely successful (to give the reverse prenyl substituent), the introduction of another electrophile followed by exchange reaction is much more reasonable. Indeed, this proposition was realized[5] when a phenylseleno group assigned the precursorial role and its replacement was effected by reaction with prenyl tributylstannane. Naturally, the replacement occurred with retention of configuration, to preserve the *cis*-5/5 fusion mode.

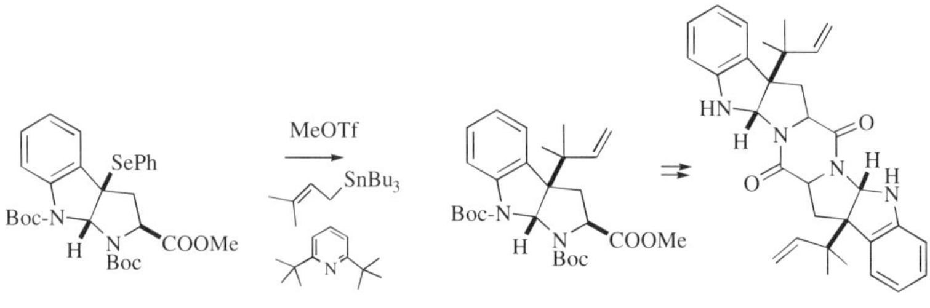

amauromine (**7-1**)

The completion of a quadrone (**7-2**) synthesis[6] was complicated by the need to regioselectively introduce a hydroxymethyl group for incorporation into the lactone ring. The problem was solved by creating a double bond to block one side of the ketone. This tactic was viable only because hydrogenation perforce reconstituted the *cis*-diquinane moiety.

There is no gainsaying the utmost importance of topography in the synthesis of dodecahedrane (**7-3**).[7] As the curvature of the intermediates increases, functional groups tend to orient themselves outward. However, they must be kept in the more congested state to be utilized in skeletal bond construction. Thus, the reductive cyclization of a diester was found to go halfway probably because direct dechlorination proceeded at a competitive rate. The unreacted ester group incapacitated for subsequent cyclization was sterically rectified by an exohedral alkylation with a detachable group.

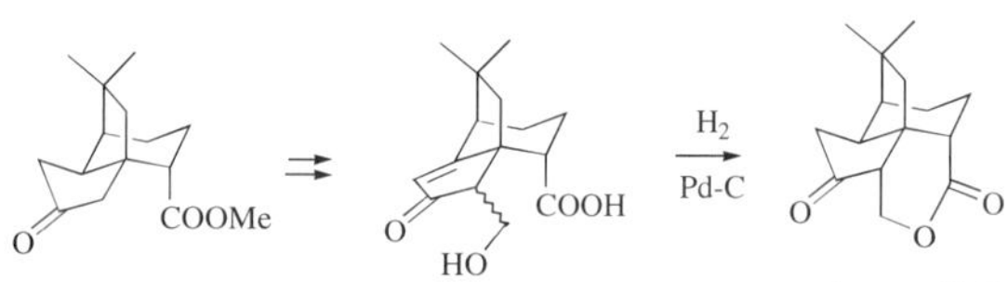

quadrone (**7-2**)

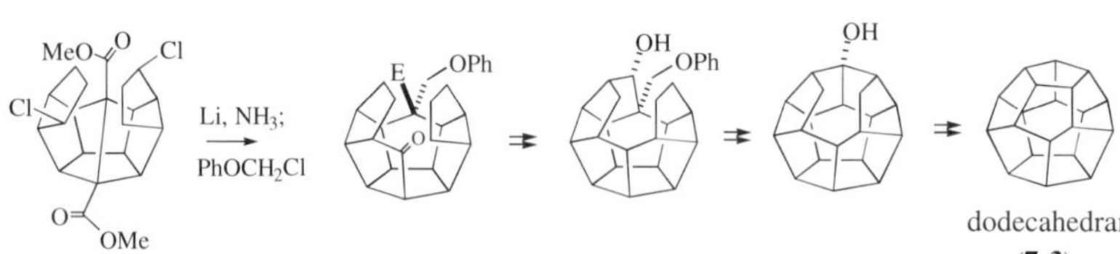

dodecahedrane (**7-3**)

7.1.1.1. Reactions at a Subangular Position. The steric courses of reactions at a subangular site are better managed in *cis*-fused ring and bridged ring systems. Generally, *cis*-fused bicyclic systems undergo reactions stereoselectively because the adjacent ring acts as an effective shield for one face.

The 1,3-dipolar cycloaddition of a fused methylenecyclopropane from the *exo* side provided a spirocyclic precursor of pumiliotoxin-C (**7-4**).[8] In this synthetic approach only the nonregioselective rearrangement of the adduct was a weakness.

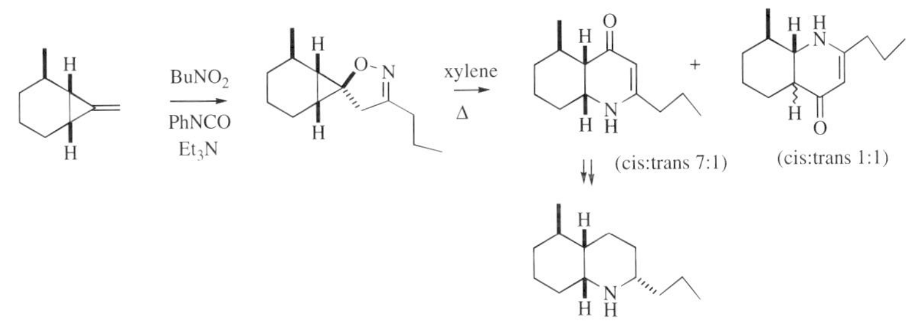

pumiliotoxin-C (**7-4**)

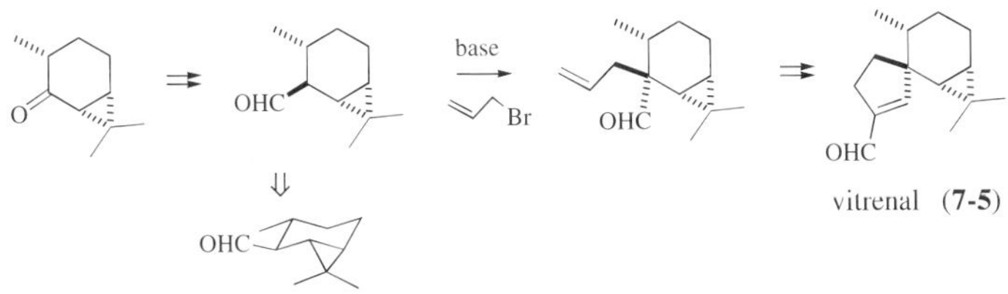

vitrenal (**7-5**)

The structural characteristic of vitrenal (**7-5**) suggests its expedient synthesis from a carane derivative.[9] 2-Carone is most appropriate since it already contains three of the four stereocenters. The spiroannulated cyclopentenecarboxaldehyde is so constituted that the subangular double bond is on the *endo* side and retrosynthetic analysis readily reveals that the aldehyde derived by homologation of 2-carone is an ideal intermediate. It all depends on attachment of a longer chain from the *exo* face of the molecule. This chain was to be used to construct the third ring.

(+)-Hanegokedial (**7-6**) was synthesized from the dimethylbicyclo[5.1.0]-octenone precursor by 1,2-dialkylation of the double bond, initiated by a stereoselective conjugate addition.[10] Unfortunately, only the minor product (*cis*) was useful. On the other hand, the initiation of a synthesis[11] of indolizomycin (**7-7**) by

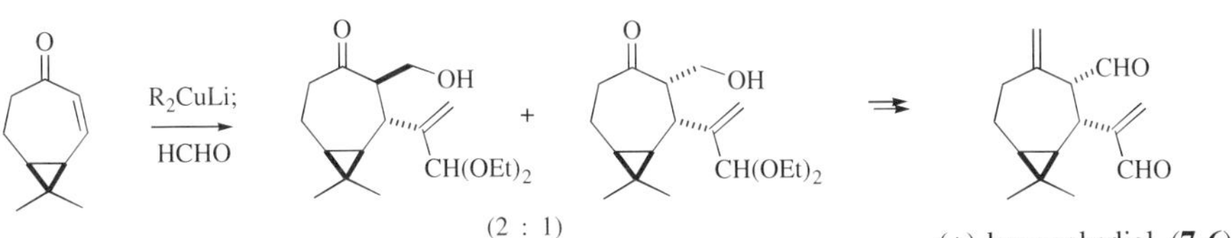

(+)-hanegokedial (**7-6**)

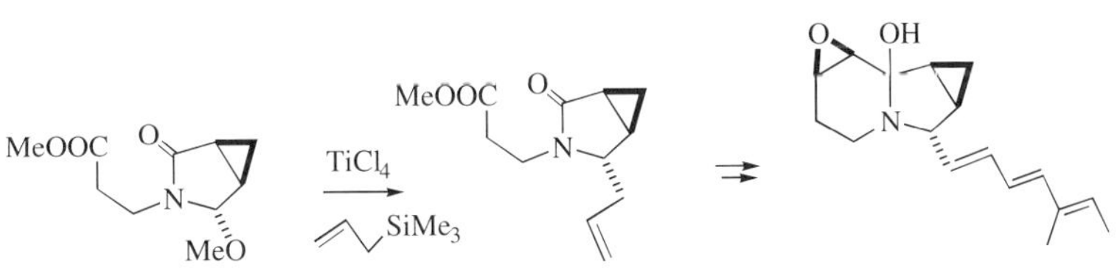

indolizomycin (**7-7**)

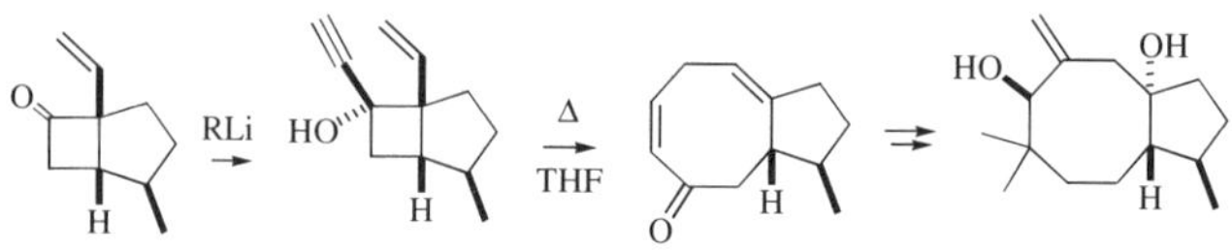

poitediol (**7-8**)

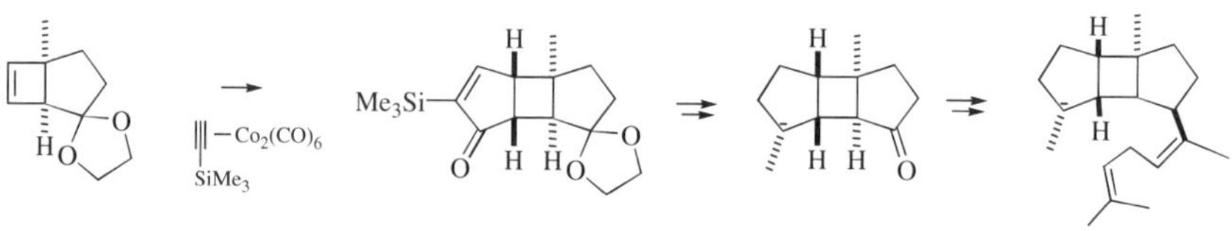

(13Z)-spata-13(15),17-diene (**7-9**)

configurationally retentive displacement of an *exo*-methoxy group of an azabicyclo[3.1.0]hexane derivative with an allyl chain was a blessed event.

The employment of a ring expansion strategy for a synthesis[12] of poitediol (**7-8**) was greatly facilitated by the addition of an acetylene unit to a fused cyclobutanone from the *exo* face. The key step was an oxy-Cope rearrangement.

The Pauson–Khand reaction[13] is a powerful method for annulation of an alkene to create a cyclopentenone unit. Because the hexacarbonyldicobalt complexes of alkynes are rather bulky, the reaction is rather sensitive to the steric environment of the alkene. For a synthesis of a spatane (**7-9**)[14] from a bicyclo[3.2.0]heptene one can use ethynyltrimethylsilane instead of ethyne. The silyl group provides steric bias in the insertion step and it can be removed from the product. Only one *cis-anti-cis* tricyclic enone was observed in this case due to complexation with the cobalt species and subsequent reaction on the *exo* side of the fused cyclobutene substrate.

A versatile method for achieving indirect Michael addition to cyclic enones with excellent stereoselectivity involves [2+2]-cycloaddition with ynamines and subsequent hydrolysis and fragmentation.[15] The configuration of the exocyclic center (α to the emerging carboxyl group) can be determined by proper choice of the hydrolytic conditions. On equilibration the double bond of the cyclobutenylamine moiety shifts into conjugation, while disposing the larger alkyl group (versus H) in an *exo*-configuration, before it is transformed into the β-diketone and suffers cleavage. On the other hand, rapid hydrolysis following kinetic protonation of the enamine from the *exo*-face leads to the diastereomeric keto acid.

From the adduct of 2-cyclohexenone and *N,N*-diethyl-1-propynamine it has been possible to synthesize juvabione (**7-10**)[16] by the kinetic route. [Note an example of employing the thermodynamic protonation for a synthesis of the D-ring synthon containing the side chain of (*1S*)-hydroxycholecalciferol.[17]]

Hydration of a cyclopentenone fused to a four-membered ring was a critical step in the conversion of the photocyclized tropolone methyl ether to prostaglandin-$F_{2\alpha}$ (**7-11**).[18]

The surge in the discovery of naturally occurring polyquinanes has prompted intense research in cyclopentannulation and related transformations. For diquinane

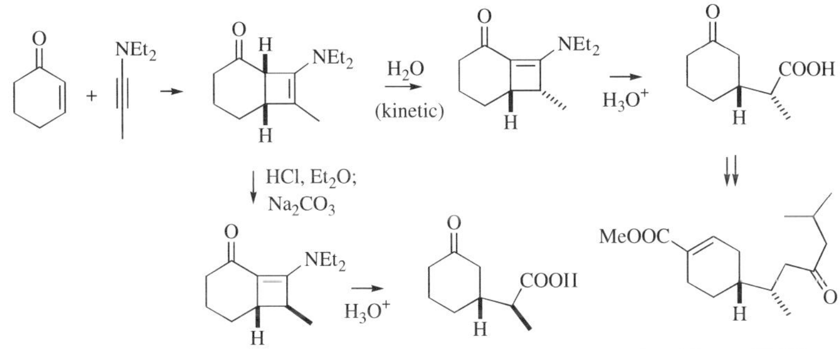

juvabione (**7-10**)

the strain energy is lower in the *cis*-isomer (12.0 kcal/mol) than the *trans*-isomer (18.4 kcal/mol),[19] and access to the former ring system is quite easy. Angular alkylation is also expected to give the *cis*-fused product, and by such steric course it is possible to elaborate (–)-α-cedrene (**7-12**) from pulegone via the fused lactone.[20]

The *exo*-selectivity for reactions of a diquinane derivative was shown in a synthesis of gymnomitrol (**7-13**).[21] Conjugate addition to introduce an angular methyl substituent as well as subangular methylation at the same site as the allyl group delivered an intermediate which embraced total stereochemical information for completion of the required skeleton.

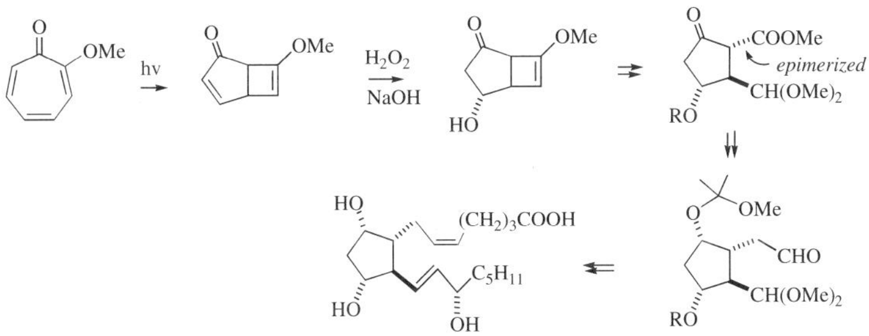

prostaglandin-$F_{2\alpha}$ (**7-11**)

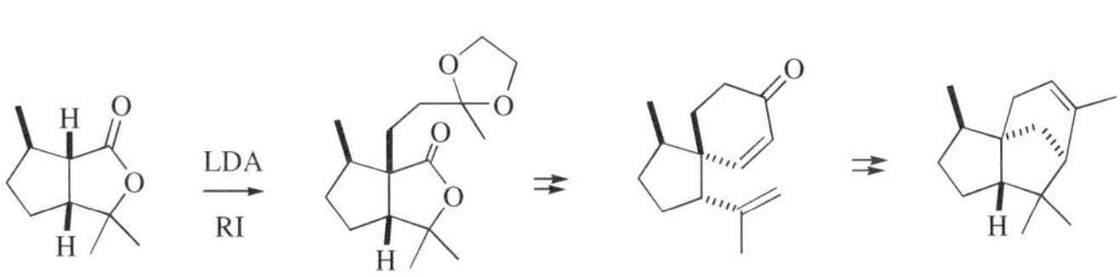

(-)-α-cedrene (**7-12**)

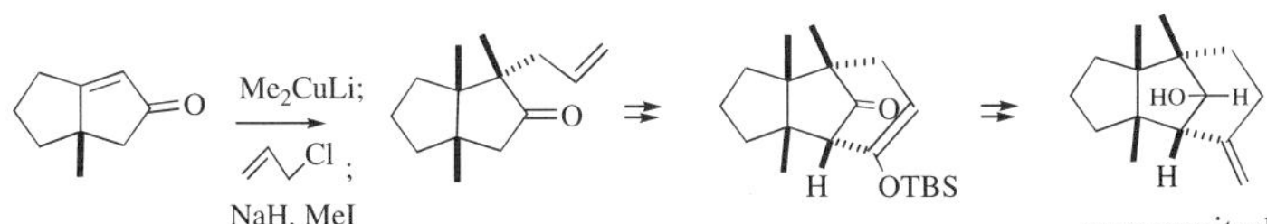

gymnomitrol

(**7-13**)

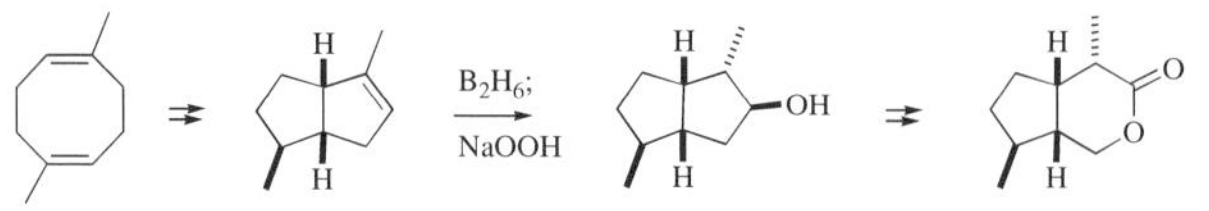

iridomyrmecin (**7-14**)

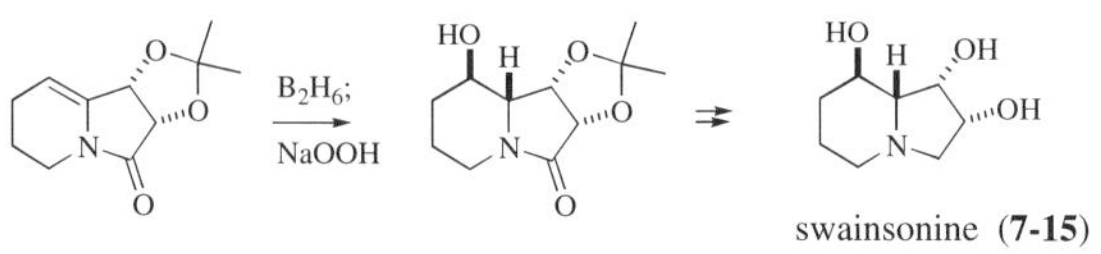

swainsonine (**7-15**)

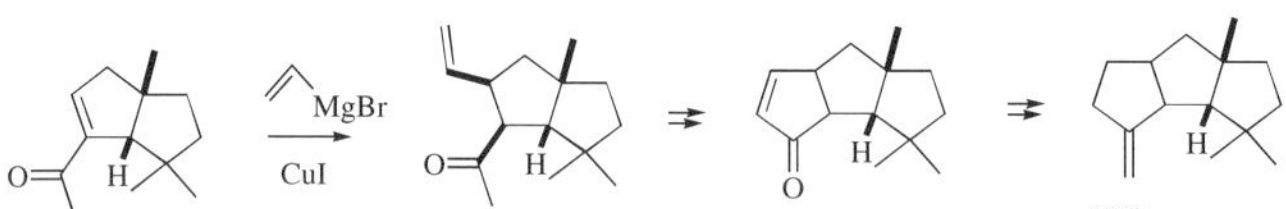

$\Delta^{9(12)}$-capnellene
(**7-16**)

Iridomyrmecin (**7-14**) has been prepared from a diquinene[22] in which the α-methyl group was established in conjunction with elaboration of the lactone ring by hydroboration and other transformations. The method is congruent to requirements of regiochemical and stereochemical aspects. The functionalization of heteradiquinanes with an exocyclic double bond begets the same stereochemical consequences and it has been exploited in a synthesis of swainsonine (**7-15**).[23]

Diquinenes are usually approached from the exo side, and this preference enabled annulation through conjugate addition and cyclization to construct $\Delta^{9,12}$-capnellene (**7-16**)[24] and silphinene (**7-17**).[25]

cis-anti-cis-Triquinanes such as hirsutene (**7-18**) are amenable to strategy involving linear addition of new cyclopentane units. Regardless of whether the reaction at a subangular position forms only one C–C bond[26] or involves cycloaddition,[27] the stereochemical consequences are the same.

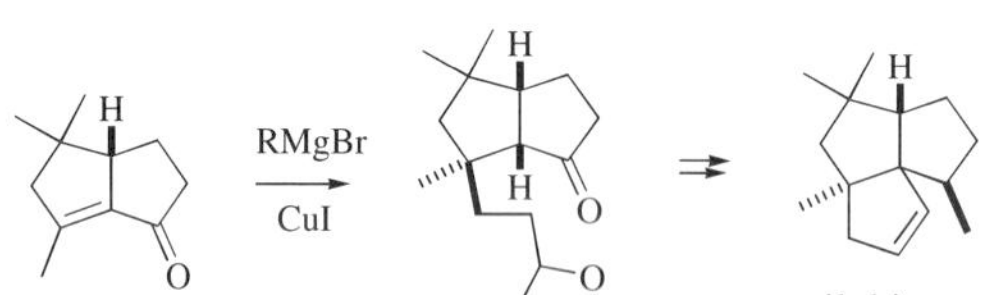

silphinene
(**7-17**)

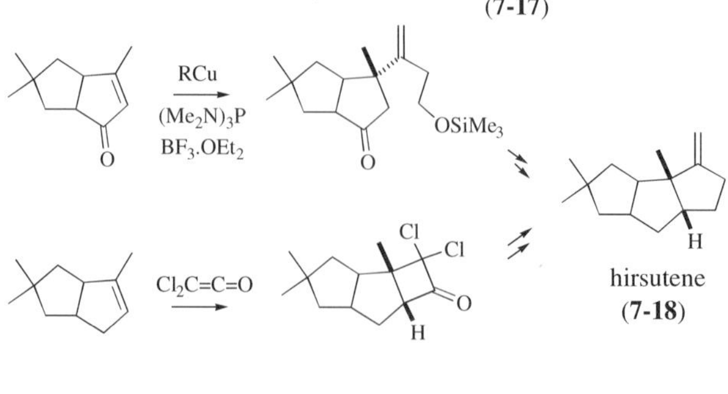

hirsutene
(**7-18**)

Exo approach of reagents to polyquinane networks and preference for generating *cis*-fused ring systems may be illustrated by a synthesis of magellanine (**7-19**).[28] From a retrosynthetic perspective the disconnection of the piperidine ring to expose a seco derivative with the reconstituting chains to be introduced by a Michael reaction–carbomethoxylation tandem appears most expedient. The Michael addend should effect bonding from the less hindered face, and the δ-lactam elaborated thereafter is susceptible to equilibration. Such expectations were fulfilled and it is interesting to note that the two annulation processes in synthesis relied solely on the topographical features of the diquinane subunit. The proper intermediates underwent Michael reactions from the *exo*-face.

The *exo*-selectivity associated with reactions at a subangular position of diquinanes dictated the order of alkylation when such intermediates were employed to complete the bridged ring system of gymnomitrol (**7-13**).[21,29,30] Three tactics evolved to accomplish the goal.

Generation of a brefeldin-A (**7-20**) intermediate from a diquinane by a fragmentation method must attend to the stereochemistry of the electrofugal group (Me_3Sn) and the carbon chain in the adjacent carbon atom. However, it was quite easily accomplished by operation on an enone.[31]

The *endo* orientation of the secondary methyl group of aplysin (**7-21**) readily suggests its generation from a dehydro derivative by hydrogenation. Indeed, two syntheses[32,33] were based on such a tactic.

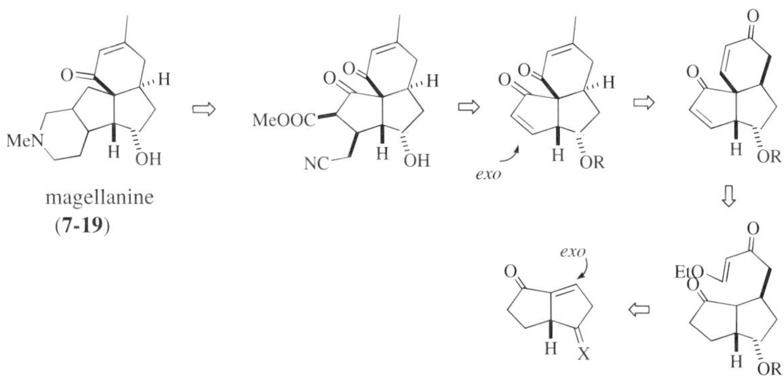

magellanine
(**7-19**)

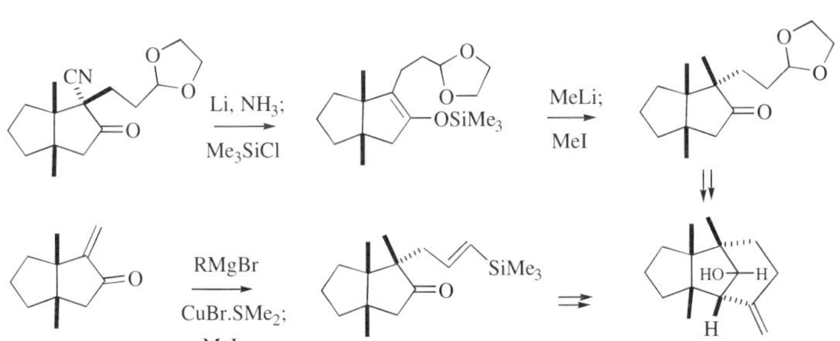

gymnomitrol
(**7-13**)

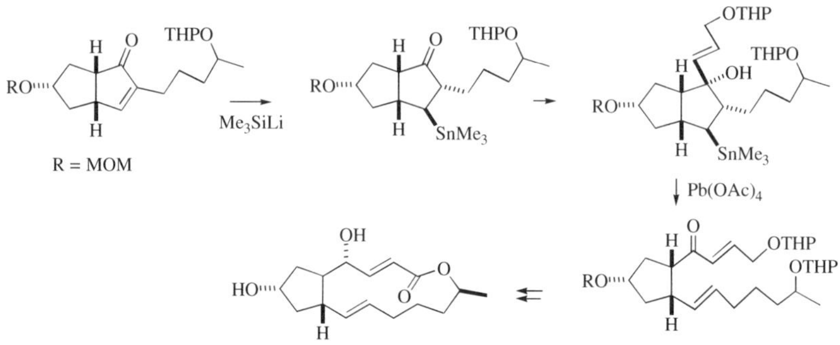

brefeldin-A (**7-20**)

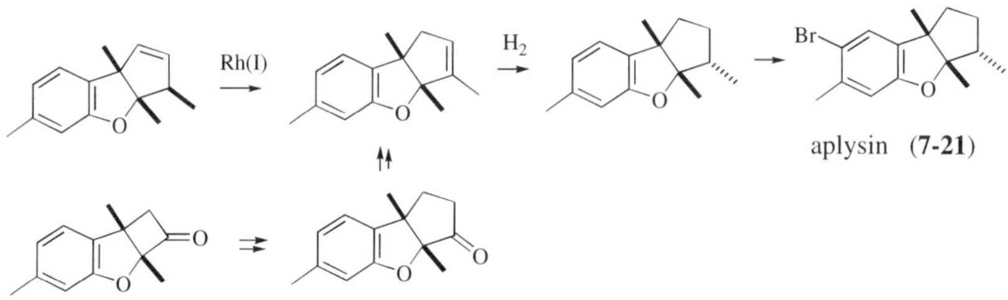

aplysin (**7-21**)

A synthesis of trichodiene (**7-22**)[34] engaged a bicyclic γ-lactone in a Diels–Alder reaction. The diene approached the exocyclic methylene group at the α-position of the lactone unit in accordance with the steric bias of the ring system. The use of 2-phenylthio-1,3-butadiene as the diene component was due to the higher regioselectivity it confers.

Radical reactions at a subangular position also occur preferentially from the *exo* face. This phenomenon has important implications to synthetic designs for prostaglandins (**7-23**),[35] showdomycin (**7-24**),[36] and (+)-pseudomonic acid-C (**7-25**).[37]

In the prostaglandin synthesis, introduction of the ω-side chain was by trapping the bicyclic radical from the *exo* face. Cyclization in the 5-*exo-trig* mode also posits the bulky group in an *exo* orientation.[38] Reductive cleavage of a cyclopropane during

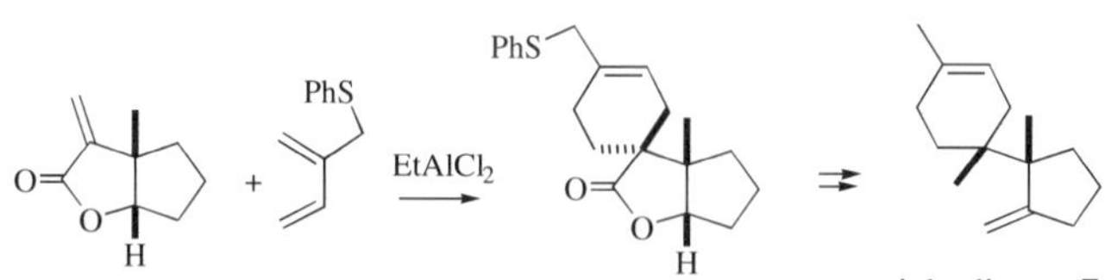

trichodiene (**7-22**)

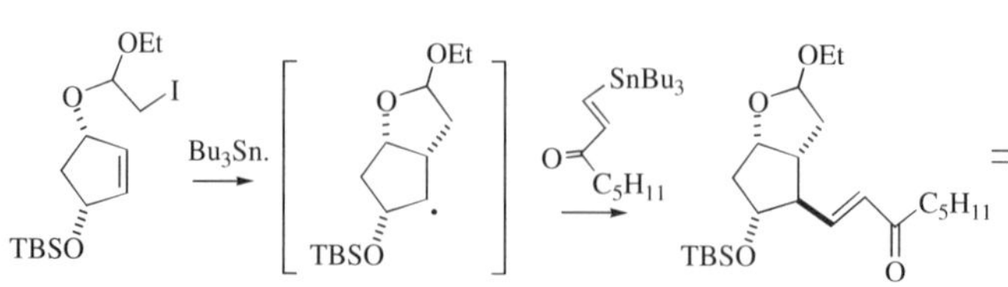

prostaglandins (**7-23**)

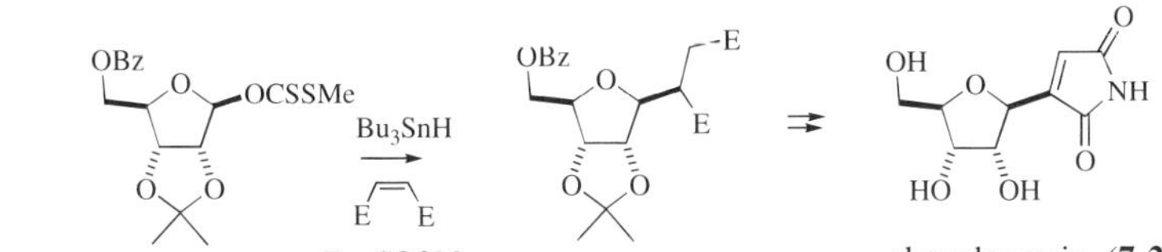

showdomycin (**7-24**)

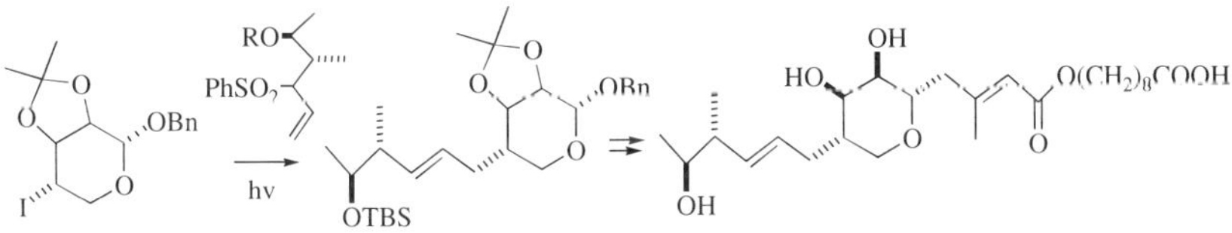

(+)-pseudomonic acid-C (**7-25**)

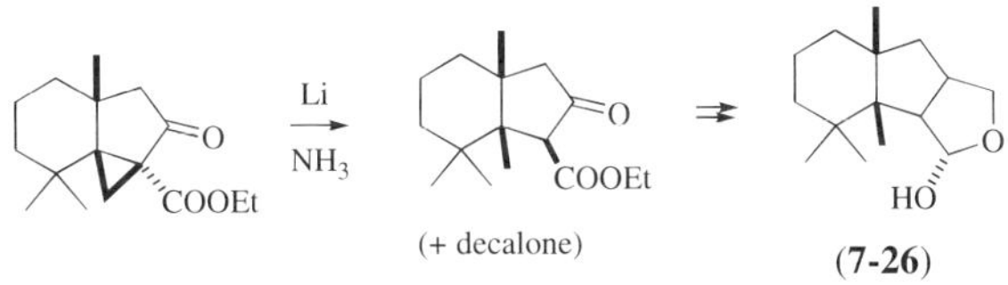

(**7-26**)

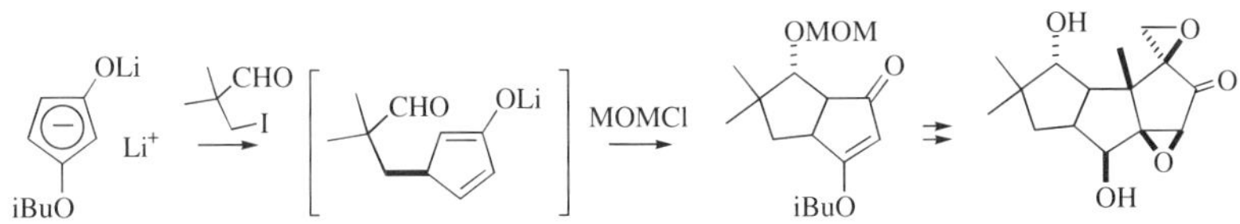

coriolin (**7-27**)

synthesis of a thapsane sesquiterpene (**7-26**)[39] left an *exo*-hydroxymethyl group which was to be incorporated into a lactol unit. One did not have to worry about the stereochemistry of the epimerizable side chain.

In many syntheses of coriolin (**7-27**) the hydroxyl group in the terminal ring arose from reduction under thermodynamically controlled conditions. However, one route[40] is different in that the diquinane precursor was formed by an intramolecular aldol condensation. Because the reaction is reversible, the more stable (*exo*-OH) product was obtained.

An intramolecular Pauson–Khand reaction has been employed successfully to construct intermediates of (–)-α-kainic acid (**7-28**).[41] Of the two diastereomeric transition states the one with a *cis-syn-cis*-heterotriquinane skeleton is in disfavor. Further bias is expected from nonbonding interactions involving an *endo*-methyl group. Significantly, an analogous substrate devoid of the oxazolidinone ring underwent cyclization to give a 1.7:1 mixture with the desired isomer predominating.

It was indicated that external reagents approach a diquinane from the *exo* face. There are exceptions, although rather rare. An example is the appearance of an *endo*-hydroxymethyl group when formaldehyde was used to trap the Michael adduct of an enone intermediate, en route to quadrone (**7-2**).[42] The statement must also be qualified: thermodynamically controlled reduction (with reagents generating solvated electrons as the reducing agent) can give rise to products in which the entered

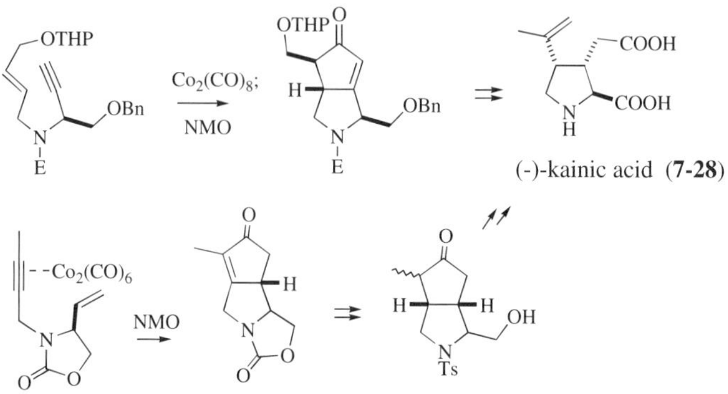

(-)-kainic acid (**7-28**)

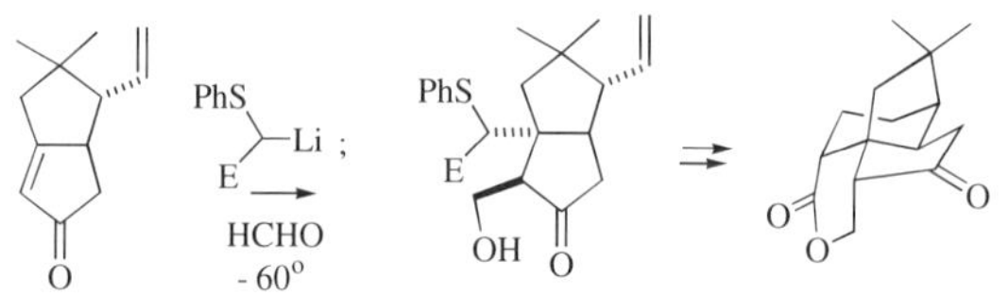

quadrone (**7-2**)

hydrogen atom is *endo*; such as in the establishment of the secondary methyl group at the bicyclic stage in a synthesis of cedrene (**7-12**)[43] as well as in the reduction of tricyclic precursors of coriolin with conforming configuration.[44,45]

endo-Dicyclopentadiene is readily available and its use as a cyclopentane/cyclopentene building block has been extensively exploited. Its attributes also include facial selectivity when reactions are carried out in the nonbridged portion of the molecule. For example, in a short synthesis of multifidene (**7-29**),[46] cycloaddition with dichloroketene achieved a formal *cis*-1,2-difunctionalization of cyclopentadiene.

Chiral dicyclopentadienones can be obtained by resolution methods,[47] and the Sharpless kinetic resolution.[48] The Diels–Alder adduct of cyclopentadiene with chiral 4-hydroxymethylbutenolide which can be formed from mannitol is also a versatile precursor since it is possible to convert the (+)-*endo*-dicyclopentadienone to the

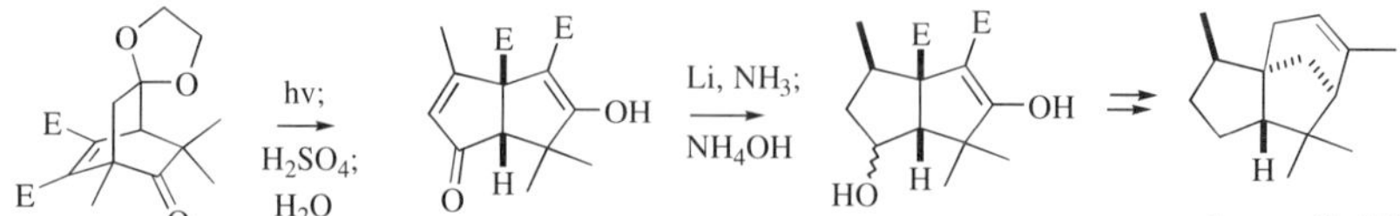

cedrene (**7-12**)

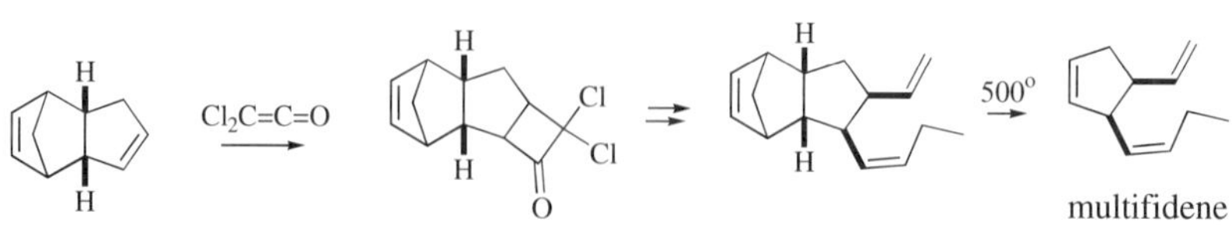

multifidene (**7-29**)

(−)-isomer by way of a Wharton rearrangement and oxidation. More recently, a report described the lipase-mediated transesterification of *cis*-4-*t*-butoxycyclopent-2-en-1-ol to separate the enantiomers and conversion of each to the chiral *endo*-dicyclopentadienone.[49] The remarkable observation is that the 4-*t*-butoxycyclopent-2-enone underwent a contrasteric Diels–Alder reaction, resulting in an adduct with an *endo*-*t*-butoxy group. Synthetic applications of these chiral building blocks have been extensively investigated,[50] and are based on *exo*-selective reactions on the cyclopentenone portion and release of the processed intermediates by a retro-Diels–Alder reaction to get rid of cyclopentadiene.

Chiral amine precursors of *S*-glutamic acid,[51] and the dihydroxycyclopentenylamine portion of queuosine have been prepared[52] using reductive amination on the (−)-enone and (+)-enone, respectively. Most other syntheses operated on the functionalization of the double bond and proceeded to give cyclopentenone or cyclopentanone subtargets. These latter compounds were then subjected to proper modification, either with or without a ring cleavage maneuver according to demand. It is needless to reiterate the fact that reactions on the *exo* face of the diquinane moiety were planned and observed.

A synthesis of (+)-α-cuparenone (**7-30**)[53] involved reaction of the (+)-enone with methyllithium, allylic transposition, and conjugate addition of the *p*-tolyl group before pyrolytic generation of the cyclopentenone precursor. The diquinane intermediate of (−)-$\Delta^{9,12}$-capnellene was formed by conjugate addition, methylation, intramolecular alkylation, and pyrolysis.[54] In the approach to (+)-equilenin,[55] the alkylation sequence that followed conjugate addition of the 6-methoxy-2-naphthyl substituent specified allylation prior to methylation so as to establish the *trans*-C/D ring junction. The allyl group was to be degraded and activated for ring closure.

Two syntheses of (−)-aphanorphine were accomplished,[56] operated on the β-carbon of the (+)-enone or the α-carbon of the (−)-enone to generate the quaternary center. Interestingly, the carbinolamine intermediate in the latter process has been converted into (−)-physostigmine (**7-31**).[57] By varying the aryl group and starting with the proper enantiomeric form of the enone to prepare the tricyclic α,α-disubstituted cyclopentanones, the elaboration of (−)-herbertene and (−)-aplysin has been demonstrated.[58]

(−)-Goniomitine (**7-32**) is another alkaloid which has been elaborated[59] via a chiral cyclopentanone derived from the tricyclic ketone.

An intriguing question arises when one considers the possibility of employing the tricyclic dienone as a D-ring synthon for a more direct approach to estrone (**7-33**), in other words, by a Diels–Alder reaction followed by angular methylation. To realize

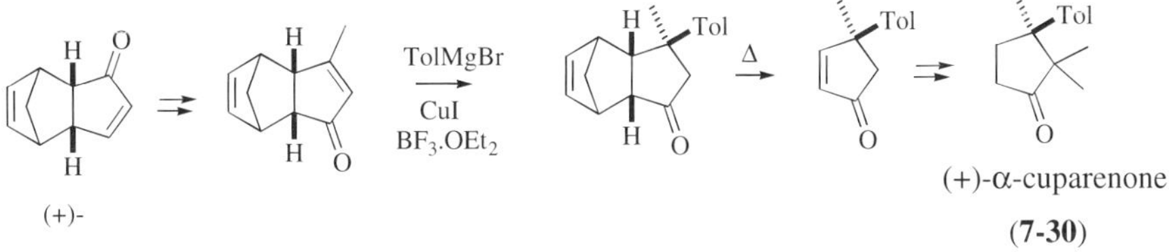

(+)-α-cuparenone

(**7-30**)

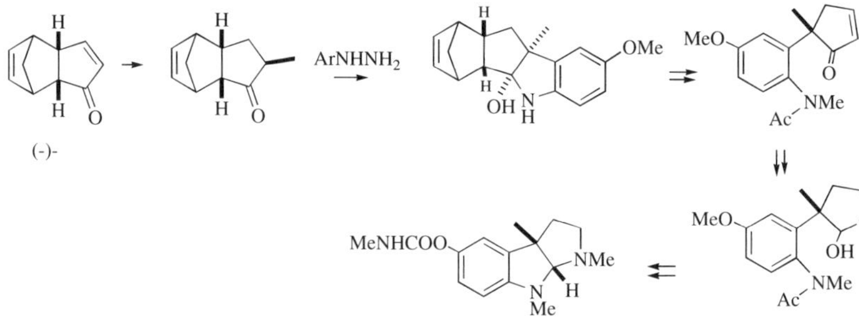

(-)-physostigmine (**7-31**)

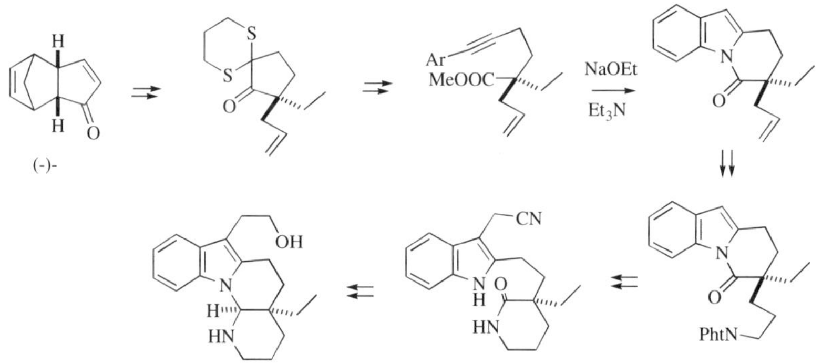

(-)-goniomitine
(**7-32**)

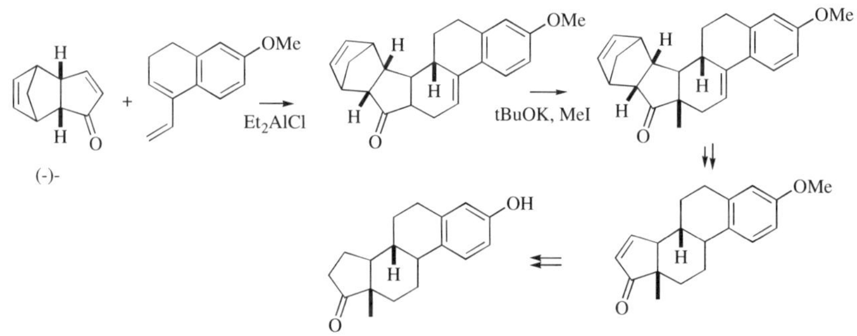

(+)-estrone (**7-33**)

the goal, two unusual features must be conformed to, namely, that the Diels–Alder reaction proceeds via an *endo* transition state and the methylation gives rise to a *trans* ring fusion. The special disposition of the dienone and that of the ensuing cycloadduct indeed favor the desired steric course.[60]

Modification of the Pauson–Khand reaction of norbornadiene gave a chiral α-alkoxy *exo*-dicyclopentadienone which has found use in a synthesis of

(+)-brefeldin-A (**7-20**).[61] Conjugate addition, reductive removal of the alkoxy group, followed by Lewis acid catalyzed retro-Diels–Alder reaction led to a cyclopentenone.

In the synthetic study toward the ginkgolides (**7-34**),[62] when the stage was set to elaborate the γ-lactone by annexation to a cyclopentenone subunit, the organometallic reaction was found to be sterically unaccommodating. However, after epoxidation of the conjugated double bond, the steric situation was reversed and the reaction with *t*-butyl α-lithiopropionate occurred predominantly from the opposite face.

Highly stereoselective reactions can be achieved on the Diels–Alder adducts of a 1,5-bridged cyclopentadiene.[63] By kinetic resolution using a chiral diene in the Diels–Alder reaction, it is possible to acquire enantioselectively products such as (–)-methyl cucurbate (**7-35**)[64] and didemnenone-C (**7-36**).[65]

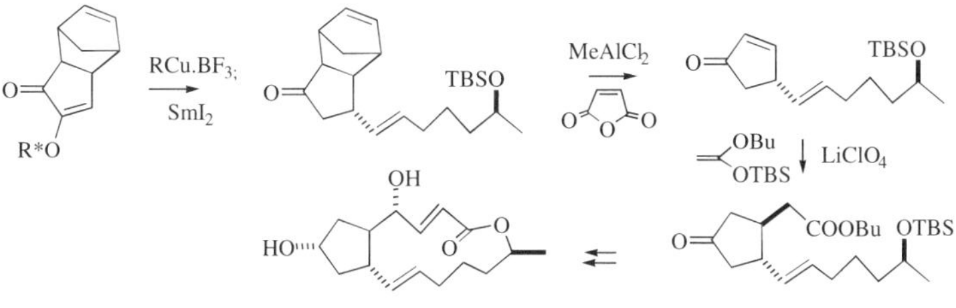

brefeldin-A (**7-20**)

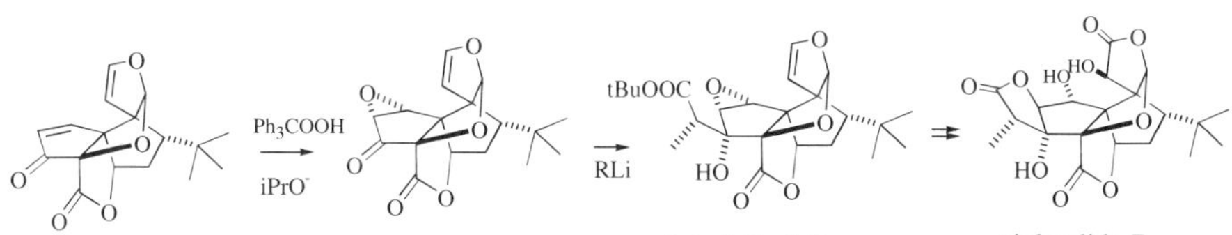

(selectivity 8:1)

ginkgolide-B
(**7-34**)

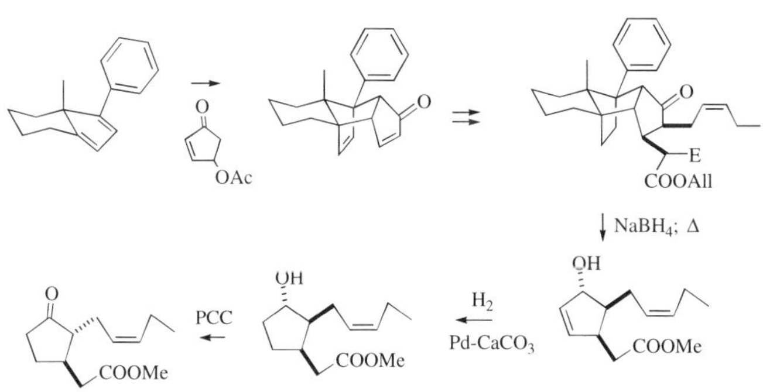

(-)-methyl jasmonate (-)-methyl cucurbate
(**7-35**)

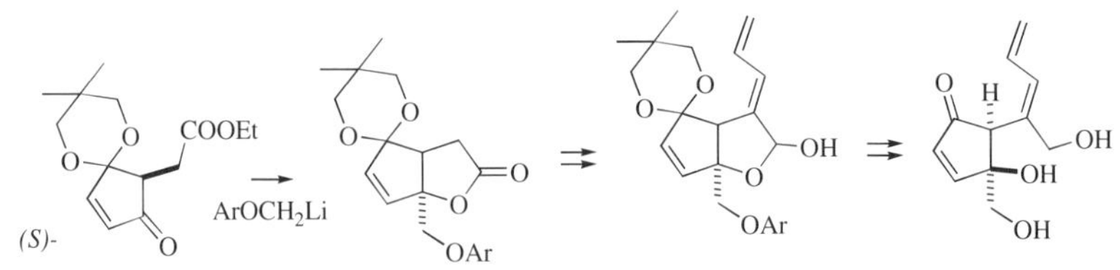

[unreacted with diene]

didemnenone-C
(**7-36**)

The 5/6-fused ring systems also show a similar trend to the lower homologues in many reactions regarding their steric courses. Thus, a useful solution to the problem in establishing the two contiguous quaternary centers of trichodiene (**7-22**) involved stepwise alkylation of a *cis*-fused lactone.[66] The properly substituted lactone was devolved to a cyclopentanone precursor of the terpene molecule by a Dieckmann condensation.

An analogous stereochemical issue is the spiroepoxy ring of phyllanthocin (**7-37**). A synthesis of the sesquiterpene[67] also took advantage of the topographical guidance in alkylation.

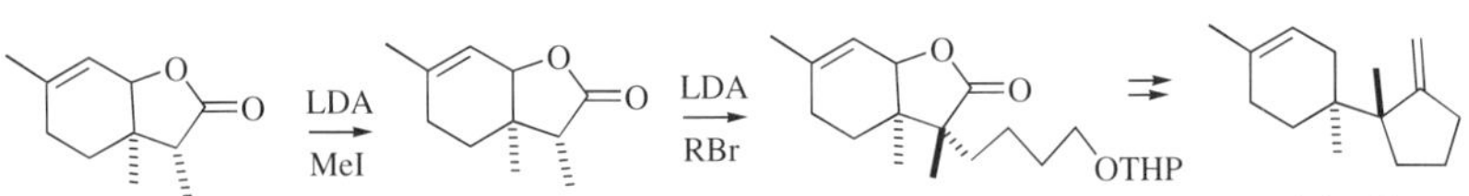

trichodiene
(**7-22**)

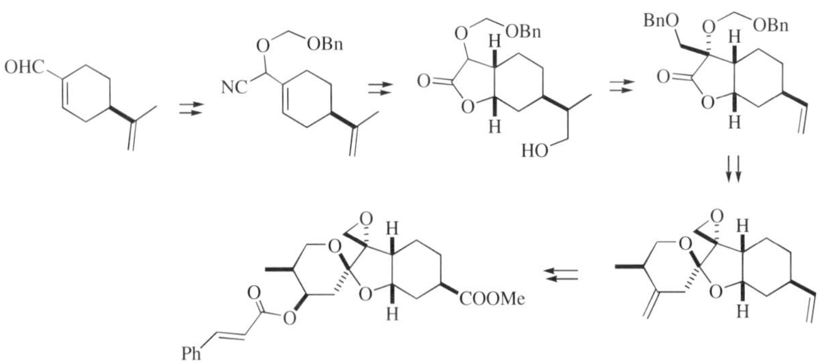

phyllanthocin (**7-37**)

An approach to gibberellin A-38 methyl ester (**7-38**)[68] took advantage of allylation of an aldehyde at a subangular position of a bridged perhydrindan from the *exo*-side. The allyl group provided the $C_{(1)}$-to-$C_{(3)}$ fragment of the target.

Although an intramolecular Diels–Alder reaction was the key step for assembling the A-ring of gibberellic acid (**7-39**),[69] it also required methylation at $C_{(4)}$. This was readily accomplished because the topography of the molecule strongly disfavored reaction at the concave side. The methylation site is both an angular atom

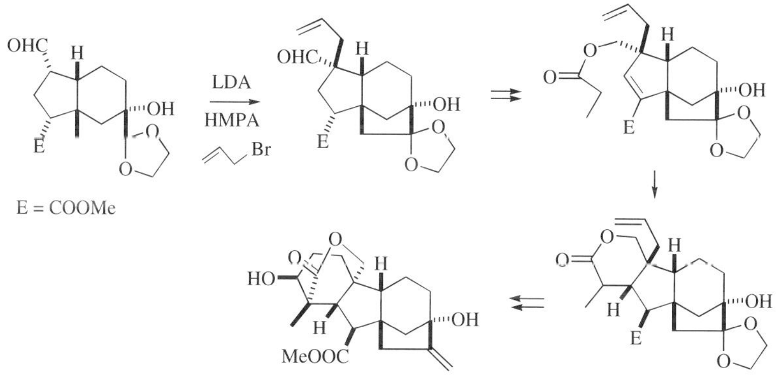

gibberellin-A_{38} methyl ester (**7-38**)

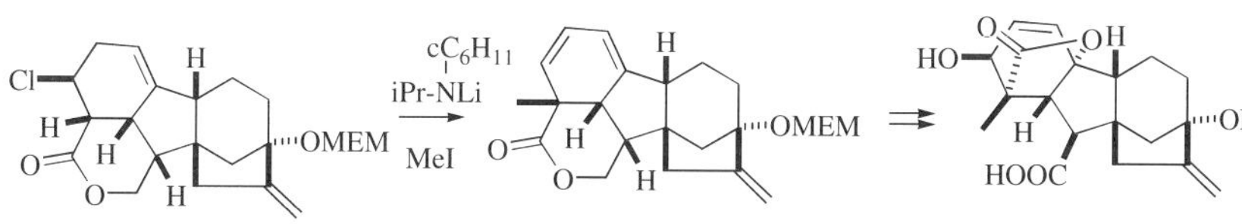

gibberellic acid (**7-39**)

(of the homophthalide system) and a subangular position (of a lactone fused to the cyclopentane).

Michael addition at a subangular position of a 5/6-*cis*-fused ring system is subject to conformational effects. For a synthesis of pinguisone (**7-40**)[70] from a bisenone, the more reactive cyclopentenone was attacked quite indiscriminately in stereochemical terms. The problem was partially solved by using isopropanethiol as Michael donor and the *endo* adduct was processed at the cyclohexenone moiety until the furan was installed. The preferred conformation of the regenerated cyclopentenone at that stage is less flat and only one side was exposed to the dimethylcuprate reagent. Pinguisone was thus obtained.

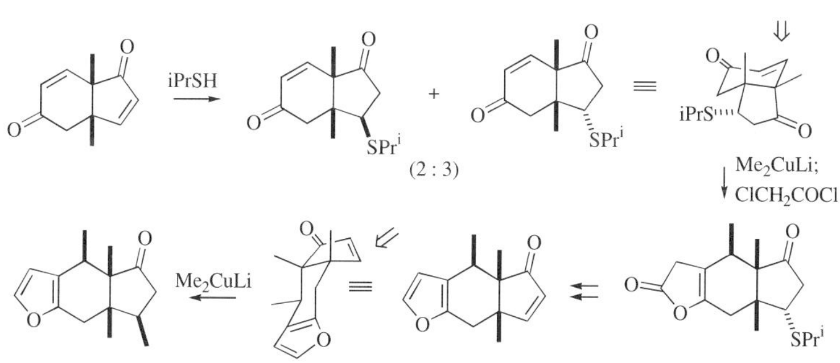

pinguisone (**7-40**)

Another tactic for ensuring the steric course is by locking the hydrindenone conformation with the help of an acetal in the six-membered ring.[71] The acetal group also blocked the *endo* side for the methyl group to enter.

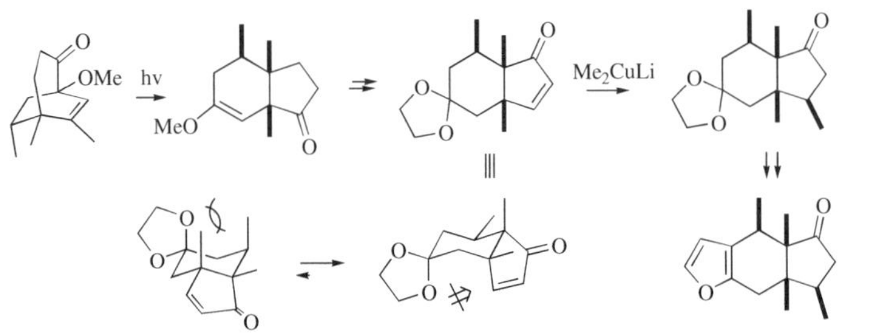

pinguisone
(**7-40**)

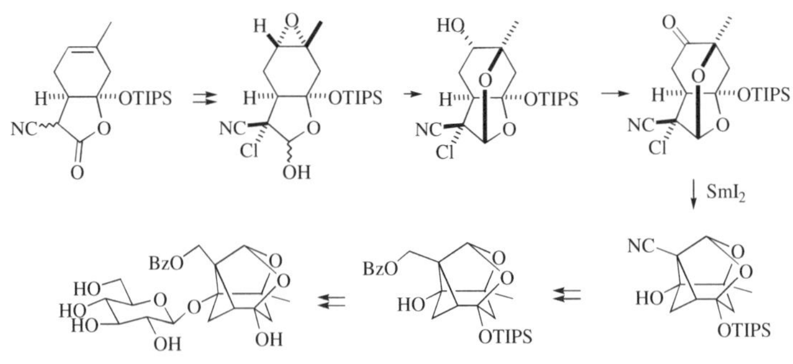

paeoniflorin (**7-41**)

Epoxidation at the *exo*-face of an unsaturated *cis*-fused γ-lactone was crucial to creation of the cyclic acetal substructure of paeoniflorin (**7-41**).[72] With the locked conformation the closure of a cyclobutane was facilitated.

The predilection for the formation of *cis*-5/5 and *trans*-5/6 fused ring systems has an accumulative influence on the stereochemical outcome when pleurotin (**7-42**) was pursued by way of radical cyclization from a bridged lactone.[73]

The success in synthesizing (–)-ptaquilosin (**7-43**)[74] depended on *exo*-attack on a hydrindenone by a Grignard reagent which established the cyclopropylcarbinol.

Thus, *cis*-5/6 fused ring systems generally exhibit the same stereoselectivity as the 5/5-counterparts. A synthesis of isovelleral (**7-44**)[75] involving cyclopropanation was planned accordingly. However, the reason for favoring an opposite steric course in a similar substrate, therefore leading to methyl isomarasmate (**7-45**) instead,[76] remains unknown.

The synthesis of 17-oxoaspidofractinine (**7-46**) is an imposing problem due to its molecular complexity. However, a Diels–Alder approach to construct the hexacyclic framework[77] proved satisfactory with the dienophile adding from the same side as the angular methine hydrogen.

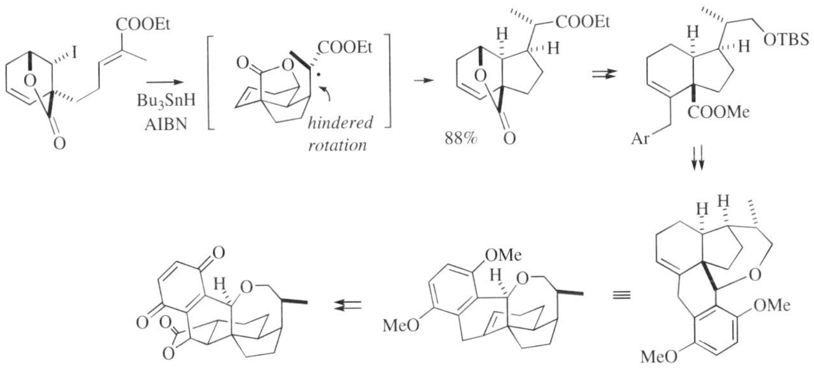

pleurotin (**7-42**)

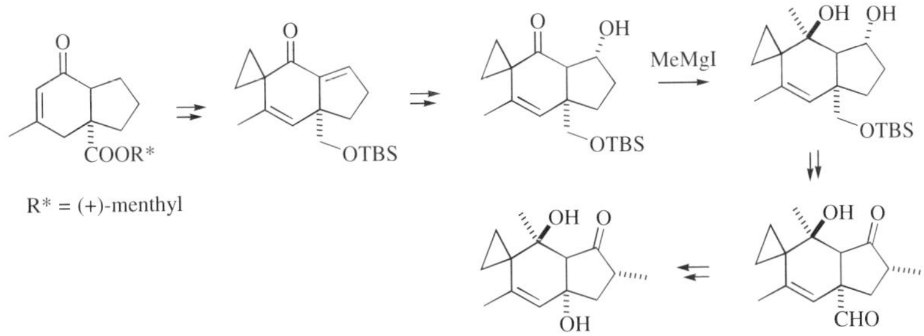

(-)-ptaquilosin
(**7-43**)

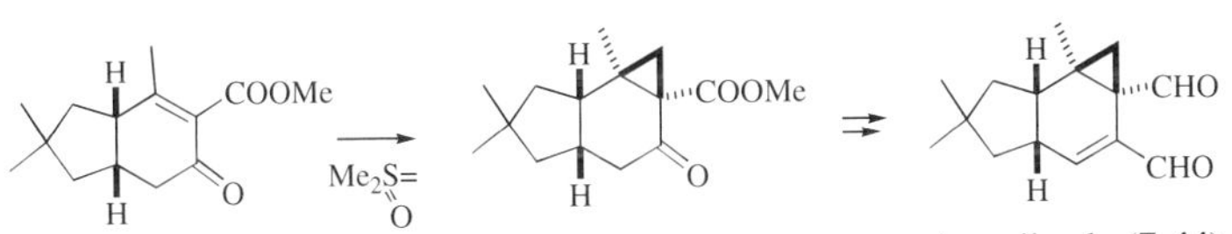

isovelleral (**7-44**)

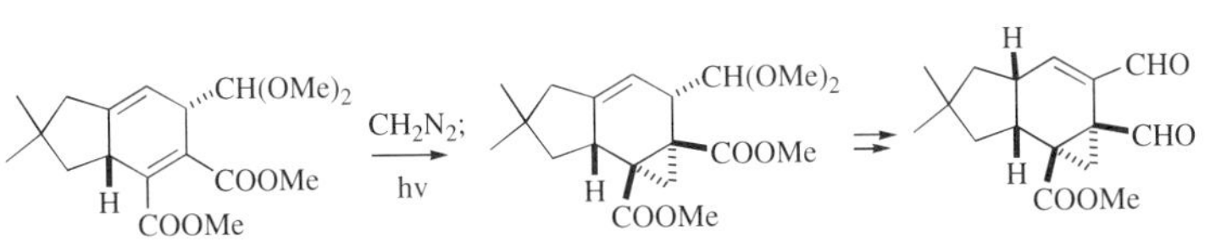

methyl isomarasmate
(**7-45**)

The extracyclic aldehyde of a fused bistetrahydrofuran was found to undergo crotylation to give mainly one diastereomer, affording a favorable situation en route to kumausallene (**7-47**).[78] It seems that the *endo*-aldehyde was incapable of forming a chelate together with the proximal ethereal oxygen atom, even in the presence of a bidentate Lewis acid, leading to simple Cram stereoselection.

An approach to seychellene (**7-48**)[79] which involved formation of the bridged ring system by intramolecular alkylation of a *cis*-decalone was hinged on the

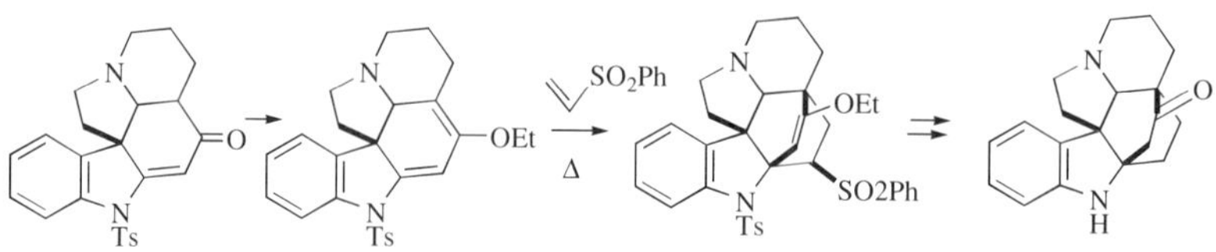

17-oxoaspidofractinine (**7-46**)

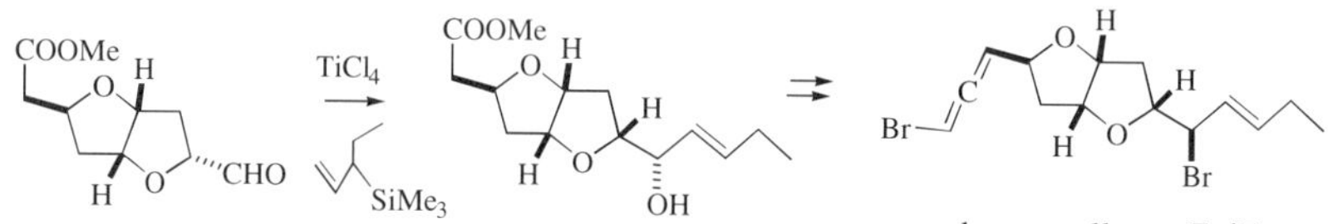

kumausallene (**7-47**)

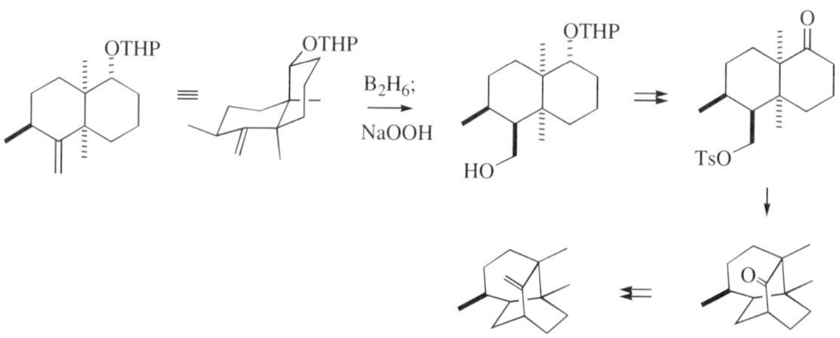

seychellene (**7-48**)

availability of the substrate. The crucial one-carbon pendant containing the leaving group was placed correctly in an *endo* orientation by means of hydroboration.

7.1.1.2. Reactions at Positions Farther from the Angle. Stereoselectivity is still observable at positions farther from the angle of certain *cis*-fused ring systems. The synthetic usefulness of such a phenomenon is demonstrated in hydrogenation which led to crucial intermediates of coronafacic acid (**7-49**)[80] and ibophyllidine (**7-50**).[81]

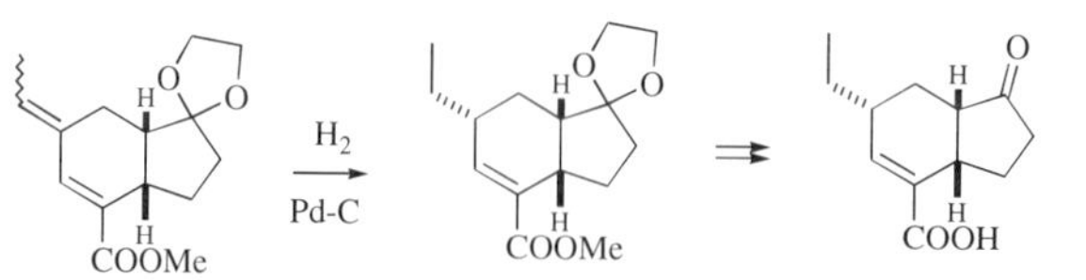

coronafacic acid (**7-49**)

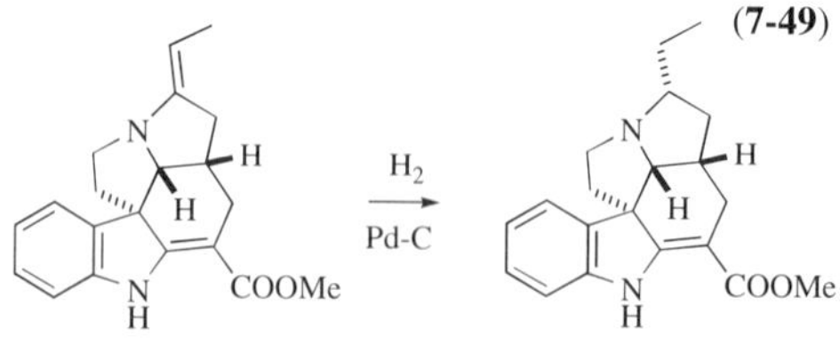

ibophyllidine (**7-50**)

Other examples pertain to syntheses of hydroxysemperoside aglucone (**7-51**),[82] sterpuric acid (**7-52**),[83] bakkenolide-A (**7-53**),[84] and isoclovene (**7-54**),[85] which involved sigmatropic rearrangement with rehybridization of an exocyclic sp^2-carbon atom. C–C Bond formation proceeded from the *exo*-side. (Asterisks in the formulas indicate the induced stereogenic centers.)

Natural bicyclic compounds in which one of the fused rings is a cyclopropane are often valuable building blocks even if the target molecules lack such structural components. For example, α-cyperone (**7-55**) is much more readily available from 3-caranone[86] or 2-carone[87] than from dihydrocarvone because the latter affords predominantly a bicyclic ketone corresponding to epi-α-cyperone, due to stereoelectronic effect which governs the Robinson annulation. By the same token, a synthesis of (–)-phytuberin (**7-56**)[88] took advantage of the directing effect of the *gem*-dimethylated cyclopropane.

It is of interest to note the effective shielding of the cyclohexane unit on an *endo*-oriented unsaturated side chain attached to $C_{(7)}$ of the bicyclo[4.1.0]hexane system.[89] The conjugate addition on this compound served to introduce the isopropenyl group with relative configuration corresponding to that of eremophilone (**7-57**).

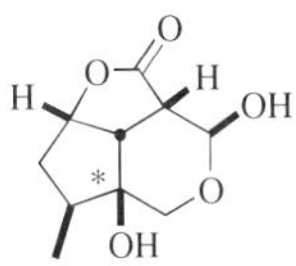

hydroxysemperoside aglucone (**7-51**)

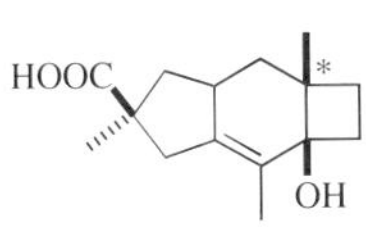

sterpuric acid (**7-52**)

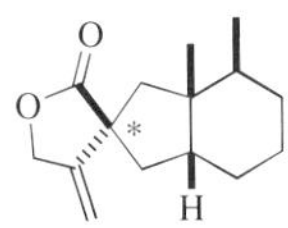

bakkenolide-A (**7-53**)

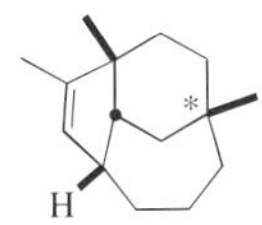

isoclovene (**7-54**)

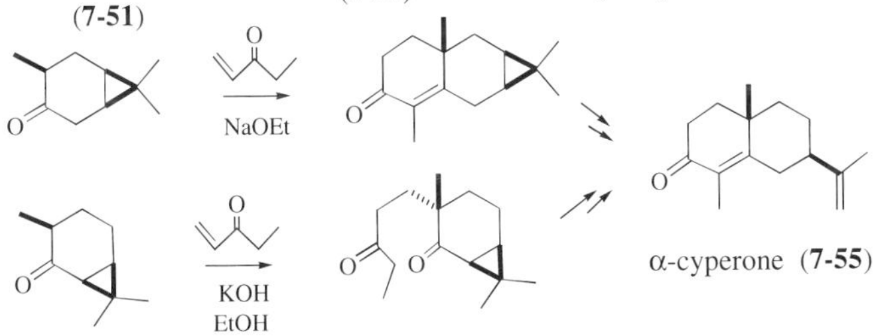

α-cyperone (**7-55**)

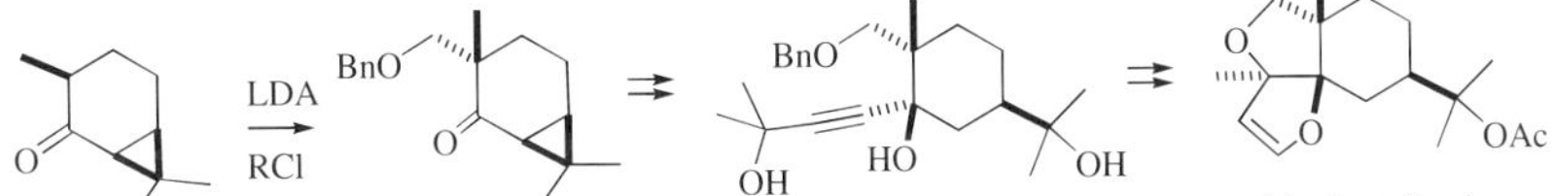

(-)-phytuberin (**7-56**)

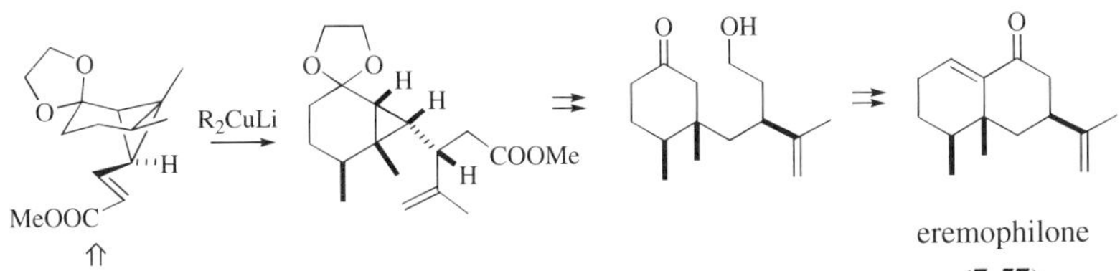

eremophilone (**7-57**)

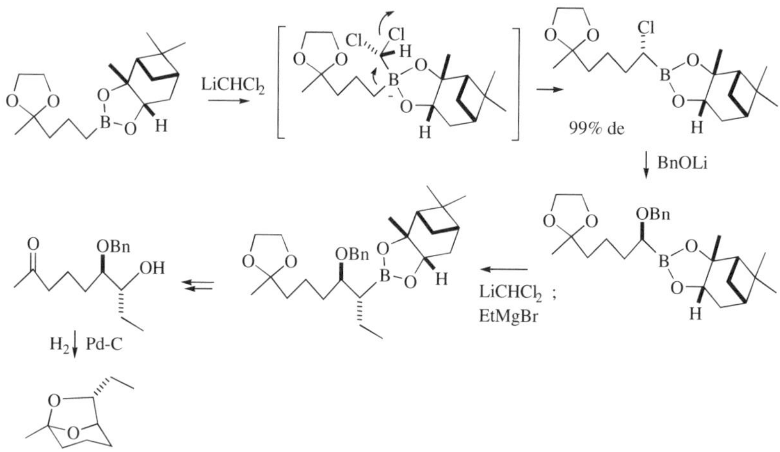

exo-brevicomin (**7-58**)

Boronic esters derived from chiral pinanediols have been investigated in terms of their ability to induce chirality to inserted fragments. To illustrate this capacity a synthesis of *exo*-brevicomin (**7-58**)[90] may be mentioned. Attack of an organolithium reagent at boron *syn* to the nearby angular substituents was followed by rearrangement. After an S_N2 reaction with lithium benzyloxide the process was repeated with the variation of chlorine atom displacement by reaction with ethylmagnesium bromide before oxidative cleavage of the C–B bond.

An astute analysis of the problem facing the synthesis of tetracycline antibiotics (**7-59**)[91] that concerns the beginning of annexing the A-ring to a tricyclic intermediate (in the classical BCD+A approach) indicated that a superior solution lies in delaying closure of the C-ring. As the direction for the $C_{(4)}$–$C_{(4a)}$ bond formation is determined by the stereochemistry at the B/C-ring junction (C_{5a}), the undesired result always appears from reactions on the linear array. This propensity for convex approach can be turned into advantage by setting the C-ring precursor to an angular fusion mode for

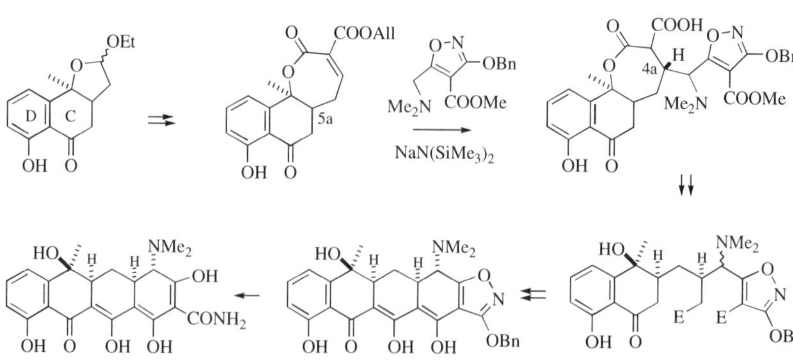

(**7-59**)

the stereoinduction, and after the proper elements for future construction of the A-ring are attached the completion of a tetracycline synthesis is a matter of release of the latent chains and their cyclization.

7.1.2. Bridged Ring Systems

Nonequivalent spatial environments on the two surfaces of the cyclohexane ring in pinane derivatives are easily appreciated. This property has prompted many investigations of stereoselective elaboration of pinanes into useful and/or interesting compounds. Nopinone has been gainfully employed in a synthesis of nootkatone (**7-60**).[92] The relative stereochemistry of the quaternary center at the ring junction was set in the methylation step which occurred exclusively away from the protruding *endo* branch of the *gem*-dimethyl subunit.

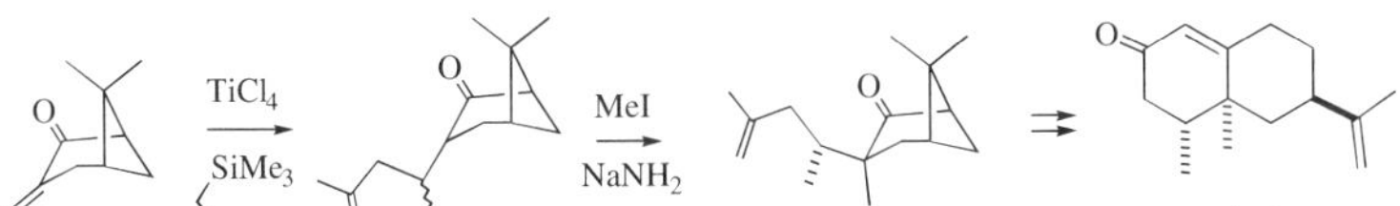

nootkatone
(**7-60**)

The marked *exo*-preference for small bridged systems such as bicyclo[2.2.1]heptane derivatives to react with either electrophiles, nucleophiles, or cycloaddends is well known. The steric propensity has been rationalized in terms of steric and torsional factors. Torsional strain is decreased on sp^2-to-sp^3 rehybridization of a carbon on the longer bridge when the attack comes from the *exo* side, but it is increased on *endo* attack.[93]

By variation of the order of alkylation, precursors of β-santalene (**7-61**) and epi-β-santalene (**7-62**) can be acquired from norcamphor at will.[94]

For an approach to pentalenene (**7-63**) according to a retrosynthetic analysis which features a diquinane intermediate, the presence of a stereocenter in a free-rotating side chain is most problematic. A solution emerged when the side chain was to be developed from a bridged ring system.[95] A viable route involved formation of a 4,5-cyclopropanonorbornene was devised, and the secondary methyl group was introduced correctly from the *exo* side.

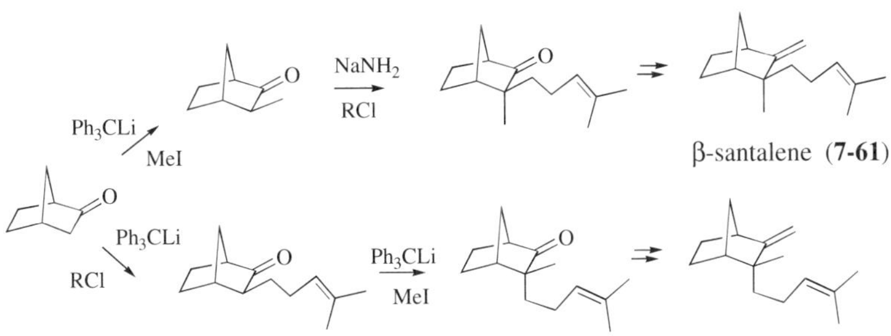

β-santalene (**7-61**)

epi-β-santalene (**7-62**)

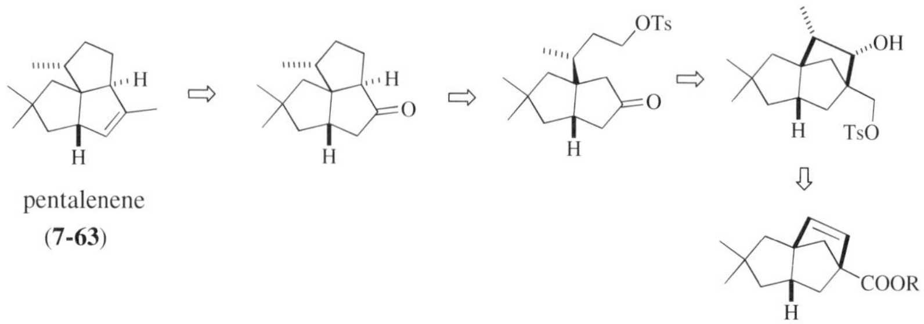

pentalenene (**7-63**)

Another example is the devolution of a bridged ring ketone to a trisubstituted cyclopentanone for elaboration of the CD-block of 1α,25-dihydroxyvitamin D_3 (**7-63a**).[96]

Claisen rearrangement of a norbornenylmethyl vinyl ether placed an acetaldehyde chain in an *exo*-configuration, thereby ensuring the access to β-santalene (**7-61**).[97]

The hydrogenation of an exocyclic unsaturated ester delivered a stereochemically appropriate intermediate for the synthesis of hirsutic acid-C

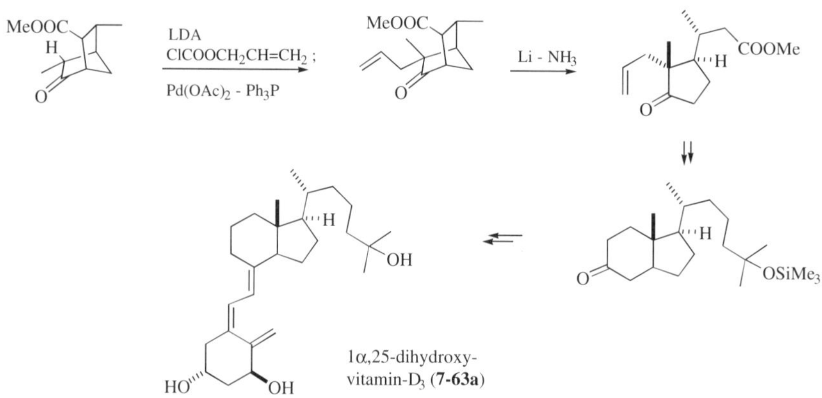

1α,25-dihydroxy-vitamin-D_3 (**7-63a**)

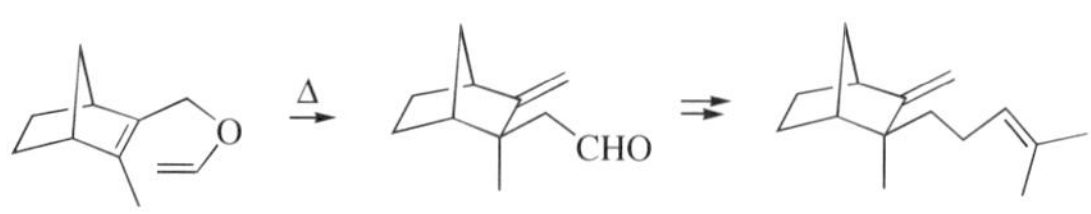

β-santalene (**7-61**)

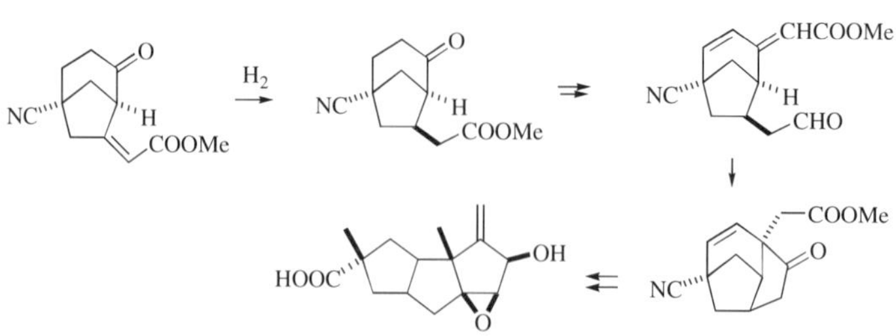

hirsutic acid-C (**7-64**)

(**7-64**).[98] Even in a more complex ring system, the hydrogenation of an alkene occurred from the *exo* side. This step was crucial in an elaboration of 9-isocyanopupukeanane (**7-65**).[99]

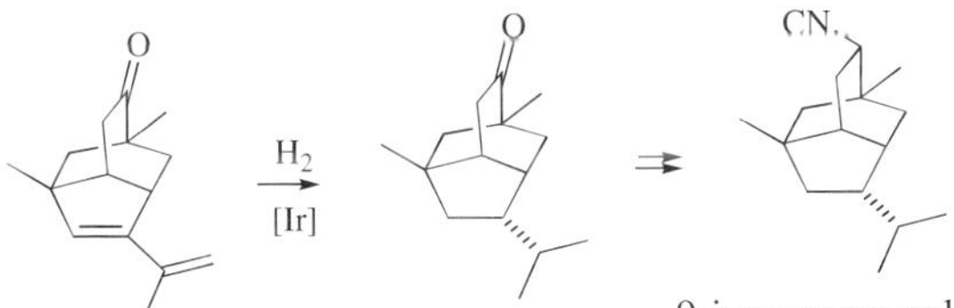

9-isocyanopupukeanane
(**7-65**)

Let us consider a development of strychnos alkaloid synthesis.[100] A very efficient route involves coupling of an indole-2-acetic ester with a 3-substituted pyridinium salt. In essentially one operation the tetracyclic nucleus can be assembled. Successive nucleophilic attack on $C_{(4)}$ and $C_{(6)}$ of the pyridinium species led to the bridged ring system. For the synthesis of tubifolidine (**7-66**), hydrogenation of the intermediate from the *exo* side was the last crucial step which needed any stereochemical attention.

It must be emphasized that the presence of a bulky *syn*-7 substituent (a methyl group is sufficient) in a bicyclo[2.2.1]heptane or heptene derivative suppresses the *exo*-selectivity to the extent of its exclusion. Reports on steric guidance of this kind are numerous and only a few cases are mentioned here. *endo*-Methylation of such norbornenones was exploited in the synthesis of damsin (**7-67**)[101] and the CD-synthon (**7-68**) for steroids.[102]

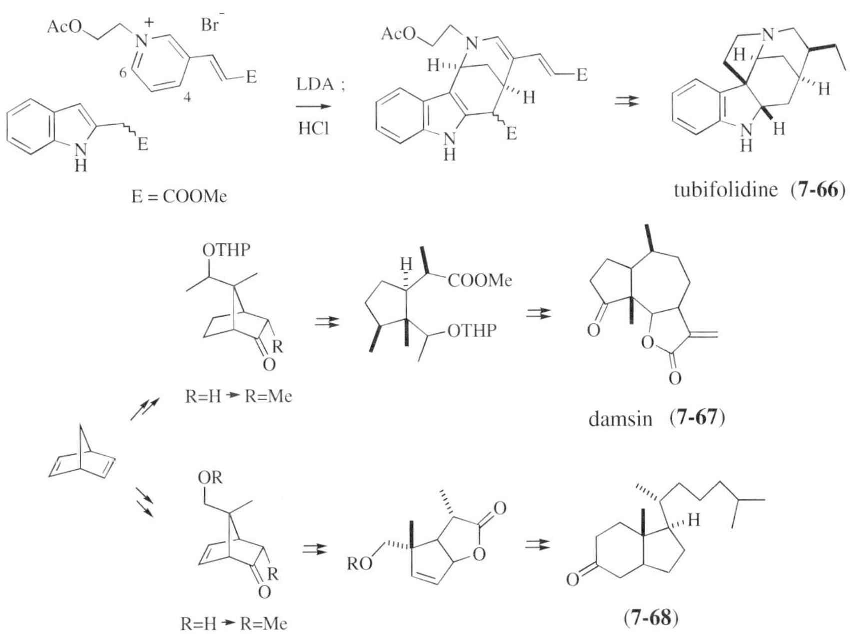

A route to coronafacic acid (**7-49**)[103] involved an oxy-Cope rearrangement of a substrate prepared from 7,7-dimethoxybicyclo[2.2.1]hept-5-en-2-one. The benzofuran-2-yl moiety was introduced from the *endo* side.

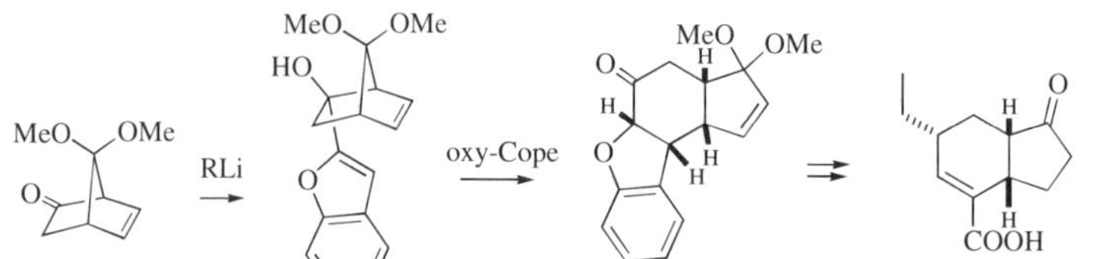

coronafacic acid
(**7-49**)

A remarkable exception to the trend of *endo* entry was witnessed in the kinetic aldol condensation of 9-bromocamphor with furaldehydes.[104] While all of the factors that cause this apparent anomaly are difficult to clarify, torsional strain which prompts the twist of the functionalized bridge away from the bridgehead methyl group (increase in dihedral angle) should make the *exo*-face accessible. This steric course of predominant reaction enabled the synthesis of furodysin (**7-69**) and furodysinin (**7-70**) in a more straightforward fashion.

Also unusual is the observation of the stereoselectivity for methylation of bicyclic allyl sulfones.[105] The results are contrary to expectation.

The *exo*-selectivity of cycloaddition to the norbornene system has been exploited in the synthesis of chasmanine (**7-71**)[106] and albene (**7-72**).[107]

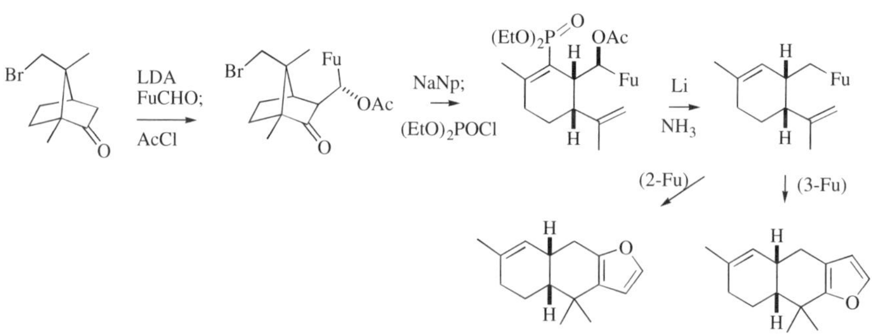

furodysinin
(**7-70**)

furodysin
(**7-69**)

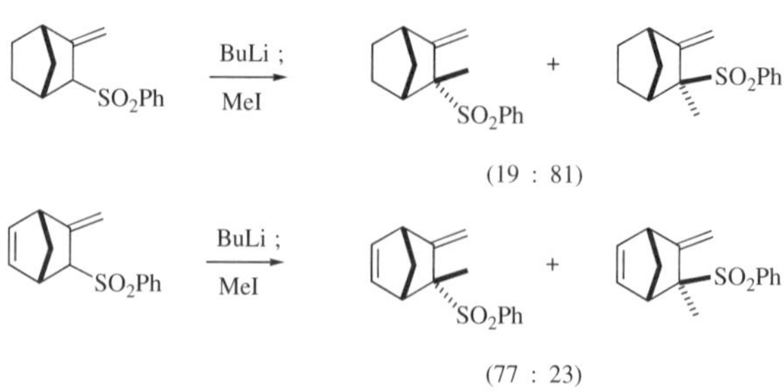

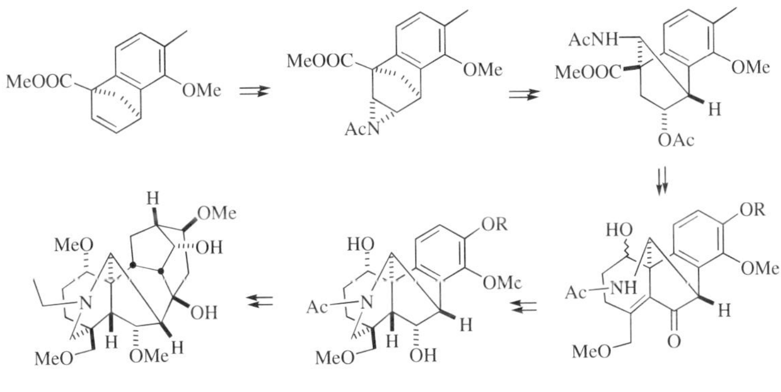

chasmanine
(**7-71**)

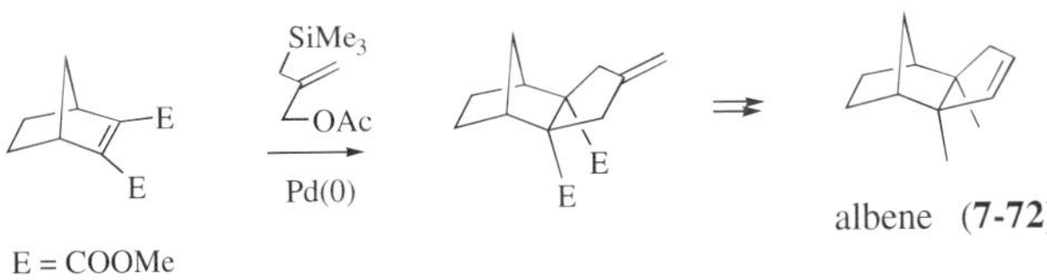

albene (**7-72**)

Pentalenolactone (**7-73**) is a rather complex sesquiterpene. A retrosynthetic analysis simplifies the structure into a bicyclic lactone diester. Since the two ester groups are *cis*-1,3-related in a cyclopentane, their derivation from a norbornene precursor is indicated. For the angular substituent of the bicyclic lactone diester, which is equated to an *endo*-oriented group of the bridged ring system, one can envision its original presence in a substituted norbornadiene and the employment of cycloaddition (from the *exo* side) to introduce the element of the δ-lactone to start the synthesis.[108] It is important to use a symmetrical anhydride to increase the dienophilic activity and to preclude any regiochemical problem.

The classical route to cantharidin (**7-74**)[109] was lengthened to address the same stereoselectivity of the cycloaddition of the 7-oxa analogue. Additional steps were needed to readjust the substituents, i.e., in the conversion of esters to the methyl groups and degradation of the cyclohexene ring of the adduct to obtain the succinic anhydride unit.

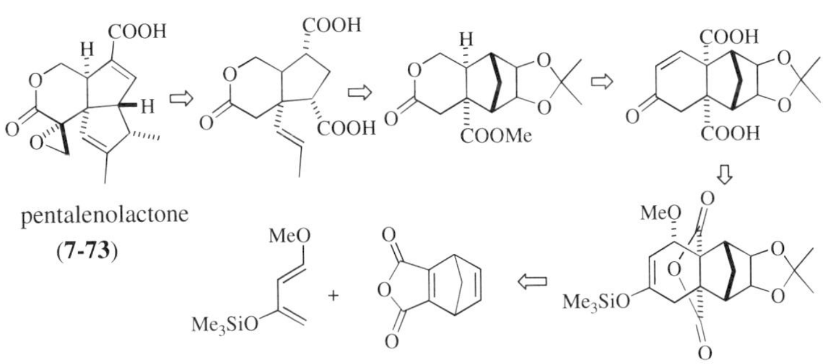

pentalenolactone
(**7-73**)

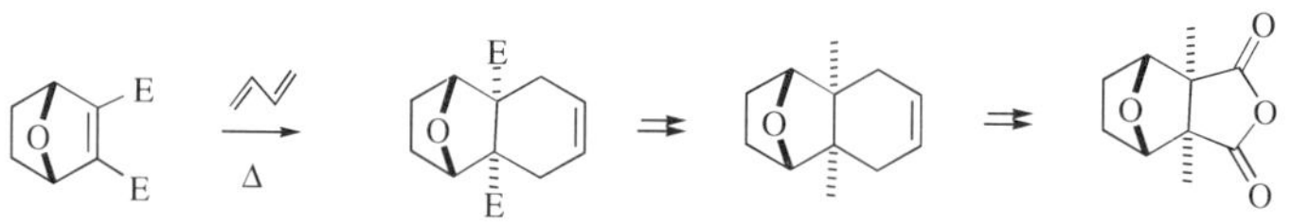

cantharidin
(**7-74**)

Using only two oxa-bridged building blocks, a synthesis of an unusual macrocycle (**7-75**) has been accomplished.[110] The dienophilic alkene reacted on the *exo* side.

An S_N2' ring opening of oxa-bridged alkenes by preferential *exo*-attack of organometallic reagents facilitated the access of methyl nonactate (**7-76**).[111] With two stereocenters latent in the bridged intermediate, the ring opening reaction generated the third and release of the secondary alcohol.

An array (**7-77**) of five contiguous stereocenters spanning $C_{(22)}$ to $C_{(26)}$ of rifamycin-S has been established by the same method, while exploiting the symmetry of an oxabicyclo[3.2.1]octenol.[112] The ring opening protocol is an effective complement to the popular aldol method because more stereocenters are set in place in short order.

Norcamphor is a valuable building block with special attribute in the stereoselectivity of reactions for modifying the bridged framework. The

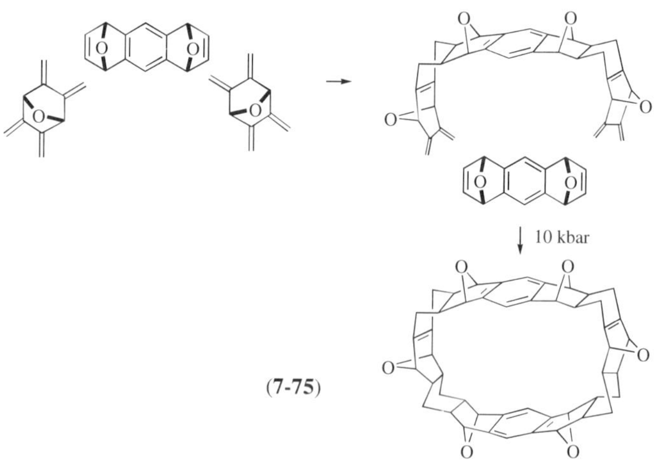

(**7-75**)

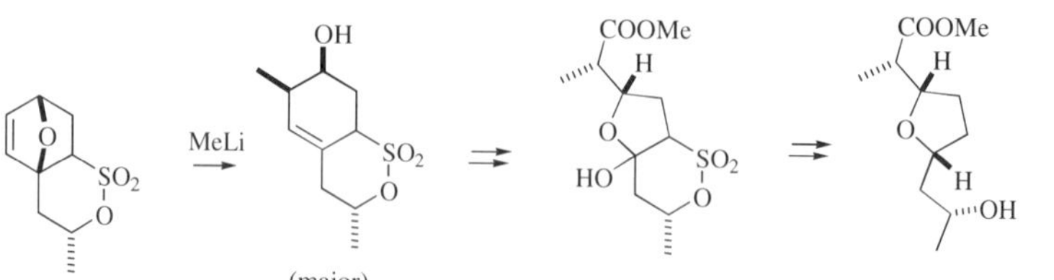

methyl nonactate
(**7-76**)

corresponding lactone has been used to elaborate antirhine (**7-78**)[113] after allylation and epimerization.

In a synthesis of *N*-benzoylmeroquinene aldehyde (**7-79**),[114] which is a precursor of quinine, the stereochemical integrity was unfortunately destroyed during hydride reduction of the glutarimide intermediate.

The conversion of (+)-limonene to a bridged lactone by way of hydroboration and oxidation afforded stereoselective elaboration of the isopropenyl group. Such was the tactic for establishing one of the secondary methyl groups toward a synthesis of (–)-pseudopterosin-A (**7-80**).[115]

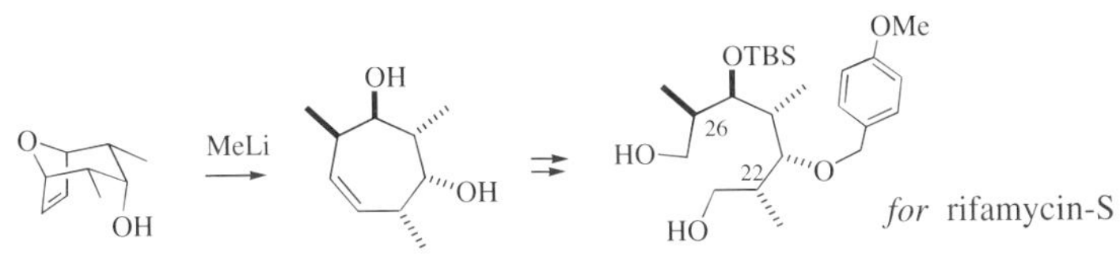

(**7-77**)

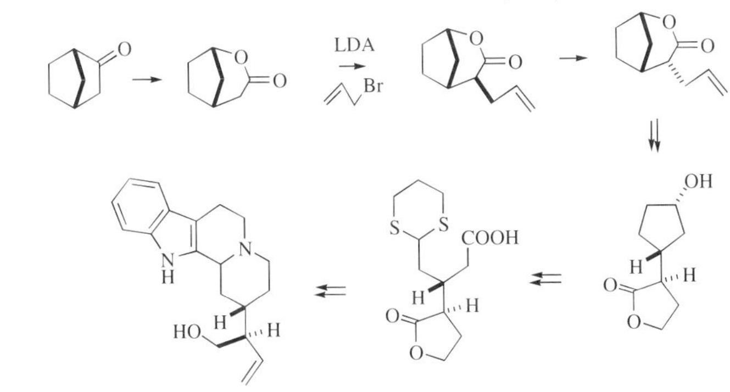

antirhine (**7-78**)

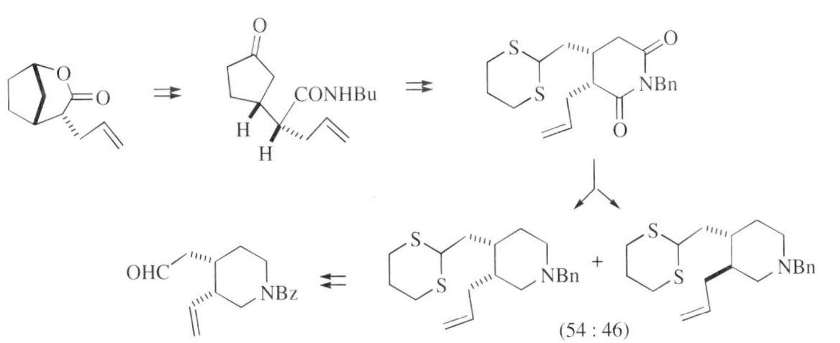

N-benzoylmeroquinene aldehyde (**7-79**)

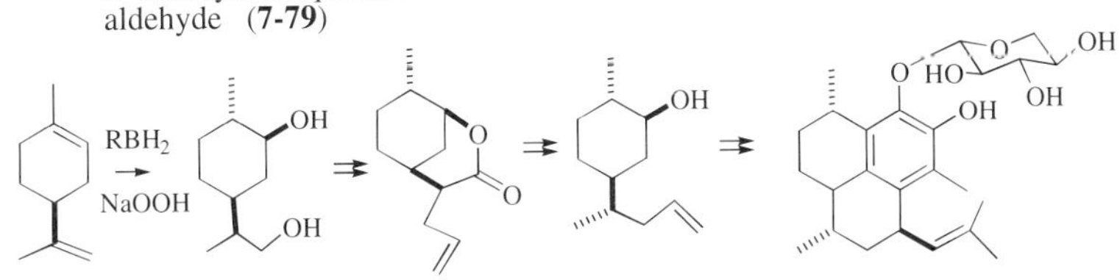

(+)-limonene

(-)-pseudopterosin-A (**7-80**)

A preparation of the enantiopure enyne synthon (**7-81**) for vitamin Ds (e.g., 1-hydroxyvitamin D_3 and calcitriol)[116] from a cyclohexenecarboxylic acid involved bromolactonization and modification of the bridged lactone. Epoxidation of the subsequently introduced double bond served to establish the configuration of $C_{(1)}$.

Analogously, several reports on the synthesis of tirandamycin-A[117] described epoxidation from the *exo* face of trioxabicyclo[3.3.1]nonene derivatives.

A retrosynthetic analysis of phyllanthocin (**7-37**) divides the molecule into a perhydrobenzofuran and a spiroacetal subunit. The norketone generated by disconnection of the epoxide provides an active site for chain extension toward construction of the spiroacetal. As the perhydrocoumarone is a β-hydroxy ketone tautomer, its preparation by nitrile oxide–alkene cycloaddition is attractive, particularly when such a reaction features the formation of *cis*-fused isoxazolines. The conversion of isoxazolines to β-hydroxy ketones on mild reduction is well known.

The ring fusion problem is not the total stereochemical issue concerning the elaboration of a phyllanthocin intermediate. An intriguing question seeking urgent answer at the start of the synthesis was how to effect the 1,3-dipolar cycloaddition in a regio- and stereoselective manner in the presence of an ester group at a *meta*-position of the ethereal oxygen atom. Identifying a bridged lactone as the dipolarophile was pivotal to a smooth progression toward the synthetic objective[118] as exposure of the *exo*-face is expected. While the regioselectivity is less apparent, the involvement of an asynchronous transition state in which formation of the C–C bond precedes the O–C bond seems to favor that in the axial direction (probably as a result of subtle interplay of steric and electronic factors, and frontier molecular orbital overlap. A similar regioselective cycloaddition en route to luciduline[119] by an intramolecular nitrone–alkene reaction was also noted).

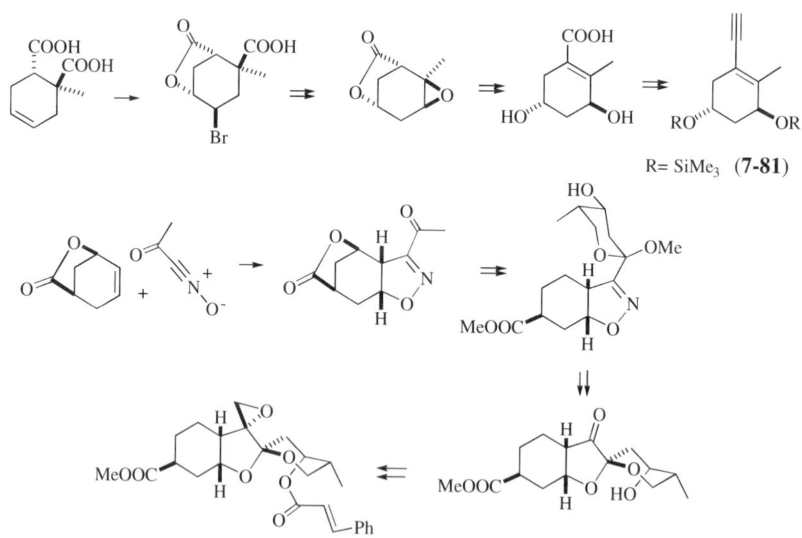

phyllanthocin (**7-37**)

The incorporation of a dienophilic double bond exocyclic to a bridged lactone in a synthesis of verrucarol (**7-82**)[120] was to take advantage of its participation in a Diels–Alder reaction In an approach to anguidine,[121] a structurally akin α-hydroxymethylene lactone was condensed with methyl vinyl ketone to create a spiroannulated cyclohexenone intermediate. The reaction proceeded according to stereochemical specification.

It is expedient to lock the hydroxymethylcyclohexenedione intermediate of (–)-phaseic acid (**7-83**) into a bicyclic acetal form because it permits regioselective introduction of the six-carbon side chain in a stereoselective manner.[122]

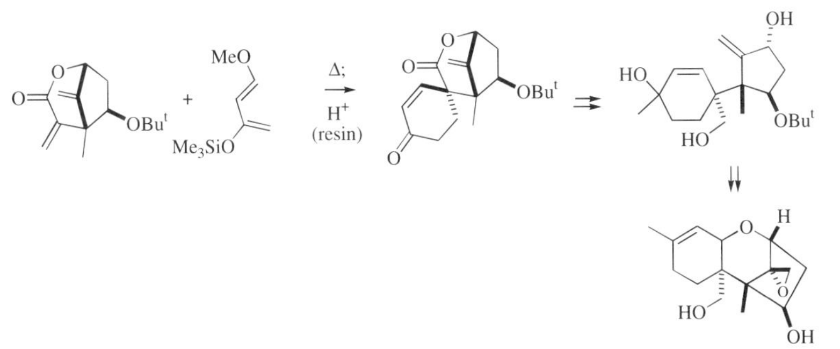

verrucarol (**7-82**)

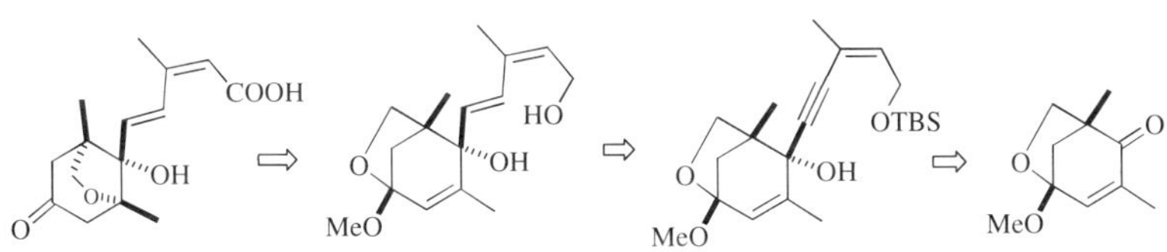

(-)-phaseic acid

(**7-83**)

An intriguing method for constructing the hydrazulene precursor of ledol (**7-84**)[123] is by intramolecular aldol condensation of a keto aldehyde derived from cleavage of a bridgehead alkene. The alkene is available from a Diels–Alder reaction, and with an enone dienophile the product is also equipped with a ketone group whose

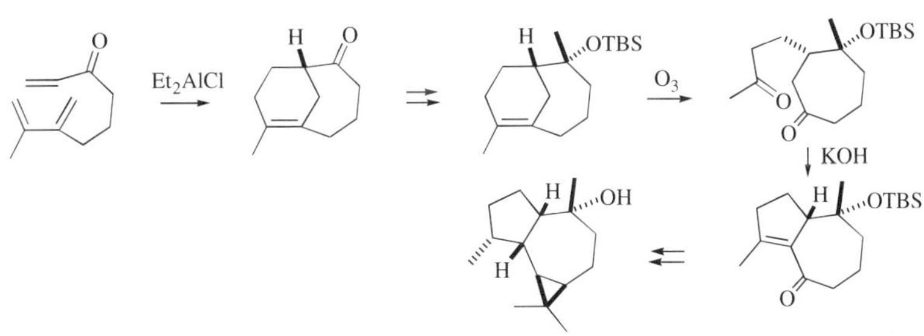

ledol (**7-84**)

reaction with methyllithium is subject to topographical influence. The *exo*-attack actually generated the correct relative configuration.

Subtle effects can upset the *exo*-selectivity of a bridged ring. Reconstitution of aphidicolin (**7-85**) from the norketone by way of methylenation and dihydroxylation is totally stereorandom because the double bond is essentially eclipsed with the bridgehead hydrogen and there is no difference in the change of torsional strain on reaction from either direction. On the other hand, an endocyclic double bond underwent *exo*-selective epoxidation, through which a solution to the stereochemical problem could be derived.[124]

An intermediate of damsin (**7-67**) has been prepared from the symmetrical *endo,endo*-2,4-dimethyl-8-oxabicyclo[3.2.1]octane-3-one.[125] The process involved alkylation of the ketone with a cobalt-coordinated propargyl cation as a prelude to the cyclopentane ring formation. (Perhaps the *exo*-alkylation in this case was of no consequence, although this steric discrimination must be of value in giving a more homogeneous product.)

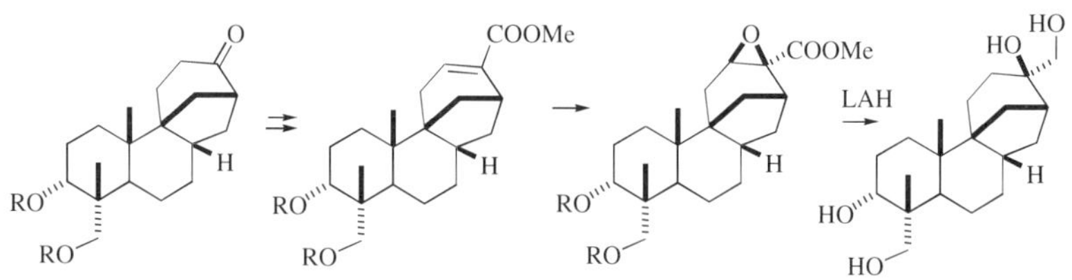

(+)-aphidicolin
(**7-85**)

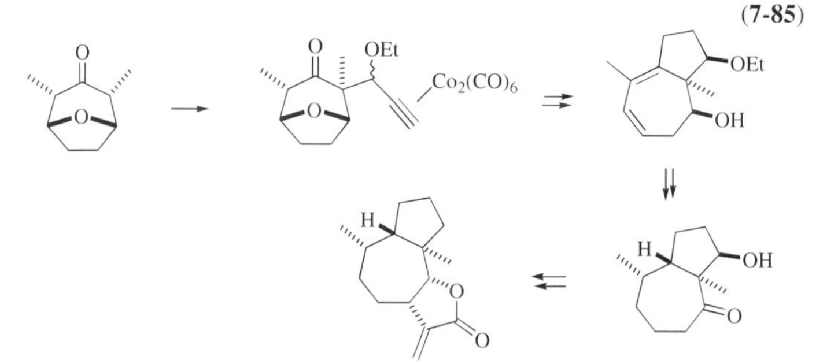

damsin (**7-67**)

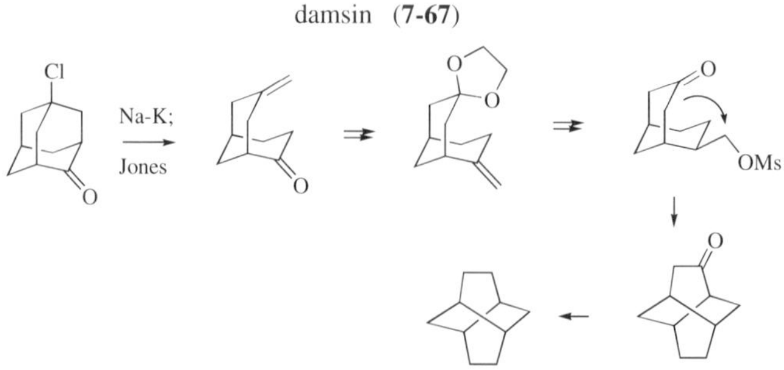

twistane (**7-86**)

Conversion of the adamantane skeleton to twistane (**7-86**) is interesting, as such contrathermodynamic reorganization necessarily involves special manipulation. The key steps of the synthesis included fragmentation and hydroboration of an exocyclic methylene to furnish an *endo*-methanol unit (selectivity 85:15) for eventual cyclization.[126]

Conjugate addition to bridged enones with organocopper reagents from the *exo* direction was instrumental to the access of valeranone (**7-87**),[127] a large segment (**7-88**) of erythronolide-B,[128] and gibberellin-A_{12} (**7-89**).[129]

The novel concept for elaborating several terpenes by intramolecular alkylation to form a cyclobutane ring from properly functionalized bicyclo[3.2.1]octenones and subsequent cleavage of the original five-membered ring has been realized. α-*trans*-Bergamotene (**7-90**) is a representative of the process.[130] The two leaving groups were installed from a *cis*-diol which must have the *exo, exo*-configuration, and it was readily prepared by dihydroxylation of an alkene.

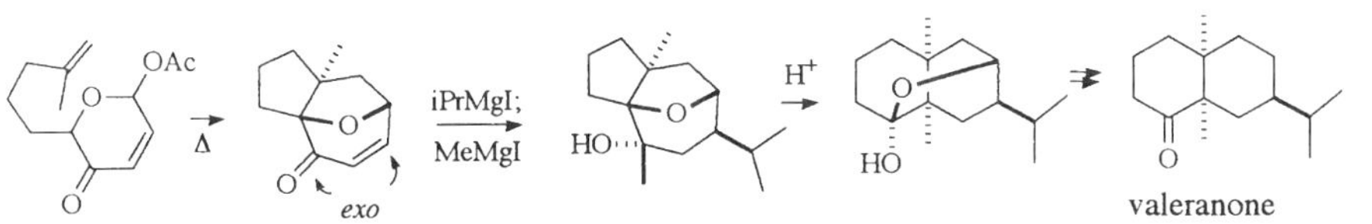

valeranone
(**7-87**)

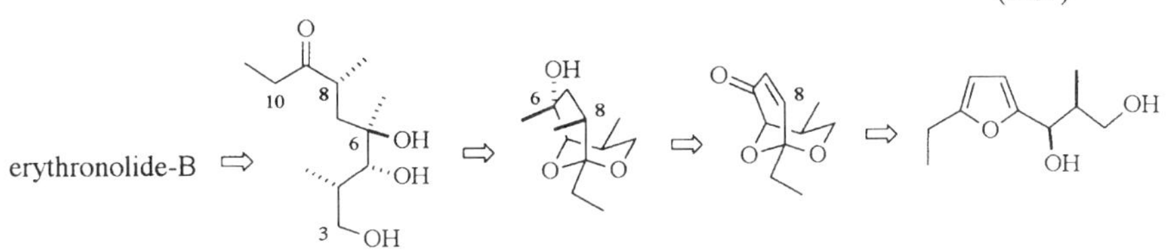

(**7-88**)

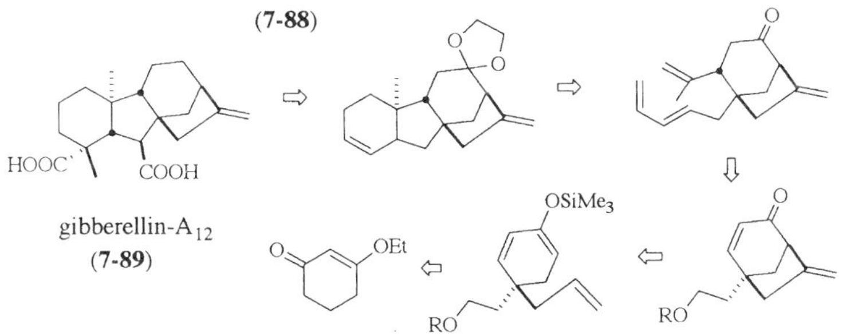

gibberellin-A_{12}
(**7-89**)

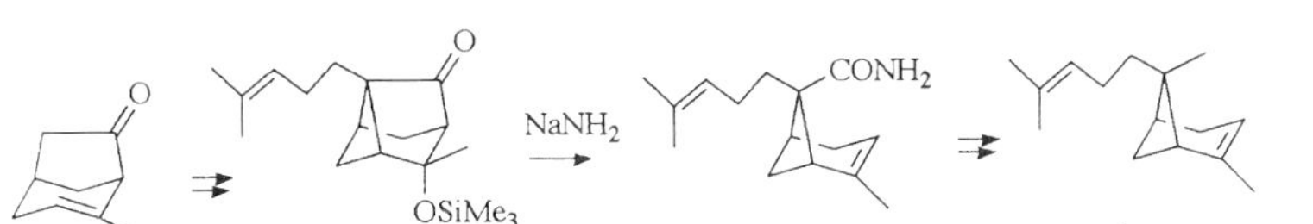

α-*trans*-bergamotene
(**7-90**)

An increasing number of natural products embodying either *syn*- or *anti*-1,3-polyol segments are being found. A priori, reiterative processes are most effective for assembling these substances and it is of interest to be able to utilize a

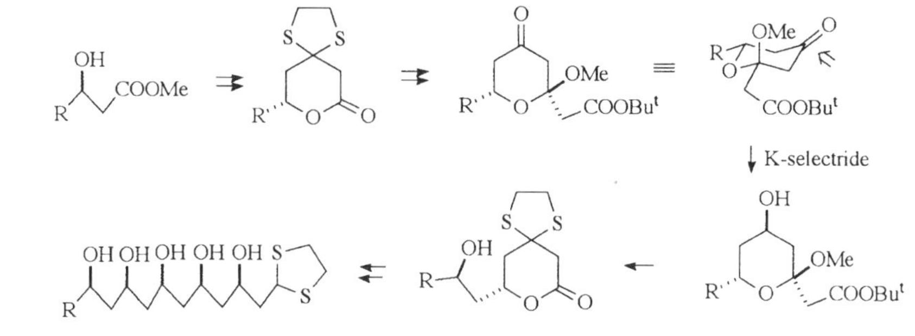

syn-1,3-polyols
(**7-91**)

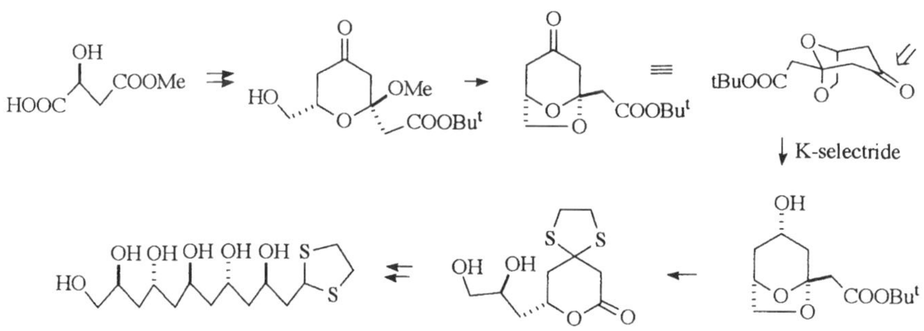

anti-1,3-polyols
(**7-92**)

reaction sequence to prepare one series and a slight variant to access the other. Actually, it is possible to achieve such an objective by relying on structural and reactivity patterns to direct the steric course. Exploiting anomeric effect to maintain a chair conformation and using a bulky metal hydride to reduce a cyclic ketone group by equatorial approach, it is easy to add one unit of *syn*-1,3-diol to a growing chain (**7-91**).[131] In a bridged acetal the axial/equatorial relationship was exchanged and reduction yielded the epimer which belongs to the *anti*-1,3-polyol series (**7-92**).[132]

7.2. TEMPLATE EFFECTS

Template effects denote the consequences of a physical barrier that restricts access of reagents to the reaction site. Such effects have great impact on steric courses of reactions and considerable efforts have been spent in molecular design to reap their benefit in synthesis. A template can be part of the substrate, a temporary attachment to it, or an external species which associates with the reactants through metal chelation, hydrogen bonding, electrostatic, or some other forces to form a sterically well-defined complex. The barrier exerted by a template often fixes the reactant in one conformation.

7.2.1. π-Stacking

Aromatic rings provide an inert and flat surface that can be enlisted to affect reactions at a distant site, by merely hindering approach of reagents from its direction. Reactions involving rehybridization of C=O and C=C bonds which, through attractive interactions to stretch above an aromatic ring are now quite useful and predictable.

Significant advances in synthetic application of π-stacking are due to investigations of the prostaglandins (**7-23**). The remarkable notion that 11-hydroxyl protection with an aromatic derivative can serve as a shield to one face of the 15-ketone to enable its stereoselective reduction was brought to fruition,[133] and the 4-biphenylcarbamate was identified as an optimal substrate [diastereoselectivity: (*S*:*R*) = 92:8]. Interestingly, the enone system adopted an *s-cis* form.

An extremely popular chiral auxiliary is the chiral 8-phenylmenthol.[134] A great variety of conjugated esters of both enantiomers have been prepared to be used in Diels–Alder reactions, intramolecular ene reaction and Michael additions with predictable absolute stereochemistry. Notable applications include synthesis of the Corey lactone intermediate for the prostaglandins (**7-23**),[135] and α-allokainic acid (**7-93**).[136]

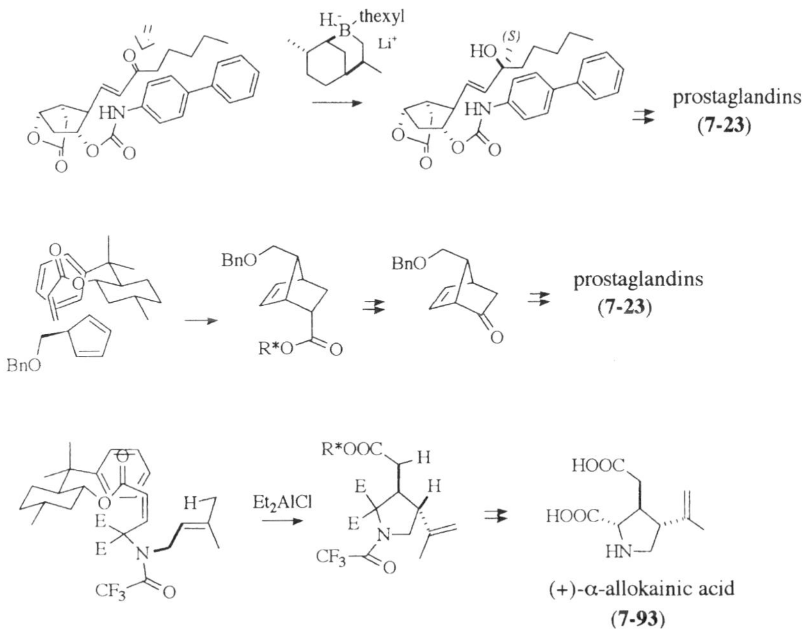

The π,π-interaction between the pyridine ring and the phenyl group of 1-(8-phenylmethoxylcarbonyl)pyridinium salt delivers an opportunity for the synthesis of alkaloids containing a piperidine ring in chiral form.[137] With unsymmetrically substituted pyridines, apparently one type of rotamers is preferred, and the attack by organometallic reagents (at $C_{(2)}$ of the pyridinium ring) leads to useful products. An example is a synthesis of (–)-porantheridine (**7-94**).[138]

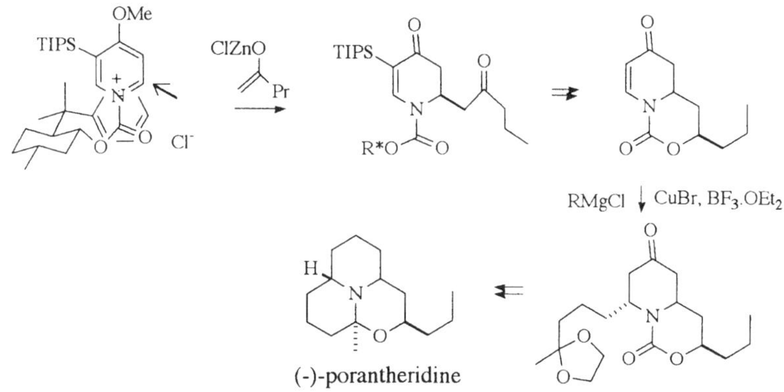

(-)-porantheridine
(**7-94**)

The phenylmenthol half esters of monosubstituted malonic acids are alkylated from a chelated transition state. Accordingly, chiral α,α-disubstituted carboxylic acid derivatives are obtainable, and the potential of this methodology has been demonstrated in approaches to (–)-eburnamonine (**7-95**) and (–)-aspidospermidine (**7-96**).[139]

The *N*-Boc-glycine ester of 8-phenylmenthol on bromination and treatment with Grignard reagents, which effect in situ dehydrobromination and addition, leads to chiral α-amino acid derivatives.[140] Grignard reaction with the glyoxalate ester is an excellent method for obtaining chiral α-hydroxy acid derivatives.[141]

Lewis acid catalyzed ene reaction of 8-phenylmenthyl glyoxalate with racemic *cis*-bicyclo[3.3.0]octa-2,6-diene gave an 8:1 mixture of adducts with the major isomer possessing all of the framework atoms of (–)-xylomollin (**7-97**) in a latent form and, most importantly, two contiguous stereogenic centers.[142] The adduct was processed accordingly to afford (–)-xylomollin, while effecting inversion of the configuration of the side chain on transforming the ester into a methyl carbinol. The ene reaction afforded the same ratio of isomers even when a 10-fold excess of the diene was used.

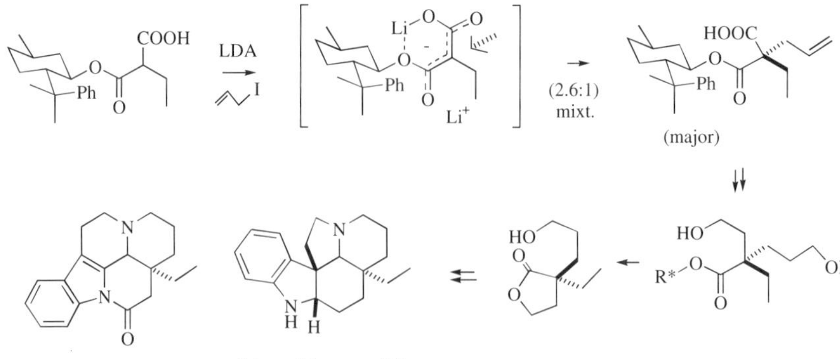

(-)eburnamonine
(**7-95**)

(-)-aspidospermidine
(**7-96**)

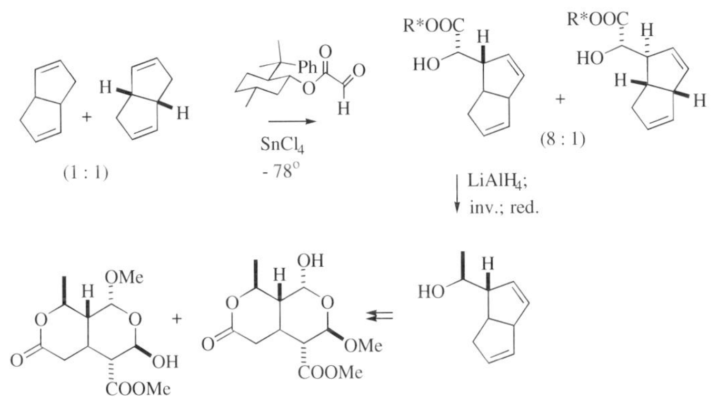

isoxylomollin xylomollin (**7-97**)

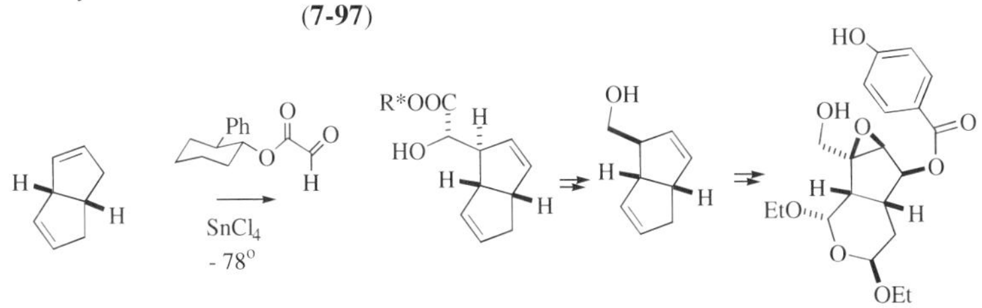

specionin
(**7-98**)

trans-2-Arylcyclohexanols serve the same purpose almost as well as the 8-phenylmenthols. Thus Lewis acid catalyzed ene reaction of (*1R*)-*trans*-2-phenylcyclohexyl glyoxalate with *cis*-bicyclo[3.3.0]octa-2,7-diene furnished a chiral adduct which was readily transformed into (–)-specionin (**7-98**).[143] The discrimination of two *meso*-related double bonds by this method is equally satisfactory. Note that the starting diene is *meso*, whereas that used in the xylomollin synthesis is racemic.

Enol ethers derived from these auxiliaries have been used in the synthesis of (–)-α-cuparenone and (+)-β-cuparenone by way of cycloaddition with dichloroketene[144] and of (+)-hirsutene on the basis of a Pauson–Khand reaction.[145]

2,2,6,6-Tetramethyl-3,5-heptanediol monobenzoate monoalkenoates are functional analogues to the above compounds. Although acyclic, the exhibited conformational preference of these molecules (**7-99a, 7-99b, 7-99c**) renders them rather useful for asymmetric synthesis.[146]

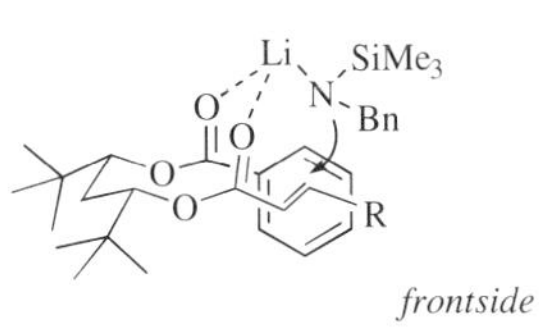

(**7-99a**)

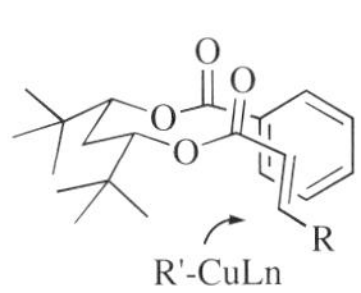

(**7-99b**)

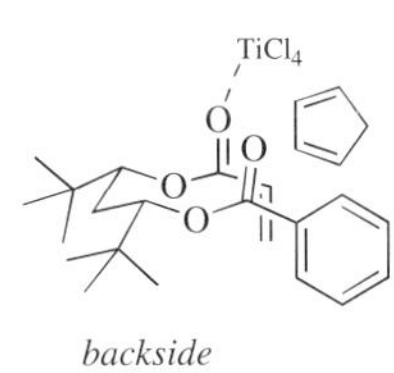

(**7-99c**)

It is possible to convert the methyl ester of fumaraldehydic acid to chiral 3-substituted succinaldehydic esters (**7-100**)[147] by condensation with a proline-derived diamine, followed by Cu(I)-catalyzed conjugate addition with Grignard reagents and acid hydrolysis. The preferred *Si*-face attack of the Grignard reagent was due to discriminate shielding by the *N*-phenyl substituent which is expected to be *exo*-oriented. (To access the enantiomeric esters, see Ref. 148.)

How can *p*-benzoquinone be converted into conduritol-A (**7-101**)? All it requires is *cis*-dihydroxylation of one double bond, and reduction of the two ketone groups from the same direction. However, reactivity problems as well as stereochemical issues negate a straightforward transformation. A useful solution is to mask one of the double bonds by forming an adduct with an anthracene derivative.[149]

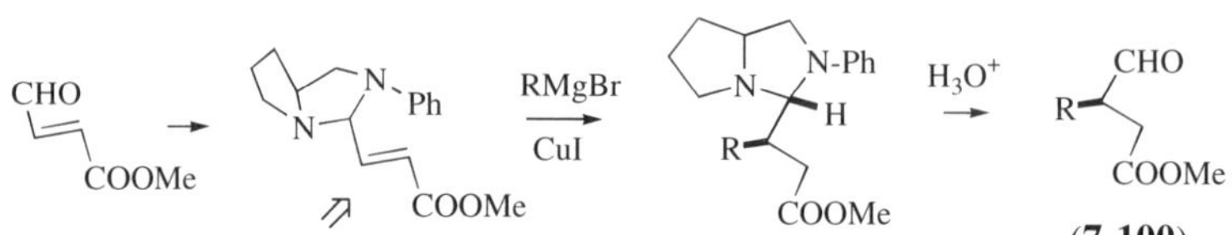

(**7-100**)

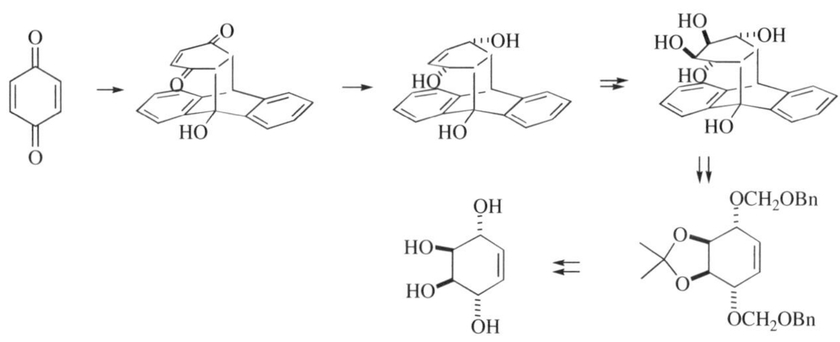

conduritol-A
(**7-101**)

7.2.2. Steric Bulk

The use of the trityl (trimethylphenyl) group for alcohol protection has a long history. The spatial occupance by this group has also been recognized as an advantageous feature when stereoselectivity is desired for reactions in its environ. The effect of a trityloxy group can be felt at least four atoms away in a conformationally restricted framework. When (*S*)-4-hydroxymethyl-4-butanolide (**7-102**) which is readily obtained from glutamic acid or the *R*-isomer from a Mitsunobu reaction of the corresponding hydroxy acid is tritylated, subsequent alkylation at $C_{(2)}$ is directed completely to the face opposite the existing side chain.

The implication of this phenomenon is that numerous natural products can be synthesized using the lactone as a building platform. Representatives include (−)-antirhine,[150] (+)-eburnamine,[151] (+)- and (−)-velbanamine,[152] (+)- and (−)-quebrachamine,[153] podorhizon,[154] (−)-megaphone,[155] an intermediate of calonectrin,[156] and swainsonine.[157]

Even more effective steric control is expected for reactions at the nearer β-carbon of the lactone. Actually, the corresponding butenolide has been employed in Michael additions with predictable stereoselectivity, as shown in its use for a synthesis of (+)-staganacin (**7-103**).[158]

These butanolides and butenolides are very valuable for elaboration of the propionate metabolites.[159] The reiteratable processes is highly efficient.

In recent times, bulky siloxy groups have been developed for protection and stereocontrol. Interestingly, a triisopropylsiloxy group at $C_{(3)}$ of a pyran-oxyacetaldehyde (**7-104**) displayed an opposite diastereoselectivity to the corresponding benzyl ether or acetate in reaction with an allylstannane.[160]

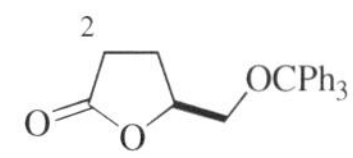

(**7-102**)

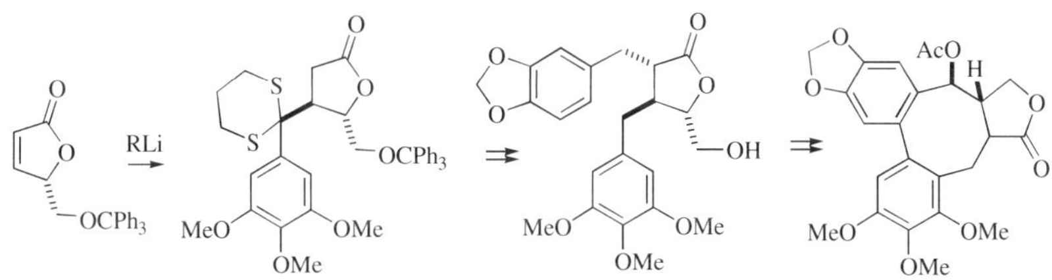

(+)-staganacin
(**7-103**)

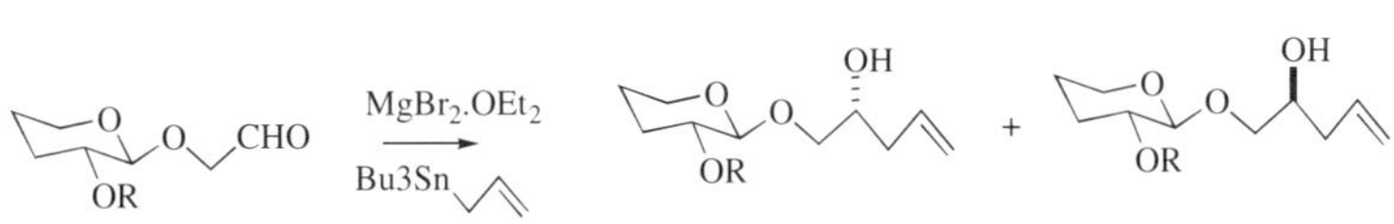

(**7-104**)

R		
R = Ac	86 :	14
R = Bn	97.6 :	2.4
R = iPrSi	7.2 :	92.8

In a synthesis of gephyrotoxin (**7-105**)[161] the stereoselective establishment of the pentenyne side chain is critical. By derivatizing the primary alcohol present in the pyrrolidine ring as a *t*-butyldiphenylsilyl ether, the exocyclic double bond of the conjugated ester could be hydrogenated stereoselectively in the desired manner (diastereoselectivity 10:1). Note that when the acetate of the alcohol was subjected to reduction with lithium in liquid ammonia, the wrong isomer was produced predominantly (ratio 35:1).

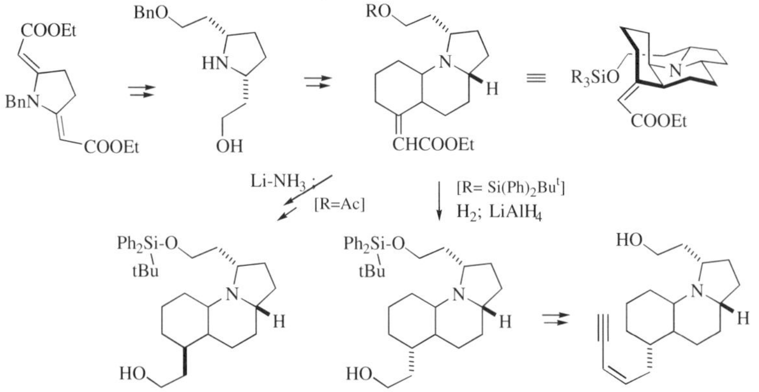

gephyrotoxin
(**7-105**)

In the conjugate addition of an alkenyl chain to 4-hydroxycyclopenten-1-yl methyl ketone during synthesis of brefeldin-A (**7-20**),[162] it was noted that the formation of a bulky alkoxide forced the addend to enter antarafacially.

One of the crucial operations in a synthesis of annotinine (**7-106**)[163] was to establish the relative configuration of the secondary methyl group. It was found that the methylenecyclobutane unit of the photocycloadduct derived from allene was hydrogenated from both sides, actually favoring the undesirable isomer because of less steric congestion in the proximity of the cyclohexanone ring. The situation was changed when the ketone was converted to an ethyleneacetal.

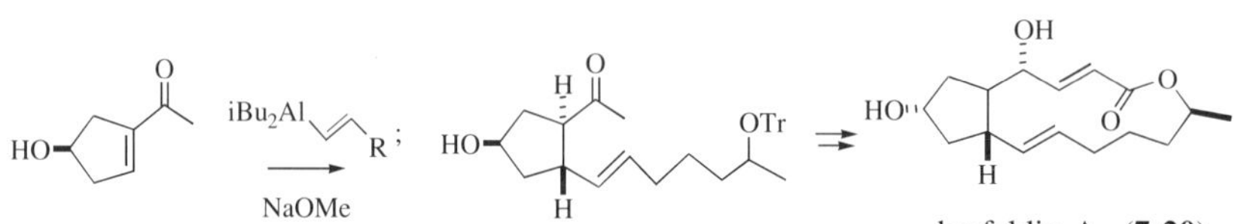

brefeldin-A (**7-20**)

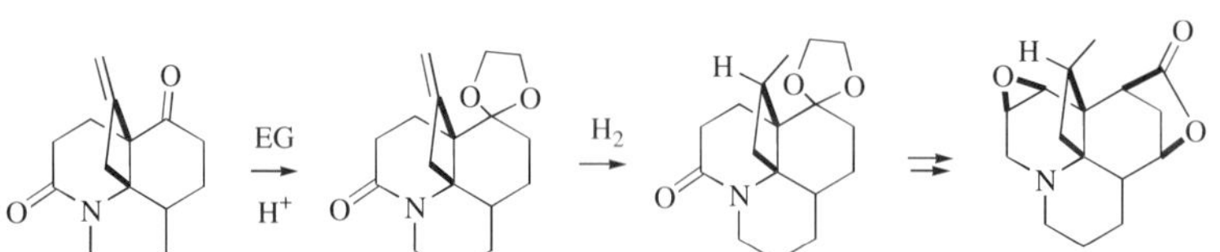

annotinine
(**7-106**)

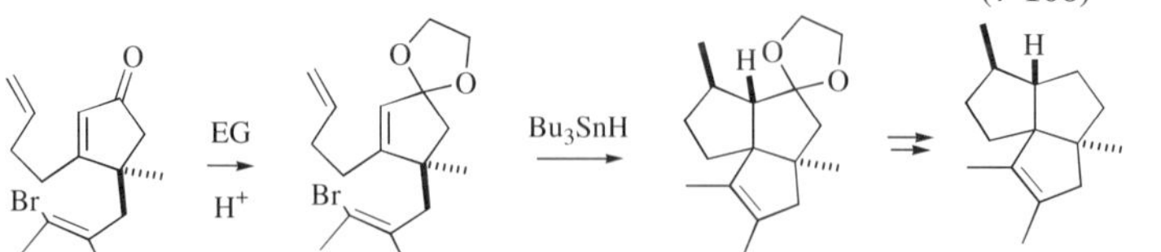

silphiperfolene
(**7-107**)

A tandem cyclization initiated by an alkenyl radical to create the tricyclic framework of silphiperfolene (**7-107**)[164] suffered from generation of the unwanted stereoisomer, when a cyclopentenone was used as the relay unit. The same tactic of acetalization, which did not affect the reactivity and regiochemistry, changed the course of cyclization to favor the *exo*-methyl isomer (ratio 3:1).

A historically significant asymmetric induction of aliphatic systems is the Grignard reaction of α-keto esters of chiral alcohols as predicted by Prelog's rule. The manifest of stereoinduction is due to the population of one particular conformation and the preference for approach of the reagent. Extension of the concept to alkenoic esters[165] is important because these compounds are useful substrates for the most versatile organic reactions such as the Diels–Alder reaction and the Michael reaction.

In the 2-neopentyl-3-acryloxy derivatives of *exo,exo*-bornane-2,3-diol, facial discrimination of the conjugated double bond reaches a high level. The 2,3-butadienoate is sufficiently reactive to give a cycloadduct with cyclopentadiene which was converted to (–)-β-santalene (**7-61**).[166]

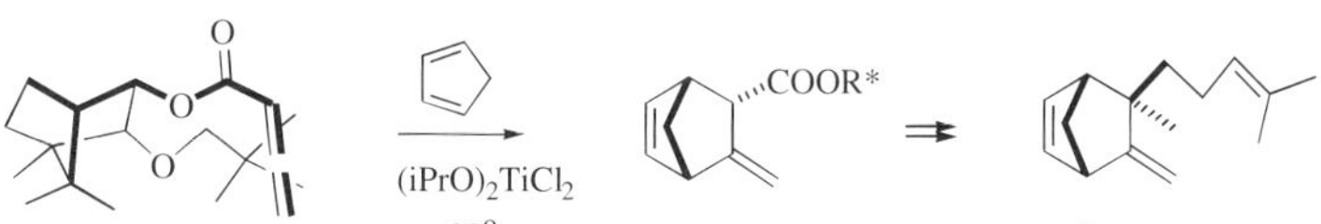

β-santalene
(**7-61**)

Rapid and stereoselective assembly of a *trans*-2,3-disubstituted cyclopentanone intermediate for (–)-khusimone (**7-108**)[167] has been realized. A chelated transition state for the Michael addition was indicated.

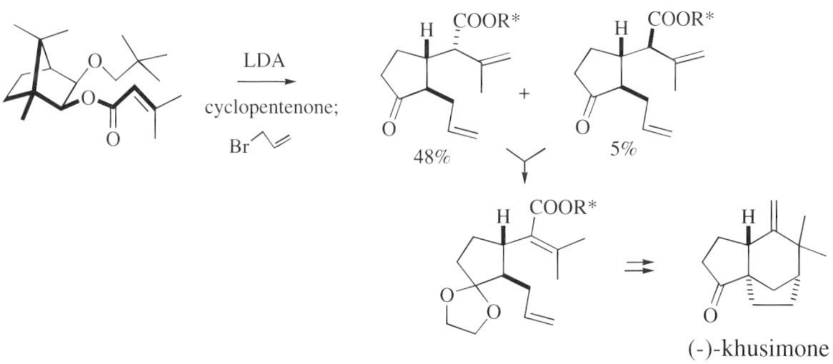

(-)-khusimone
(**7-108**)

One of the various chiral auxiliaries based on the camphor skeleton is bornane-10,2-sultam which can be prepared from (+)-camphor-10-sulfonyl chloride. In its application to a β-necrodol (**7-109**) synthesis,[168] a conjugate addition with methylcopper in the presence of boron trifluoride was executed. Due to rigidity of the molecular conformation only the *Si*-face was exposed to attack.

Cyclization of 2-octenediol through enamine formation can be rendered asymmetric by using a chiral oxazolidine.[169] The three substituents on the heterocycle blocked off one face completely so that the chain containing the enal curled only in one direction in the active transition state. It is also important that the more stable rotamer underwent cyclization. The product was used for a synthesis of (+)-brefeldin-C (**7-110**).

One must not neglect the possibility of achieving asymmetric induction on very simple substructures. A case in point is a relatively straightforward synthesis of (–)-ajmalicine (**7-111**)[170] which relied heavily on $NaBH_4$ reduction of a methyl ketone. The shielding of one face by an ester side chain emanating from an adjacent carbon atom was sufficient to give the desired result. The restricted conformation of the methyl ketone was also a contributing factor. Furthermore, the accessibility of the (*Z*)-alkene from the secondary alcohol by controlled dehydration opened a pathway to (–)-tetrahydroalstonine (**7-112**).

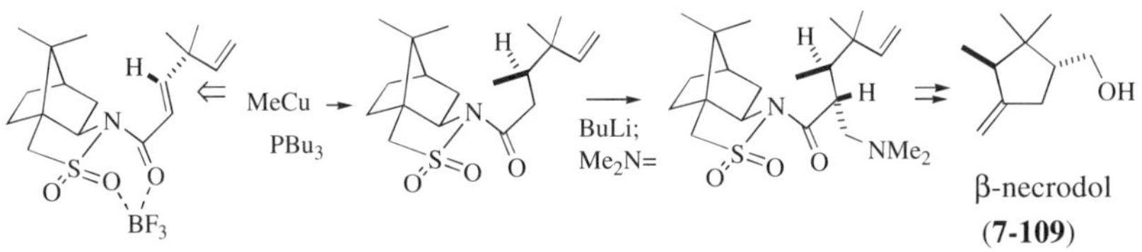

β-necrodol
(**7-109**)

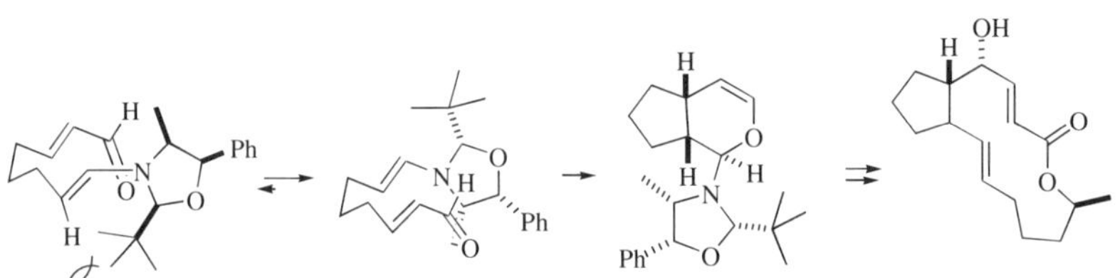

(+)-brefeldin-C
(**7-110**)

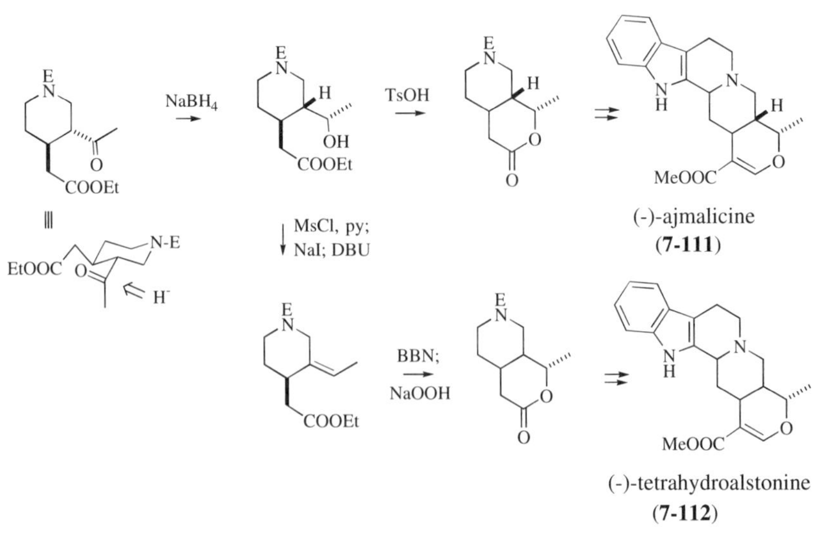

(-)-ajmalicine
(**7-111**)

(-)-tetrahydroalstonine
(**7-112**)

7.2.3. Ligated Metallic Components for Diastereofacial Selection

The formation and judicious use of organometallic complexes constitute a very popular trend in organic synthesis. Methodologies involving palladium, chromium, iron, and molybdenum seem to be the most extensively developed. They all share the common feature of directing reaction of the π-ligand antarafacially, although a change of reagents to those that prefer attack on the metal center before being transferred to the ligand would give different results. Of course this phenomenon can also be exploited.

Ring junction stereochemistry is always a vexing problem in the synthesis of polycyclic compounds, particularly highlighted in steroids. One way to solve this problem is by Pd-catalyzed, transpositional reduction of allylic formates.[171] The preparation of the stereoisomeric allylic alcohols (**7-113, 7-114**) demonstrates the genuine value of the method.

The stereochemistry of the reductive cleavage of alkenylcyclopropanes with formic acid depends on the geometry of the double bond. Thus, the configurational control of the new stereogenic center becomes possible. A synthesis of clavukerin-A (**7-115**) and isoclavukerin-A (**7-116**)[172] was achieved based on this method.

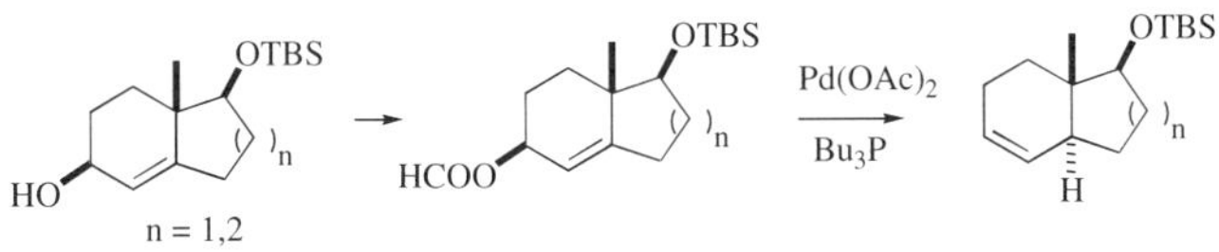

(**7-113**)

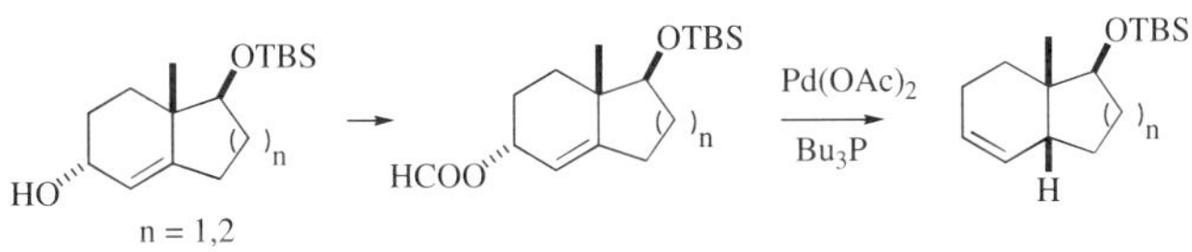

(**7-114**)

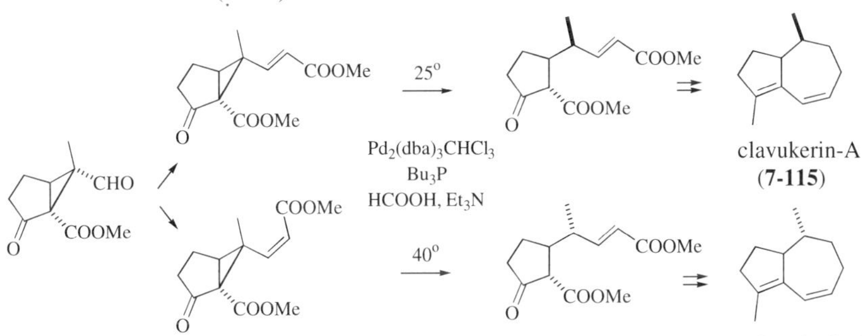

clavukerin-A (**7-115**)

isoclavukerin-A (**7-116**)

Allylic esters such as acetates and carbonates undergo Pd(0)-catalyzed substitutions with retention of configuration, due to intervention of π-allylpalladium complexes. Carbon and heteroatom nucleophiles can be used making the process very

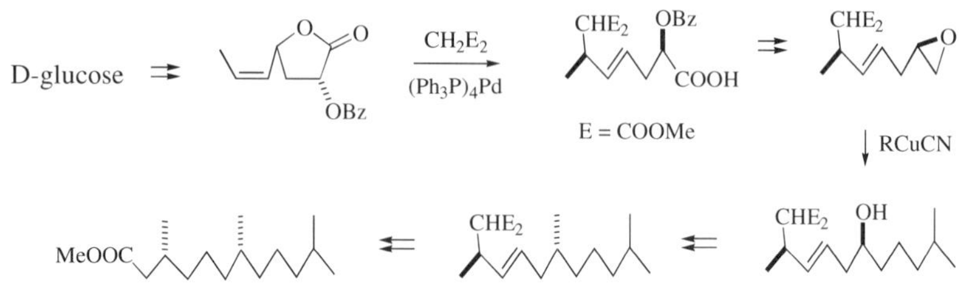

(**7-117**)

versatile. Assembly of the side chain (**7-117**) of vitamin E from D-glucose with complete control of relative and absolute stereochemistry has been realized.[173]

Conjugate dienes such as 1,3-cyclohexadiene and 1,3-cycloheptadiene can be 1,4-difunctionalized by means of the Pd-catalyzed reaction. Furthermore, the counterion of the added lithium salt influences the stereochemistry of the product.[174] By a proper choice of substrate and reaction sequence, the syntheses of both tropine (**7-118**) and pseudotropine (**7-119**) were achieved.[175]

A synthesis of ibogamine (**7-120**)[176] highlights a concise assembly of the bridged ring system using Diels–Alder reaction and the Pd(0)-mediated intramolecular S_N2' reaction. The latter involved a secondary amine as the nucleophile.

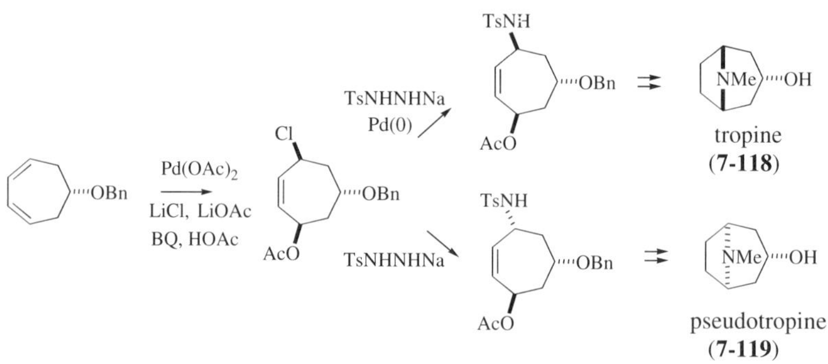

tropine
(**7-118**)

pseudotropine
(**7-119**)

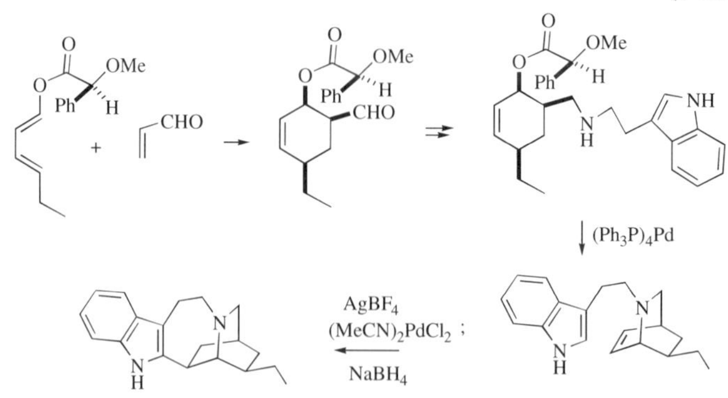

ibogamine (**7-120**)

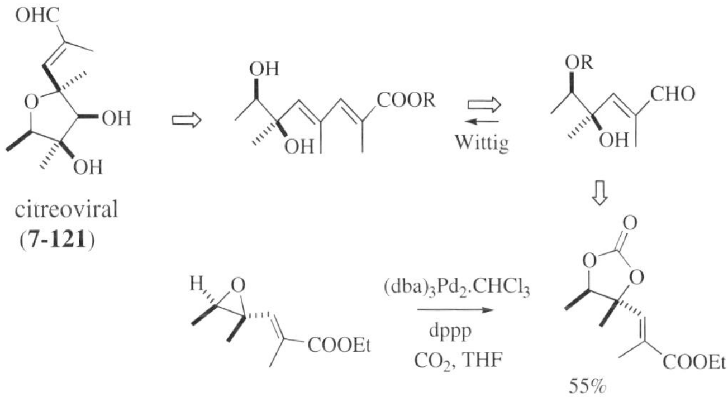

citreoviral
(**7-121**)

An alternative method for dihydroxylation of a diene is by retentive insertion of a C–O bond of the monoepoxide. As illustrated in a formal synthesis of citreoviral (**7-121**),[177] Pd(0)-catalyzed cleavage of the epoxide in the presence of carbon dioxide led to the cyclic carbonate.

3-Dimethylaminocyclopentene undergoes functionalization through chelation to palladium. Two different azapalladocycle intermediates thus participated in the stereoselective introduction of three pendants which enabled the access of a well-known intermediate of prostaglandin-$F_{2\alpha}$.[178]

Metal-complexed pentadienyl cations are electrophilic. Reaction of such species with nucleophiles occurs antarafacial to the metal.[179] Illustrated in an approach to the Prelog–Djerassi lactone (**7-122**) from 1,3-cycloheptadiene[180] is the critical dependence of the template effect of iron during attachment of two new secondary methyl groups.

Molybdenum has the same capacity in terms of exerting template effects. For example, *cis*-2,5-disubstituted 5,6-dihydro-2*H*-pyrans can be prepared from Mo complexes of 2*H*-pyran.[181] By this protocol a precursor (**7-123**) of monic acid is readily made available.

Nucleophiles are expected to enter the allylic position antarafacially to tricarbonyliron-complexed dienes. The conformation of the allylic cation is maintained by interaction with the metal. Elaboration of the *as*-indacene unit (**7-124**)

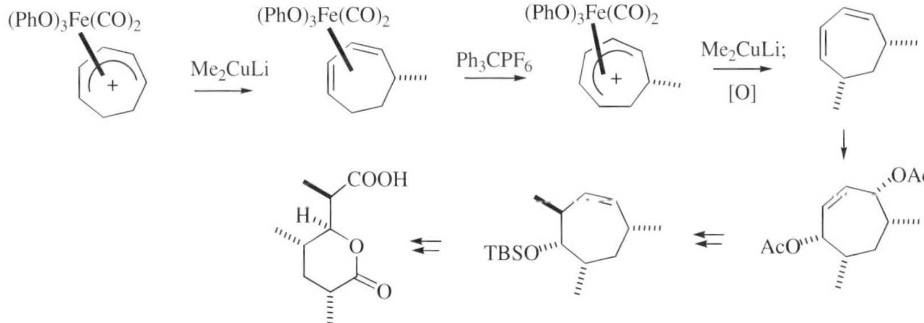

Prelog-Djerassi
lactone (**7-122**)

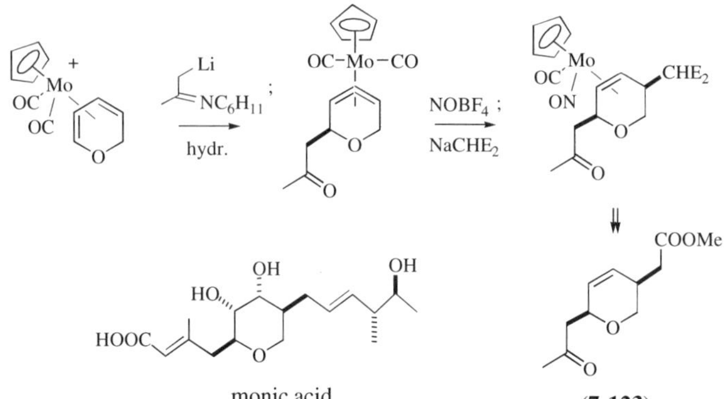

(**7-123**)

of ikarugamycin[182] exploited fully the asymmetric induction by the diene complex, establishing three stereocenters which subsequently influenced the mode of intramolecular Diels–Alder reaction.

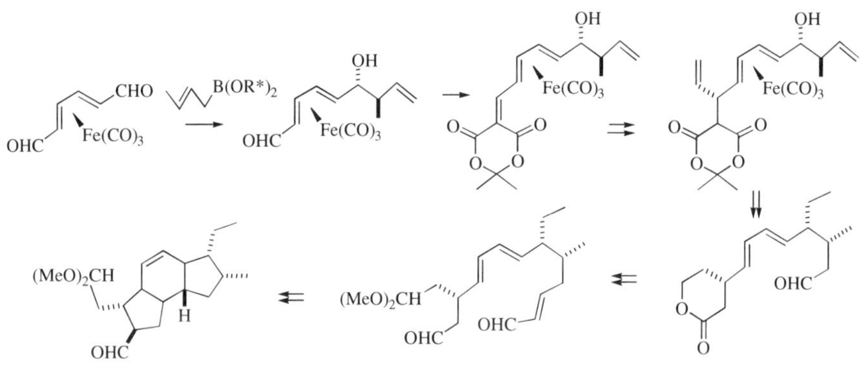

(**7-124**)

An important factor governing the stereoselectivity of these reactions is the conformation of the chain (i.e., the electrophilic C=O). With (π-allyl)-tricarbonyliron-ligated butadienyl carbonyl compounds, the *s-cis*-conformers give φ-*exo* products, whereas the *s-trans*-conformers furnish the φ-*endo* isomers. The usefulness of this correlation is further demonstrated in a synthesis of the intermediate of leukotriene-B_4 (**7-125**).[183] The $Fe(CO)_3$ group also played a protective role as in its absence the yield in the chain homologation step plunged from 65% (>98% diastereomeric excess) to less than 10% (racemic).

The Ireland-Claisen rearrangment involving an uncomplexed double bond of a conjugated triene is subject to steric influence of the $Fe(CO)_3$ group associated with the diene unit.[183a] The result can be ascribed to a chair transition state.

The stoichiometric usage of the $Fe(CO)_3$ is uneconomical. Remedy for this defect is now on the horizon[184] as it is possible to shift the unit to a new position while a new double bond is being created and the other terminus of the original diene is

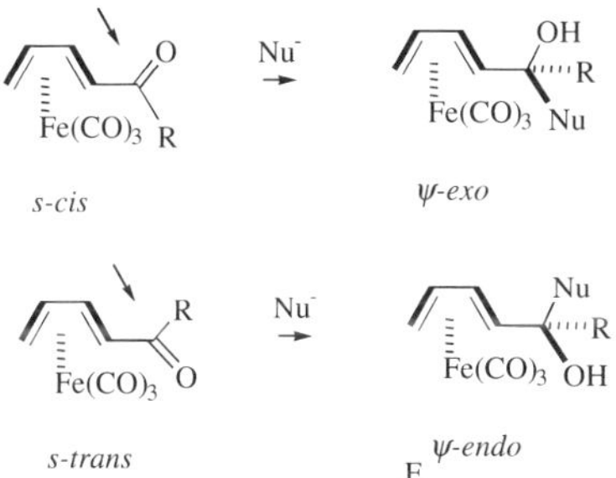

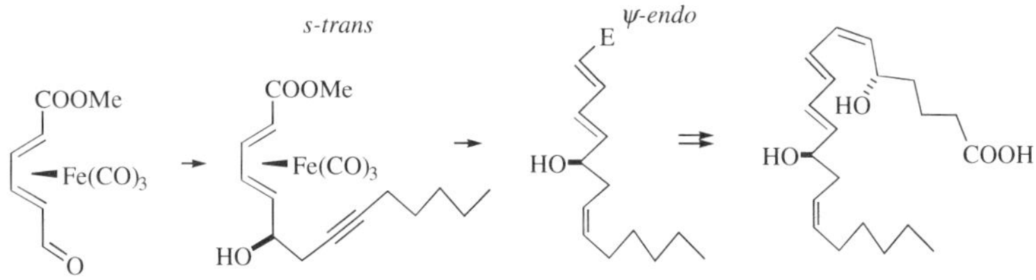

leukotriene-B_4
(**7-125**)

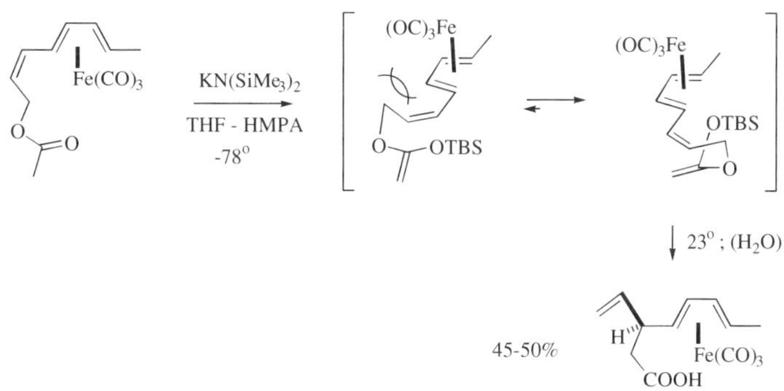

(**7-125a**)

under attack (after activation). Effectually, it accomplishes an S_N2' reaction and, in principle, the process can be repeated.

As indicated above, soft nucleophiles may attack the metal of a complex. The reverse stereoselectivity in the reaction of π-coordinated 2-formyl-1,3-butadiene with methyllithium versus lithium dimethylcuprate[185] should serve as a reminder.

Diastereofacial selection figured prominently in 1,3-dipolar cycloadditions involving dipolarophiles substituted with metal-complexed diene moieties. Thus, the stereoselectivity is instrumental to the success in an (*S*)-[6]-gingerol (**7-126**) synthesis.[186]

An allylic epoxide is involved in 1,4-asymmetric induction for ketone reduction when treated with $Fe_2(CO)_9$, as demonstrated in a synthesis of β-dimorphecolic acid (**7-127**).[187]

The synthetic utility of (π-allyl)tricarbonyliron lactone complexes is increased by the discovery of a transpositional atom insertion process. There is a report on the

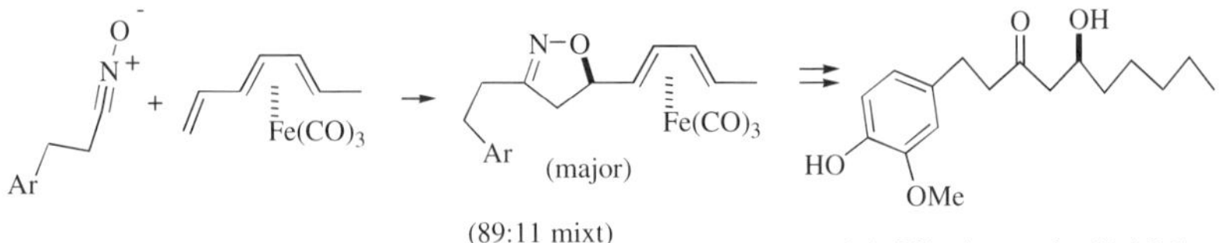

(+)-[6]-gingerol (**7-126**)

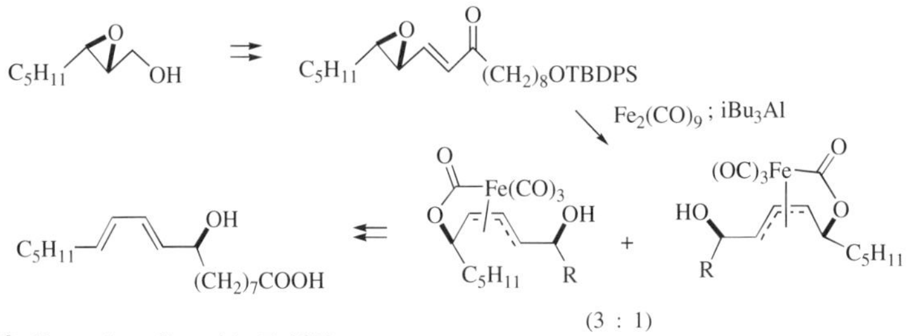

β-dimorphecolic acid (**7-127**)

elaboration of (+)-thienamycin (**7-128**)[188] based on this chemistry. Direct formation of β-lactones is possible as shown in a synthesis of valilactone (**7-129**).[189]

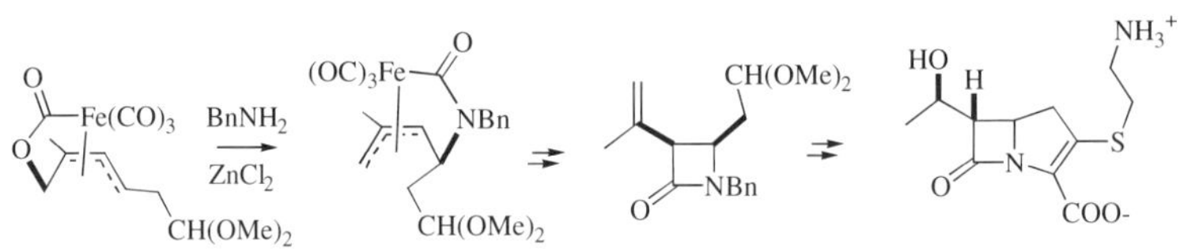

(+)-thienamycin
(**7-128**)

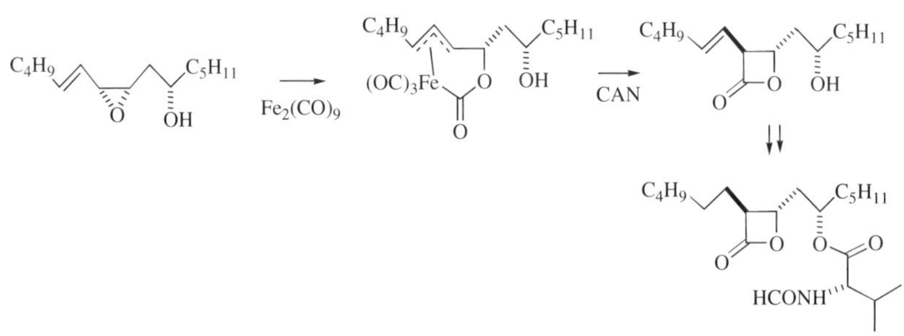

valilactone
(**7-129**)

By complexation of an aromatic ring with the tricarbonylchromium group, reactions at a benzylic position can be controlled sterically. The activation of benzylic and certain nuclear positions by the metallic residue is particularly valuable for synthesis. Thus, such processes were on prominent display in the serial reactions leading to dihydroxyserrulatic acid (**7-130**).[185] In a synthesis of mutisianthol

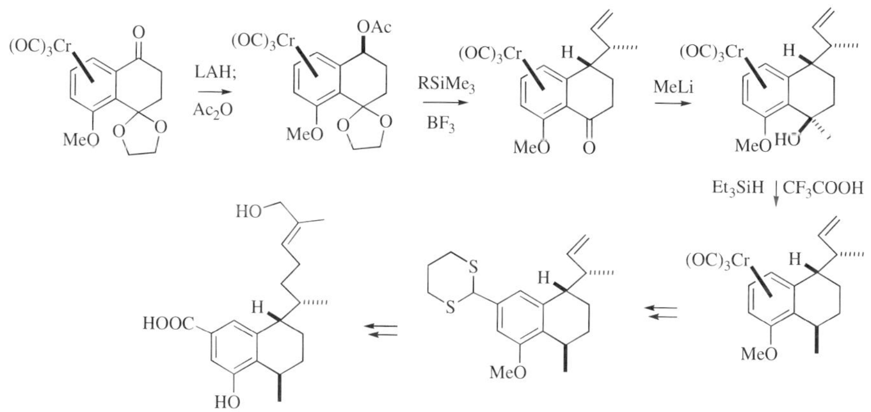

dihydroxyserrulatic acid (**7-130**)

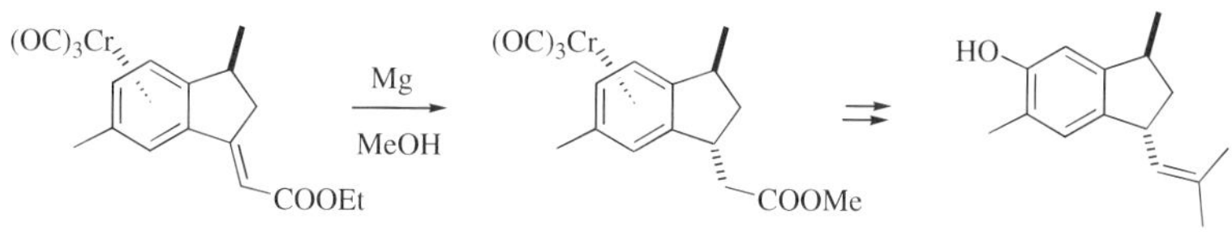

mutisianthol (**7-131**)

(**7-131**),[190] the *trans*-1,3-disubstituted indane skeleton evolved from chemical reduction of a tricarbonylchromium-complexed indanylidenecarboxylic ester. Various other methods were unsuccessful.

Even more elegant control is witnessed in the improved synthesis of acorenone (**7-132**) and acorenone-B (**7-133**)[191] which was based on previous work.[192] Instead of generating a mixture of diastereomeric complexes, each isomer was formed from an individual silylanisole. This involved OH-directed reaction with chromium hexacarbonyl, nuclear methylation, desilylation, replacement of the benzylic alcohol (via the acetate) by a Sakurai–Hosomi reaction, and modification of the side chain. The tricarbonylchromium group served to activate the *ortho* position (of the methoxy group) for methylation and the benzylic ionization, while maintaining the steric integrity of the benzylic center. The silyl group in the aromatic ring had the role of establishing the proper conformation of the neighboring side chain, enforcing its proximity to a hydrogen atom instead of the isopropyl group, thereby causing the metal complexation on different surfaces of the benzene ring in two silylanisoles.

A very interesting cyclization was observed during a synthetic study of clavicipitic acid.[193] Only one (η^6-arene)chromium complex diastereomer was found to undergo cyclization. It was also fortunate that the cyclized product has the desired stereochemistry to be converted to clavicipitic alcohol (**7-134**).

Enantioselectivity of asymmetric allylation of aromatic and propargylic aldehydes with allylborates is greatly enhanced by metal complexation of the aldehydes.[194]

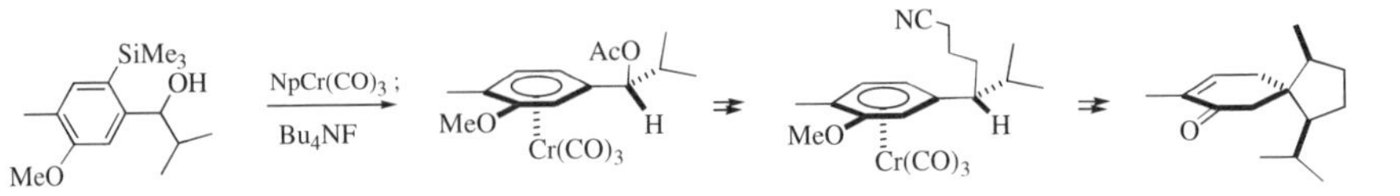

acorenone (**7-132**)

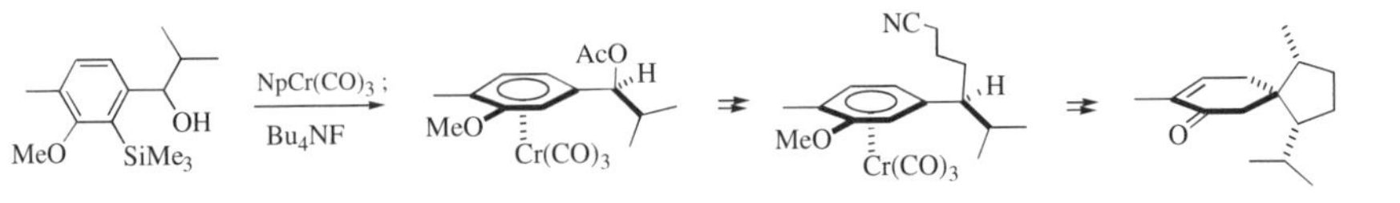

acorenone-B (**7-133**)

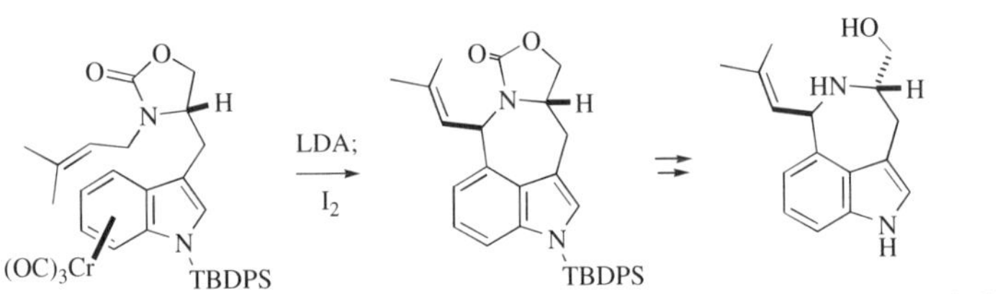

clavicipitic alcohol (**7-134**)

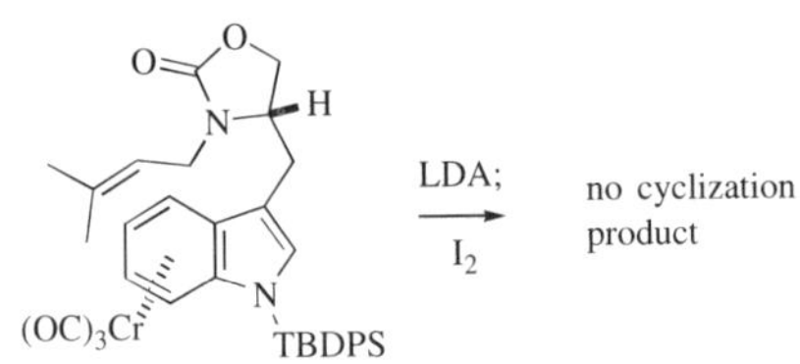

REFERENCES

1. Confalone, P.N.; Pizzolato, G.; Confalone, D.L.; Uskokovic, M.R. *J. Am. Chem. Soc.* **1980**, *102*, 1954.
2. Takano, S.; Inomata, K.; Ogasawara, K. *Chem. Lett.* **1992**, 443.
3. Curran, D.P.; Raciewicz, D.M. *J. Am. Chem. Soc.* **1985**, *107*, 1448.
4. Woodward, R.B. *Pure Appl. Chem.* **1968**, *17*, 519.
5. Marsden, S.P.; Depew, K.M.; Danishefsky, S.J. *J. Am. Chem. Soc.* **1994**, *116*, 11143.
6. Danishefsky, S.; Vaughan, K.; Gadwood, R.C.; Tsuzuki, K. *J. Am. Chem. Soc.* **1981**, *103*, 4136.
7. Paquette, L.A.; Ternansky, R.J.; Balogh, D.W.; Kentgen, G. *J. Am. Chem. Soc.* **1983**, *105*, 5446.
8. Brandi, A.; Cordero, F.M.; Goti, A.; Guarna, A. *Tetrahedron Lett.* **1992**, *33*, 6697.
9. Magari, H.; Hirota, H.; Takahashi, T.; Matsuo, A.; Uto, S.; Nozaki, H.; Nakayama, M.; Hayashi, H. *Chem. Lett.* **1982**, 1143.
10. Taylor, M.D.; Smith, A.B, III. *Tetrahedron Lett.* **1983**, *24*, 1867.
11. Kim, G.; Chu-Moyer, M.Y.; Danishefsky, S.J. *J. Am. Chem. Soc.* **1990**, *112*, 2003.
12. Gadwood, R.C.; Lett, R.M.; Wissinger, J.E. *J. Am. Chem. Soc.* **1986**, *108*, 6343.

13. Schore, N.E. *Chem. Rev.* **1988**, *88*, 1081.
14. Dauben, W.G.; Kowalczyk, B.A. *Tetrahedron Lett.* **1990**, *31*, 635.
15. Ficini, J. *Tetrahedron* **1976**, *32*, 1449.
16. Ficini, J.; d'Angelo, J.; Noire, J. *J. Am. Chem. Soc.* **1974**, *96*, 1213.
17. Desmaële, D.; Ficini, J.; Guingant, A.; Kahn, P. *Tetrahedron Lett.* **1983**, *24*, 3079; Desmaële, D.; Ficini, J.; Guingant, A.; Touzin, A.M. *Tetrahedron Lett.* **1983**, *24*, 3083.
18. Greene, A.E.; Teixeira, M.A.; Barreiro, E.; Cruz, A.; Crabbé, P. *J. Org. Chem.* **1982**, *47*, 2553.
19. Chang, S.; McNally, D.; Shary-Tehrany, S.; Hickey, M.J.; Boyd, R.H. *J. Am. Chem. Soc.* **1970**, *92*, 3109.
20. Solas, D.; Wolinsky, J. *J. Org. Chem.* **1983**, *48*, 670.
21. Welch, S.C.; Chayabunjonglerd, S.; Rao, A.S.C.P. *J. Org. Chem.* **1980**, *45*, 4086.
22. Matthews, R.S.; Whitesell, J.K. *J. Org. Chem.* **1975**, *40*, 3312.
23. Rodriguez, R.; Bermejo, F. *Tetrahedron Lett.* **1996**, *37*, 5581.
24. Stevens, K.E.; Paquette, L.A. *Tetrahedron Lett.* **1981**, *22*, 4393.
25. Tsunoda, T.; Kodama, M.; Ito, S. *Tetrahedron Lett.* **1983**, *24*, 83.
26. Magnus, P.; Quagliato, D.A. *Organometallics* **1982**, *1*, 1243.
27. Greene, A.E. *Tetrahedron Lett.* **1980**, *21*, 3059.
28. Williams, J.P.; St. Laurent, D.R.; Friedrich, D.; Pinard, E.; Roden, B.A.; Paquette, L.A. *J. Am. Chem. Soc.* **1994**, *116*, 4689.
29. Paquette, L.A.; Han, Y.-K. *J. Am. Chem. Soc.* **1981**, *103*, 1831.
30. Coates, R.M.; Shah, S.K.; Mason, R.W. *J. Am. Chem. Soc.* **1982**, *104*, 2198.
31. Nakatani, K.; Isoe, S. *Tetrahedron Lett.* **1985**, *26*, 2209.
32. Ronald, R.C.; Gewali, M.B.; Ronald, B.P. *J. Org. Chem.* **1980**, *45*, 2224.
33. Ghosh, A.; Biswas, S.; Venkateswaran, R.V. *Chem. Commun.* **1988**, 1421.
34. Schlessinger, R.H.; Schultz, J.A. *J. Org. Chem.* **1983**, *48*, 407.
35. Stork, G.; Sher, P.M. *J. Am. Chem. Soc.* **1986**, *108*, 303.
36. Araki, Y.; Endo, T.; Tanji, M.; Nagasawa, J.; Ishido, Y. *Tetrahedron Lett.* **1988**, *29*, 351.
37. Keck, G.E.; Tafesh, A.M. *J. Org. Chem.* **1989**, *54*, 5845.
38. Hart, D.J.; Tsai, Y.-M. *J. Am. Chem. Soc.* **1984**, *106*, 8209.
39. Srikrishna, A.; Krishnan, K. *J. Org. Chem.* **1993**, *58*, 7751.
40. Koreeda, M.; Mislankar, S.G. *J. Am. Chem. Soc.* **1983**, *105*, 7203.
41. Takano, S.; Inomata, K.; Ogasawara, K. *Chem. Commun.* **1992**, 169; Yoo, S.-e.; Lee, S.H. *J. Org. Chem.* **1994**, *59*, 6968.
42. Bornack, W.K.; Bhagwat, S.S.; Ponton, J.; Helquist, P. *J. Am. Chem. Soc.* **1981**, *103*, 4467.
43. Yates, P.; Burnell, D.J.; Freer, V.J.; Sawyer, J.F. *Can. J. Chem.* **1987**, *65*, 69.
44. Danishefsky, S.; Zamboni, R.; Kahn, M.; Etheredge, S.J. *J. Am. Chem. Soc.* **1980**, *102*, 2097.
45. Tatsuta, K.; Akimoto, K.; Kinoshita, M. *J. Am. Chem. Soc.* **1979**, *101*, 6116.
46. Boland, W.; Jaenicke, L. *Chem. Ber.* **1978**, *111*, 3262.
47. Woodward, R.B.; Katz, T.J. *Tetrahedron* **1959**, *5*, 70.

49. Sugahara, T.; Kuroyanagi, Y.; Ogasawara, K. *Synthesis* **1996**, 1101.
50. Ogasawara, K. *J. Synth. Org. Chem.* **1996**, *54*, 15.
51. Takano, S.; Kamikubo, T.; Moriya, M.; Ogasawara, K. *Synthesis* **1994**, 601.
52. Tanaka, K.; Ogasawara, K. *Synthesis* **1996**, 219.
53. Takano, S.; Inomata, K.; Ogasawara, K. *Chem. Commun.* **1989**, 271.
54. Tanaka, K.; Ogasawara, K. *Chem. Commun.* **1996**, 1839.
55. Takano, S.; Inomata, K.; Ogasawara, K. *Chem. Commun.* **1990**, 1544.
56. Takano, S.; Sato, T.; Inomata, K.; Ogasawara, K. *Chem. Commun.* **1991**, 462.
57. Takano, S.; Moriya, M.; Ogasawara, K. *J. Org. Chem.* **1991**, *56*, 5982.
59. Takano, S.; Sato, T.; Inomata, K.; Ogasawara, K. *Chem. Commun.* **1991**, 402.
60. Takano, S.; Moriya, M.; Ogasawara, K. *Tetrahedron Lett.* **1992**, *33*, 329.
61. Bernardes, V.; Kann, N.; Riera, A.; Moyano, A.; Pericas, M.A.; Greene, A.E. *J. Org. Chem.* **1995**, *60*, 6670.
62. Corey, E.J.; Kang, M.-c.; Desai, M.; Ghosh, A.K.; Houpis, I.N. *J. Am. Chem. Soc.* **1988**, *110*, 649; Desai, M.; Ghosh, A.K.; Kang, M.-c.; Houpis, I.N. *Strat. Tact. Org. Synth.* **1991**, *3*, 89.
63. Borm, C.; Meibom, D.; Winterfeldt, E. *Chem. Commun.* **1996**, 887.
64. Borm, C.; Winterfeldt, E. *Liebigs Ann. Chem.* **1996**, 1209.
65. Göres, M.; Winterfeldt, E. *J. Chem. Soc. Perkin Trans. 1* **1994**, 3525.
66. Welch, S.C.; Rao, A.S.C.P.; Gibbs, C.G. *Synth. Commun.* **1976**, *6*, 485; Welch, S.C.; Rao, A.S.C.P.; Wong, R.Y. *Synth. Commun.* **1976**, *6*, 443.
67. McGuirk, P.R.; Collum, D.B. *J. Am. Chem. Soc.* **1982**, *104*, 4496.
68. Lombardo, L.; Mander, L.N. *J. Org. Chem.* **1983**, *48*, 2298.
69. Corey, E.J.; Danheiser, R.L.; Chandrasekaran, S.; Keck, G.E.; Gopalan, B.; Larsen, S.D.; Siret, P.; Gras, J.-L. *J. Am. Chem. Soc.* **1978**, *100*, 8034; Corey, E.J.; Danheiser, R.L.; Chandrasekaran, S.; Siret, P.; Keck, G.E.; Gras, J.-L. *J. Am. Chem. Soc.* **1978**, *100*, 8031.
70. Bernasconi, S.; Ferrari, M.; Gariboldi, P.; Jommi, G.; Sisti, M. *J. Chem. Soc. Perkin Trans. 1* **1981**, 1994; Bernasconi, S.; Gariboldi, P.; Jommi, G.; Montanari, S.; Sisti, M. *J. Chem. Soc. Perkin Trans. 1* **1981**, 2394.
71. Uyehara, T.; Kabasawa, Y.; Kato, T. *Bull. Chem. Soc. Jpn.* **1986**, *59*, 2521.
72. Corey, E.J.; Wu, Y.-J. *J. Am. Chem. Soc.* **1993**, *115*, 8871.
73. Hart, D.J.; Huang, H.-C.; Krishnamurthy, R.; Schwartz, T. *J. Am. Chem. Soc.* **1989**, *111*, 7507.
74. Kigoshi, H.; Imamura, Y.; Mizuta, K.; Niwa, H.; Yamada, K. *J. Am. Chem. Soc.* **1993**, *115*, 3056.
75. Thompson, S.K.; Heathcock, C.H. *J. Org. Chem.* **1990**, *55*, 3004.
76. Greenlee, W.J.; Woodward, R.B. *J. Am. Chem. Soc.* **1976**, *98*, 6075.
77. Le Menez, P.; Sapi, J.; Kunesch, N.; Angell, E.C.; Wenkert, E. *J. Org. Chem.* **1989**, *54*, 3216.
78. Grese, T.A.; Hutchinson, K.D.; Overman, L.E. *J. Org. Chem.* **1993**, *58*, 2468.
79. Piers, E.; de Waal, W.; Britton, R.W. *J. Am. Chem. Soc.* **1971**, *93*, 5113.
80. Mehta, G.; Praveen, M. *Chem. Commun.* **1993**, 1573.
81. Bornmann, W.; Kuehne, M.E. *J. Org. Chem.* **1992**, *57*, 1752.

82. Kraus, G.A.; Thurston, J. *J. Am. Chem. Soc.* **1989**, *111*, 9203.

83. Paquette, L.A.; Lin, H.S.; Gunn, B.P.; Coghlan, M.J. *J. Am. Chem. Soc.* **1988**, *110*, 5818.

84. Evans, D.A.; Sims, C.L. *Tetrahedron Lett.* **1973**, 4691.

85. Baraldi, P.G.; Barco, A.; Benetti, S.; Pollini, G.P.; Simoni, D. *Tetrahedron Lett.* **1983**, *24*, 5669.

86. Fringuelli, F.F.; Taticchi, A.; de Giuli, G. *Gazz. Chim. Ital.* **1969,** *99*, 219; Fringuelli, F.F.; Taticchi, A.; Traverso, G. *Gazz. Chim. Ital.* **1969,** *99*, 231.

87. Caine, D.; Gupton, J.T., III. *J. Org. Chem.* **1974**, *39*, 2654.

88. Caine, D.; Smith, T.L. *J. Am. Chem. Soc.* **1980**, *102*, 7568.

89. Ziegler, F.E.; Reid, G.R.; Studt, W.L.; Wender, P.A. *J. Org. Chem.* **1977**, *42*, 1991.

90. Matteson, D.S.; Sadhu, K.M.; Peterson, M.L. *J. Am. Chem. Soc.* **1986**, *108*, 810.

91. Stork, G.; La Clair, J.J.; Spargo, P.; Nargund, R.P.; Totah, N. *J. Am. Chem. Soc.* **1996**, *118*, 5304.

92. Yanami, T.; Miyashita, M.; Yoshikoshi, A. *J. Org. Chem.* **1980**, *45*, 607.

93. Schleyer, P.v.R. *J. Am. Chem. Soc.* **1967**, *89*, 699; Houk, K.N.; Paddon-Row, M.N.; Rondan, N.G.; Wu, Y.-D.; Brown, F.K.; Spellmeyer, D.C.; Metz, J.T.; Li, Y.; Loncharich, R.J. *Science* **1986**, *231*, 1108.

94. Corey, E.J.; Hartmann, R.; Vatakencherry, P.A. *J. Am. Chem. Soc.* **1962**, *84*, 2611.

95. Hatanaka, M.; Ueno, F.; Ueda, I. *Tetrahedron Lett.* **1996**, *37*, 89.

96. Nagasawa, K.; Matsuda, N.; Noguchi, Y.; Yamanashi, M.; Zako, Y.; Shimizu, I. *J. Org. Chem.* **1993**, *58*, 1483.

97. Gibson, T.; Barneis, Z.J. *Tetrahedron Lett.* **1972**, 2207.

98. Trost, B.M.; Shuey, C.D.; DiNinno, F. *J. Am. Chem. Soc.* **1979**, *101*, 1284.

99. Yamamoto, H.; Sham. H.L. *J. Am. Chem. Soc.* **1979**, *101*, 1609.

100. Bosch, J.; Bennasar, M.-L. *Synlett* **1995**, 587.

101. Grieco, P.A.; Ohfune, Y.; Majetich, G. *J. Am. Chem. Soc.* **1977**, *99*, 7393.

102. Grieco, P.A.; Takigawa, T.; Moore, D.R. *J. Am. Chem. Soc.* **1979**, *101*, 4380.

103. Jung, M.E.; Hudspeth, J.P. *J. Am. Chem. Soc.* **1980**, *102*, 2463.

104. Vaillancourt, V.; Agharahimi, M.R.; Sundram, U.N.; Richou, O.; Faulkner, D.J.; Albizati, K.F. *J. Org. Chem.* **1991**, *56*, 378.

105. Trost, B.M.; Schmuff, N.R. *J. Am. Chem. Soc.* **1985**, *107*, 396.

106. Tsai, T.R.Y.; Nambiar, K.P.; Krikorian, D.; Botta, B.; Marini-Bettolo, R.; Wiesner, K. *Can. J. Chem.* **1979**, *57*, 2124.

107. Trost, B.M.; Renaut, P. *J. Am. Chem. Soc.* **1982**, *104*, 6668.

108. Danishefsky, S.; Hirama, M.; Gombatz, K.; Harayama, T.; Berman, E.; Schuda, P.F. *J. Am. Chem. Soc.* **1979**, *101*, 7020.

109. Stork, G.; van Tamelen, E.E.; Friedman, L.J.; Burgstahler, A.W. *J. Am. Chem. Soc.* **1953**, *75*, 384.

110. Ashton, P.R.; Brown, G.R.; Isaacs, N.S.; Giuffrida, D.; Kohnke, F.H.; Mathias, J.P.; Slawin, A.M.Z.; Smith, D.R.; Stoddart, J.F.; Williams, D.J. *J. Am. Chem. Soc.* **1992**, *114*, 6330.

111. Metz, P.; Meiners, U.; Cramer, E.; Fröhlich, R.; Wibbeling, B. *Chem. Commun.* **1996**, 431.

112. Lautens, M. *Pure Appl. Chem.* **1992**, *64*, 1873.
113. Takano, S.; Takahashi, M.; Ogasawara, K. *J. Am. Chem. Soc.* **1980**, *102*, 4282.
114. Takano, S.; Takahashi, M.; Hatakeyama, S.; Ogasawara, K. *Chem. Commun.* **1979**, 556.
115. Broka, C.A.; Chan, S.; Peterson, B. *J. Org. Chem.* **1988**, *53*, 1584.
116. Dixon, J.; Lythgoe, B.; Siddiqui, I.A.; Tideswell, J. *J. Chem. Soc. C* **1971**, 1301; Harrison, R.G.; Lythgoe, B.; Wright, P.W. *J. Chem. Soc. Perkin Trans. 1* **1974**, 2654.
117. DeShong, P.; Ramesh, S.; Elango, V.; Perez, J.J. *J. Am. Chem. Soc.* **1985**, *107*, 5219; Schlessinger, R.H.; Bebernitz, G.R.; Lin, P.; Poss, A.J. *J. Am. Chem. Soc.* **1985**, *107*, 1777.
118. Martin, S.F.; Dappen, M.S.; Dupre, B.; Murphy, C.J. *J. Org. Chem.* **1987**, *52*, 3706.
119. Oppolzer, W.; Petrzilka, M. *J. Am. Chem. Soc.* **1976**, *98*, 6722; *Helv. Chim. Acta* **1978**, *61*, 2755.
120. Schlessinger, R.H.; Nugent, R.A. *J. Am. Chem. Soc.* **1982**, *104*, 1116.
121. Brooks, D.W.; Grothaus, P.G.; Mazdiyasni, H. *J. Am. Chem. Soc.* **1983**, *105*, 4472.
122. Kitahara, T.; Touhara, K.; Watanabe, H.; Mori, K. *Tetrahedron* **1989**, *45*, 6387.
123. Gwaltney, S.L., II; Sakata, S.T.; Shea, K.J. *J. Org. Chem.* **1996**, *61*, 7438.
124. Rizzo, C.J.; Smith, A.B., III. *J. Chem. Soc. Perkin Trans. 1* **1991**, 969.
125. Montana, A.M.; Nicholas, K.M. *J. Org. Chem.* **1990**, *55*, 1569.
126. Hamon, D.P.G.; Young, R.N. *Aust. J. Chem.* **1976**, *29*, 145.
127. Sammes, P.G.; Street, L.J.; Whitby, R.J. *J. Chem. Soc. Perkin Trans. 1* **1986**, 666.
128. Martin, S.F.; Pacofsky, G.J.; Gist, R.P.; Lee, W.-C. *J. Am. Chem. Soc.* **1989**, *111*, 7634.
129. Toyota, M.; Wada, T.; Nishikawa, Y.; Yanai, K.; Fukumoto, K.; Kabuto, C. *Tetrahedron* **1995**, *51*, 6927.
130. Larsen, S.D.; Monti, S.A. *J. Am. Chem. Soc.* **1977**, *99*, 8015.
131. Nakata, T.; Hata, N.; Iida, K.; Oishi, T. *Tetrahedron Lett.* **1987**, *28*, 5661.
132. Nakata, T.; Nagao, S.; Mori, N.; Oishi, T. *Tetrahedron Lett.* **1985**, *26*, 6461; Nakata, T.; Nagao, S.; Oishi, T. *Tetrahedron Lett.* **1985**, *26*, 6465.
133. Corey, E.J.; Becker, K.B.; Varma, R.K. *J. Am. Chem. Soc.* **1972**, *94*, 8616.
134. Ensley, H.E.; Parnell, C.A.; Corey, E.J. *J. Org. Chem.* **1978**, *43*, 1610.
135. Corey, E.J.; Ensley, H.E. *J. Am. Chem. Soc.* **1975**, *97*, 6908.
136. Oppolzer, W.; Robbiani, C.; Bättig, K. *Tetrahedron* **1984**, *40*, 1391
137. Comins, D.L.; Joseph, S.P.; Goehring, R.R. *J. Am. Chem. Soc.* **1994**, *116*, 4719.
138. Comins, D.L.; Hong, H. *J. Am. Chem. Soc.* **1993**, *115*, 8851.
139. Ihara, M.; Takahashi, M.; Taniguchi, N.; Yasui, K.; Niitsuma, H.; Fukumoto, K. *J. Chem. Soc. Perkin Trans. 1* **1991**, 525.
140. Ermert, P.; Meyer, J.; Stucki, C.; Schneebeli, J.; Obrecht, J.-P. *Tetrahedron Lett.* **1988**, *29*, 1265.
141. Whitesell, J.K.; Bhattacharya, A.; Henke, K. *Chem. Commun.* **1982**, 988.
142. Whitesell, J.K.; Minton, M.A. *J. Am. Chem. Soc.* **1986**, *108*, 6802.
143. Whitesell, J.K.; Allen, D.E. *J. Am. Chem. Soc.* **1988**, *110*, 3585.
144. Greene, A.E.; Charbonnier, F.; Luche, M.-J.; Moyano, A. *J. Am. Chem. Soc.* **1987**, *109*, 4752.

145. Castro, J.; Sorensen, H.; Riera, A.; Morin, C.; Moyano, A.; Pericas, M.A.; Greene, A.E. *J. Am. Chem. Soc.* **1990**, *112*, 9388.

146. Suzuki, I.; Kin, H.; Yamamoto, Y. *J. Am. Chem. Soc.* **1993**, *115*, 10139.

147. Asami, M.; Mukaiyama, T. *Chem. Lett.* **1979**, 569.

148. Bernardi, A.; Cardani, S.; Pilati, T.; Poli, G.; Scolastico, C.; Villa, R. *J. Org. Chem.* **1988**, *53*, 1600.

149. Knapp, S.; Ornaf, R.M.; Rodriques, K.E. *J. Am. Chem. Soc.* **1983**, *105*, 5494.

150. Takano, S.; Tamura, N.; Ogasawara, K. *Chem. Commun.* **1981**, 1155.

151. Takano, S.; Yonaga, M.; Morimoto, M.; Ogasawara, K. *J. Chem. Soc. Perkin Trans. 1* **1985**, 305.

152. Takano, S.; Yonaga, M.; Chiba, K.; Ogasawara, K. *Tetrahedron Lett.* **1980**, 3697; Takano, S.; Uchida, W.; Hatakeyama, S.; Ogasawara, K. *Chem. Lett.* **1982**, 733.

153. Takano, S.; Yonaga, M.; Ogasawara, K. *Chem. Commun.* **1981**, 1153.

154. Tomioka, K.; Koga, K. *Tetrahedron Lett.* **1979**, 3315.

155. Tomioka, K.; Kawasaki, H.; Iitaka, Y.; Koga, K. *Tetrahedron Lett.* **1985**, *26*, 903.

156. Tomioka, K.; Sugimori, M.; Koga, K. *Chem. Pharm. Bull.* **1987**, *35*, 906.

157. Ikota, N.; Hanaki, A. *Chem. Pharm. Bull.* **1987**, *35*, 2140.

158. Tomioka, K.; Ishiguro, T.; Koga, K. *Tetrahedron Lett.* **1980**, 2973.

159. Hanessian, S.; Murray, P.J.; Sahoo, S.P. *Tetrahedron Lett.* **1985**, *26*, 5623, 5627.

160. Charette, A.B.; Mellon, C.; Rouillard, L.; Malenfant, E. *Synlett* **1993**, 81.

161. Fujimoto, R.; Kishi, Y.; Blount, J.F. *J. Am. Chem. Soc.* **1980**, *102*, 7154.

162. Bartlett, P.A.; Greene, F.R., III. *J. Am. Chem. Soc.* **1978**, *100*, 4858.

163. Wiesner, K.; Poon, L.; Jirkovsky, I.; Fishman, M. *Can. J. Chem.* **1969,** *47*, 433.

164. Curran, D.P.; Kuo, S.C. *Tetrahedron* **1987**, *43*, 5653.

160. Oppolzer, W. *Angew. Chem. Int. Ed. Engl.* **1984**, *23*, 876.

165. Oppolzer, W.; Chapuis, C.; Dupuis, D.; Guo, M. *Helv. Chim. Acta* **1985**, *68*, 2100.

166. Oppolzer, W.; Pitteloud, R.; Bernardinelli, G.; Bättig, K. *Tetrahedron Lett.* **1983**, *24*, 4975.

167. Oppolzer, W.; Schneider, P. *Helv. Chim. Acta* **1986**, *69*, 1817.

168. Schreiber, S.L.; Meyers, H.V. *J. Am. Chem. Soc.* **1988**, *110*, 5198.

169. Hirai, Y.; Momose, T. *Stud. Nat. Prod. Chem.* **1994**, *14*, 551.

170. Mandai, T.; Matsumoto, T.; Kawada, M.; Tsuji, J. *J. Org. Chem.* **1992**, *57*, 1326.

171. Shimizu, I.; Ishikawa, T. *Tetrahedron Lett.* **1994**, *35*, 1905.

172. Trost, B.M.; Klun, T.P. *J. Am. Chem. Soc.* **1981**, *103*, 1864.

173. Bäckvall, J.E.; Byström, S.E.; Nordberg, R.E. *J. Org. Chem.* **1984**, *49*, 4619.

174. Bäckvall, J.E.; Renko, Z.D.; Byström, S.E. *Tetrahedron Lett.* **1987**, *28*, 4199.

175. Trost, B.M.; Godleski, S.A.; Genet, J.P. *J. Am. Chem. Soc.* **1978**, *100*, 3930.

176. Trost, B.M.; Lynch, J.K.; Angle, S.R. *Tetrahedron Lett.* **1987**, *28*, 375.

177. Holton, R.A. *J. Am. Chem. Soc.* **1977**, *99*, 8083.

178. Pearson, A.J. *Synlett* **1990**, 10.

179. Pearson, A.J.; Lai, Y.S. *Chem. Commun.* **1988**, 442.

180. Hansson, S.; Miller, J.F.; Liebeskind, L.S. *J. Am. Chem. Soc.* **1990**, *112*, 9660.

181. Roush, W.R.; Wada, C.K. *J. Am. Chem. Soc.* **1994**, *116*, 2151.
182. Nunn, K.; Mosset, P.; Gree, R.; Saalfrank, R.W. *Angew. Chem. Int. Ed.* **1988**, *27*, 1188.
183. Iwata, C.; Takemoto, Y. *Chem. Commun.* **1996**, 2497.
183a. Roush, W.R.; Works, A.B. *Tetrahedron Lett.* **1997**, *38*, 351.

184. Franck-Neumann, M.; Martina, D.; Heitz, M.-P. *J. Organomet. Chem.* **1986**, *301,* 61.
185. Le Gall, T.; Lellouche, J.-P.; Beaucourt, J.-P. *Tetrahedron Lett.* **1989**, *30*, 6521.
186. Ley, S.V.; Meek, G. *Chem. Commun.* **1995**, 1751.
187. Hodgson, S.T.; Hollinshead, D.M.; Ley, S.V. *Tetrahedron* **1985**, *41*, 5871.
188. Bates, R.W.; Fernandez-Moro, R.; Ley, S.V. *Tetrahedron* **1991**, *47*, 9929.
189. Jiang, Y.; Hu, W.; Deng, J.; Liu, G. *Synth. Commun.* **1991**, *21*, 1755; Uemura, M.; Nishimura, H.; Minami, T.; Hayashi, Y. *J. Am. Chem. Soc.* **1991**, *113*, 5402.
190. Ho, T.-L.; Lee, K.-Y.; Chen, C.-K. *J. Org. Chem.* **1997**, *62*, 3365
191. Uemura, M.; Kobayashi, T.; Isobe, K.; Minami, T.; Hayashi, Y. *J. Org. Chem.* **1986**, *51*, 2859.
192. Semmelhack, M.F.; Yamashita, A. *J. Am. Chem. Soc.* **1980**, *102*, 5924.
193. Semmelhack, M.F.; Knochel, P.; Singleton, T. *Tetrahedron Lett.* **1993**, *34*, 5051.
194. Roush, W.R.; Park, J.C. *J. Org. Chem.* **1990**, *55*, 1143.

8

STERIC, ELECTROSTATIC, AND STEREOELECTRONIC EFFECTS

8.1. STERIC EFFECTS

Steric effects due to van der Waals forces between two groups are well known to organic chemists. Although there are attractive (at long distances) and repulsive (at short distances) interactions, it is sufficient to consider the latter in our evaluation of their influences in synthetic circumstances.

Conformations are of the utmost importance in determining ways molecules react in terms of stereochemical consequences. Because of lower degrees of freedom, cyclic compounds, especially smaller rings (<six-membered), usually prefer certain conformations even if there are no additional bonding constraints. Conformational analysis of six-membered ring compounds has been developed since the time Barton enunciated the basic principles in the early 1950s.

For cyclohexane and cyclohexene derivatives in the chairlike form, the less hindered equatorial substituents are more reactive. This behavior is witnessed in esterification or saponification and oxidation, fruitfully exploited in a synthesis of yohimbine,[1] to resolve the quandary of seperating two diacetoxy lactams (**8-1a, 8-1b**). Thus, saponification of the acetates and pivaloylation led to two separable products without affecting the original ratio. It was found that only the equatorial hydroxyl group in each diol underwent acylation with the bulky reagent. The major compound (55% yield) was processed into (+)-yohimbine. The minor (25% yield) pivalate is, in principle, convertible to the alkaloid by a different set of reactions.

In the same work the last step involved reduction of yohimbinone. Since yohimbine contains an axial alcohol, it requires reduction from the equatorial side. Extremely bulky metal hydrides such as L-Selectride [$LiB(i\text{-}Bu)_3H$] are the reagents of choice.

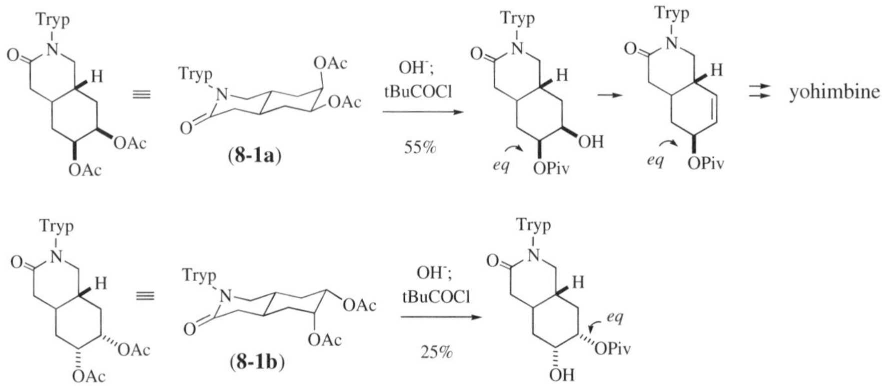

Telesteric effects are those arising from nonbonding interactions of a reagent with remote part(s) of the molecule during a reaction. Such an effect determined the course of the organoyttrium-catalyzed cyclization that led to a synthesis of epilupinine (**8-2**).[2] One of the methylated cyclopentadienyl ligands interfered with methylene groups of the piperidine ring in the alternative transition state, which was promptly avoided.

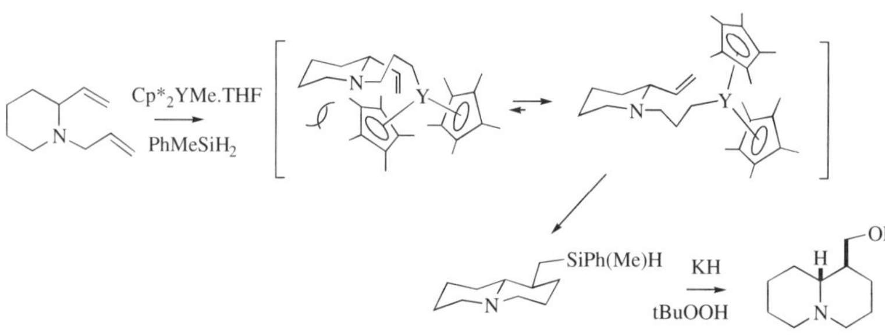

epilupinine
(**8-2**)

Stereoinduction of tetraene cyclization during a photoinduced electron transfer process[3] occurs when the terminal alkene group is incorporated into a 1,3-dioxin-4-one unit which was in turn spiroannulated to $C_{(3)}$ of *p*-menthane. Each of the two diastereomeric dioxinones derived from (–)-menthone generated tricarbocyclic alcohols with >99% enantiomeric excess and diastereoselectivity arising from telesteric effects. Six stereocenters were created, and the major and minor products in each series are enantiomeric at the five new centers. The disfavored transition states (**8-3, 8-4**) are presented to emphasize the steric interactions.

The assembly of the BCDE-ring system (**8-5**) of (+)-conessine[4] utilized an *endo*-selective 1,3-dipolar cycloaddition to an unsaturated oxazolidine synthon. The

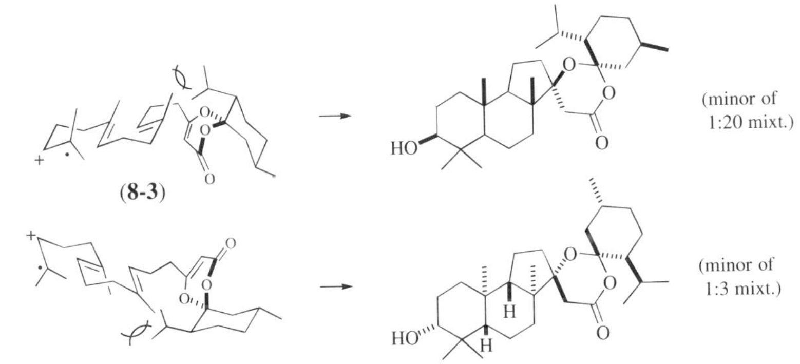

(8-3)

(8-4)

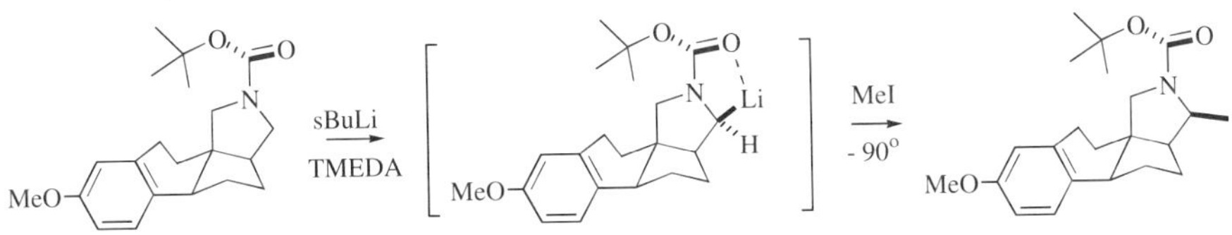

(8-5)

last stage concerning introduction of the secondary methyl substituent at the α-carbon of the pyrrolidine ring was extremely critical. The success in using an *N*-(*t*-butoxycarbonyl) group to direct lithiation was due to chelation of *s*-butyllithium prior to deprotonation. The bulky group is expected to stay away from interaction with the methylene groups of the CD-ring component, and only an 20-*endo*-hydrogen was reachable by the complexed reagent.

The *t*-Boc group on the nitrogen atom of 2-piperidineacetic esters steers the side chain to an axial position. When the allyl ester was enolized and treated with trimethylsilyl chloride, an Ireland–Claisen rearrangement led to a precursor (**8-6**) of (+)-epilupinine. The corresponding methyl ester underwent allylation to afford the precursor (**8-7**) of (–)-lupinine.[5]

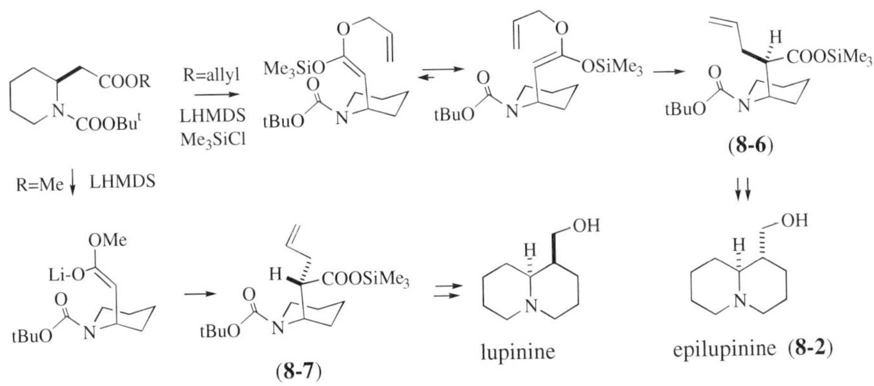

The intriguing stereoselectivity for alkylation of Meyers's bicyclic lactam (**8-8a**)[6] has been ascribed to preferential solvation of the lithium enolate from the *exo* face,[7] leaving the *endo* side more accessible.

Stereocontrol of the Pauson–Khand reaction for cyclopentenone assembly by allylic and propargylic substituents of the substrate(s) has direct bearing on syntheses such as that of quadrone (**8-8**).[8] The stereoinduction was the result of an option for an intermediate in which spatial interactions between the siloxymethyl and the alkoxyethyl chains are minimized. The importance of telesteric interactions was also manifested in the formation of angular triquinanes and, as a result, a facile synthesis of pentalenene[9] could be developed.

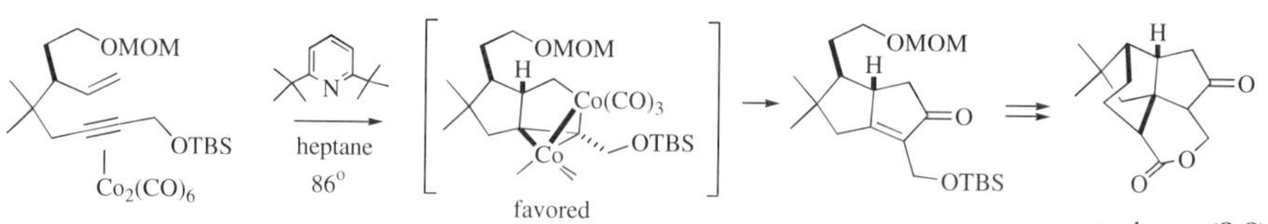

quadrone (**8-8**)

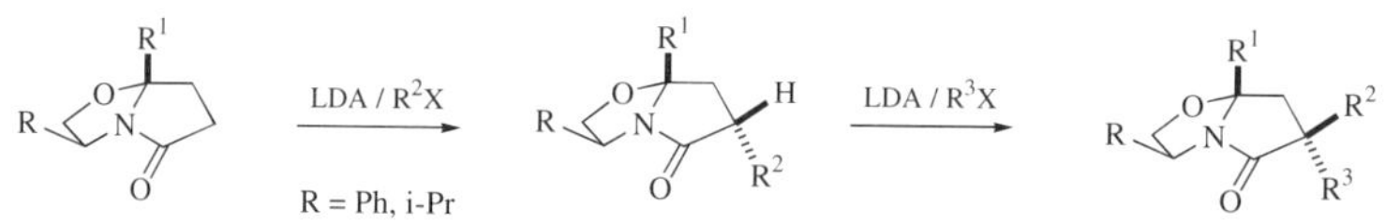

(**8-8a**)

Many intramolecular Diels–Alder reactions do not obey the *endo* rule. They are concerted but asynchronous, with formation of one of the two C–C bonds at a more advanced stage in the transition state. Such a reaction was observed in a forskolin (**8-9**) synthesis[10] from which three stereocenters (especially the *trans* ring junction) were created while leaving a double bond at the desired position for dihydroxylation.

The acceleration of the Diels–Alder reactions in aqueous environments has been attributed to pressure exerted on the reactant pairs. Consequently, components of such

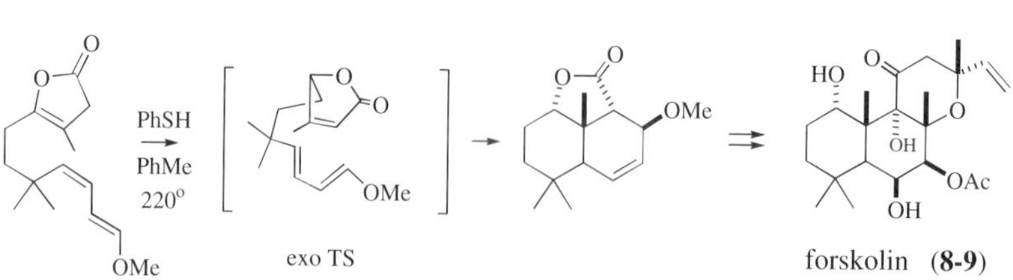

forskolin (**8-9**)

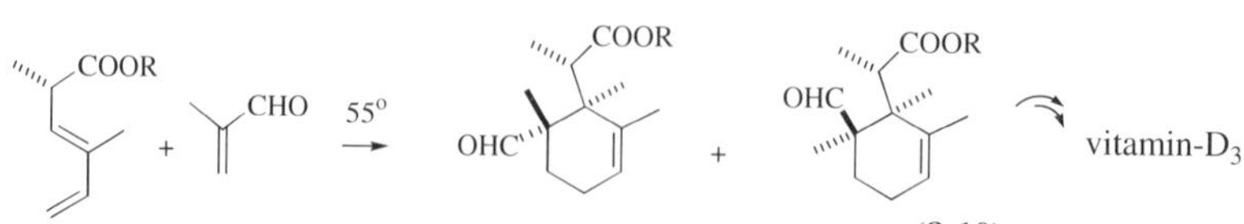

(**8-10**)

R = Me	neat, 63 h (10%)	1	:	1
R = Na	H_2O, 16 h (85%)	1	:	4.6

reactant pairs tend to align themselves to form a more compact entity. The increase in the *endo* adduct from the reaction of methacrolein with the sodium salt of a substituted hexadienoic acid is an example of the effect. This change greatly facilitated the preparation of a vitamin D_3 intermediate (**8-10**).[11]

8.2. ELECTROSTATIC EFFECTS

Electrostatic interactions among various parts of a molecule can influence its conformation profoundly and, therefore, the outcome of reactions. In 1959 the Cornforth model for addition to α-halocarbonyl compounds was formulated,[12] and a more recent evaluation of the electrostatic factor from the theoretical viewpoint[13] has been prompted by such observations as the opposite facial attack on norbornen-7-one by methylmagnesium iodide and pentafluoroethyllithium.[14] Surprising as it may seem at first glance, the Mn(III) oxidative cyclization of a β-ketoester produced a precursor (**8-11**) of *O*-methylpodocarpic acid.[15] The *anti* conformation to minimize dipole–dipole interaction between the ketone and the ester group must have been maintained before the first C–C bond formation. The result also implied an ester participation during the intramolecular Friedel–Crafts alkylation.

4,4-Disubstituted 2,5-cyclohexadienones in which one of the substituents is an oxy group were found to react with Grignard reagents predominantly from the side away from that group.[16] The cause was probably electrostatic in nature, but in any case the results suited the convenience of an aranorosin (**8-11a**) synthesis because the tertiary alcohol could be used to direct a stereoselective epoxidation.

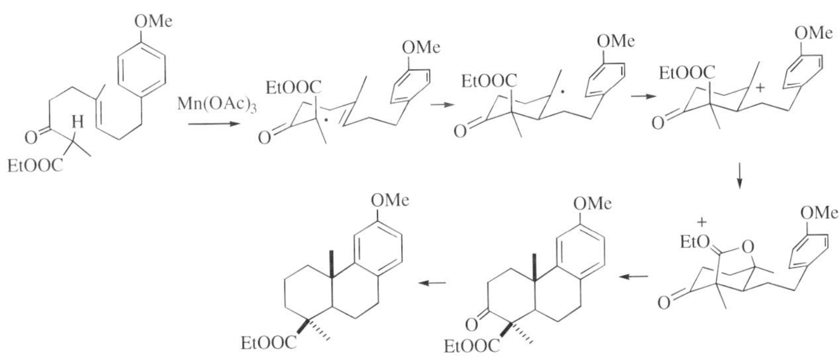

ethyl *O*-methylpodocarpate (**8-11**)

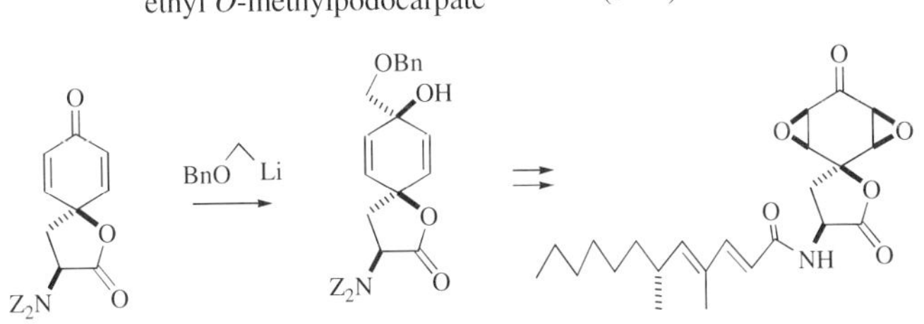

(major) aranorosin (**8-11a**)

A synthesis of *ent*-grindelic acid (**8-12**)[17] involving a pinacolic rearrangement for ring expansion was related to some telesteric effects during the organometallic reaction which provided the substrate. The dihydrofuranylcerium reagent subtly discriminated (with 10.4:1 selectivity) the substituted *trans*-perhydroindanone by virtue of the chirality at the quaternary center. The subangular site of the ketone was greatly influenced by the presence of the axial methyl groups, particularly that at the adjacent carbon. Apparently, in the favored transition state the less sterically demanding oxygen atom of the reagent was closer to the cyclohexane and also the vinyl group instead of the methyl projected over the ring system. When rearrangement started, the oxonium ion would have swung around to dispose the oxygen atom toward the cyclopentane so as to minimize dipole–dipole interaction with the hydroxyl group.

Electrostatic attraction of the mercuric ion to the face of a cyclopentene unit of a sesbanine (**8-13**) precursor by the benzylic ester is likely the cause of a very satisfactory hydratomercuration reaction. This solved the stereochemical problem in the synthesis.[18]

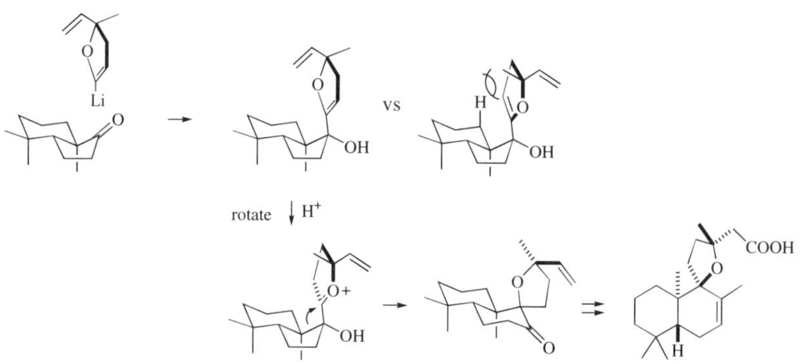

ent-grindelic acid
(**8-12**)

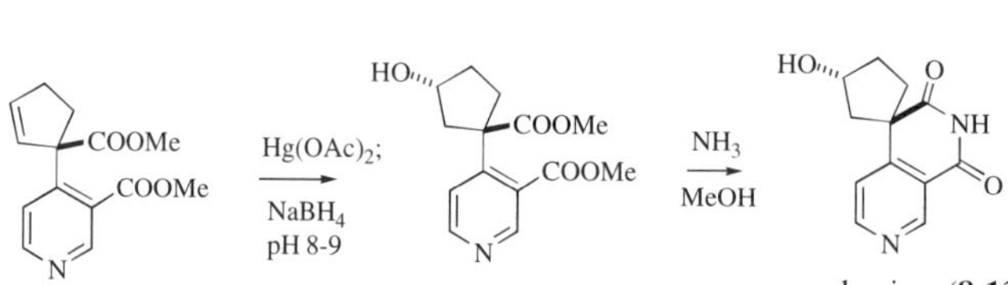

sesbanine (**8-13**)

In the course of a synthetic investigation of kopsane alkaloids [e.g., kopsanone (**8-14**)],[19] an extremely helpful reaction was found. The exclusive *endo*-allylation is seemingly contrasteric. While evidence ruled out pathways involving reaction from the *exo*-side and subsequent isomerization by ring cleavage and reclosure, the most reasonable explanation appears to be carbanion pyramidalization toward the *endo* direction to avoid a dipole–dipole interaction with the lone electron-pair of the nitrogen atom.

A synthetic route to byssochlamic acid (**8-15**)[20] involved a [2+2]-photocycloaddition and thermolysis to generate the mesocycle. The cycloreversion that took place in refluxing toluene, which is an exceptionally mild condition, suggests enormous internal strain of the cycloadduct. The eclipsing lactone units probably contributed to the instability of the compound.

A transannular cyclization (intramolecular Michael addition) tactic for synthesis of cephalotaxine (**8-16**)[21] was plagued by unfavorable equilibrium. However, when the Michael acceptor was set in a higher oxidation state, the problem vanished. The reason is that the 3-cyclopentene-1,2-dione unit has a low tendency to exist in the antiaromatic cyclopentadienone form, and the diketone has strong dipole interactions which can be lowered on cyclization. The resulting 1,2-cyclopentanedione can then exist as a hydroxy enone.

Aldol condensation can be rendered *anti*-selective (98:2) by exploiting dipole alignment.[22] Thus, the (*Z*)-boron enolate attached to a bicyclic lactam (**8-17**) prefers a chairlike transition state in which the lactam carbonyl group and the enolic oxygen are as far apart as possible. It must be emphasized that the presence of dipole–dipole interactions favors open transition states when reactions are conducted in polar solvents and/or involve highly dissociating counterions (e.g., R_4N^+). However, situations change with strong chelation by cations.

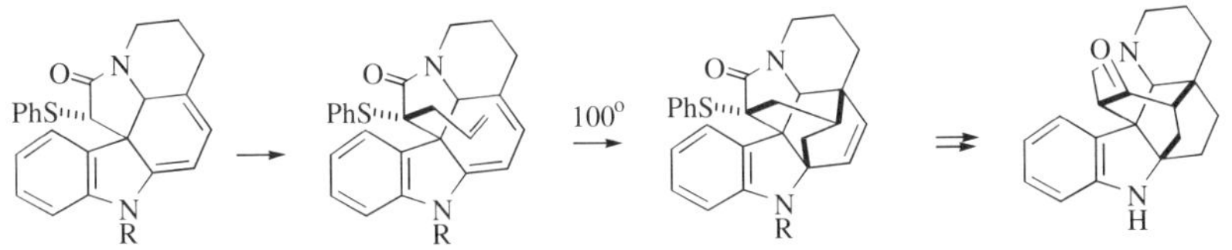

kopsanone (**8-14**)

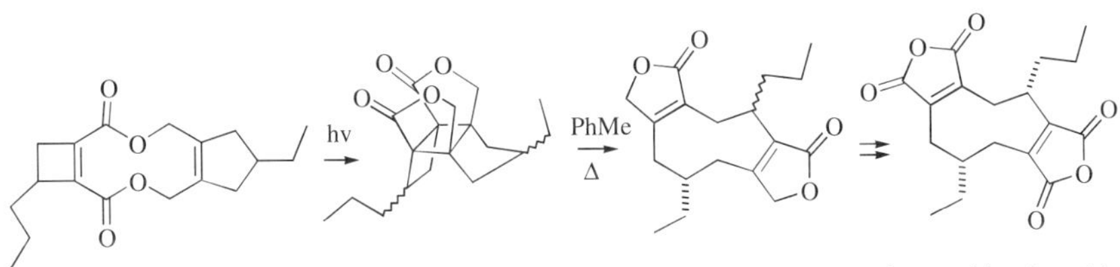

byssochlamic acid (**8-15**)

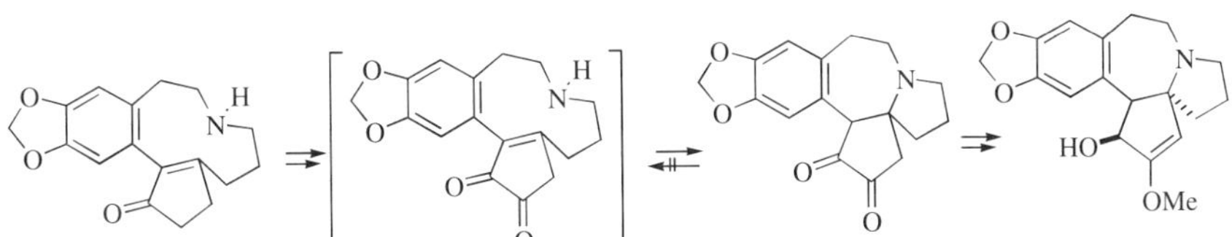

cephalotaxine (**8-16**)

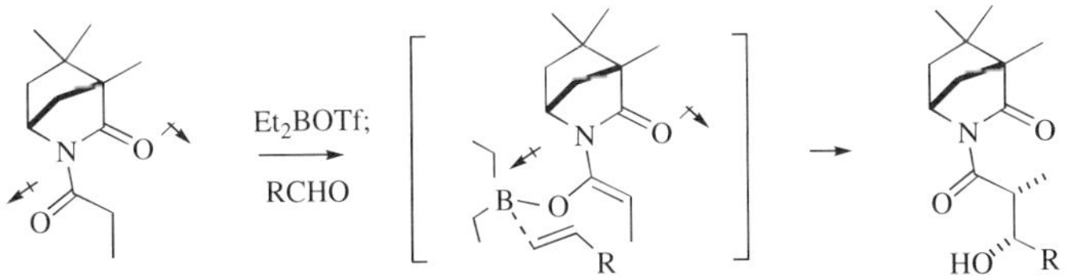

(**8-17**)

syn: anti > 98: 2

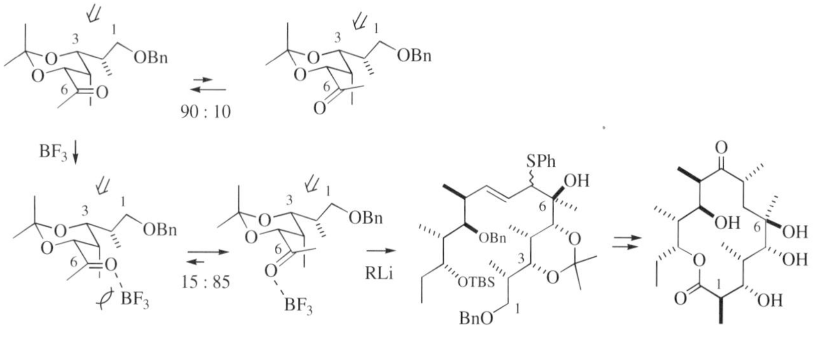

(-)-erythronolide-B
(**8-18**)

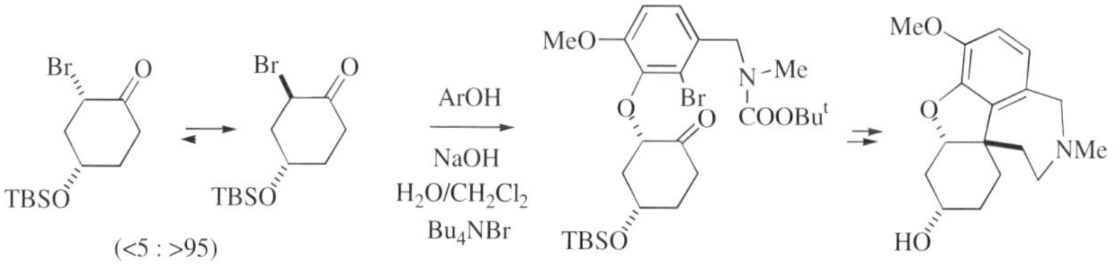

lycoramine (**8-18a**)

The methyl ester of *meso*-2,4-dimethylglutaric hemialdehyde has been found to react with allyltin reagents to afford mainly the *erythro* (anti-Cram) products. The results indicate the existence of a stable conformation which is maintained by electrostatic interaction of the two carbonyl groups.[23]

In a synthesis of (–)-erythronolide-B (**8-18**),[24] the assembly of two large segments [$C_{(6)}$–$C_{(7)}$ bond formation] which created a quaternary stereocenter was critically dependent on reaction conditions. Due to dipole–dipole interactions which decided a transoid conformation for the ketone with respect to the proximal oxygen atom of the dioxane ring, the correct isomer could only be formed by precoordination with boron trifluoride which destabilizes the conformation by introducing steric interference with other parts of the substrate.

It cannot be overemphasized that a set of simple rules are sufficient to predict or account for the vicissitudes of various chemical systems. The predominant existence (>95%) of *trans*-2-bromo-4-(*t*-butyldimethyl)siloxy-cyclohexanone in solution[25] indicates the importance of solvent–dipole interactions to favor a conformer with a larger net dipole. The acquisition of this *trans* bromoketone was critical to a convergent synthesis of lycoramine (**8-18a**).

8.3. STEREOELECTRONIC EFFECTS

8.3.1. Stereoelectronic Effects in Carbocycles

Stereoelectronic effect[26] relates reactivity or properties of a chemical species with the orientation of electron orbitals (filled or unfilled). Its manifestation is most unambiguous when a molecule is fixed in one conformation. For example, the $sp^2 \rightarrow sp^3$ rehybridization occurring in reactions of a chairlike six-membered ring is favored by axial bond formation.

The discussion in this chapter is focused on the consequences of stereoelectronic effects in the synthetic context. However, their dominance is not always welcomed in the course of a synthesis, and one should also bear in mind that stereoelectronic effects may be subverted by steric effects.

When the stereoelectronic requirement for groups participating in reactions to be antiperiplanar is met, these reactions are generally very facile. Thus, diaxial opening of cyclohexene oxides is preferred whenever possible. In a rather complicated situation related to the synthesis of haemanthamine (**8-19**),[27] the epoxide ring opening with methanol followed this tenet. The ring system involved is conformationally mobile but fortunately the reactive conformer was the more favorable, owing to the resident aryl group (bulky) being equatorial.

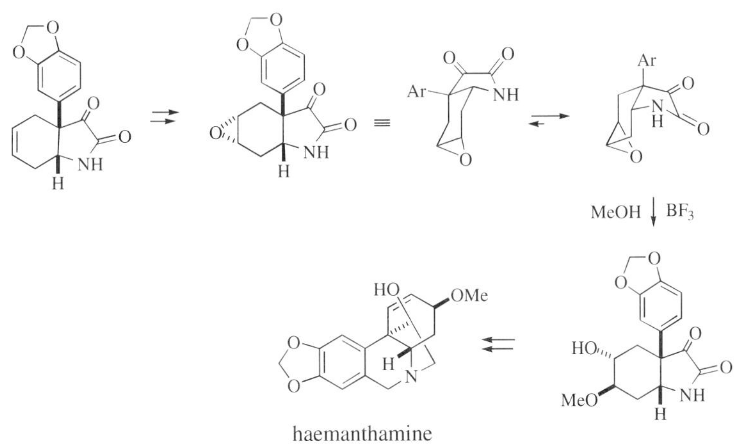

haemanthamine
(**8-19**)

For synthesis of vernolepin (**8-20**) and vernomenin,[28] the γ-lactone construction was initiated by reaction of an epoxide with lithioacetate anion. To direct the reaction at $C_{(7)}$ of the *cis*-oxadecalin framework, the A-ring lactone was transformed into an orthoester. The conformer with which axial attack occurs at $C_{(7)}$ would be predominant. Two other favoring factors are that the nucleophile would be hindered in its approach to $C_{(6)}$ in the alternative conformer and that the alkoxide species can form a chelate with the epoxy oxygen only in the case of the desired conformation.

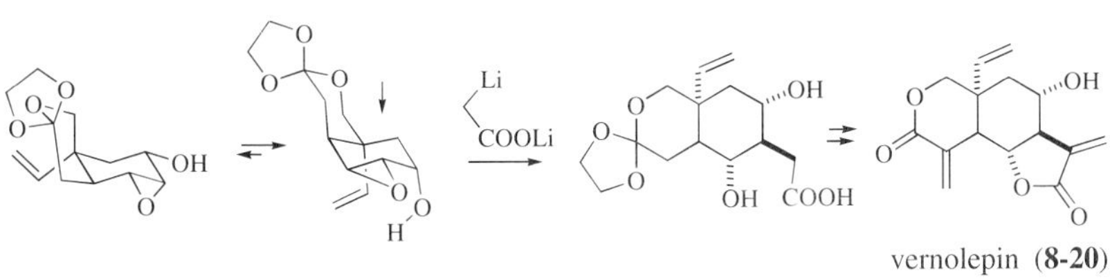

vernolepin (**8-20**)

The operation of stereoelectronic effect for diaxial opening of epoxides was extremely helpful during a synthesis of actinobolin (**8-21**).[29] Although epoxidation was practically nonstereoselective, both epoxides gave the same diol after reaction with formic acid and hydrolysis.

Difficulties were encountered in lactam ring closure of a cyclohexene intermediate in a synthesis of pancratistatin (**8-22**).[30] In one set of conditions, the ring formation was accompanied by double bond migration to the benzylic position. Fortunately, the tetrasubstituted cyclohexene could be epoxidized and analysis

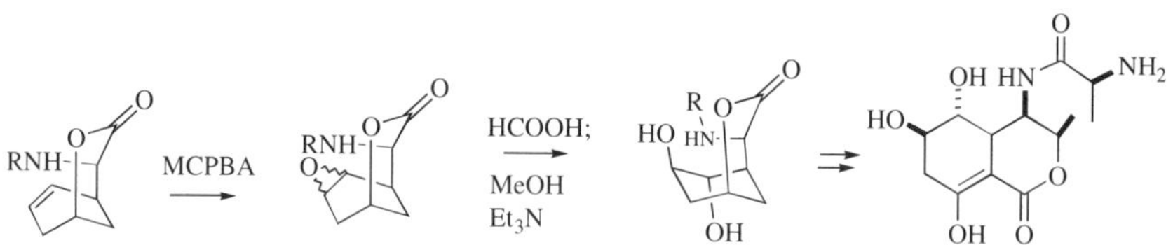

actinobolin (**8-21**)

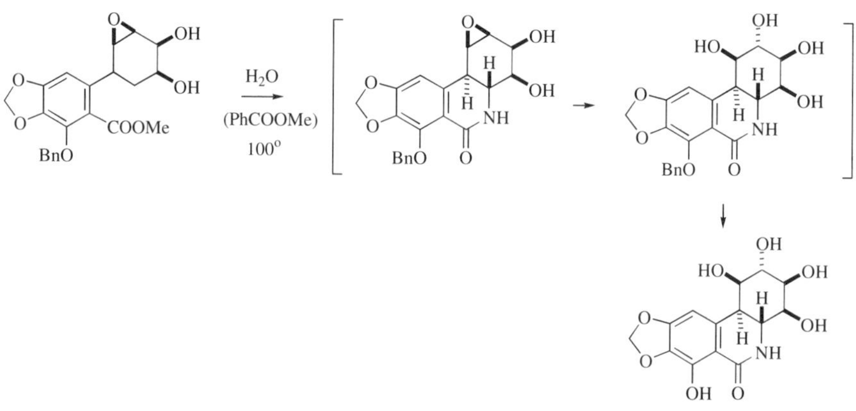

pancratistatin (**8-22**)

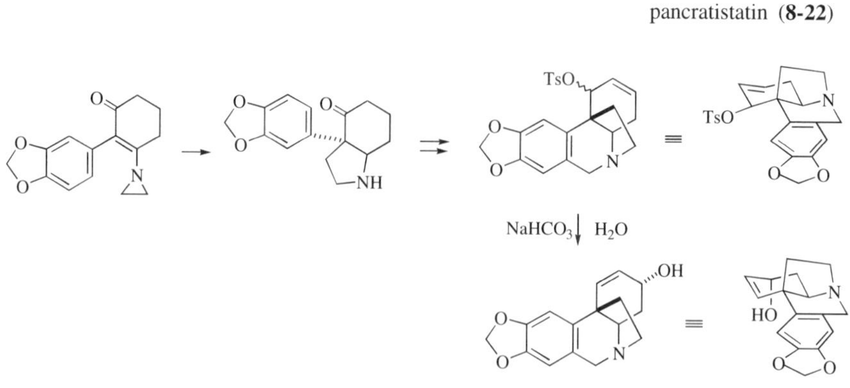

crinine (**8-23**)

indicated that both the α-epoxide and the β-epoxide would lead to the same product on diaxial opening. The OH-directed epoxidation gave only the β-epoxide which indeed showed a compliant reactivity. Actually, it was slowly transformed into the target compound by heating with water in the presence of a catalytic amount of sodium benzoate. Events that took place in this operation involved carbamate cleavage, lactam ring formation, epoxide opening, and debenzylation.

In hindsight, the last step for the synthesis of crinine (**8-23**),[31] involving solvolysis of an allylic tosylate, is interesting. Neutralization of an allyl cation from the less hindered position (secondary instead of neopentyl) was also dictated by the stereoelectronic factor favoring axial entry by the nucleophile.

Axial attack on cyclohexanones is virtually always preferred with a small and linear nucleophile such as the acetylide anion. Thus, it is not surprising that a tertiary alcohol which is *cis* to an ester group at $C_{(3)}$ was obtained on reaction of a cyclohexenone ester (**8-24**).[32] Interestingly, these two polar functional groups were thought to be involved in a $Mo(CO)_6$-catalyzed epoxidation, for which the two groups must be 1,3-diaxial. The resulting epoxide was converted to daunomycinone.

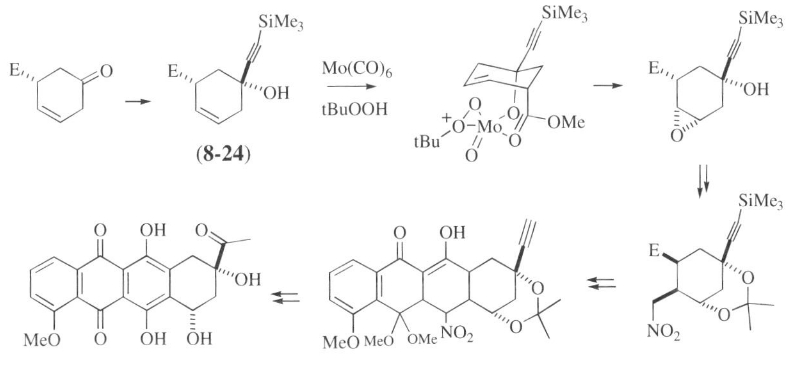

(**8-24**)

daunomycinone

The preferred axial attack by small nucleophiles on cyclohexanones is ascribable to, at least partly, the decrease in torsional strain between the equatorial hydrogen atoms and the oxygen. On the contrary, C–H and C=O eclipse is along the pathway for bond formation from the equatorial side. A model[33] which considers the relative stabilities of two diastereoisomeric transition states has been proposed. It concerns the interactions between the σ* orbital of the developing bond in such a way that hyperconjugative σ–σ* interactions with C–H bonds (in the case of axial attack) are greater than with C–C bonds (in the case of equatorial attack). However, there are controversies concerning this model.

From a tricyclic enone the formation of a tertiary alcohol constituted a critical step toward synthesis of (+)-digitoxigenin (**8-25**).[34] A Grignard reaction occurring from the α-side of the rigid enone was used to establish the 14β-hydroxyl group.

In the context of assembing the spiroannulated tetronic acid fragment (**8-26**) of kijanolide,[35] the axial addition of a metal acetylide reagent to a model compound resulted in a *cis* relationship with the alkenyl substituent present. However, when a

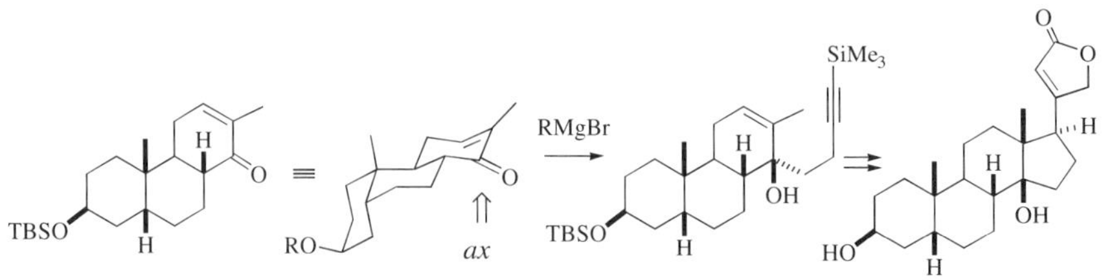

digitoxigenin (**8-25**)

more fully equipped cyclohexenone with *cis*-disposed allylic methyl and alkenyl groups, was similarly treated, a stereochemically conformable product was obtained. Either equatorial addition to one conformational isomer bearing an equatorial alkenyl chain and axial methyl group or axial addition to the other alternative with interchanged substituent conformation occurred.

Also quite essential to the synthetic success toward (+)-streptolic acid (**8-27**)[36] starting from D-glucose is the axial attack of a pyranone by benzyloxymethyllithium. (In this case the formation of the epimer would not have mattered, although it would require different manipulation to construct the epoxide. However, it would be disastrous to achieve a nonstereoselective reaction.)

Pulegone is a cyclohexane chiron of great popularity.[37] For elaborating mevinolin (**8-28**)[38] the unnatural but synthetically readily available (*S*)-pulegone is a more convenient source. However, one must address the stereoselective introduction of a side chain to the α′-position of the ketone which is not likely accomplishable by direct alkylation because of the *cis*-relationship to the existing methyl group. A

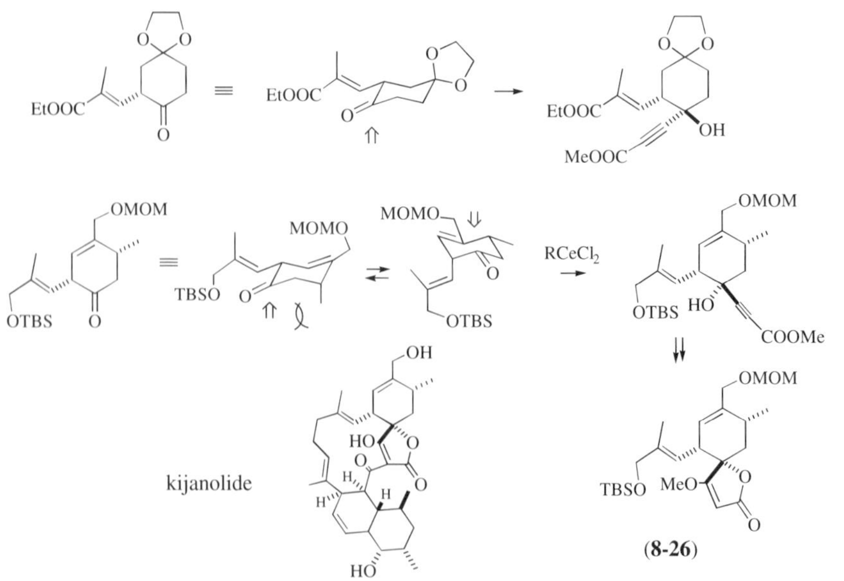

kijanolide

(**8-26**)

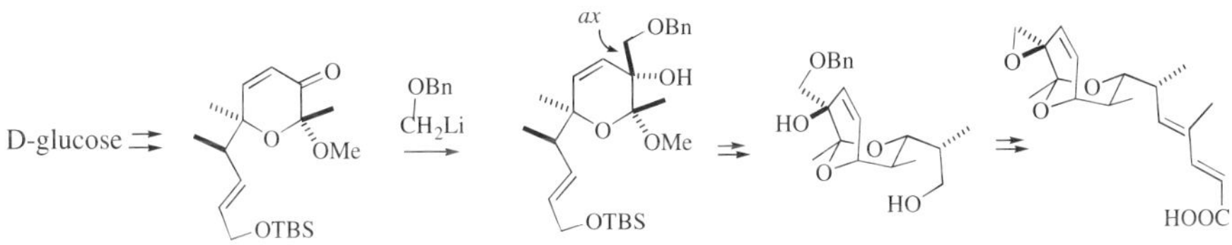

(+)-streptolic acid (**8-27**)

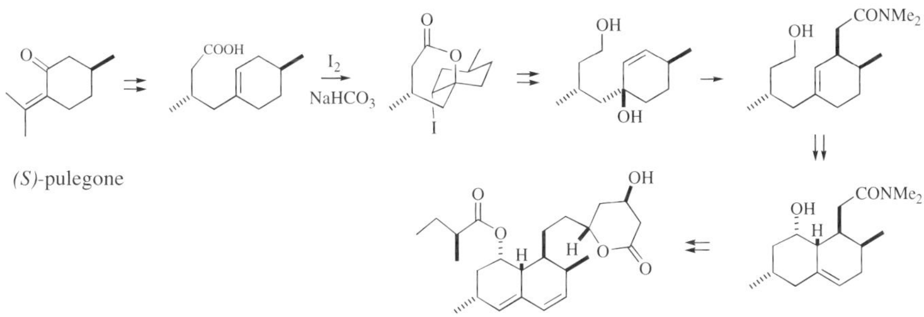

mevinolin (**8-28**)

solution that emerged consisted of degradative modification of the enone unit side chain construction by an orthoester-Claisen rearrangement, and iodolactonization. The last reaction occurred in a half chain cyclohexene containing an equatorial methyl group, giving a product which was readily processed to an allylic alcohol designed for stereoselective attachment of the carbon chain by another Claisen rearrangement.

A classical problem pertained to the synthesis of α-cyperone (**8-29**)[39,40] from dihydrocarvone. Axial approach of the electrophile to the pseudochair form of the enolate is on the same face as the isopropenyl pendant if the latter assumes a pseudoequatorial orientation, leading mainly to the epi-α-cyperone stereochemistry.

The stereoelectronic effect which frustrates the access to α-cyperone is ideally suited for the synthesis of valeranone (**8-30**),[41,42] occidentalol (**8-31**),[43] and a few other terpenes. Note that 2,4-disubstituted cyclohexanones undergo Robinson annulation to give octalones in which the original *cis*-relationship of the two substituents is retained. An excellent exploitation of this situation resulted in a stereoselective synthesis of eremophilone (**8-32**).[44]

A very interesting solvent dependence of diastereoselection has been detected in the Robinson annulation of cyclohexanones with 3-penten-2-one.[45] The major product has the *vic-cis*- and *vic-trans*-pattern in dioxane and DMSO, respectively. The role of the metal counterion is also important: the more covalent Li enolate favors the formation of the *trans* compound, whereas the more ionic Na and K species give more *cis* product, due to higher degree of solvation. For a synthesis of isonootkatone (**8-33**)[46] the proper reaction conditions were evident.

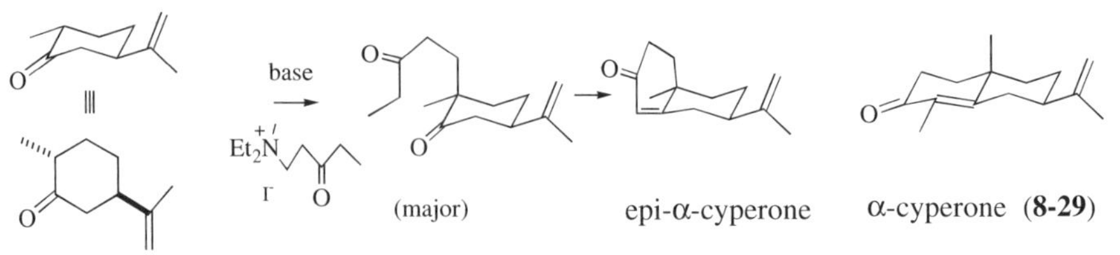

epi-α-cyperone α-cyperone (**8-29**)

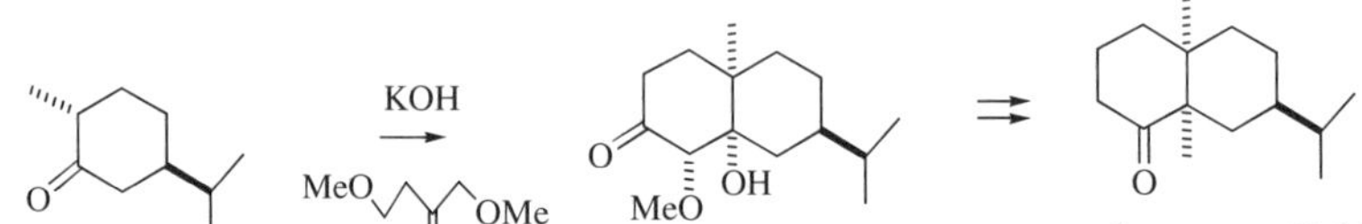

valeranone (**8-30**)

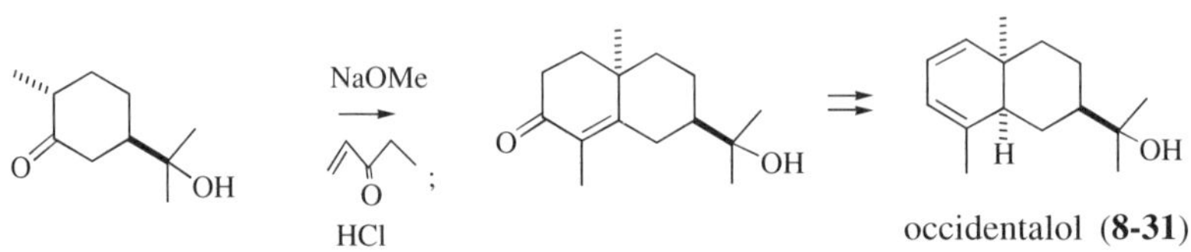

occidentalol (**8-31**)

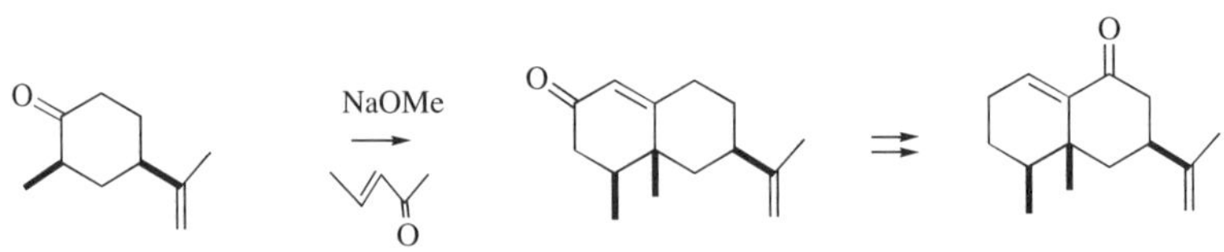

eremophilone (**8-32**)

Steroid synthesis based on the CD′ → BCD′ → ABCD approach[47] suffers from adversary stereoelectronic effect in the A-ring formation. The problem can be alleviated by using an intermediate with a *cis*-B/C ring junction and a route to conessine (**8-34**)[48] was developed along this line. Thus, cyanoethylation of such a tetracyclic ketone proceeded from the α-side is both sterically and stereoelectronically (axial bond formation) sound. Unfortunately, this gain was somewhat offset by the additional steps to adjust the configuration of $C_{(8)}$ (changing the B/C-ring junction to *trans*).

The cyclohexane subunit of hapalindole-Q (**8-35**) contains four contiguous stereocenters. It is fortunate that three of these centers are constituted by equatorial groups. When equating the isocyanate with a ketone, the adjacent quaternary carbon site becomes amenable to establishment by alkylation methods. Since the vinyl group is axial, its introduction should be subsequent to the methyl group. By this analysis the synthetic problem is reduced to the preparation of a 3-substituted dihydrocarvone enolate. Reductive cleavage of the proper derivative of 8-bromocamphor to provide regiochemical control seems most expedient.[49]

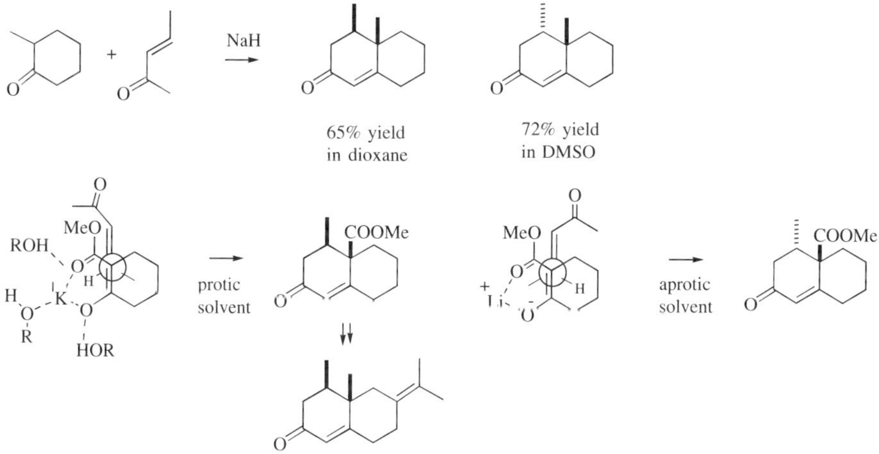

isonootkatone (**8-33**)

A remarkable difference in reactivity was observed in an intramolecular alkylation of an iodoacetaldehyde acetal to form the AB-ring system of vernolepin (**8-20**).[50] Apparently the reaction occurred with the electrophilic chain oriented axially to the enolate. The reactive isomer had a pseudoequatorial methoxy group. The epimer probably preferred the alternative conformation also containing an equatorial

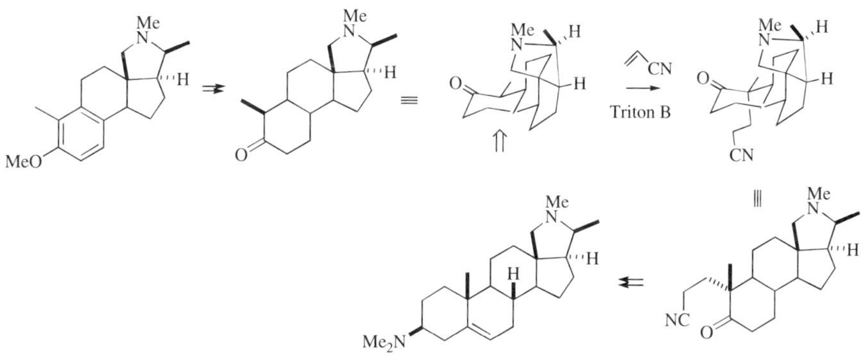

conessine (**8-34**)

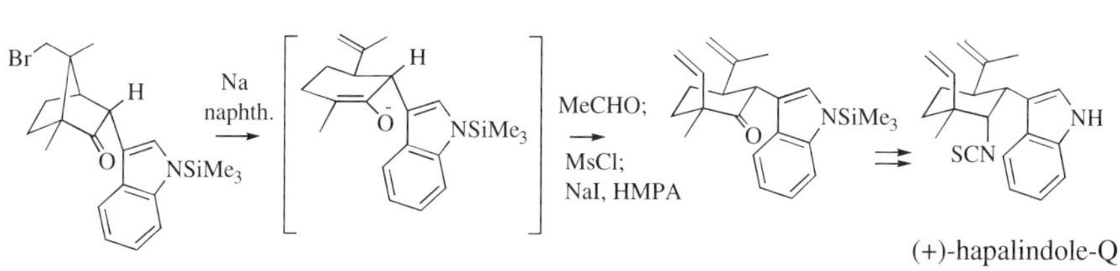

(+)-hapalindole-Q
(**8-35**)

methoxy, but the reacting centers of this conformer are stereoelectronically mismatched. Note that the analogous reaction of a cyclopentanone did not show any stereoselection.[51] The ketone lactol ether mixture can be used in a synthesis of methyl epijasmonate (**8-36**).

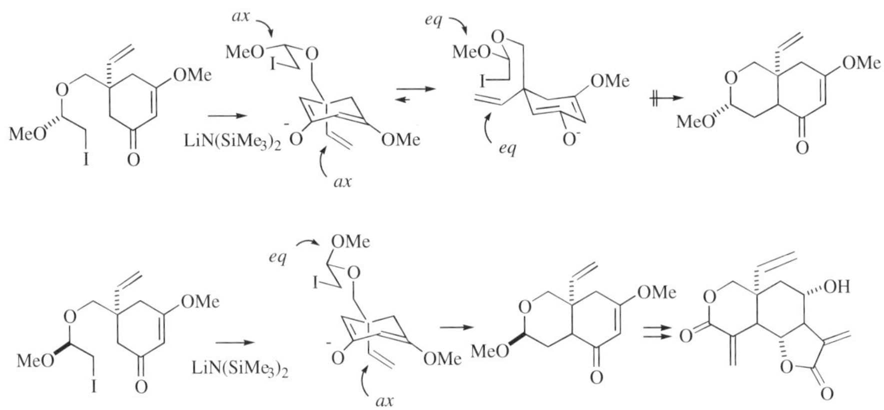

vernolepin (**8-20**)

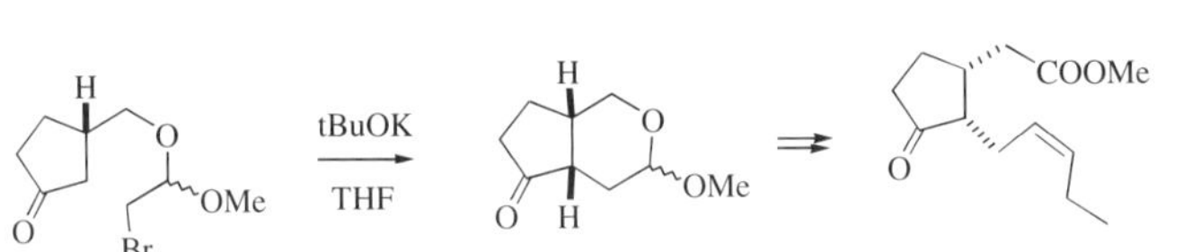

71%

methyl epijasmonate (**8-36**)

Chiral 3-substituted piperidines are available by alkylation of 2-piperidones bearing a chiral auxiliary at the nitrogen atom. The nitrogen in amide enolates is highly pyramidalized, and behaves as a chelator to stabilize a certain conformation of the piperidone. Alkylation from the axial direction follows. This method was used to synthesize (+)-stenusine (**8-37**).[52]

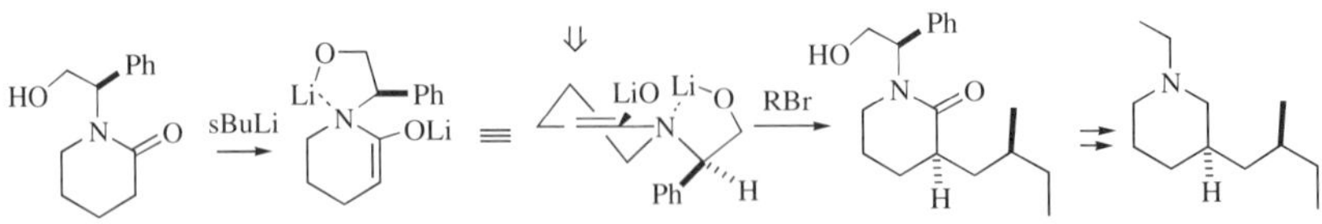

(+)-stenusine (**8-37**)

For the transformation of a *cis*-octalone to fukinone (**8-38**),[53] the stereoselective conjugate addition of a methyl group was rendered quite facile because the acetoxy substituent occupied an equatorial position and the nonsteroidal conformation was favored. However, a totally different situation arose when a cross-conjugated cyclohexadienone was submitted to the same treatment.[54] A *vic-trans*-dimethyl product was obtained. To use it in a synthesis of nootkatone (**8-39**), it was necessary to return the product to the dehydro form. Because on $NaBH_4$ reduction, the delivery of a hydride ion from the same direction led to the desired *vic-cis*-dimethyl isomer.

Axial attack by cuprate reagent to a cyclohexenone was also exploited in a synthesis of (–)-cordiaquinone-B (**8-40**).[55] It resulted in a *cis*- 3,4-dimethylcyclohexanone as required. The secondary methyl group in the C-ring of methyl vouacapenate (**8-41**) was also introduced by such a conjugate addition.[56] It is interesting that in the diterpene the methyl group is pseudoaxial and more stable. In this orientation $A^{1,3}$-strain with the furan ring is minimized.

In the area of resin acid synthesis previously it was indicated that in order to juxtapose the acid to the angular methyl group (to derive the podocarpic acid stereochemistry) one needs to methylate a certain compound at $C_{(4)}$ to bring up the existing latent carboxyl function (an ester being the most direct surrogate). However, it is also possible to construct a useful precursor (**8-41a**) by conjugate hydrocyanation of a Δ^3-2-one.[57] Apparently, the axial entry of the cyano group is favored for both

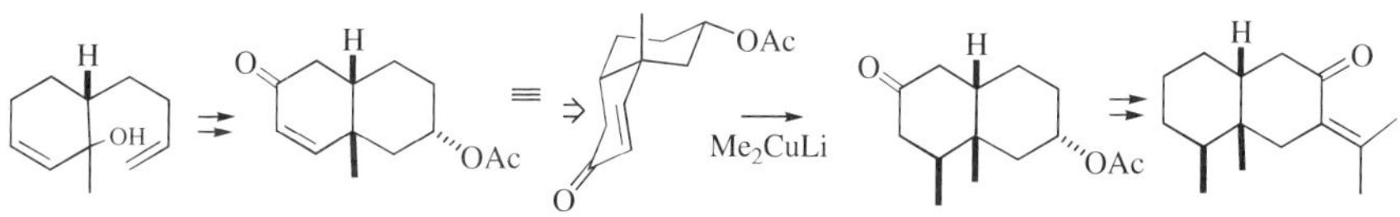

fukinone (**8-38**)

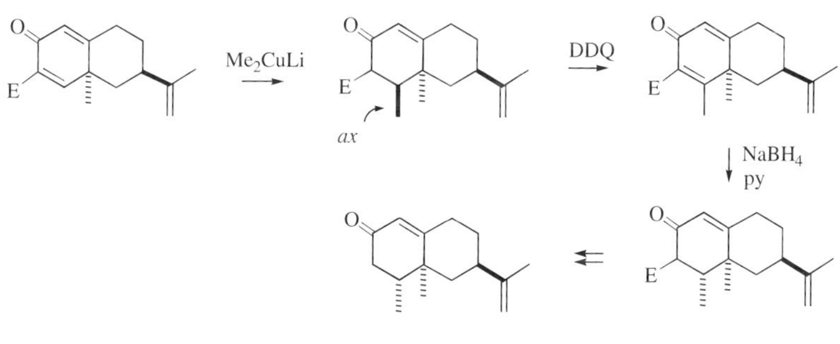

nootkatone (**8-39**)

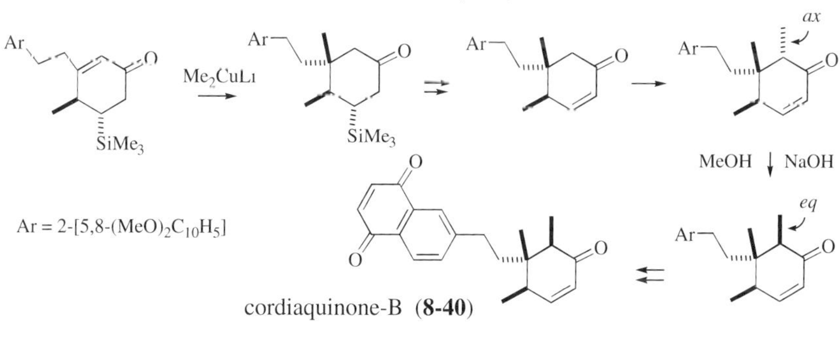

cordiaquinone-B (**8-40**)

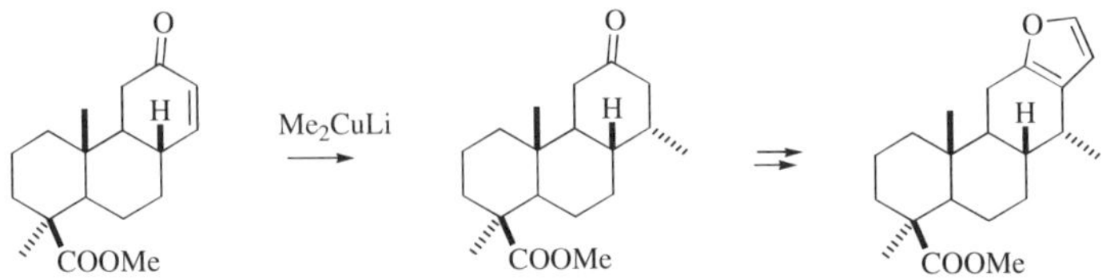

methyl vouacapenate
(**8-41**)

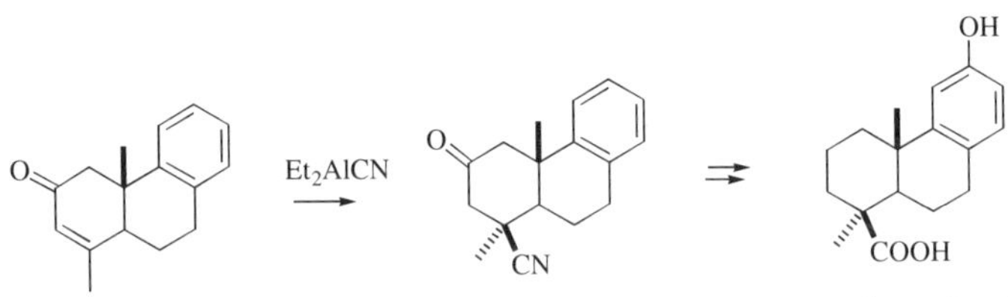

(**8-41a**) podocarpic acid

stereoelectronic and steric reasons, the cyanide ion being linear and far less demanding of space.

Griseofulvin (**8-42**) was formed as the major product from a double Michael reaction.[58] It has been attributed to a kinetic pathway with better orbital overlap in the transition state for the cyclization.

Many intramolecular Michael additions also follow axial attack (for most efficient orbital overlap). Two examples are found in the synthesis of vernolepin (**8-20**)[59] and lycopodine (**8-43**).[60] In the latter case the favored pathway involved axial

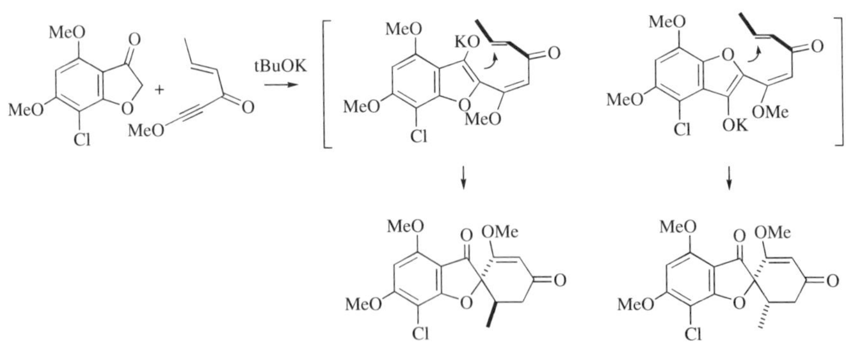

griseofulvin (**8-42**) epigriseofulvin

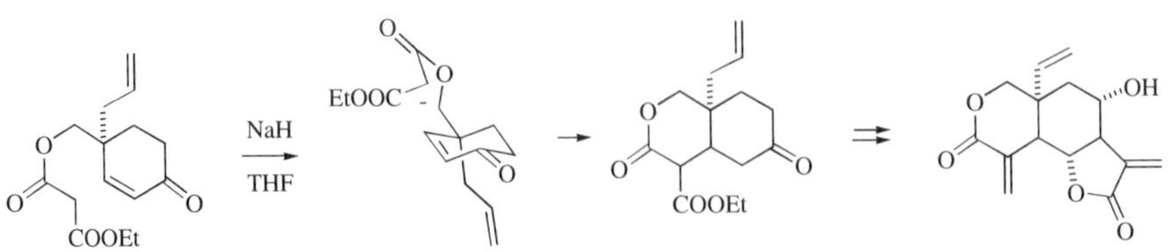

vernolepin (**8-20**)

addition on the side opposite the equatorial methyl group, and subsequent aldol condensation to form the bridged system.

Methods for the formation of *trans*-hydrindanones have implications in steroid synthesis [CD component (**8-44**)]. Thus, the development of an intramolecular Michael–aldol reaction tandem[61,62] was an important contribution. An excellent selectivity (*trans:cis* = 25:1) was obtained when it was promoted by $(n\text{-PrO})_4Zr$ in benzene. This was ascribed to the favoring of the chelated transition state in which electron repulsion and eclipsing interactions were minimal. The protocol was used to prepare a difficult intermediate for retigeranic acid-A.[63]

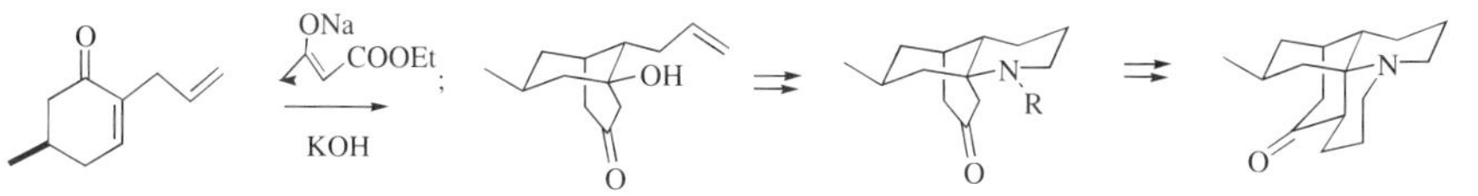

lycopodine (**8-43**)

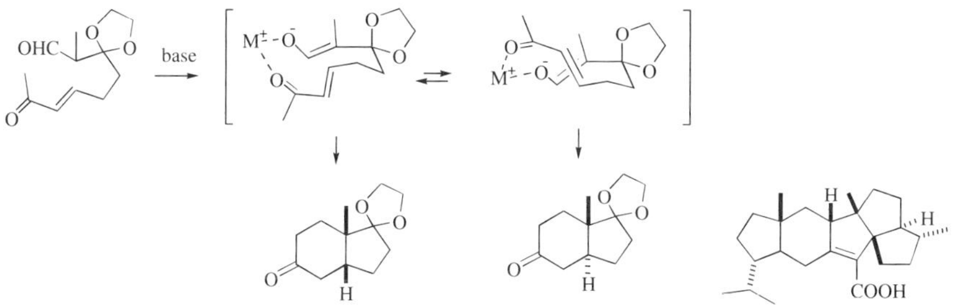

(**8-44**)

retigeranic acid-A

base	solvent			
$Ba(OH)_2$	MeOH	1	:	3
$Ca(OH)_2$	MeOH	1	:	10
$Mg(OMe)_2$	MeOH	1	:	12
$Al(OBu^t)_3$	MeOH	1	:	20
$Zr(OPr^n)_4$	PhH	1	:	25

Fragmentation reactions often play an important role in synthesis.[64] Since application of powerful methods such as cycloadditions may generate overly complex skeletons, these must be unraveled. Concerted fragmentations proceed from a transition state that places the leaving group and the β,γ-bond in an antiperiplanar array are most valuable.

A classic example is found in a synthesis of caryophyllene (**8-45**) and isocaryophyllene (**8-46**).[65] The tricyclic substrates were each prepared by controlled reduction of a ketone followed by tosylation. Base-promoted fragmentation led to olefin products of defined geometry. It is of interest to note that the cyclopentane ring which contained the leaving group can assume two different envelope conformations to accommodate the stereoelectronic requirements, with each of the active

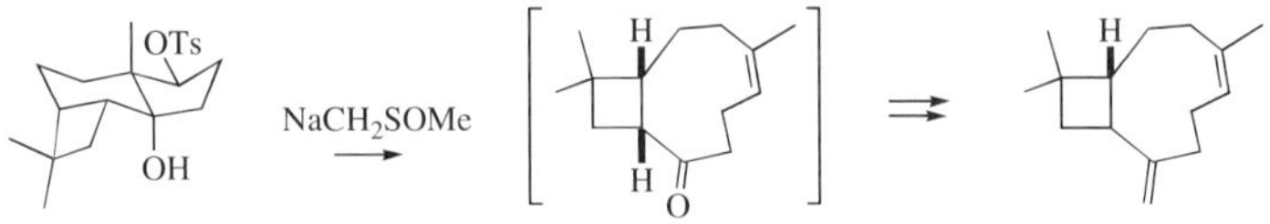

caryophyllene
(**8-45**)

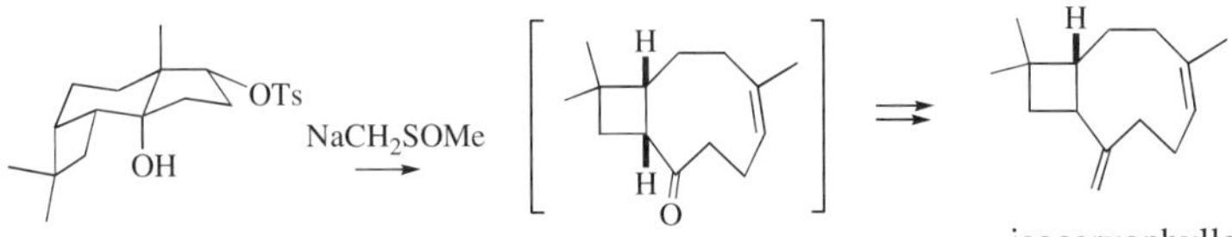

isocaryophyllene
(**8-46**)

conformations projecting a pseudoequatorial tosyl group which should be thermodynamically more stable.

Two fragmentation reactions were used in a synthesis of the cecropia juvenile hormone JH-I (**8-47**).[66] Proper configuration of the first substrate was established during methylation and hydride reduction steps.

Conversion of an eudesmane derivative to hedycaryol (**8-48**)[67] which has a germacrane skeleton was achieved by hydroboration and alkali decomposition of the organoborane. Cleavage of the intercyclic bond was triggered by deboronation and simultaneous expulsion of the tosylate ion.

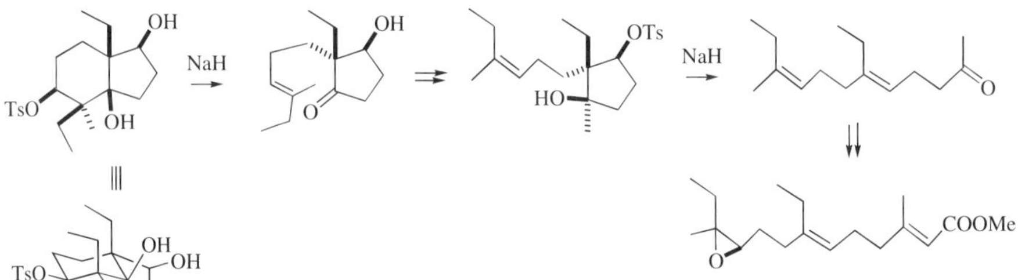

JH-I (**8-47**)

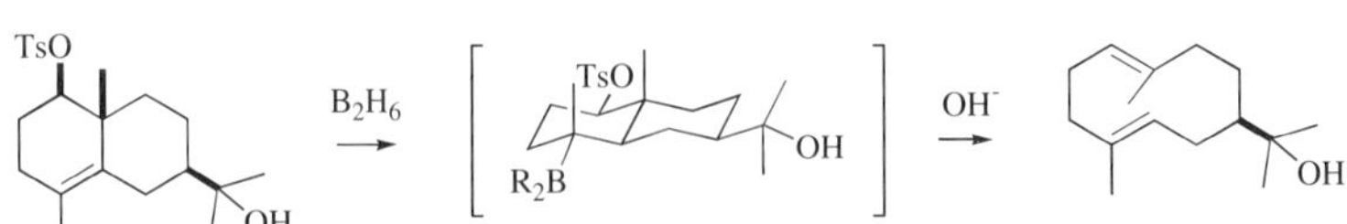

hedycaryol (**8-48**)

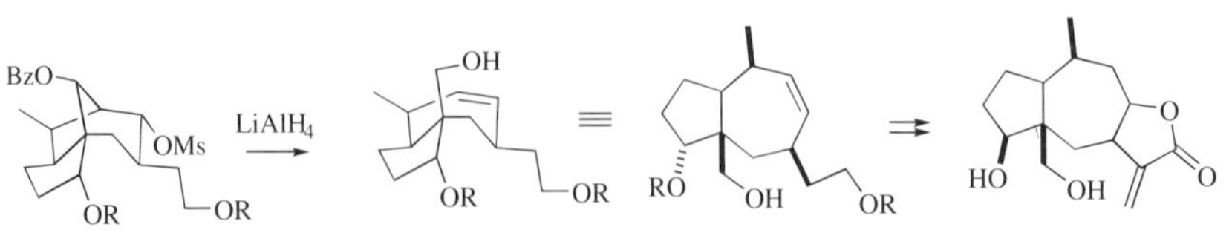

rudmollin (**8-49**)

A bridged ring system was disassembled to give a functionalized cycloheptene in a synthesis of rudmollin (**8-49**).[68] Originally this substrate was derived from an intramolecular arene–alkene *meta*-cycloadduct.

8.3.2. Stereoelectronic Effects in Heterocycles

There are many other variants of fragmentation substrates and conditions. In a synthesis design for isochanoclavine-I (**8-50**)[69] a tetracyclic amine was transformed into an *O*-mesylate to trigger fragmentation. A [3,3]-sigmatropic rearrangement followed.

Conjugated enones in six-membered unsaturated heterocycles behave very similarly in Michael reactions to the all-carbon counterparts, and therefore the operation of stereoelectronic effects is expected. A synthetic route to ceroplastol-I (**8-51**)[70] involved an axial 1,4-addition to an unsaturated enollactone. The newly created stereocenter was to become a ring junction upon formation of a new cyclopentane ring.

A synthesis of akuammigine (**8-52**)[71] from a tetracyclic enone depended on axial addition of a malonate anion. Equilibration of the acetyl group to an equatorial orientation ensured a *cis* disposition of two vicinal side chains which were to be incorporated into the dihydropyran ring. A similar situation was encountered in a synthesis of hirsuteine (**8-53**).[72] In this tetracyclic alkaloid the two side chains are *trans*, but they can become diequatorial by a conformational change of the C-ring.

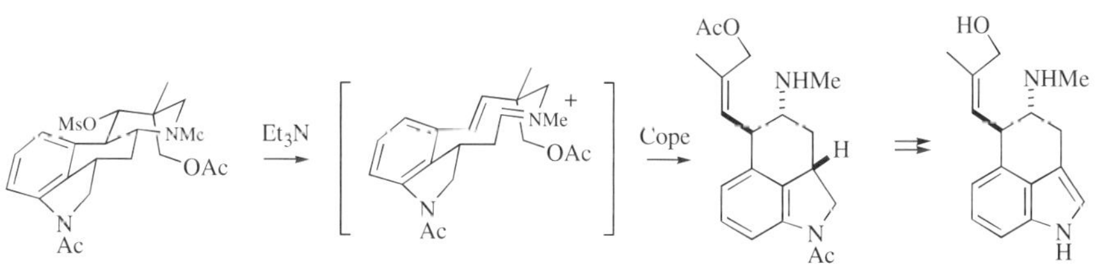

isochanoclavine-I
(**8-50**)

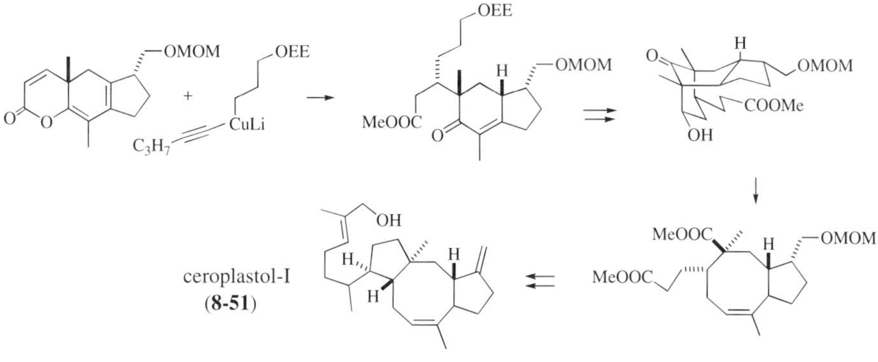

ceroplastol-I
(**8-51**)

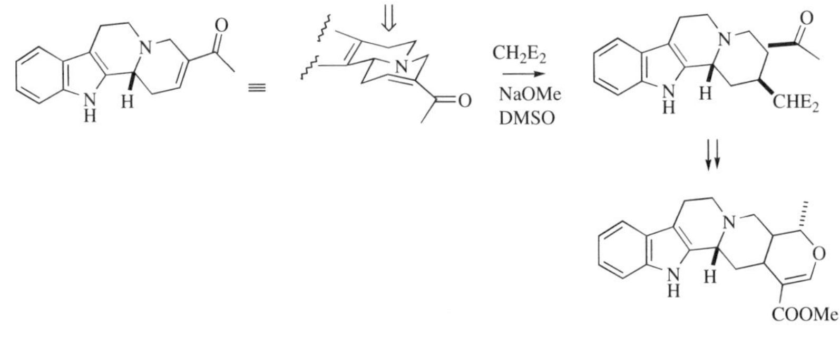

akuammigine (**8-52**)

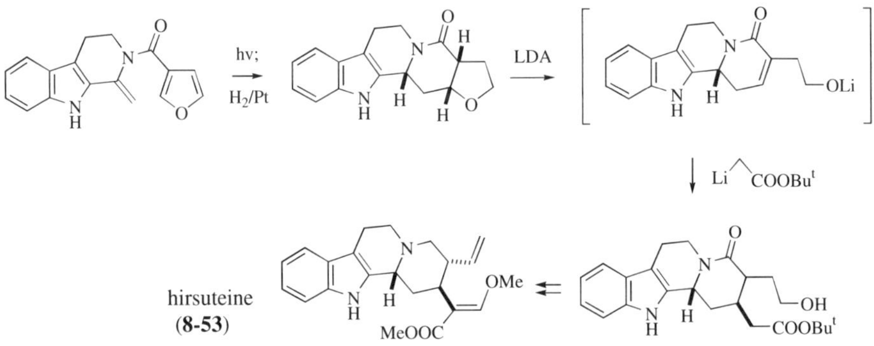

hirsuteine (**8-53**)

The preparation of a potential hexacyclic intermediate (**8-54**) for strychnine[73] featured as key steps an intramolecular Diels–Alder reaction, alkylation, conjugate allylation, and reductive amination. Introduction of new stereocenters at each stage relied on thermodynamic factors (*cis*-5/6 fusion) and a stereoelectronic effect in the Sakurai–Hosomi allylation.

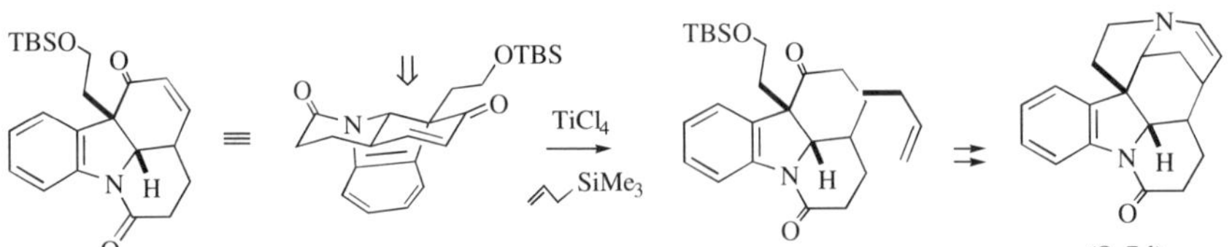

(**8-54**)

With an anomeric *t*-butoxy group to confine conformational mobility of a pyranone, the conjugate introduction of a vinyl group at the β-position became unequivocal and it ensured the production of meroquinene (**8-55**) after proper manipulations.[74]

The anomeric effect is often observed in carbohydrates. It denotes the axial preference of the $C_{(1)}$-substituent in glycopyranosides. It is expected that 1,6-dioxaspiro[5.5]undecane formation from ketodiols results in product(s) with one oxygen axial to the other tetrahydropyran ring, and vice versa, while placing a maximal number of other substituents in equatorial positions. This pattern has profound significance to synthesis of natural products containing such structural subunits, e.g., the avermectins and milbemycins.[75,76]

Talaromycin-A (**8-56**) and talaromycin-B (**8-57**) are two metabolites differing in the configuration of one secondary alcohol. Talaromycin-B is relatively easy to form from a linear keto tetraol, as all of the ring substituents are equatorial. The task is rendered easier still if the ethyl group is replaced with a hydroxymethyl, because spirocyclization would leave a 1,3-diol unit and a remote primary alcohol free. Differentiation of the two systems by selective acetal formation is obvious.[77]

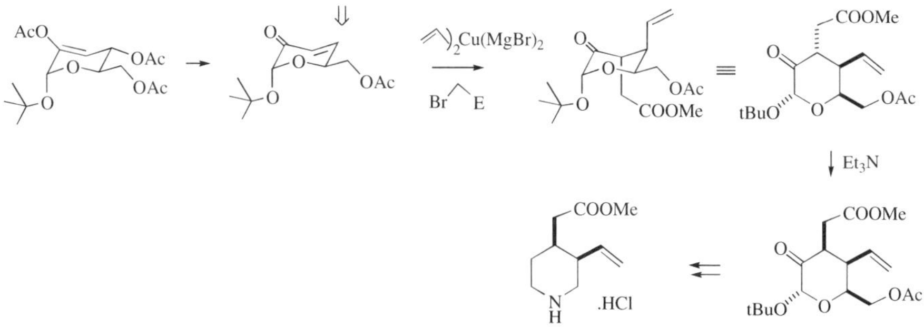

meroquinene.HCl
(**8-55**)

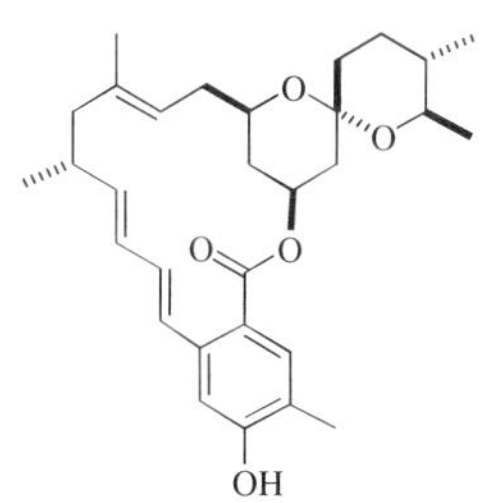

milbemycin-β_3

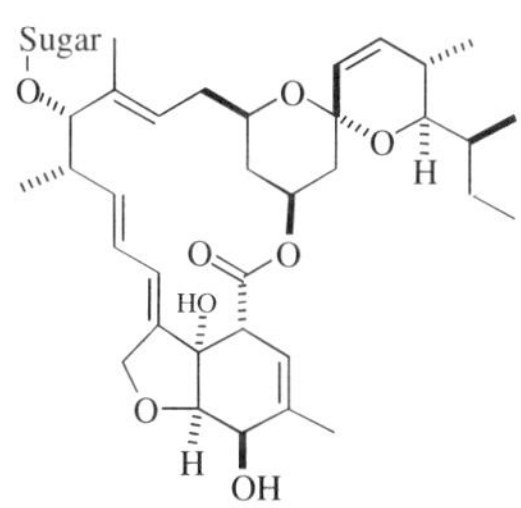

avermectin-B_{1a}

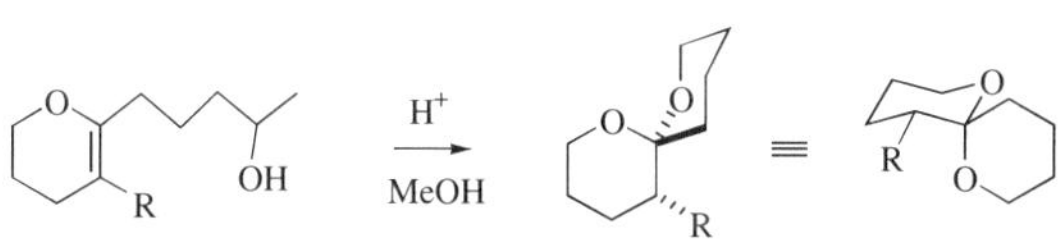

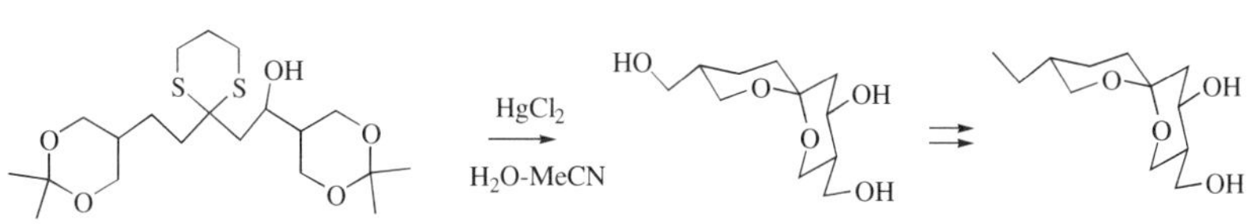

talaromycin-B
(**8-57**)

Talaromycin-A is susceptible to rearrangement, e.g., via the open chain compound to give talaromycin-B and possibly also another isomer with an *R*-configuration at the spirocyclic center which has similar energy to talaromycin-B. Thus it is unwise to pursue a synthesis of talaromycin-A by an equilibration route. However, it is known in carbohydrate chemistry that an axial polar group (e.g., OH) usually directs α-glycosidation, therefore the presence of such a group would determine the spiroacetalization to give a precursor of talaromycin-A. This stereoelectronic effect indeed supported the development of a synthetic route starting from D-glucofuranose 1,2-acetonide.[78]

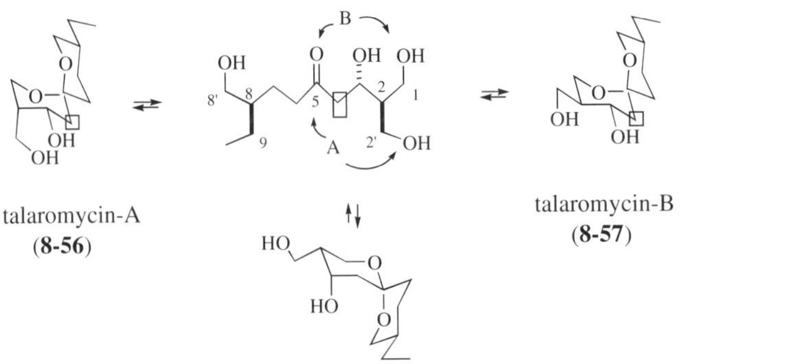

talaromycin-A
(8-56)

talaromycin-B
(8-57)

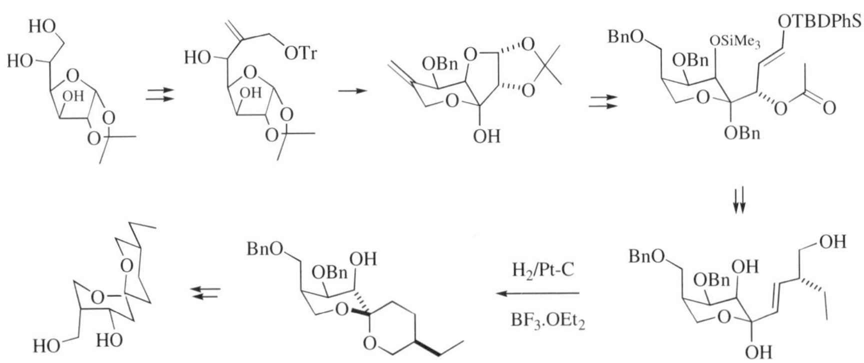

talaromycin-A
(8-56)

The first synthesis of reserpine[79] suffered slightly in that the reduction of the iminium salt after C-ring closure led to a compound of the iso series. It required epimerization of a derived γ-lactone. It was thought,[80] on the basis of stereoelectronic considerations, that reaction of 6-methoxytryptamine with a tosyloxy aldehyde which contained the full complement of the E-ring substituents would give reserpine directly. Precedents supporting this notion include the cyclization step leading to pseudoyohimbine[81] with the 3β-H configuration. Interestingly, the synthesis of yohimbine (**8-58**) also required a change of configuration.

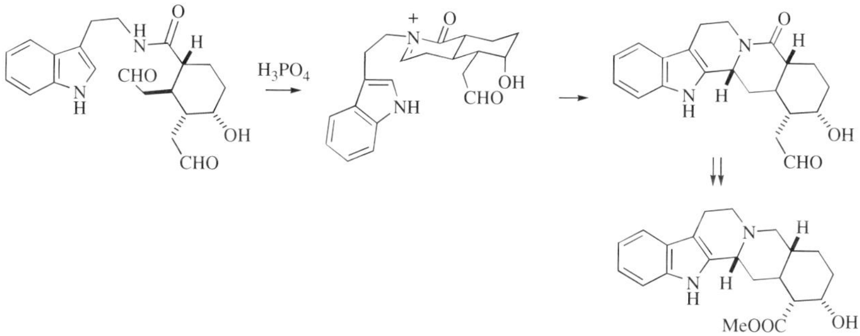

yohimbine (**8-58**)

The attempt to derive reserpine (**8-59**) stereoselectively was thwarted due to the fact that the Pictet–Spengler cyclization proceeded at a faster rate, while fulfillment of the original premise depended on the formation of the D-ring prior to the said cyclization. Tactical changes were implemented by modifying the substrate to one already having a D-ring, and activating the proper position with a cyano group by adding silver ion to the reaction medium. It was shown that thermolysis of the cyclic α-aminonitrile alone in dimethylformamide still furnished predominantly (15:1) isoreserpine, indicating a persistent steric shielding by cyanide ion (intimate ion-pair), which was cleared by the added silver ion. The axial space was then left free for the indole moiety to react.

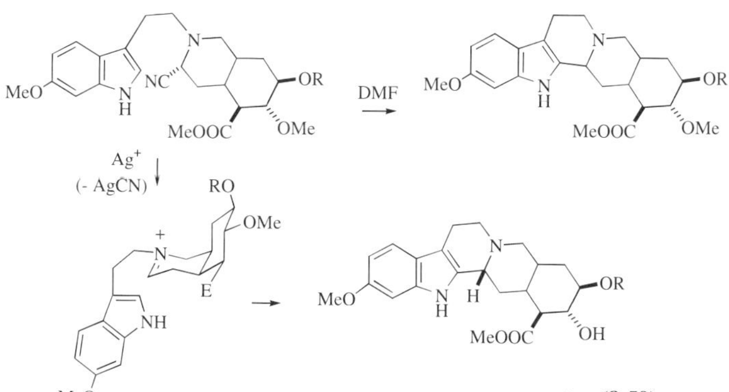

R= 3,4,5-$(MeO)_3C_6H_2CO$ reserpine (**8-59**)

It should be noted that in another approach to deserpidine (reserpine without the Ar-methoxyl group),[82] a counter-stereoelectronic cyclization was observed. The virtually exclusive formation of the 3α-H isomer suggests the E-ring in a conformation with three axial substituents. The result was totally unexpected and unexplainable.

A synthetic study of ladybug alkaloids[83] has shed much light on the operation of stereoelectronic effects in the Robinson–Schöpf condensation. The amino dialdehyde liberated in the presence of dimethyl 3-oxoglutarate formed a tetrahydropyridinium species which underwent a Mannich reaction to afford a *trans*-2,6-disubstituted

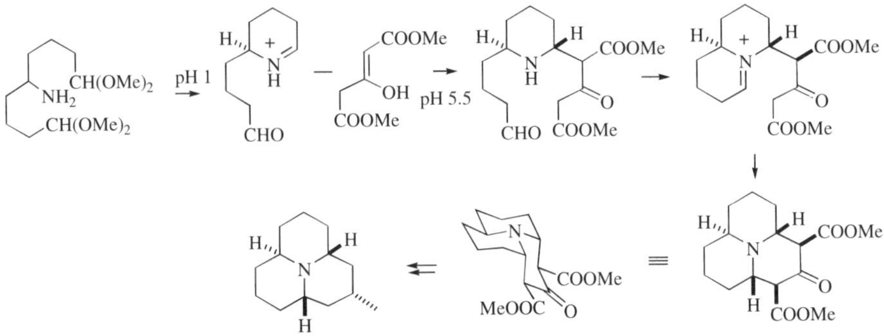

precoccinelline
(**8-60**)

piperidine, which was the result of axial attack on the iminium ion (bearing an equatorial α′-chain). Subsequent cyclization to give a dehydroquinolizidinium ion initiated the intramolecular Mannich reaction which also occurred axially. The reaction led to a perhydro[9*b*]azaphenalene derivative with five stereocenters, three of which corresponding to that of precoccinelline (**8-60**). Actually, the alkaloid was obtained after a few more steps.

This remarkable result warrants a more detailed analysis of the reaction pathway. In the first intermolecular condensation, the Mannich acceptor has two conformations each possessing a pseudoequatorial or pseudoaxial allylic (α′-position) substituent. There were four possible transition states which can maintain maximum orbital overlap between the approaching nucleophile and the developing lone electron-pair of the nitrogen atom. Two of these transition states are kinetically disfavored because of being boatlike, and one of the chairlike transition states featured a pseudoaxial group which can interact strongly with the incoming nucleophile. By comparison, one transition state is highly favored and its adoption was observed (by inference on the basis of the tricyclic product structure).

For synthesis of monomorine-I (**8-61**),[84] a dehydroindolizidine was prepared. The reduction under acidic conditions proceeded via the iminium ion, and stereoelectronic control was apparent. Again, two boatlike transition states were

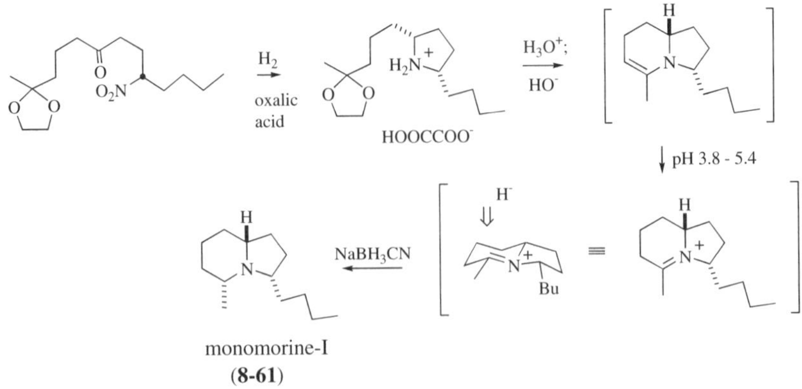

monomorine-I
(**8-61**)

disqualified, and one of the chairlike transition states was suppressed by the presence of serious steric hindrance from a methylene group of the pyrrolidine portion of the molecule. It should be noted that diastereomeric structures may be constructed by Grignard reactions on the α-unsubstituted indolizidinium ions.[85]

A highly stereoselective reaction sequence leading to the alkaloid indolizidine-223AB (**8-62**) included an intramolecular Diels–Alder reaction of a dienophilic acylnitroso component and reduction of a bicyclic iminium species.[86] A similar principle was invoked in the design of the critical operation.

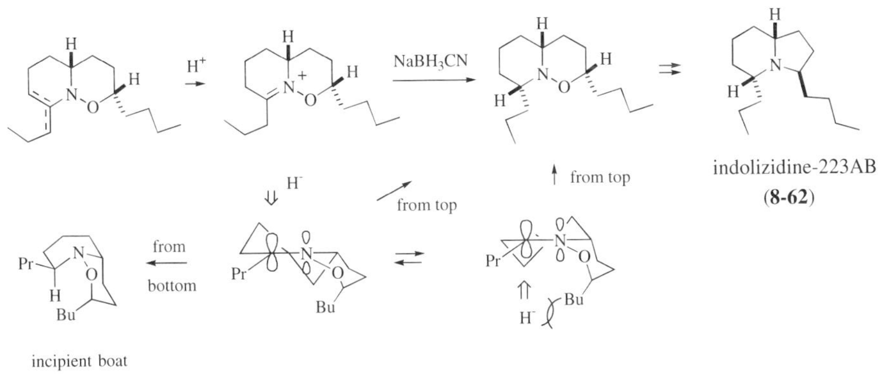

indolizidine-223AB
(**8-62**)

An approach to deoxoprosopinine (**8-63**) by addition of a nucleophile to a tetrahydropyridinium precursor is not advisable, as the *cis*-2,6-disubstituted piperidine is expected to form.[87] Reversal of this steric course is realized by changing the substrate to a bicyclic oxazolidinone.[88]

An elegant approach to makomakine (**8-64**) and hobartine (**8-64a**)[89] attests to the fruitful manipulation of stereoelectronic effects. In the final step reduction of the bridged imines followed the axial course. In two intramolecular Mannich reactions the formation of two axial bonds were a necessity during a synthesis of porantherine (**8-65**).[90]

Similarly, acid treatment of a diacetal (1:1 mixture of *cis,trans*- and *trans,trans*-isomers) gave only one tricyclic amino ketone which corresponded to lycopodine (**8-43**).[91] Isomerization of the uncyclizable bicyclic iminium species must have occurred in situ.

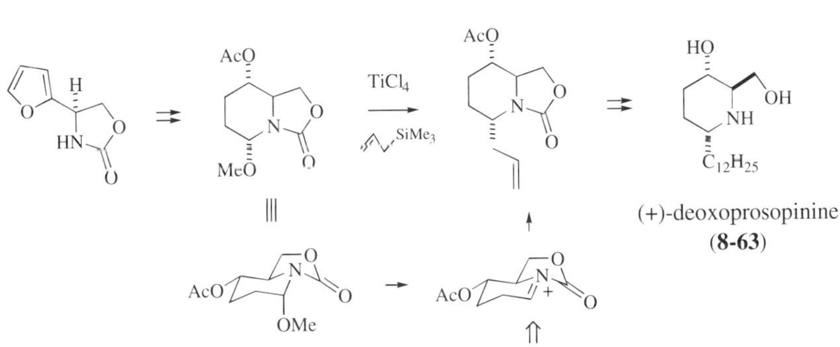

(+)-deoxoprosopinine
(**8-63**)

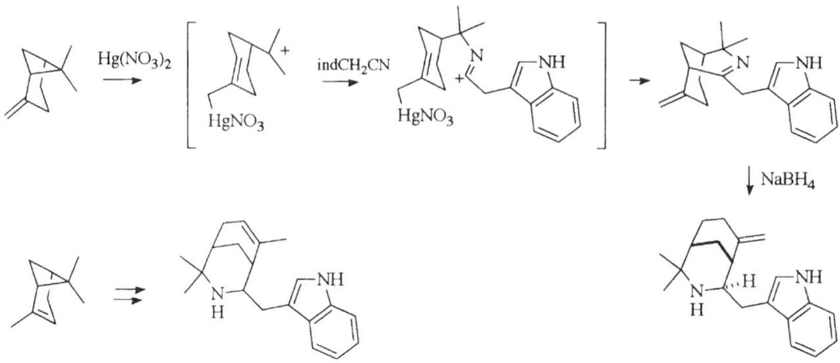

hobartine (**8-64a**)

(+)-makomakine (**8-64**)

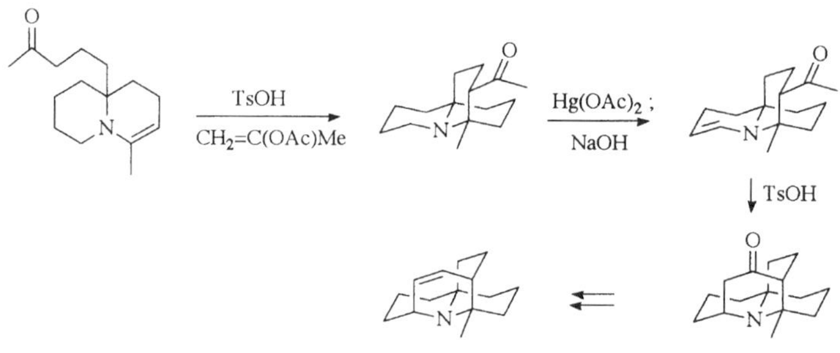

porantherine (**8-65**)

The employment of an aza-Claisen rearrangement–Mannich cyclization tandem in a synthesis of α-allokainic acid (**8-66**)[92] is interesting. The first stage was accomplished when the double bond in the side chain approached the iminium ion axially.

A deeper understanding of stereoelectronic effects emerged from a failed attempt to synthesize the *cis*-hydroindolone intermediate of *N*-methyllycodine.[93] Analysis showed the inevitability that, for the cyclization to occur, the Michael adduct derived

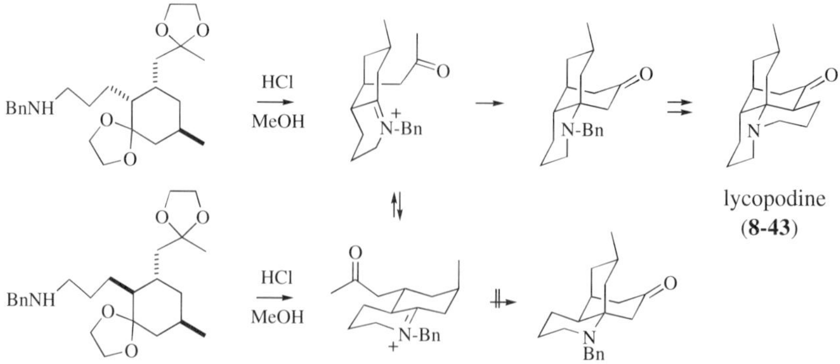

lycopodine (**8-43**)

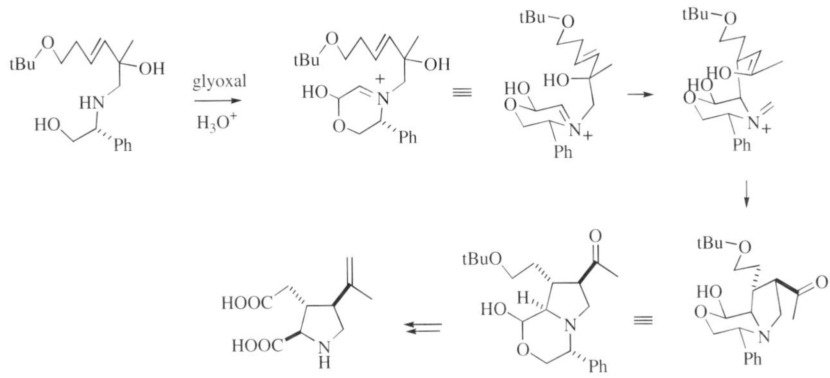

(-)-α-allokainic acid (**8-66**)

from the endocyclic enamine must go through a double boatlike transition state. The chairlike transition state being strongly destabilized by a severe $A^{1,2}$-interaction between the aryl substituent and the *N*-methyl group. On the other hand, the geometrical demand of such a transition state for the synthesis of karachine (**8-67**)[94] had already reached in the intermediate and the ring closure was a foregone conclusion.

Reduction of a 2,6-disubstituted 1-piperideine can be diverted to the formation of the *trans*-piperidine[95] by adding an excess of an organoaluminum reagent. In combination with the alkylative Beckmann rearrangement promoted by organoaluminums, the reduction leads to various important alkaloids including solenopsin-A (**8-68**). The reduction follows pathways favored by stereoelectronic

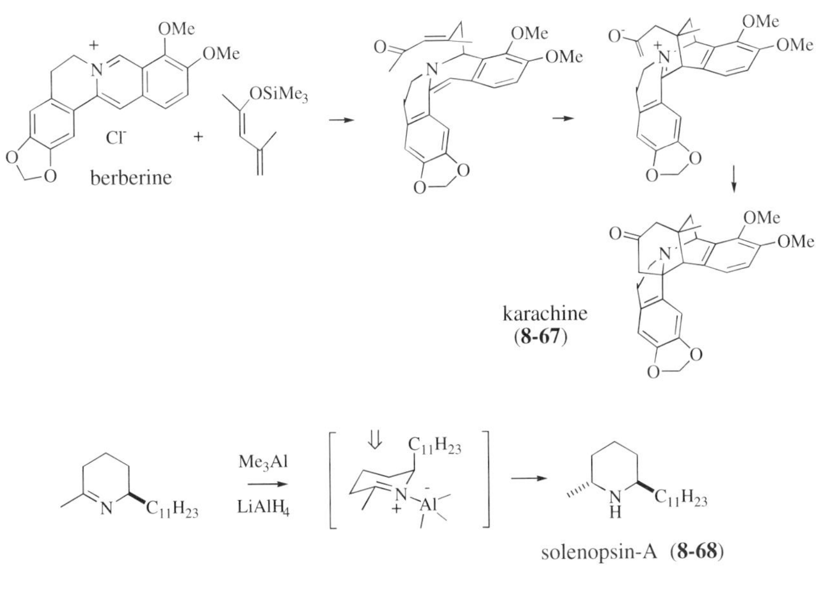

karachine (**8-67**)

solenopsin-A (**8-68**)

factors which apparently overwhelm steric interference by the pseudoaxial group at the α′-position.

In situ reduction of an iminium species was involved in a synthesis of apovincamine (**8-69**)[96] after hydroxymethylation of a cyclic enamine. Apparently the hydrogen bonding between the OH group and the indolic nitrogen atom stabilized the conformation with an axial ethyl substituent, causing the reduction to give the undesired isomer. Fortunately, stereochemical rectification was achieved by equilibration of the derived aldehyde.

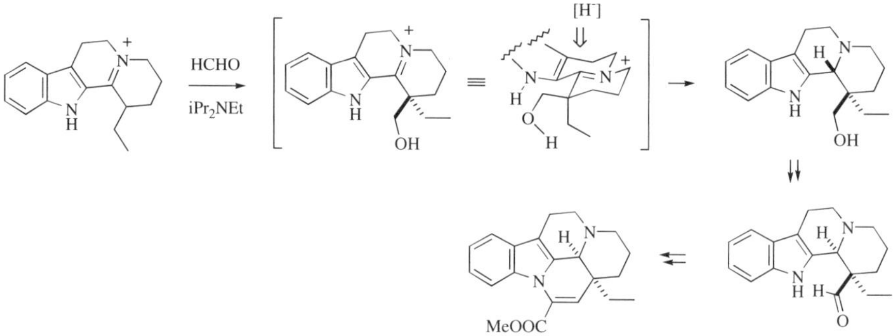

apovincamine (**8-69**)

A general method for synthesis of aspidosperma alkaloids [e.g., minovine (**8-70**)] involves union of an indole derivative and a (tetra)hydropyridine building block, followed by intramolecular alkylation to form the pyrrolidine ring. The formation of both a *cis*-hydroquinoline in the former step and a *cis*-hydrindole unit by the latter reaction are favored stereoelectronically.[97] A more stable transition state with the best orbital overlap was selected en route to the observed products.

Doubly activated cyclopropanes are susceptible to ring opening by nucleophiles. Cleavage occurs with the the ring bond whose constituting orbital can overlap effectively with the π-bond of the activating group. Several synthetic situations can benefit from transformation of this type. In the cases of isoretronecanol (**8-71**) and trachelanthamidine (**8-72**) these compounds are derivable from pyrrolidines with an ester side chain at $C_{(2)}$. By ring exchange (pyrrolidine and cyclopropane) a more rigid system whose preparation is under better stereocontrol can be identified.[98] The general consensus is that stereochemical issues concerning a free rotating group are more difficult to address, and in the present case the hydroxymethyl branch can be incorporated as a lactone to facilitate the synthetic process.

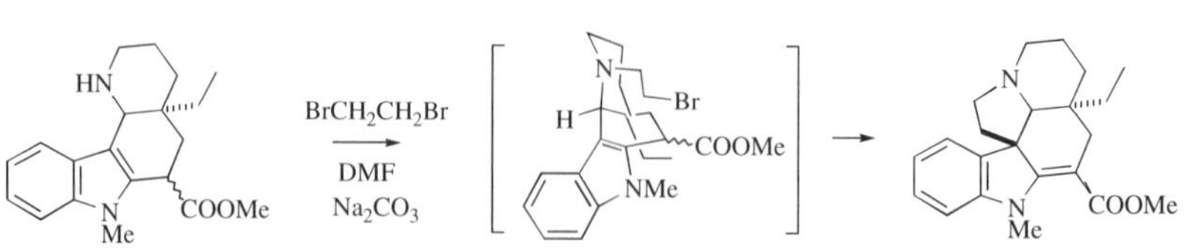

minovine (**8-70**)

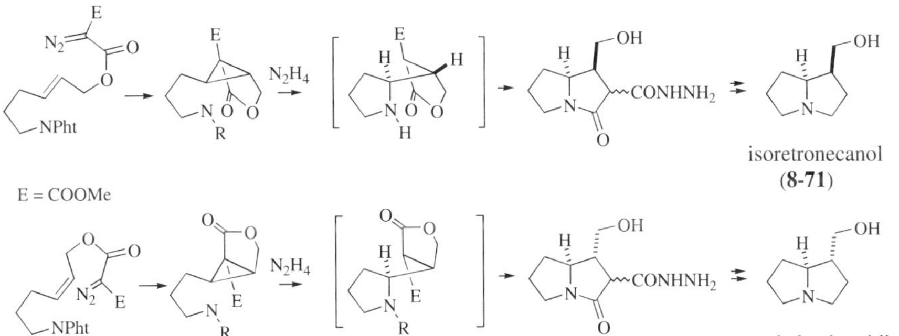

isoretronecanol (**8-71**)

trachelanthamidine (**8-72**)

Stereoselective construction of a prostaglandin intermediate (**8-73**)[99] and a D-ring component (**8-74**) of cholesterol containing the complete side chain[100] can be realized by the same method of cyclopropane ring opening using external carbon nucleophiles.

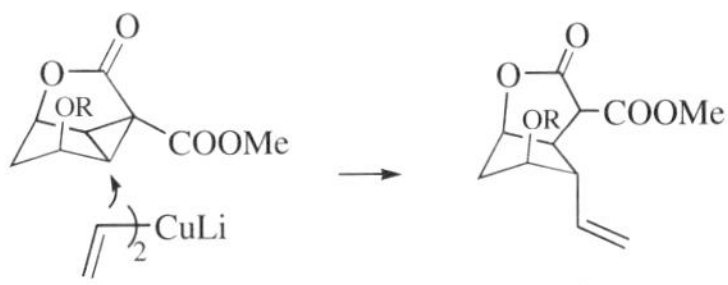

(**8-73**)

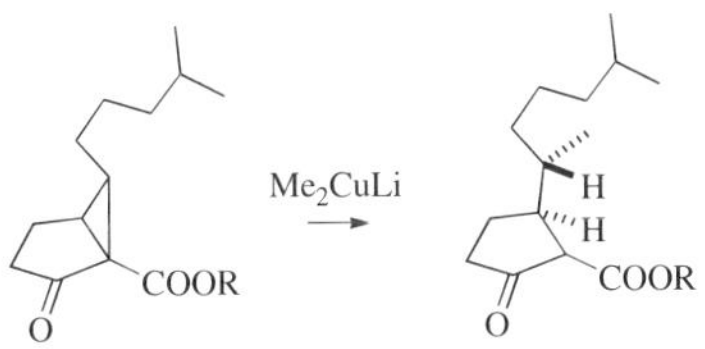

(**8-74**)

8.4. PERICYCLIC REACTIONS

Pericyclic reactions follow pathways which are guided by changes in frontier orbital interactions.[101] The enormous volume of the literature and numerous reviews on the subject, even when limited to the pertinent stereochemical aspects, cannot be treated adequately in a monograph, and therefore this section deals with several individual cases whose selection is quite arbitrary.

The photochemical transformation of a homonuclear *trans*-hexalin derivative results in a conjugated cyclodecatriene. However, on warming, the latter compound is converted to two *cis*-hexalins possessing the same constitution as the starting material but differing in the relative stereochemistry at the ring junction.[102] The two steps involving conrotatory ring opening and disrotatory ring closure in sequence are photoinduced and thermally induced, respectively. For a synthesis of occidental (**8-31**, **8-31ent**) through such a sequence, each diastereomer was converted to an enantiomeric form the of terpene.

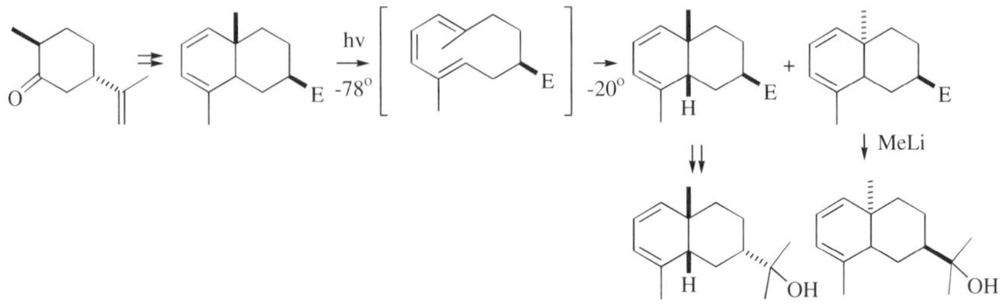

(-)-occidentalol (**8-31ent**) (+)-occidentalol (**8-31**)

The 5C-4e electrocyclization of cross-conjugated dienones, known as the Nazarov reaction, has a well-defined stereochemical pathway as predicted by the Woodward–Hoffmann rules. Very interesting applications of this method to the synthesis of trichodiene (**8-75**)[103] and nootkatone (**8-39**)[104] have been reported.

The Diels–Alder reaction is usually *endo*-selective due to secondary orbital overlap of the diene and dienophile, and other factors. The intramolecular version may not obey this rule if steric constraints disfavor the *endo* transition state. An intriguing synthetic problem concerns antibiotics such as vermisporin and MBP 049-13 (**8-76**) which contain a *cis*-octalin nucleus. As *endo*-selective Diels–Alder reactions of (*E*)-enoic esters would lead to *trans*-octalins, a more elaborate design is needed to divert this trend. Replacing the unsaturated ester with an unsaturated lactone appears to provide the simplest solution.[105]

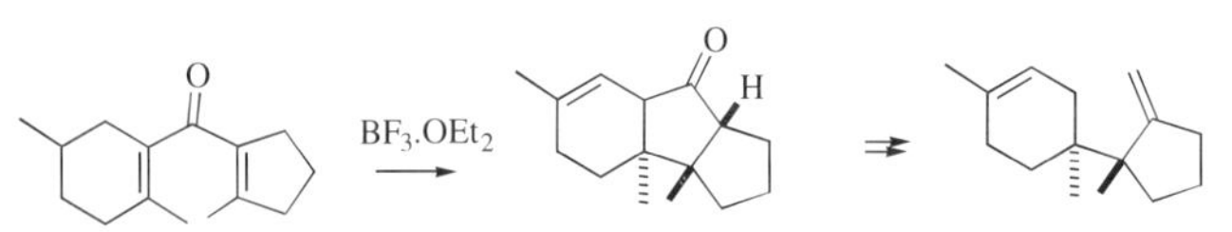

trichodiene (**8-75**)

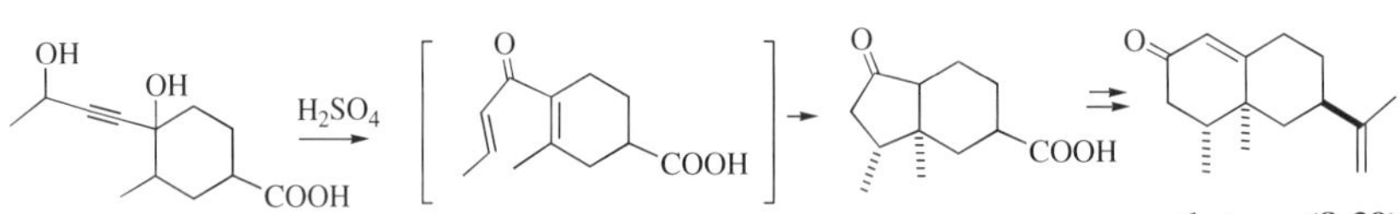

nootkatone (**8-39**)

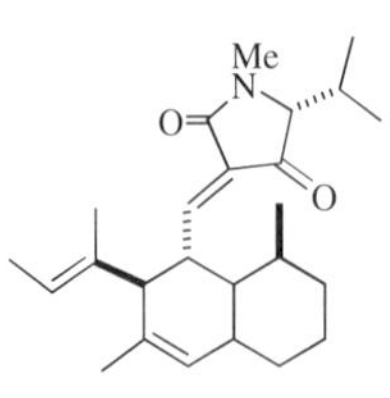

MBP 049-13
(**8-76**)

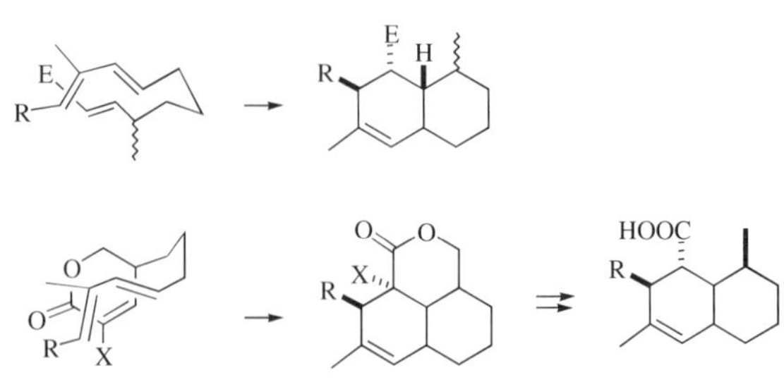

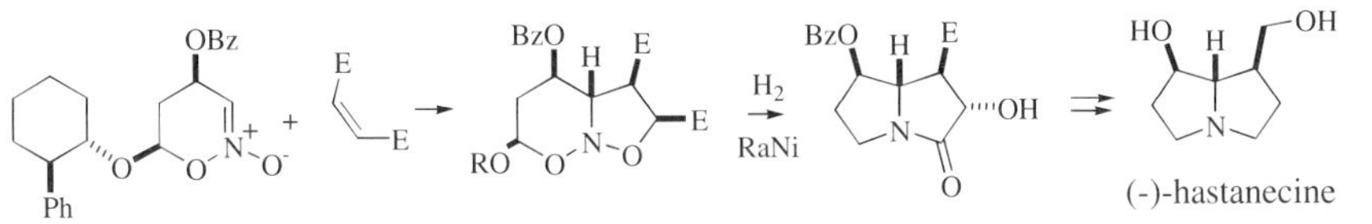

(-)-hastanecine
(**8-77**)

Stereoelectronic factors account for the formation of a bicyclic dioxaindolizidine with all of the functionalities suitable for manipulation to give hastanecine (**8-77**).[106] The 1,3-dipolar cycloaddition apparently occurred with the nitrone in a conformation featuring two pseudoaxial substituents. Strong bias due to the anomeric effect is recognized.

Sigmatropic rearrangement is suitable for suprafacial 1,3-chirality transfer[107] because of their concerted nature. Claisen, Cope, and [2,3]-Wittig rearrangements and their many variants have received intensive attention with respect to mechanistic interpretations and synthetic applications. In [3,3]-sigmatropic rearrangements the chair transition states disposing the maximum number of pseudoequatorial substituents are normally adopted, and therefore the geometry of the double bonds in the substrates can determine the configuration of new stereogenic centers. Formation of (*E*)- disubstituted and (*E*)- trisubstituted double bonds is also favored. From a study of the thermal behavior of four propenyl crotyl ether isomers,[108] it was shown that the (*E,E*)- and (*Z,Z*)-isomers rearranged to a mixture majoring (95–96%) in the *threo*-aldehyde, whereas the (*Z,E*)- and (*E,Z*)-isomers underwent *erythro*-selective (95%) reaction. There is a correlation of rates [fastest for the (*E,E*)- isomer, slowest for the (*Z,Z*)- isomer] to the number of pseudoaxial substituents in the transition state conformations. Of course, the steric courses of the rearrangements are also influenced by existing chirality elements proximal to but outside of the circuit involved in the rearrangement.

A method for construction of the vitamin E side chain (**8-78**)[109] demonstrates the full utility of enantiomeric propargyl alcohols by proper manipulation of the double bond geometry. Such an enantioconvergent approach is particularly significant in the potential development of an industrial route.

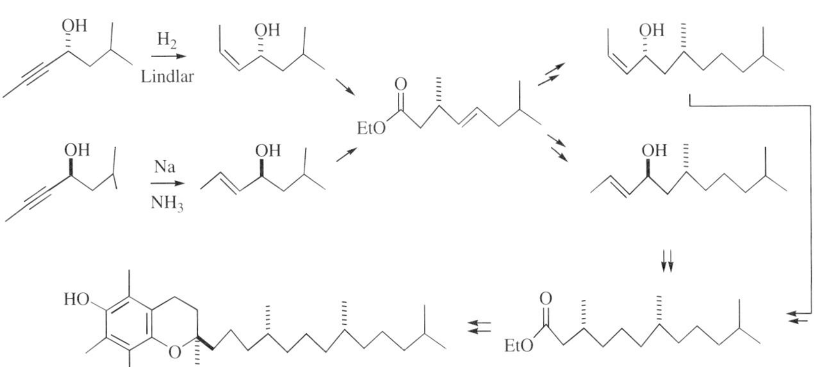

(**8-78**)

α-tocopherol

Analogous manipulation of *R,E*- and *S,Z*-allylic alcohols into the proper silyl ketene acetals to effect the Ireland–Claisen rearrangement enabled the enantioconvergent preparation of a pseudomonic acid-C intermediate.[110]

When kaurene (**8-79**) was considered for synthesis in the 1960s, the most obvious intermediate was a trimethyloctahydrophenanthrene.[111] Birch reduction and reduction of the ensuing conjugated enone readily gave a pseudoequatorial alcohol which was the desired substrate for a Claisen rearrangement. At this point the complete stereochemical issue was settled because with proper functionalization the D-ring would be formed correctly (i.e., it is part of a bridged system).

Gascardic acid (**8-80**) represents a more complex problem synthetically. It is interesting to note that in the accomplished work[112] the seven-membered ring was created by a Dieckmann cyclization, and the ester chain emanating from the hydrindan moiety was affixed by a Claisen rearrangement. The relative ease of obtaining a pseudoequatorial alcohol by sodium borohydride reduction of a cyclohexenone again made the task quite easy.

The double usage of the Ireland version of the Claisen rearrangement involving silyl ketene acetals in a synthesis of lasalocid-A (**8-81**)[113] attests to its effectiveness. First, its appearance was in a stereoselective attachment of a side chain to the furanoid

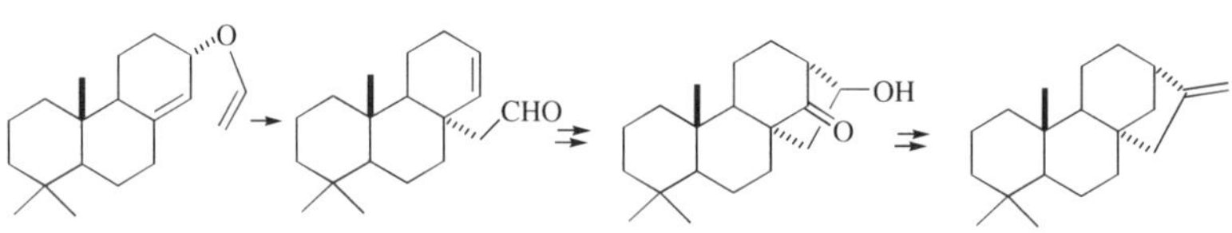

kaurene (**8-79**)

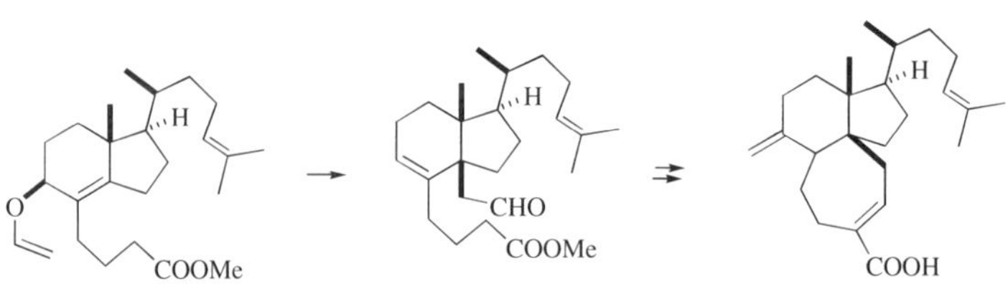

gascardic acid (**8-80**)

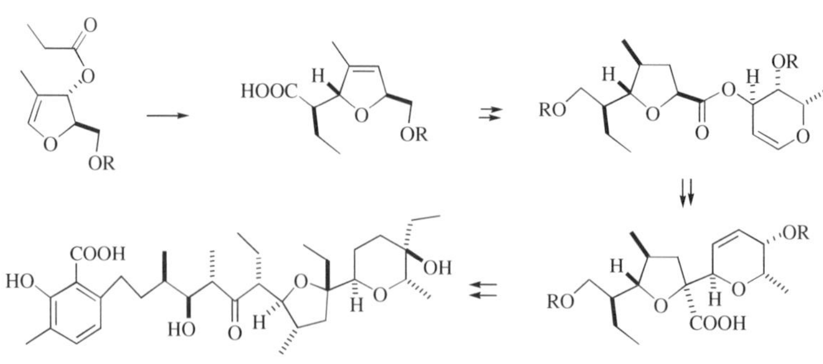

lasalocid-A (**8-81**)

ring, followed by coupling of that nucleus to a pyran. The much easier and better control of oxygenation pattern can thus be exploited indirectly for C–C bond formation under extremely mild conditions. Available protocols for generating (*E*)- and (*Z*)- silylketene acetals contributed to the value of this method.

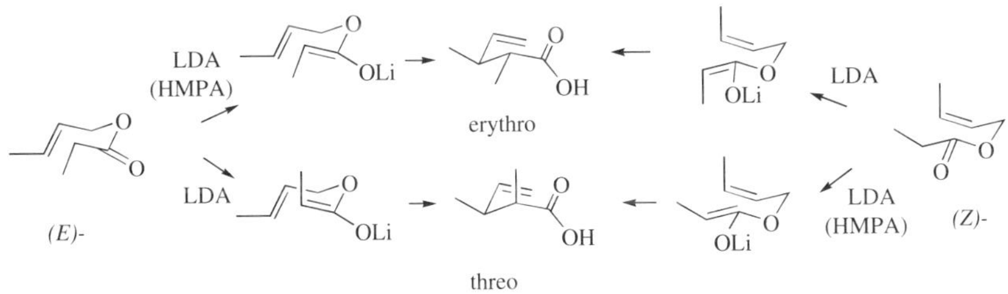

The presence of a C_2-symmetric chiral bromoborane can render the rearrangement highly enantioselective. Such a protocol has been employed in a synthesis of (+)-β-elemene (**8-82**) and (+)-fuscol (**8-83**).[114]

When unsaturated lactones undergo Claisen rearrangement, they are forced to proceed via boat transition states. Ring-contracted carbocycles of well-defined and predictable stereochemistry are accessible.

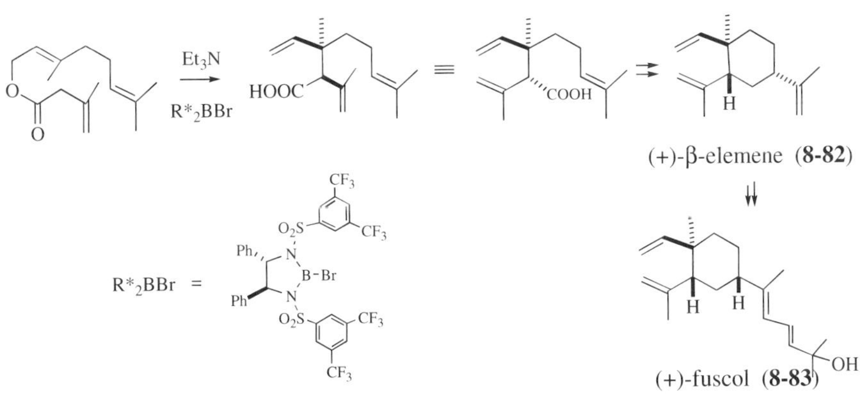

Initially, the reversible Cope rearrangement did not enjoy the popularity of its cousin (Claisen rearrangement) in synthetic utility. However, the advent of the oxy-Cope type, especially the anionic version, which destroys the equilibrium and lowers the activation energy, has led to explosive amplification of its employment. Thus, an expedient access to a precursor of juvabione (**8-84**)[115] indicates how simple it is to control two contiguous stereocenters in a freely rotating carbon chain by this method.

The combination of Cope and Claisen rearrangements is a powerful tool for molecular framework construction. When the Cope rearrangement sets up an allyl vinyl ether, which is subject to thermal reorganization, the reaction pathway becomes

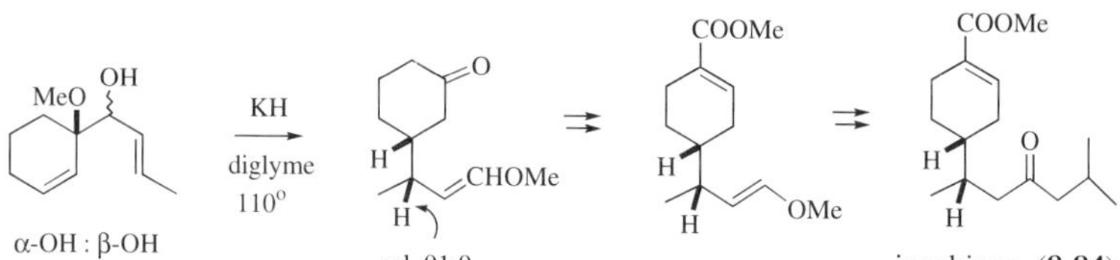

juvabione (**8-84**)

irreversible. The following analysis should convince us of the reward of the tandem processes[116] to synthetic adventures, particularly when the steric courses can be delineated.

Suppose we are interested in a synthesis of confertin (**8-85**) which has two sets of contiguous stereocenters. While it is reasonable to assume that the annexed α-methylene-γ-lactone unit can be installed relatively easily at the end, the perhydroazulene system with a *trans*-ring junction and a subangular methyl group is much more difficult to construct stereoselectively, owing to the pseudo-1,3-axial relationship of the methyl substituents. With the intriguing possibility of assembling a monocyclic precursor which contains the desired pattern of relative stereochemistry, focus can be directed to the ketone group. Application of the combined Claisen–Cope transform to such a structure reveals the proper substrate, although stereoselectivity for the two steps depends on transition state geometry (chairlike versus boatlike and nonbonding interactions).[117] A useful intermediate was indeed accessible.[118]

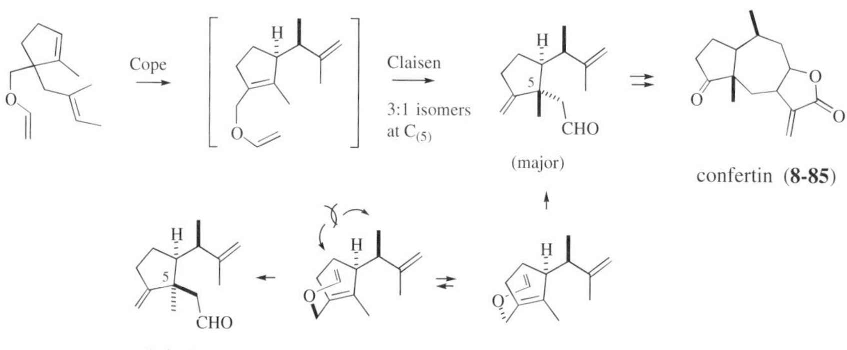

confertin (**8-85**)

The conformational preference for the anionic oxy-Cope rearrangement of an (*R*)-alcohol leading to the (*S*)-aldehyde and thence to (+)-dihydromayurone (**8-86**) is easily understood. In the alternative conformation the anionic oxygen is pseudoaxial.[119] Actually, in the availability of two chair conformations for anionic oxy-Cope rearrangement to a substrate with no apparent difference in steric compression, there is a strong equatorial preference for the oxy anion.[120] This behavior is quite different from the thermal oxy-Cope rearrangement involving nonionized species and it reflects the stabilization by the electron supply from an aligned C–O bond.

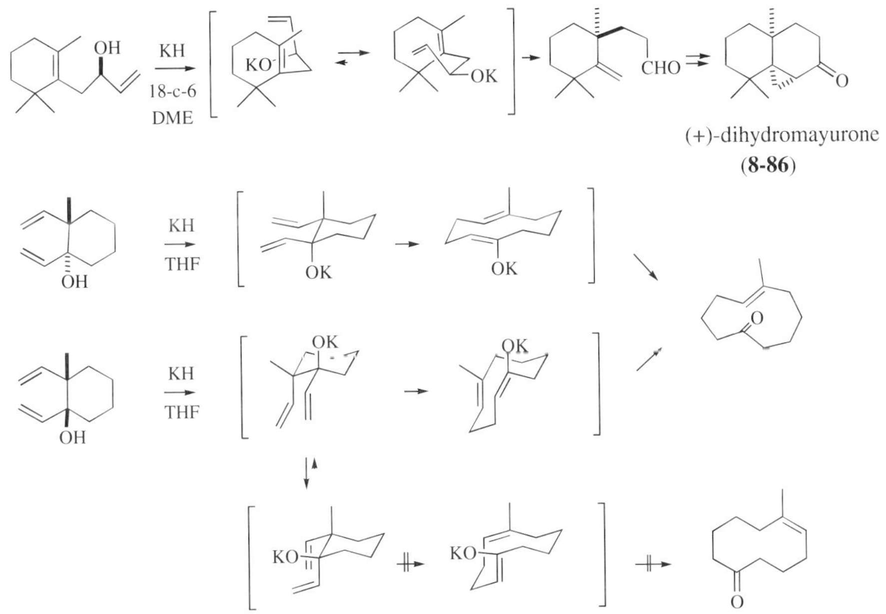

(+)-dihydromayurone (**8-86**)

[2,3]-Sigmatropic rearrangements are comparatively less familiar to the synthetic chemist. Fortunately, they have attracted much attention in recent years and the higher frequency of appearance of the [2,3]-Wittig rearrangement[121] and sulfenate–sulfoxide rearrangement in the chemical literature is evident.

Anion generation is most critical for the Wittig rearrangement. Alternatives to direct deprotonation are methods involving reductive lithiation of *O,S*-acetals and tin–lithium exchange. The latter protocol is most effective for suprafacial introduction of a hydroxymethyl group by transposition of allylic alcohols, as exemplified in a synthesis of punctatin-A (**8-87**).[122]

For acyclic substrates there has been observed the trend of *erythro*- and *threo*-selection from (*Z*)- and (*E*)- allylic ethers, respectively. Cyclic variants can exhibit a different pattern due to steric influence and conformational bias exerted by the intervening chain, although a 13-membered ring ether behaved normally. Such a compound has been converted to a cyclodecenyne on treatment with a chiral amide and ultimately to (+)-aristolactone (**8-88**).[123] The absolute configuration of the propenyl group of the rearranged product was correct but a Mitsunobu inversion of the secondary alcohol was necessary.

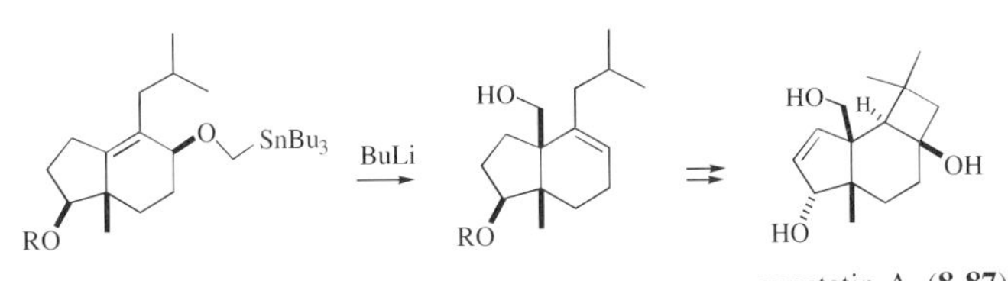

punctatin-A (**8-87**)

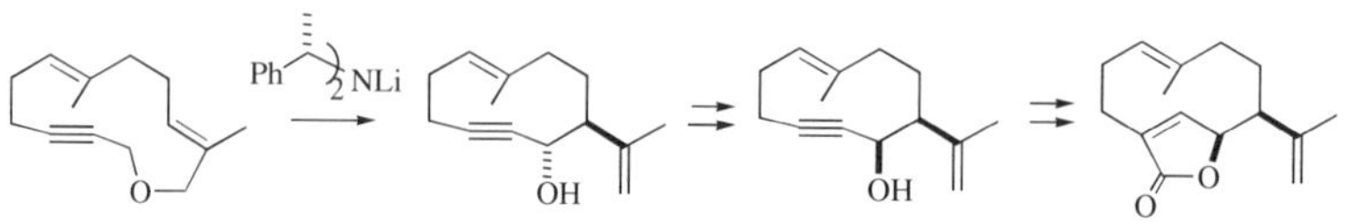

(+)-aristolactone
(**8-88**)

In recent years much work has been directed toward metal-catalyzed decomposition of α-diazo ketones with participation of heteroatoms. Given the opportunity for [2,3]-sigmatropic rearrangement, the ylides derived from α-diazo ketones would follow the expected pathways, and transfer of chirality accompanying new C–C bond formation can be achieved. A synthesis of (+)-griseofulvin (**8-42**)[124] demonstrates the potential of the process.

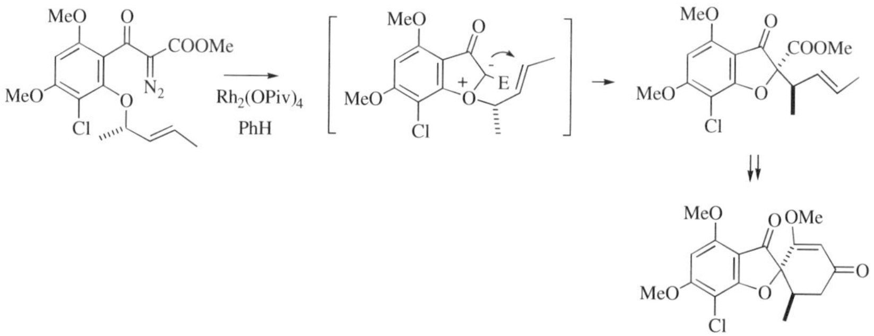

(+)-griseofulvin
(**8-42**)

For stereoselective allylic sulfenate–sulfoxide interconversion, the predominance of the thermodynamically more stable sulfoxide can be shifted by decomposing the sulfenate (allylic alcohol is usually the desired product) in situ.

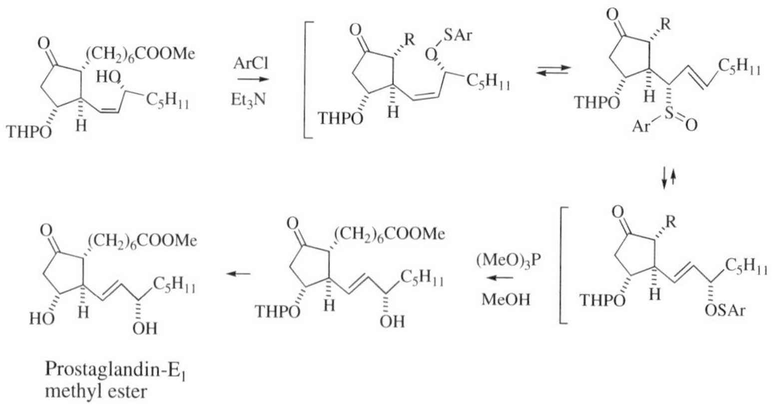

Prostaglandin-E_1
methyl ester
(**8-89**)

Addition of an organic phosphite, which has no effect on the sulfoxide, achieves the transformation. The stereochemical manipulation of a more readily accessible (*13Z*)-15α-ol intermediate of prostaglandin-E_1 (**8-89**)[125] by this method was extremely satisfying.

REFERENCES

1. Aubé, J.; Ghosh, S.; Tanol, M. *J. Am. Chem. Soc.* **1994**, *116*, 9009.
2. Molander, G.A.; Nichols, P.J. *J. Org. Chem.* **1996**, *61*, 6040.
3. Heinemann, C.; Demuth, M. *J. Am. Chem. Soc.* **1997**, *119*, 1129.
4. Kopach, M.E.; Fray, A.H.; Meyers, A.I. *J. Am. Chem. Soc.* **1996**, *118*, 9876.
5. Morley, C.; Knight, D.W.; Share, A.C. *Tetrahedron Asymmetry* **1990**, *1*, 147.
6. Romo, D.; Meyers, A.I.; *Tetrahedron* **1991**, *47*, 9503; Meyers, A.I.; Brengel, G.P. *Chem. Commun.* **1997**, 1.
7. Durkin, K.A.; Liotta, D. *J. Am. Chem. Soc.* **1990**, *112*, 8162.
8. Magnus, P.; Principe, L.M.; Slater, M.J. *J. Org. Chem.* **1987**, *52*, 1483.
9. Rowley, E.G.; Schore, N.E. *J. Org. Chem.* **1992**, *57*, 6853.
10. Hashimoto, S.; Sakata, S.; Sonegawa, M.; Ikegami, S. *J. Am. Chem. Soc.* **1988**, *110*, 3670.
11. Brandes, E.; Grieco, P.A.; Garner, P. *Chem. Commun.* **1988**, 500.
12. Cornforth, J.W.; Cornforth, R.H.; Mathews, K.K. *J. Chem. Soc.* **1959**, 112.
13. Wu, Y.-D.; Tucker, J.A.; Houk, K.N. *J. Am. Chem. Soc.* **1991**, *113*, 5018.
14. Gassman, P.G.; O'Reilly, N.J. *J. Org. Chem.* **1987**, *52*, 2481.
15. Mohan, R.; Kates, S.A.; Dombroski, M.; Snider, B.B. *Tetrahedron Lett.* **1987**, *28*, 845.
16. Wipf, P.; Kim, Y.; Fritch, P.C. *J. Org. Chem.* **1993**, *58*, 7195.
17. Paquette, L.A.; Wang, H.-L. *J. Org. Chem.* **1996**, *61*, 5352.
18. Wada, M.; Nishihara, Y.; Akiba, K. *Tetrahedron Lett.* **1985**, *26*, 3267.
19. Magnus, P.; Gallagher, T.; Brown, P.; Huffman, J.C. *J. Am. Chem. Soc.* **1984**, *106*, 2105.
20. White, J.D.; Dillon, M.P.; Butlin, R.J. *J. Am. Chem. Soc.* **1992**, *114*, 9673.
21. Lin, X.; Kavash, R.W.; Mariano, P.S. *J. Org. Chem.* **1996**, *61*, 7335.
22. Boeckman, R.K.; Connell, B.T. *J. Am. Chem. Soc.* **1995**, *117*, 12368.
23. Yamamoto, Y.; Nemoto, H.; Kikuchi, R.; Komatsu, H.; Suzuki, I. *J. Am. Chem. Soc.* **1990**, *112*, 8598.
24. Mulzer, J.; Kirstein, H.M.; Buschmann, J.; Lehmann, C.; Luger, P. *J. Am. Chem. Soc.* **1991**, *113*, 910.
25. Parker, K.A.; Kim, H.-J. *J. Org. Chem.* **1992**, *57*, 752.
26. Deslongchamps, P. *Stereoelectronic Effects in Organic Chemistry*; Pergamon Press: Oxford, 1983; Kirby, A.J. *Stereoelectronic Effects*; Oxford University Press: London, 1996.
27. Tsuda, Y.; Isobe, K. *Chem. Commun.* **1971**, 1555.
28. Danishefsky, S.; Schuda, P.F.; Kitahara, T.; Etheredge, S.J. *J. Am. Chem. Soc.* **1977**, *99*, 6066.
29. Garigipati, R.S.; Tschaen, D.M.; Weinreb, S.M. *J. Am. Chem. Soc.* **1990**, *112*, 3475.

30. Hudlicky, T.; Tian, X.; Königsberger, K.; Maurya, R.; Rouden, J.; Fan, B. *J. Am. Chem. Soc.* **1996**, *118*, 10752.
31. Whitlock, H.W.; Smith, G.L. *J. Am. Chem. Soc.* **1967**, *89*, 3600.
32. Wheeler, D.M.S.; Wheeler, M.M. *Stud. Nat. Prod. Chem.* **1994**, *14*, 3.
33. Cieplak, A.S. *J. Am. Chem. Soc.* **1981**, *103*, 4540.
34. Stork, G.; West, F.; Lee, H.Y.; Isaacs, R.C.A.; Manabe, S. *J. Am. Chem. Soc.* **1996**, *118*, 10660.
35. Takeda, K.; Yano, S.; Sato, M.; Yoshii, E. *J. Org. Chem.* **1987**, *52*, 4135; Takeda, K.; Yano, S.; Yoshii, E. *Tetrahedron Lett.* **1988**, *29*, 6951.
36. Ireland, R.E.; Smith, M.G. *J. Am. Chem. Soc.* **1988**, *110*, 854.
37. Ho, T.-L. *Stereoselective Synthesis. Natural Products from Chiral Terpenes*, Wiley: New York, 1992.
38. Wovkulich, P.M.; Tang, P.C.; Chadha, N.K.; Batcho, A.D.; Barrish, J.C.; Uskokovic, M.R. *J. Am. Chem. Soc.* **1989**, *111*, 2596.
39. Adamson, P.S.; McQuillin, F.J.; Robinson, R.; Simonsen, J.L. *J. Chem. Soc.* **1937**, 1576.
40. Howe, R.; McQuillin, F.J. *J. Chem. Soc.* **1955**, 2423.
41. Wenkert, E.; Berges, D.A. *J. Am. Chem. Soc.* **1967**, *89*, 2507.
42. Marshall, J.A.; Bundy, G.L.; Fanta, W.I. *J. Org. Chem.* **1968**, *33*, 3913.
43. Heathcock, C.H.; Amano, Y. *Can. J. Chem.* **1972**, *50*, 340.
44. McMurry, J.E.; Musser, J.H.; Ahmad, M.S.; Blaszczak, L.C. *J. Org. Chem.* **1975**, *40*, 1829.
45. Scanio, C.J.V.; Starrett, R.M. *J. Am. Chem. Soc.* **1971**, *93*, 1539.
46. Marshall, J.A.; Warne, T.M. *J. Org. Chem.* **1971**, *36*, 178.
47. Woodward, R.B.; Sondheimer, F.; Taub, D.; Heusler, K.; McLamore, W.M. *J. Am. Chem. Soc.* **1952**, *74*, 4223.
48. Stork, G.; Darling, S.D.; Harrison, I.T.; Wharton, P.S. *J. Am. Chem. Soc.* **1962**, *84*, 2018.
49. Vaillancourt, V.; Albizati, K.F. *J. Am. Chem. Soc.* **1993**, *115*, 3499.
50. Kieczykowski, G.R.; Schlessinger, R.H. *J. Am. Chem. Soc.* **1978**, *100*, 1938.
51. Kitahara, T.; Warita, Y.; Abe, M.; Seya, M.; Takagi, Y.; Mori, K. *Agric. Biol. Chem.* **1991**, *55*, 1013.
52. Micouin, L.; Varea, T.; Riche, C.; Chiaroni, A.; Quiron, J.-C.; Husson, H.-P. *Tetrahedron Lett.* **1994**, *35*, 2529.
53. Marshall, J.A.; Cohen, G.M. *J. Org. Chem.* **1971**, *36*, 877.
54. Pesaro, M.; Bozzatto, G.; Schudel, P. *Chem. Commun.* **1968**, 1152.
55. Kuramochi, T.; Asaoka, M.; Ohkubo, T.; Takei, H. *Tetrahedron Lett.* **1996**, *37*, 7075.
56. Spencer, T.A.; Smith, R.A.J.; Storm, D.L.; Villarica, R.M. *J. Am. Chem. Soc.* **1971**, *93*, 4856.
57. Meyer, W.L.; Maheshwari, K.K. *Tetrahedron Lett.* **1964**, 2175.
58. Stork, G.; Tomasz, M. *J. Am. Chem. Soc.* **1964**, *86*, 471.
59. Iio, H.; Isobe, M.; Kawai, T.; Goto, T. *Tetrahedron* **1979**, *35*, 941; *J. Am. Chem. Soc.* **1979**, *101*, 6076.
60. Kraus, G.A.; Hon, Y.-S. *J. Am. Chem. Soc.* **1985**, *107*, 4341.
61. Stork, G.; Winkler, J.D.; Shiner, C.S. *J. Am. Chem. Soc.* **1982**, *104*, 3767.

62. Stork, G.; Saccomano, N.A. *Tetrahedron Lett.* **1987**, *28*, 2087.
63. Attah-Poku, S.K.; Chau, F.; Yadav, V.K.; Fallis, A.G. *J. Org. Chem.* **1985**, *50*, 3418.
64. Ho, T.-L. *Heterolytic Fragmentation of Organic Molecules*; Wiley: New York, 1993.
65. Corey, E.J.; Mitra, R.B.; Uda, H. *J. Am. Chem. Soc.* **1964**, *86*, 485.
66. Zurflüh, R,; Wall, E.N.; Siddall, J.B.; Edwards, J.A. *J. Am. Chem. Soc.* **1968**, *90*, 6224.
67. Wharton, P.S.; Sundin, C.E.; Johnson, D.W.; Kluender, H.C. *J. Org. Chem.* **1972**, *37*, 34.
68. Wender, P.A.; Fisher, K. *Tetrahedron Lett.* **1986**, *27*, 1857.
69. Kiguchi, T.; Kuninobu, N.; Naito, T.; Ninomiya, I. *Heterocycles* **1989**, *28*, 19.
70. Boeckman, R.K.; Arvanitis, A.; Voss, M.E. *J. Am. Chem. Soc.* **1989**, *111*, 2737.
71. Winterfeldt, E.; Radunz, H.; Korth, T. *Chem. Ber.* **1968**, *101*, 3172.
72. Naito, T.; Miyata, O.; Ninomiya, I. *Heterocycles* **1987**, *26*, 1739.
73. Kraus, G.A.; Thomas, P.J.; Bougie, D.; Chen, L. *J. Org. Chem.* **1990**, *55*, 1624; Kraus, G.A.; Bougie, D. *Synlett* **1992**, 279.
74. Hanessian, S.; Faucher, A.M.; Leger, S. *Tetrahedron* **1990**, *46*, 231.
75. Crimmins, M.T.; Hollis, W.G.; O'Mahony, R. *Stud. Nat. Prod. Chem.* **1988**, *1*, 435.
76. Davies, H.G.; Green, R.H. *Chem. Soc. Rev.* **1991**, *20*, 211, 271.
77. Schreiber, S.L.; Sommer, T.J. *Tetrahedron Lett.* **1983**, *24*, 4781.
78. Redlich, H.; Lenfers, J.B.; Thormählen, S. *Synthesis* **1992**, 1112.
79. Woodward, R.B.; Bader, F.E.; Bickel, H.; Frey, A.J.; Kierstead, R.W. *Tetrahedron* **1958**, *2*, 1.
80. Stork, G. *Pure Appl. Chem.* **1989**, *61*, 439.
81. van Tamelen, E.E.; Shamma, M.; Burgstahler, A.W.; Wolinsky, J.; Tamm, R.; Aldrich, P.E. *J. Am. Chem. Soc.* **1958**, *80*, 5006.
82. Baxter, E.W.; Labaree, D.; Ammon, H.L.; Mariano, P.S. *J. Am. Chem. Soc.* **1990**, *112*, 7682.
83. Stevens, R.V.; Lee, A.W.M. *J. Am. Chem. Soc.* **1979**, *101*, 7032.
84. Stevens, R.V.; Lee, A.W.M. *Chem. Commun.* **1982**, 102.
85. Stevens, R.V.; Lee, A.W.M. *Chem. Commun.* **1982**, 103.
86. Watanabe, Y.; Iida, H.; Kibayashi, C. *J. Org. Chem.* **1989**, *54*, 4088.
87. Shono, T.; Matsumura, Y.; Onomura, O.; Sato, M. *J. Org. Chem.* **1988**, *53*, 4118.
88. Ciufolini, M.A.; Hermann, C.W.; Whitmire, K.H.; Byrne, N.E. *J. Am. Chem. Soc.* **1989**, *111*, 3473.
89. Stevens, R.V.; Kenney, P.M. *Chem. Commun.* **1983**, 384.
90. Ryckman, D.M.; Stevens, R.V. *J. Am. Chem. Soc.* **1987**, *109*, 4940.
91. Heathcock, C.H.; Kleiman, E.F.; Binkley, E.S. *J. Am. Chem. Soc.* **1982**, *104*, 1054.
92. Agami, C.; Cases, M.; Couty, F. *J. Org. Chem.* **1994**, *59*, 7937.
93. Stevens, R.V. In *Strategies and Tactics in Organic Synthesis*; Lindberg, T., Ed., Academic Press: Orlando, 1984; Chapter 10.
94. Stevens, R.V.; Pruitt, J.R. *Chem. Commun.* **1983**, 1425.
95. Maruoka, K.; Miyazaki, T.; Ando, M.; Matsumura, Y.; Sakane, S.; Hattori, K.; Yamamoto, H. *J. Am. Chem. Soc.* **1983**, *105*, 2831.
96. Danieli, B.; Lesma, G.; Palmisano, G. *Tetrahedron Lett.* **1981**, *22*, 1827.

97. Ziegler, F.E.; Spitzner, E.B. *J. Am. Chem. Soc.* **1973**, *95*, 7146.
98. Danishefsky, S.; McKee, R.; Singh, R.K. *J. Am. Chem. Soc.* **1977**, *99*, 4783.
99. Corey, E.J.; Fuchs, P.L. *J. Am. Chem. Soc.* **1972**, *94*, 4014.
100. Trost, B.M.; Taber, D.F.; Alper, J.B. *Tetrahedron Lett.* **1976**, 3857.
101. Woodward, R.B.; Hoffmann, R. *The Conservation of Orbital Symmetry*; Verlag Chemie: Weinheim, 1971.
102. Hortmann, A.G.; Daniel, D.S.; Martinelli, J.E. *J. Org. Chem.* **1973**, *38*, 728.
103. Harding, K.E.; Clement, K.S.; Tseng, C.-Y. *J. Org. Chem.* **1990**, *55*, 4403.
104. Hiyama, T.; Shinoda, M.; Saimoto, H.; Nozaki, H. *Bull. Chem. Soc. Jpn.* **1981**, *54*, 2747.
105. Hasegawa, T.; Kido, M.; Bando, M.; Kitahara, T. *Synlett* **1995**, 909.
106. Denmark, S.E.; Thorarensen, A. *J. Org. Chem.* **1994**, *59*, 5672.
107. Hill, R.K. In *Asymmetric Synthesis*; Morrison, J.D., Ed.; Academic Press: Orlando, 1984, Vol. 3B, p 503.
108. Vittorelli, P.; Hansen, H.-J.; Schmid, H. *Helv. Chim. Acta* **1975**, *58*, 1293.
109. Chan, K.K.; Cohen, N.; De Noble, J.P.; Specian, A.C.; Saucy, G. *J. Org. Chem.* **1976**, *41*, 3497.
110. Barrish, J.C.; Lee, H.L.; Baggiolini, E.G.; Uskokovic, M.R. *J. Org. Chem.* **1987**, *52*, 1372.
111. Bell, R.A.; Ireland, R.E.; Partyka, R.A. *J. Org. Chem.* **1966**, *31*, 2530.
112. Boeckman, R.K.; Blum, D.M.; Arthur, S.D. *J. Am. Chem. Soc.* **1979**, *101*, 5060.
113. Ireland, R.E.; Anderson, R.C.; Badoud, R.; Fitzsimmons, B.J.; McGarvey, G.J.; Thaisrivongs, S.; Wilcox, C.S. *J. Am. Chem. Soc.* **1983**, *105*, 1988.
114. Corey, E.J.; Roberts, B.E.; Dixon, B.R. *J. Am. Chem. Soc.* **1995**, *117*, 193.
115. Evans, D.A.; Nelson, J.V. *J. Am. Chem. Soc.* **1980**, *102*, 774.
116. Ho, T.-L. *Tandem Organic Reactions*; Wiley: New York, 1992.
117. Ziegler, F.E.; Piwinski, J.J. *J. Am. Chem. Soc.* **1982**, *104*, 7181.
118. Ziegler, F.E.; Fang, J.-M. *J. Org. Chem.* **1981**, *46*, 825.
119. Lee, E.; Shin, I.-J.; Kim, T.-S. *J. Am. Chem. Soc.* **1990**, *112*, 260.
120. Clive, D.L.J.; Russell, C.G.; Suri, S.C. *J. Org. Chem.* **1982**, *47*, 1632.
121. Mikami, K.; Nakai, T. *Synthesis* **1991**, 594.
122. Sugimura, T.; Paquette, L.A. *J. Am. Chem. Soc.* **1987**, *109*, 3017.
123. Marshall, J.A.; Lebreton, J. *J. Am. Chem. Soc.* **1988**, *110*, 2925.
124. Pirrung, M.C.; Brown, W.L.; Rege, S.; Laughton, P. *J. Am. Chem. Soc.* **1991**, *113*, 8561.
125. Miller, J.G.; Kurz, W.; Untch, K.G.; Stork, G. *J. Am. Chem. Soc.* **1974**, *96*, 6774.

9

THERMODYNAMIC CONTROL AND KINETIC TRAPPING

If a synthetic target molecule or intermediate is the most stable compound (stabilomer) among several stereoisomers and its generation is under reaction conditions that permit equilibration, then there will not be any problem in reaching the objective. On the other hand, if the reaction is kinetically controlled and leads to a stereoisomer other than the desired one, recourse must be made to readjust the structure of the product. In synthesis design it is essential to pay attention to this aspect and in case an untoward steric course is unavoidable additional step(s) to correct the wrong configuration are needed. Some handle must also be provided to initiate this corrective action. It should be emphasized that most of the equilibration protocols used in synthesis are trivial, but their importance is hardly less than other transformations.

9.1. CONFIGURATION INVERSION BY EQUILIBRATION

cis-2,8-Disubstituted 4-cyclooctanones are more stable than the *trans* isomers, and therefore the generation of the latter compounds by kinetic alkylation of the monosubstituted cyclooctenones is not a serious problem when the *cis* isomers are needed. Such is the case in a synthesis of pentalenene (**9-1**)[1] which employed a 5/8 fused intermediate for transannular cyclization to elaborate the angular triquinane skeleton.

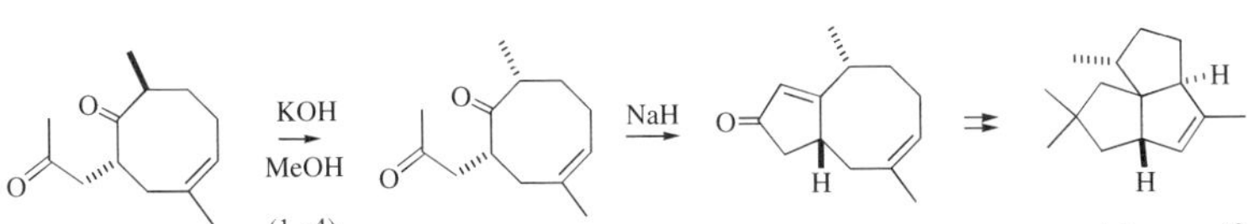

pentalenene (**9-1**)

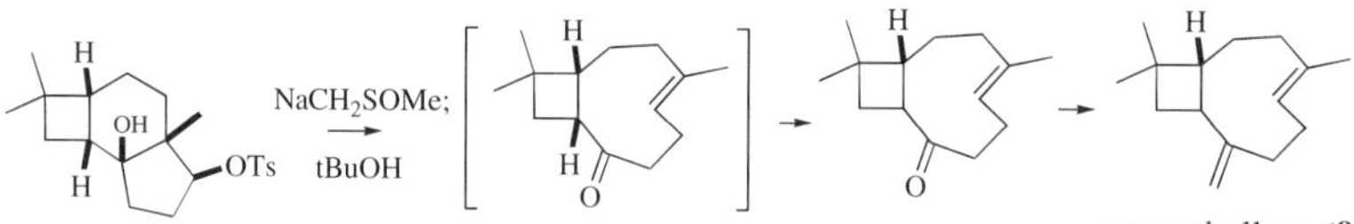

caryophyllene (**9-2**)

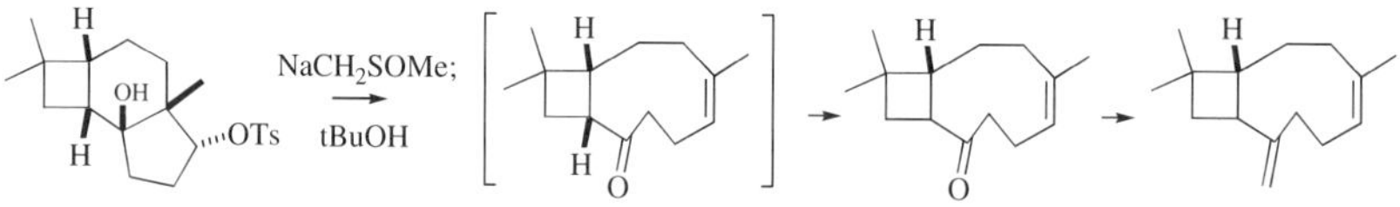

isocaryophyllene (**9-3**)

Cis fusion is preferred between a small ring and a common ring or another small ring. The trend is reversed when the second ring is larger (> nine-membered). Thus, the first synthesis of caryophyllene and isocaryophyllene[2] which started from a [2+2]-photocycloaddition of 2-cyclohexenone with isobutene actually proceeded with the *cis* bicyclo[4.2.0]octanone. The photocycloaddition gave rise to a mixture of predominantly two stereoisomers (*cis*+*trans*) but for the sake of maximizing the amount of homogeneous material, isomerization of the *trans* isomer was carried out. The stereochemical aspect of the synthesis was never in doubt because the plan called for a Grob fragmentation of the tricyclic intermediate to generate the norketone of caryophyllene (**9-2**) [also fragmentation of a stereoisomer to the norketone of isocaryophyllene (**9-3**)] in which one of the angular methine hydrogens is epimerizable, and 4/9 fused ketones are stable in the *trans* form.

The perhydroquinoline nucleus of pumiliotoxin-C (**9-4**) is conceivably derivable by way of a Beckmann rearrangement of the corresponding indanone. The ketone can be prepared by an intramolecular Diels–Alder reaction of a 1,3,8(*E*)-decatrien-5-ol derivative.[3] Two stereocenters were established in the cycloaddition step and the third one on oxidation and equilibration. The *cis*-isomer of a 5/6 fused ring system is generally more stable.

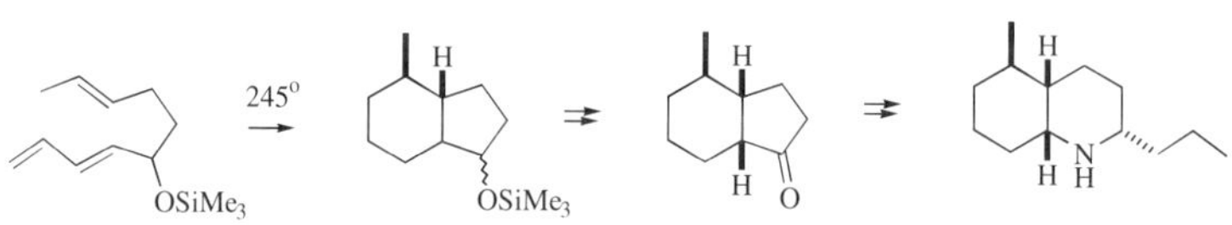

(2:1 cis:trans) at equil.

pumiliotoxin-C (**9-4**)

A retrosynthetic analysis of $\Delta^{9(12)}$-capnellene (**9-5**) indicates the desirability of skeletal assembly by an intramolecular Diels–Alder reaction which would leave a cyclohexene moiety for further elaboration (ring contraction, etc.).[4] Execution of this plan was uneventful. The configuration of the first ring-junction atoms was established during angular methylation of the fused δ-lactone, and even though a 1:2 stereoisomeric mixture of the tricarbocyclic intermediate was produced in the intramolecular Diels–Alder reaction favoring the *cis-anti-trans* ketone, the desired *cis-anti-cis* isomer was obtained exclusively by treatment of the major ketone with DBU in hot benzene.

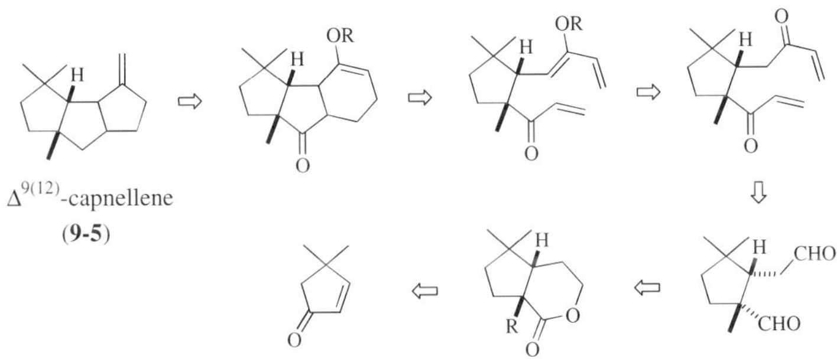

$\Delta^{9(12)}$-capnellene
(**9-5**)

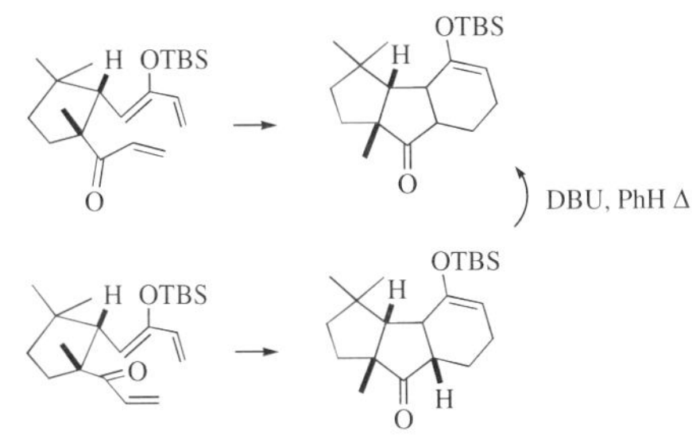

The advantages of using a hydroxydiketohydrazulene for synthesis of damsin (**9-6**)[5] are apparent. The compound, with oxygenation functionality embodying two differentiable carbonyl groups due to their respective environments, is ideally situated to permit systematic elaboration of the missing elements and is readily available by way of a [2+2]-photocycloaddition and α-glycol cleavage.[6] However, the methodology imposes the generation of a *cis*-fused diketone, and it is necessary to epimerize the ring junction. The recognition of this fact paved the proper way to reach the synthetic target.

Ecdysones are characterized by a *cis*-A/B ring junction. In many syntheses of ecdysones, intermediates featured the *trans*-stereochemistry. However, once the 2β,3β-dihydroxy group is introduced, the *cis*-ring junction becomes more favorable, and mildly alkaline conditions are sufficient to rectify the undesirable configuration, as exemplified in a route to rubrosterone (**9-7**).[7]

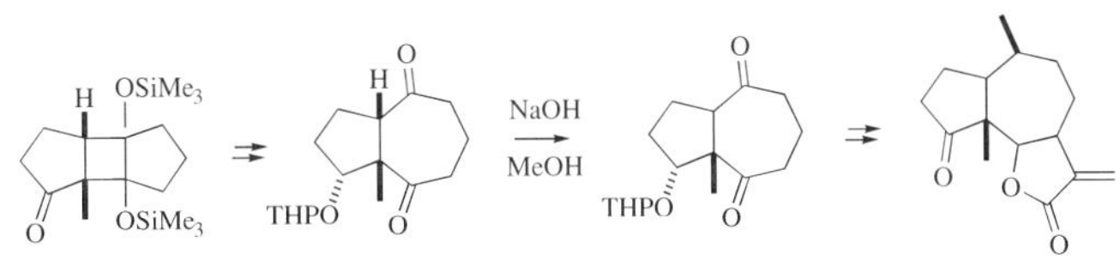

(syn + anti)

damsin (**9-6**)

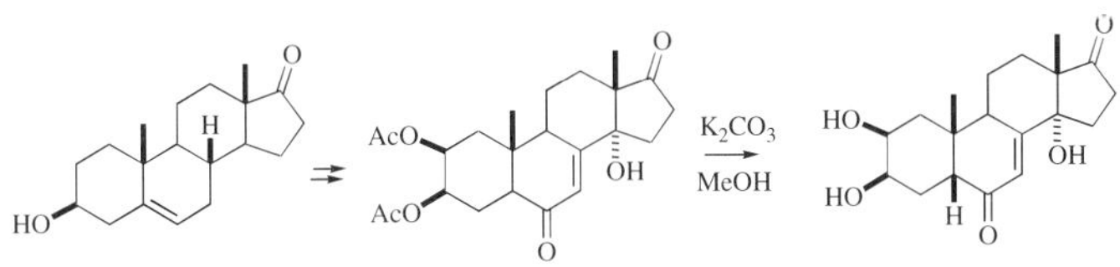

rubrosterone (**9-7**)

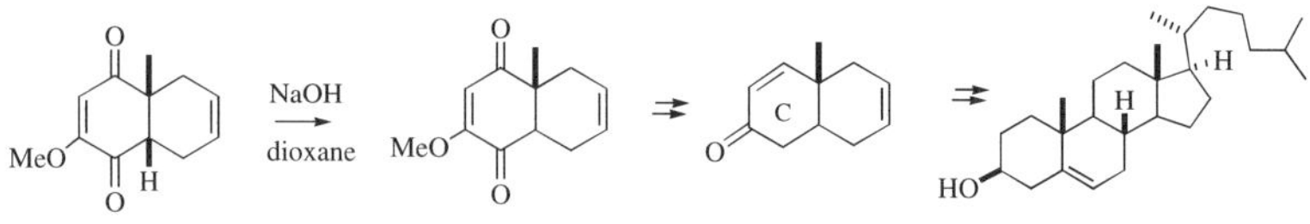

cholesterol (**9-8**)

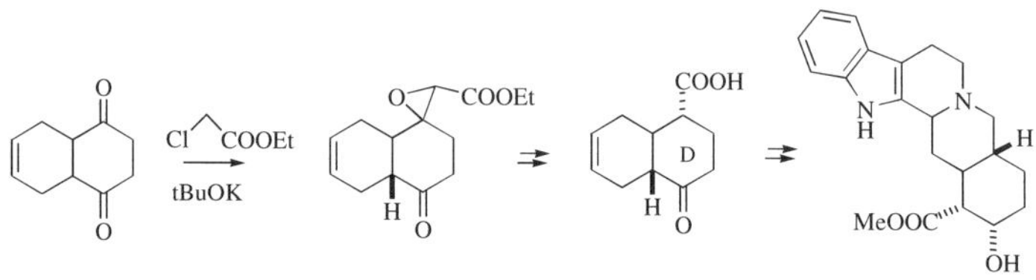

yohimbine (**9-9**)

Generally, 6/6 fusion favors the *trans*-isomer. This characteristic was exploited to ensure the correct C/D ring juncture of cholesterol (**9-8**) in a synthesis featuring a Diels–Alder reaction to form a *cis*-hexalindione in which the cyclohexene ring was destined for contraction.[8]

The ring junction epimerization of the octalindione obtained by partial reduction of the Diels–Alder adduct of butadiene and *p*-benzoquinone accompanied a Darzens reaction.[9] Three of the five stereogenic centers of yohimbine (**9-9**) emerged in this step.

In many instances the synthesis of lupinine and epilupinine involves reduction of $\Delta^{1(10)}$-hexahydroquinolizine-1-carboxylic esters. The saturated ester obtained by sodium borohydride reduction indicates *syn*-entry of two hydrogen atoms, and further reduction (e.g., with lithium aluminum hydride) gives epilupinine (**9-10**). Access to

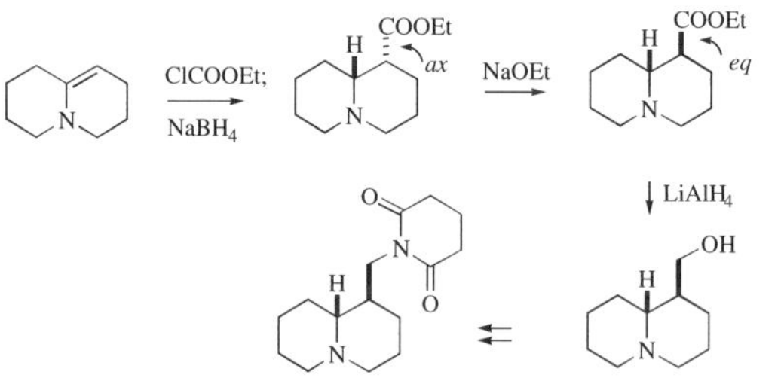

lamprolobine (**9-11**) epilupinine (**9-10**)

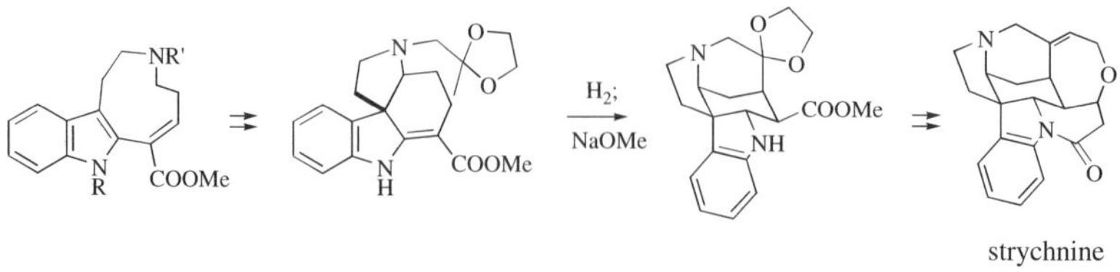

strychnine (**9-12**)

epilupinine only required equilibration of the saturated ester with an alkoxide which reoriented the ester group from an axial to an equatorial position. The synthesis of lamprolobine (**9-11**)[10] was accomplished in such a manner. Note that the same maneuver can be adopted in the preparation of epimeric pyrrolizidine bases.

A recent synthesis of strychnine (**9-12**)[11] via a pentacyclic intermediate involved hydrogenation of the unsaturated ester. Due to the mandatory situation of having at least one equatorial bond to a cyclohexane, one of the new stereocenters was incorrectly disposed. Fortunately, the axial ester group was readily epimerized and then used to elaborate the missing rings.

A synthetic approach to the xenicane skeleton (**9-13**)[12] was patterned after Corey's method in the elaboration of the *trans*-cyclononene moiety from a hydrindan precursor. For the δ-lactone ring the appropriate components were attached to the six-membered ring of the hydrindan in a stepwise manner. The carbinolic unit was derived from a cyano group which was readily introduced by an electrophilic diaxial opening of an epoxide. This was a well-orchestrated reaction sequence because it delivered the fragmentation trigger (a tertiary hydroxyl group) at the ring junction, while at the same time the cyano group could undergo epimerization before its conversion into a primary alcohol.

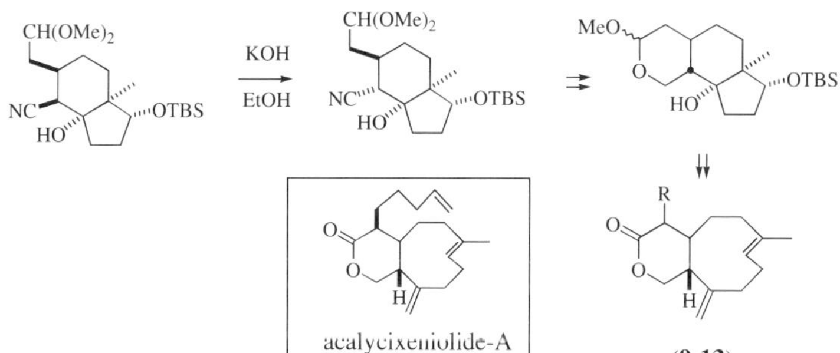

(**9-13**)

The employment of an intramolecular 1,3-dipolar cycloaddition for the formation of a tricyclic intermediate for saxitoxin (**9-14**) would solve the stereochemical problem concerning the ring junctions.[13] Reaction of the *N′*-substituted hydrazide with methyl glyoxalate hemiacetal led to an adduct with an *endo* ester group. The configuration was corrected by equilibration with NaOMe in methanol. Reduction of the resulting *exo* ester gave the desired alcohol.

A spirocyclic ketoester intermediate together with its epimer were formed by degradation of a methyl ketone during a synthesis of vetispiranes.[14] It was gratifying

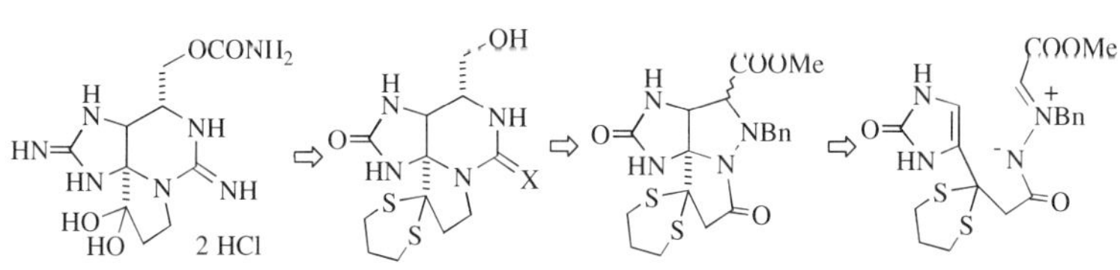

saxitoxin (**9-14**)

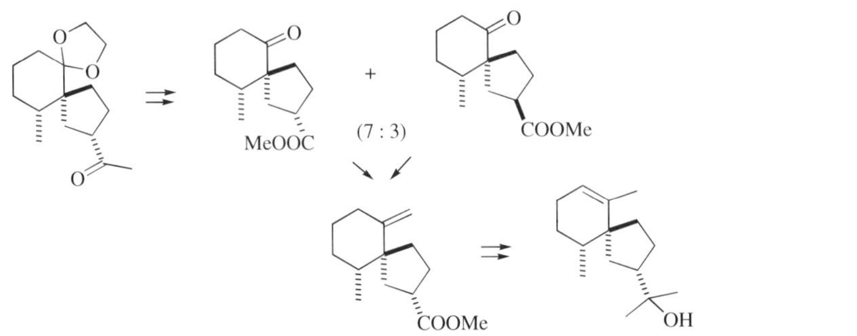

(-)-agarospirol (**9-15**)

to be able to equilibrate the esters during Wittig reaction of the cyclohexanone to obtain a homogeneous product which was converted into (–)-agarospirol (**9-15**).

After using carvomenthone to construct a bridged ring system, completion of a synthesis of helminthosporal (**9-16**)[15] required homologation of the ketone group to an aldehyde. To this end, a Wittig reaction with methoxymethylene-triphenylphosphorane was employed, and the resulting enol ether was converted to an ethyleneacetal in refluxing benzene in the presence of an acid catalyst. The conditions effected equilibration to produce the more stable isomer with the acetal pointing toward the cyclohexene ring where it encountered less severe diaxial 1,3-interaction.

A potential synthetic route to copacamphene (**9-17**) involves solvolytic cyclization (π-participation).[16] For the establishment of a two-carbon chain with the correct configuration, the noraldehyde is an appropriate intermediate. Even if the epimeric aldehyde were formed, equilibration would rectify the configuration because the desired product contains the larger formyl group (versus hydrogen atom) in a less crowded environment.

When the aldehyde is further shortened by one carbon to reach a bridged ketone, correlation with carvomenthone is readily made. A successful synthesis was carried out on the basis of this analysis, and indeed an epimerization step was involved.

An interesting scheme for the synthesis of calcimycin (**9-18**)[17] called for elaboration of a readily available norbornenone derivative to a bicyclic lactone with

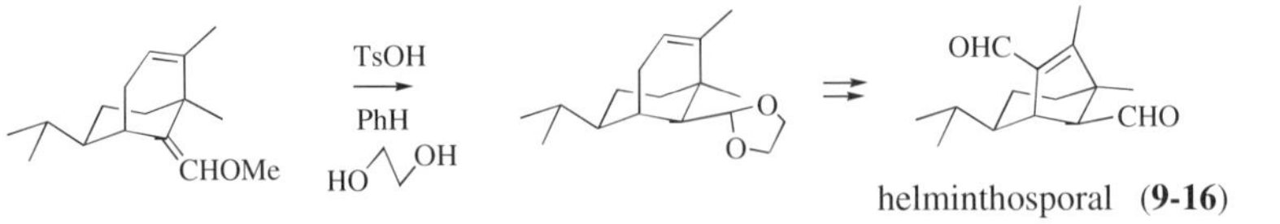

helminthosporal (**9-16**)

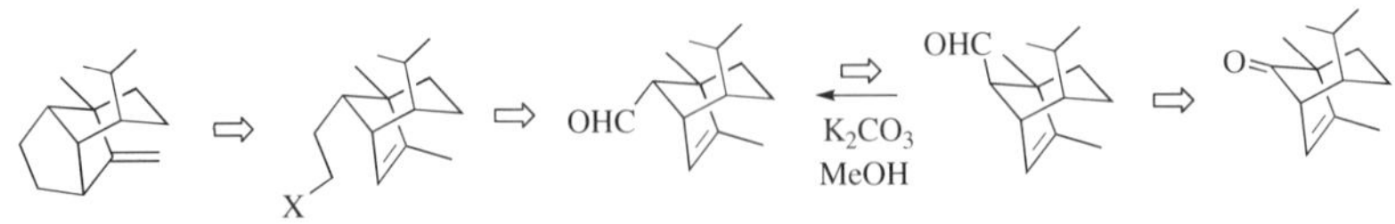

copacamphene
(**9-17**)

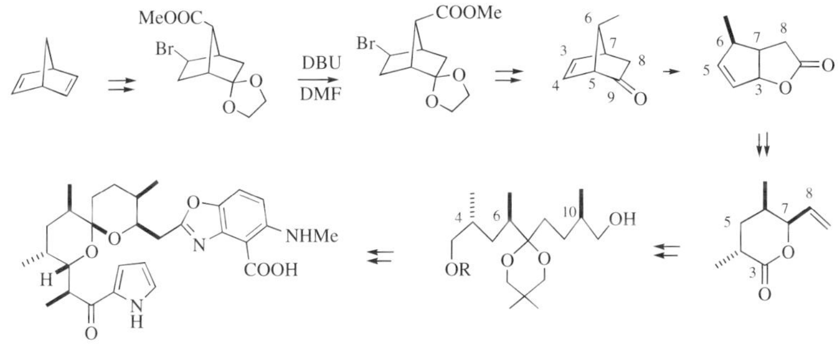

calcimycin (**9-18**)

an *endo*-oriented subangular methyl group from which a chiral chain segment would be developed. The origin of this methyl group is an epimeric ester. It seems that steric repulsion favored epimerization. Importantly, the subsequent dehydrobromination was greatly facilitated in the epimerized product.

The synthesis of pederin usually combines two building blocks which are conveniently pursued individually and simultaneously. A route developed for a precursor of (+)-pedamide[18] [part of pederin (**9-19**)] involving conjugate addition of a substituted nitromethane to an unsaturated δ-lactone passed through a furolactone with two *endo*-methyl groups. Acid-catalyzed isomerization of the unstable furolactone was promptly carried out to avoid inconvenience and complications in subsequent steps, although the configuration of the latent aldehyde was inconsequential as the stereocenter would be destroyed later.

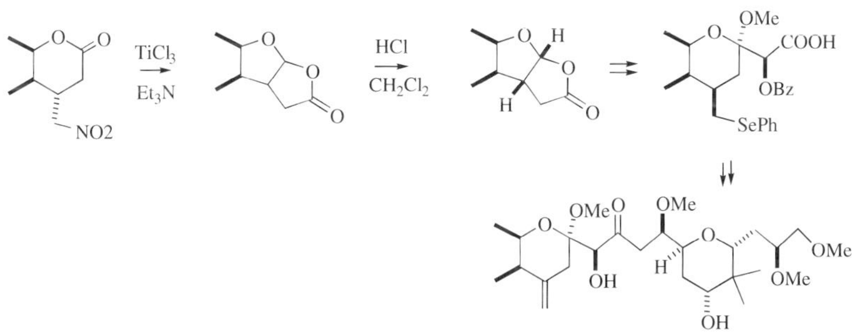

(+)-pederin (**9-19**)

Phomactin-D (**9-20**) is a macrocyclic ketone bridged with a cyclohexane unit in which four out of five substituents are equatorial. It is therefore advantageous to pursue synthesis of this compound by constructing a cyclohexane with stereochemically pliable functionalities. Considering the Diels–Alder reaction as the tool to provide a bridged ring intermediate for degradative modification, the dialdehyde generated by cleavage of one branch would be required to undergo epimerization at one center. The actual execution of such a scheme[19] involved a reflexive Michael reaction which often provides an equivalent of the Diels–Alder reaction. The presence of an ester in the product to capture one of the pendants released from ring cleavage of the bicyclo[2.2.2]octane skeleton rendered the required functional differentiation quite easy.

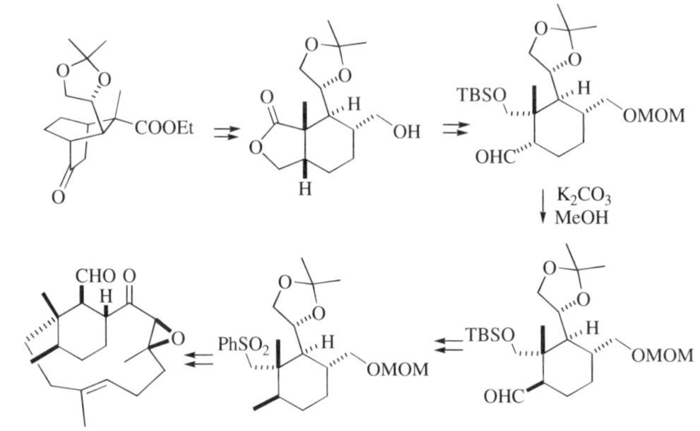

phomactin-D (**9-20**)

In the first synthesis of gibberellic acid (**9-21**),[20] the A-ring construction was delegated to an intramolecular Diels–Alder reaction. The crucial stereochemical feature that could not allow error in its establishment was the configuration at $C_{(5)}$. Thus, the dienophile had to be delivered to the α-face. The point of attachment of this dienophilic chain is $C_{(6)}$ which bears a β-oriented carboxyl group in the target molecule; the temporarily inverted configuration in the precursors was not a matter of great concern. Previous knowledge of the thermodynamic stability of gibberellic acid and derivatives at that position was a reassurance. Indeed, that stereochemical aspect was never at issue during the synthesis.

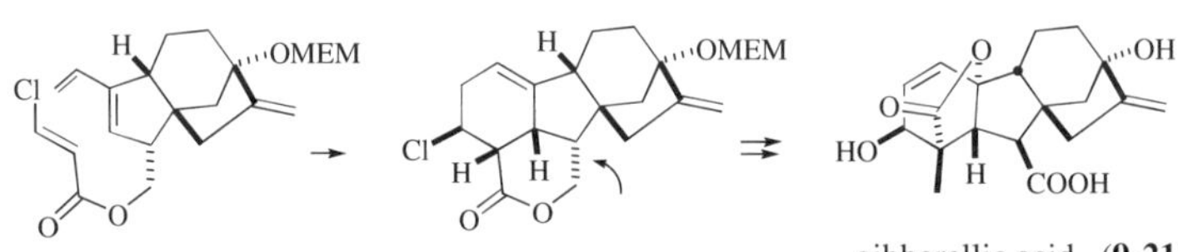

gibberellic acid (**9-21**)

The access to 11-ketosteroids (**9-22**) is of perennial interest to synthetic chemists. Novel concepts and/or methodologies that eventuate in a solution of this problem have always had enormous impact on the development of synthetic organic chemistry in general. What special structural features in 11-ketosteroids are synthetically exploitable?

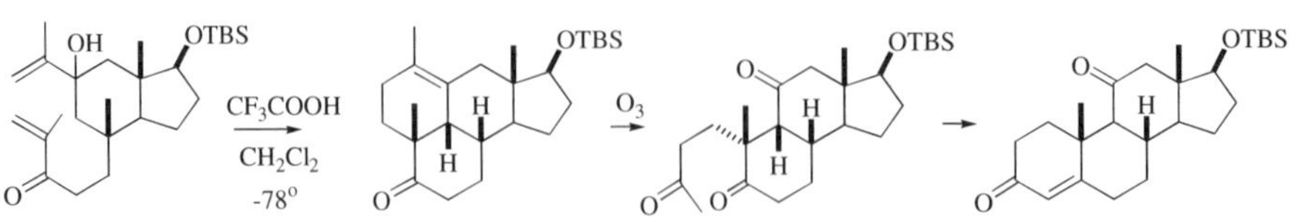

(**9-22**)

The two oxygenated sites at $C_{(3)}$ and $C_{(11)}$ are 1,6-related, but these carbon atoms cannot form a six-membered ring because of steric factors. What advantages are there in having such a ring in a synthetic precursor? This question is moot if the precursor is incapable of existence or unattainable, and therefore one must first find a modified structure in which the imposed steric constraints are removed. An A-seco triketone fits the requirement perfectly.

The most important revelation on deoxygenative association of $C_{(3)}$ and $C_{(11)}$ creates a cyclohexene subunit which is immediately suggestive of a Diels–Alder approach.[21] Based on this conceptual insight further analysis is relatively easy and the prospect for assembly of the intermediate is excellent. On introduction of the diene and dienophile components to a steroid CD-ring synthon, the intramolecular Diels–Alder reaction is sure to follow.

The useful tetracyclic intermediate is an *endo*-adduct; it was actually obtained as the major product (94:6 diastereomeric mixture) from a reaction at –78°C. With improvement of substrate preparation, the process is very efficient. It should be noted that the configuration at $C_{(9)}$ was singularly errant, but on cleavage of the double bond, it became epimerizable. The conditions for aldol condensation to close the A-ring were sufficient to amend the stereochemical malfeasance.

It has been demonstrated repeatedly that the Diels–Alder reaction is very beneficial to the preparation of intermediates for synthesis of Amaryllidaceae alkaloids. For example, in an approach to lycoricidine (**9-23**)[22] ethyl acrylate was used as dienophile to condense with a 1-arylbutadiene (generated by dehydration in situ). The regioselectivity was perfect but the *cis* disposition of the aryl and the ester groups was unsuitable. The aberrant relative configuration was easily corrected because the *trans*-diequatorial isomer was more stable.

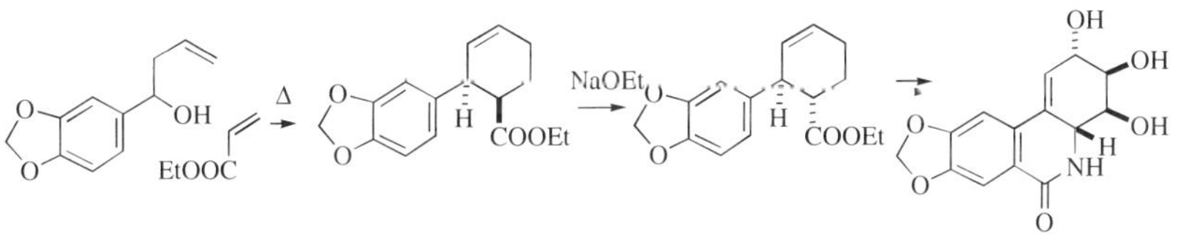

lycoricidine (**9-23**)

It is apparent that both clividine (**9-24**) and clivonine (**9-25**) can be elaborated from a common intermediate by *cis*-dihydroxylation of the cyclohexene moiety from one side or the other.[23] A tetrahydrophthalic anhydride presumably obtainable from a Diels-Alder reaction is well suited for functional differentiation because the bulky aryl ring tends to hinder reagents from approaching the nearby carbonyl group. Thus, methanolysis gave predominantly one monoacid which allowed introduction of an amino group in the desired position (with retention of configuration) by a Curtius rearrangement. Closure of the *cis*-hydroindole ring then involved saponification of the ester followed by an Arndt–Eistert homologation.

The caveat of the scheme is that the synthesis enlisted a tetrahydrophthalic anhydride (**9-26**) which is diastereomeric to the Diels–Alder adduct of the 1-arylbutadiene and maleic anhydride. In other words, the important intermediate was

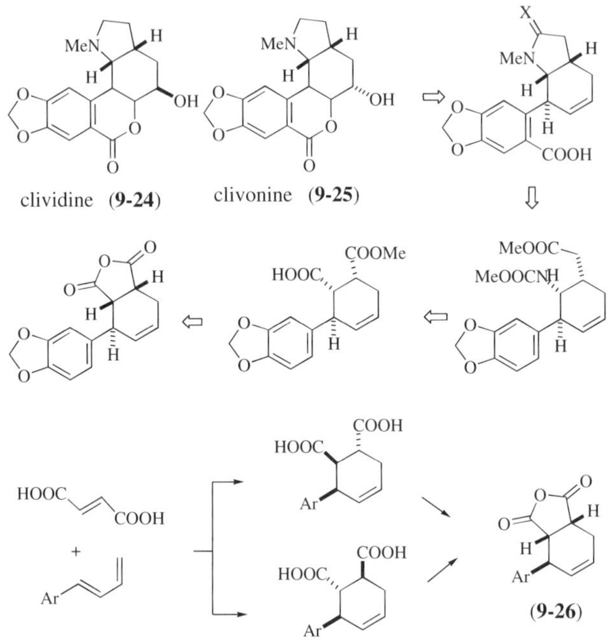

not available directly. Fortunately, when the Diels–Alder adducts derived from fumaric acid instead of maleic anhydride were subjected to equilibration, the desired anhydride was obtained. Note that in this compound the aryl group occupied a more stable *exo* position.

The eremophilane-type sesquiterpenes feature a *cis-vic*-dimethyl pattern at an angular and adjacent sites. This substitution pattern is more stable than the *trans*-dimethyl isomer when incorporated in a *trans*-decalin. Thus, it is possible to acquire eremophilanes by means of equilibration, provided that a ketone is left on the carbon atom neighboring the subangular position. For compounds such as isopetasol (**9-27**),[24] the situation is particularly conducive to their synthesis in view of the presence of a hydroxyl group in that locality. Interestingly, the same tactic can be applied to certain *cis*-β-decalones, as shown in a synthesis of furanoeremophilane (**9-28**).[25]

Stereochemical adjustments made during a synthesis of occidentalol[26] from (+)-carvone depended on the relationship between intermediates and the target molecule. The homonuclear diene ester having a *trans* ring junction underwent photoinduced conrotatory opening at low temperature to an (*E,Z,E*)-1,3,5-cyclodecatriene which reclosed at 20°C to afford two diastereomers because

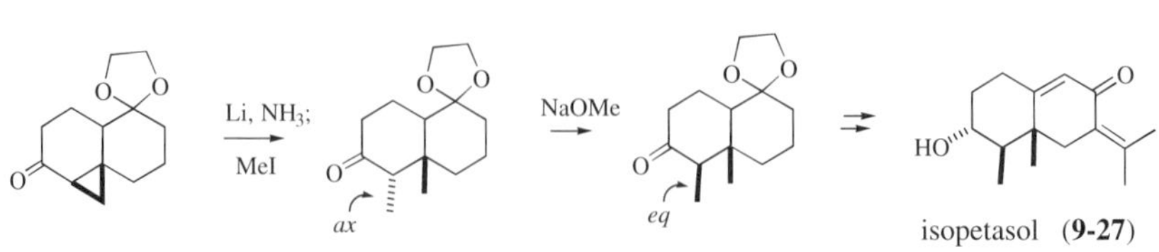

isopetasol (**9-27**)

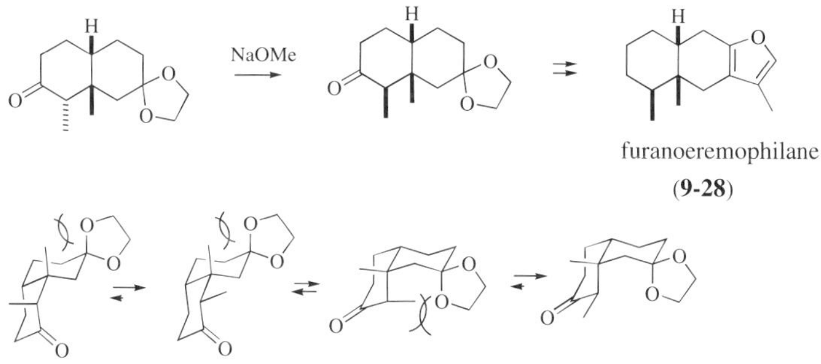

furanoeremophilane
(**9-28**)

disrotatory electrocyclization could proceed in two different directions. While one of the products could be tranformed to (+)-occidentalol (**9-29**) on reaction with methyllithium, the other diastereomer yielded 7-epi-(–)-occidentalol. The latter diastereomer could also be converted to (–)-occidentalol (**9-30**) by treatment with potassium *t*-butoxide prior to the reaction with methyllithium. Thus, it seems that the ester enolate adopted a nonsteroid conformation in which the angular methyl group was equatorial. Protonation of the enolate ion from the convex face led to the thermodynamically stable equatorial ester.

The all-*cis* stereochemistry in an octalincarboxylic ester inherited from an intramolecular Diels–Alder reaction of ethyl 9,10-dimethyl-6-oxo-2,4,10-undecatrienoate is alterable.[27] The β-ester epimer is a known precursor of valencene and eremophilene. Interestingly, the initial Diels–Alder adduct possessing a *trans* ring fusion underwent epimerization in situ. This did not affect transformations planned for the product, such as its use in a synthesis of eremophilone (**9-31**).

The $C_{(1)}/C_{(12)}$ segment (**9-32**) of amphotericin-B, which embodies five of the nine stereocenters in the aglycone portion, has been the focus of many synthetic efforts. A construction scheme based on $C_{(6)}/C_{(7)}$ bond formation is deemed reliable

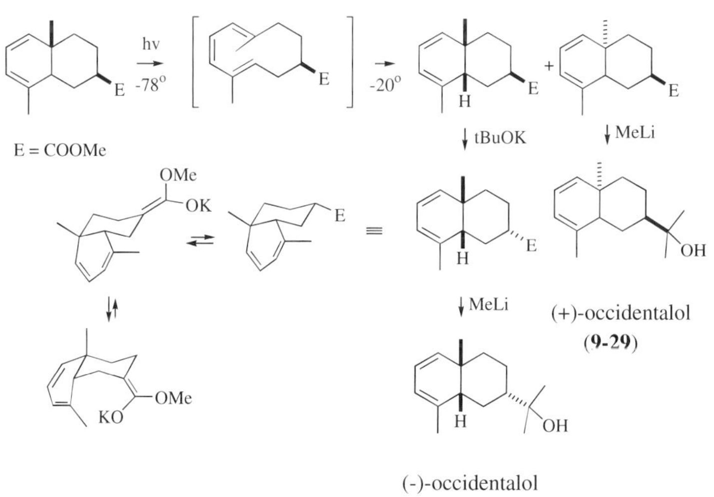

(+)-occidentalol
(**9-29**)

(-)-occidentalol
(**9-30**)

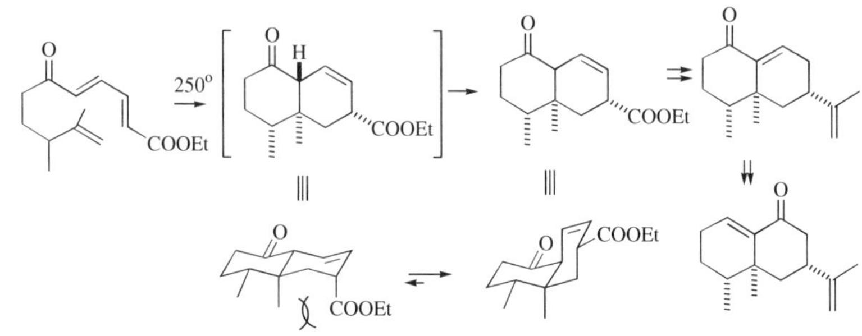

eremophilone
(**9-31**)

as it does not involve the creation of new stereocenters. The proper $C_{(1)}/C_{(6)}$ and $C_{(7)}/C_{(12)}$ components would have $C_{(6)}$ bearing a leaving group and $C_{(7)}$ activated to play the role of a nucleophile. The two components were separately derived from a trihydroxypentanal.[28]

Although several methods are conceivable for the preparation of this building block, a chiral glycidyl ether is an obvious choice as starting material in view of its availability and enantiopurity. The one-carbon homologation protocol achieved by reaction with 2-lithio-1,3-dithiane, after protection of the newly generated hydroxyl group, can be repeated after hydrolysis of the heterocycle. Acetonide formation from the diols and equilibration of the released aldehyde rectified any stereochemical impropriety of the intermediates. In a 4,6-disubstituted 2,2-dimethyl-1,3-dioxane, the two secondary substituents prefer a *cis*-1,3-diequatorial orientation, and the actual operation involved merely treatment of the aldehydes with potassium carbonate in methanol.

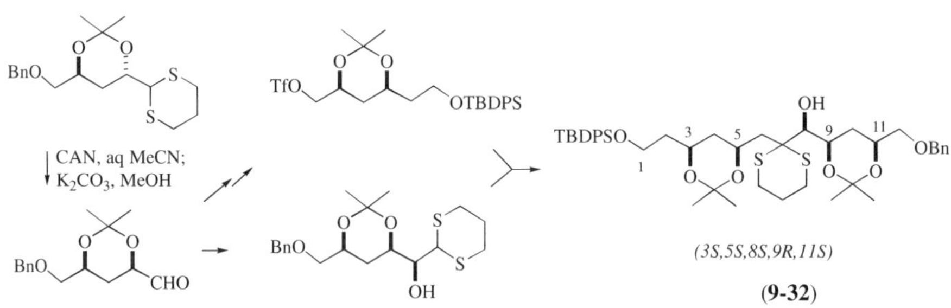

(**9-32**)

The intramolecular magnesio-ene reaction can achieve cyclization with simultaneous functional group transfer to a remote site. With mechanistically

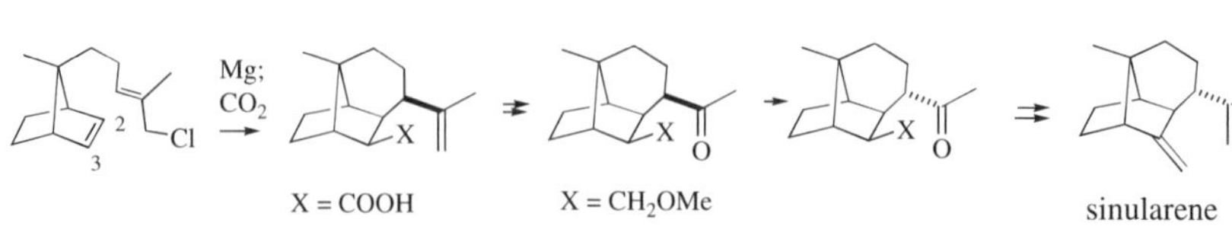

sinularene
(**9-33**)

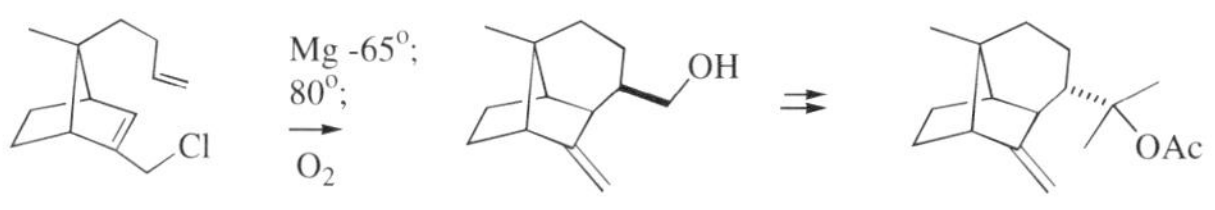

12-acetoxysinularene
(**9-34**)

predictable stereoselectivity the magnesio-ene reaction served as a key step in an approach to sinularene (**9-33**).[29] A six-membered ring was closed by enlisting the *syn*-7 functionalized side chain of a norbornene to unite with the ene portion from the *exo* face, while forming a Grignard reagent at the adjacent carbon atom ($C_{(3)}$ of the norbornane skeleton) for carboxylation. The cyclic transition state of the magnesio-ene reaction resulted in the undesired orientation of the isopropenyl group so that hydrogenation would give an epimer of the proper sinularene precursor. The problem was corrected by oxidative cleavage of the double bond to afford an epimerizable methyl ketone. The isopropyl group was then rebuilt after the epimerization.

An analogous protocol was implemented in the synthesis of 12-acetoxysinularene (**9-34**)[30] designed on a different set of ene and enophile components.

Pseudopterosin-A (**9-35**) and pseudopterosin-E (**9-36**) are tricyclic diterpenes possessing remarkable anti-inflammatory activity. With (+)-menthol identified as starting material for synthesis,[31] $C_{(8)}$ must undergo desymmetrization while functionalizing $C_{(9)}$ for an annulation process.

Activation of $C_{(9)}$ by the Barton reaction gave an oxime. The desired configuration was affixed in the γ-lactone intermediate through enolization and quenching the lithium enolate with ammonium chloride at 0°C.

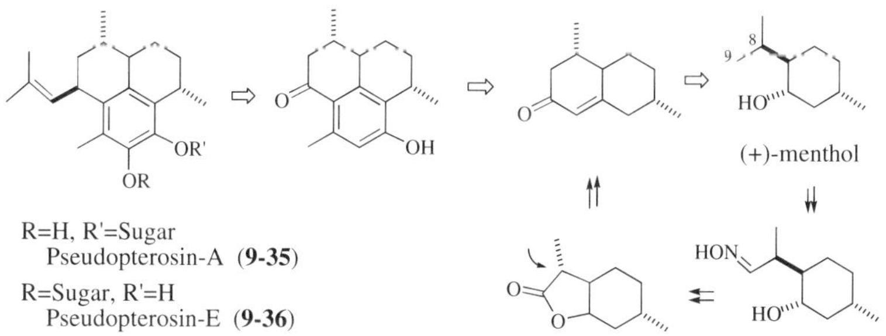

R=H, R'=Sugar
Pseudopterosin-A (**9-35**)
R=Sugar, R'=H
Pseudopterosin-E (**9-36**)

The ready formation of bridged ring systems by tandem intramolecular Michael and aldol reactions was the key to a design for synthesis of picrotoxinin (**9-37**) and coriamyrtin (**9-38**).[32] The rigidity of bridged frameworks and exploitable nonbonding interactions ensured steric adjustments in a predictable manner. Thus, a comparison of a tricyclic intermediate with the target molecule(s) indicated modifications involving C–C bond cleavage and epimerization of two groups.

To force the methyl ester to adopt an *exo* orientation, the bicyclic ketone was acetalized to increase steric compression in the *endo* domain. Cleavage of the functionalized ethano bridge was achieved by α-hydroxylation of the corresponding

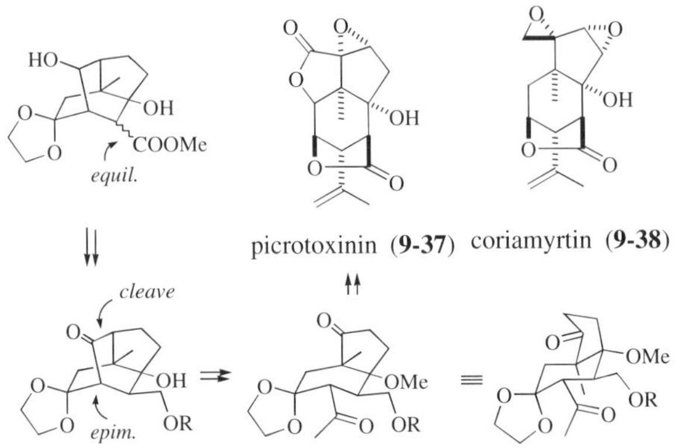

picrotoxinin (**9-37**) coriamyrtin (**9-38**)

ketone, reaction with methyllithium, and treatment with lead tetraacetate. Epimerization of the acetyl group from the concave side to the convex face on its release was also induced by a *cis*-substituent at the adjacent carbon atom.

The major difference between two major classes of pseudoguaianolide sesquiterpenes is in the configuration of $C_{(10)}$ (β-methyl: ambrosanolides; α-methyl: helenanolides). Synthetically, it is always desirable to use a common intermediate to reach both series, and one of the most obvious approaches is to invert the pertinent stereocenter of a certain intermediate. Isomerization of a 10β-methyl-7-ketoperhydroazulene, acquired in a synthesis of damsinic acid (**9-39**),[33] to the 10α-methyl epimer was performed after inserting a double bond between $C_{(8)}$ and $C_{(9)}$. The product was then processed into aromaticin (**9-40**).[34] Actually, a phenylthio group was also introduced into the unsaturated intermediate, but it was "accidental" and not a necessity, although it did not affect the outcome of the proposed epimerization.

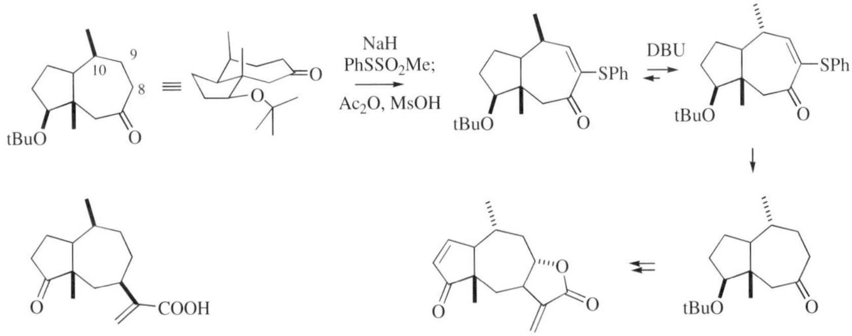

damsinic acid (**9-39**)

aromaticin (**9-40**)

Inversion of the configuration at $C_{(8)}$ was a necessity in the intermediate (**9-41**) en route to conessine.[35] It was a compromise for the stereoselectivity at establishment of the correct configuration of $C_{(19)}$ in a previous step. Creation of an enone was required for reaching the ring junction.

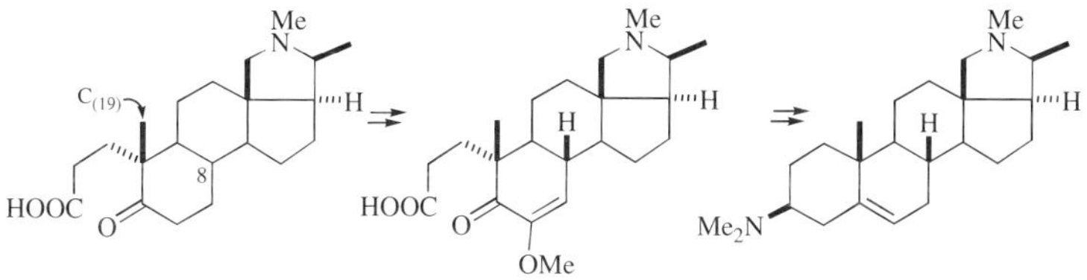

(9-41)

conessine

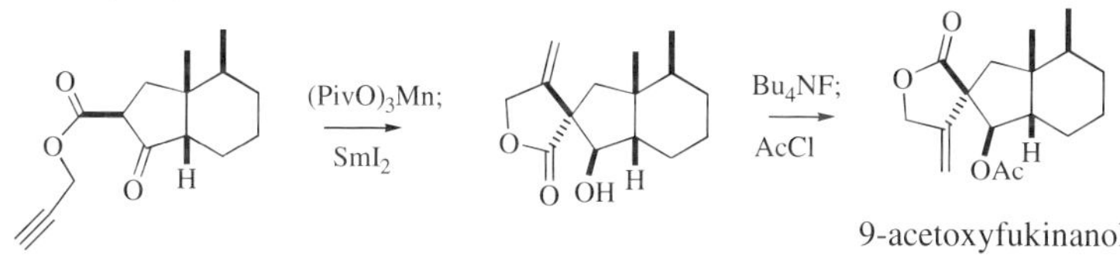

9-acetoxyfukinanolide

(9-42)

The intramolecular oxidative cyclization of a propargyl ester occurred from the convex face of a hydrindan system.[36] This result would have been disastrous to a synthesis of 9-acetoxyfukinolide (**9-42**) were it not amended by a fortuitous retroaldol–aldol reaction sequence mediated by fluoride ion.

A reversible Michael reaction served to equilibrate a tricyclic ketone during a synthesis of (–)-norsecurinine (**9-43**),[37] thereby rendering it much more efficient. A mechanistically akin process rehabilitated the minor and otherwise useless adduct of a hetero-Diels–Alder reaction which was designed to pivot a new route to cephalotaxine (**9-44**).[38]

For a synthesis of luciduline (**9-45**),[39] one of the critical aspects was to acquire a *cis*-aminooctalin. Lewis acid catalyzed Diels–Alder reaction between butadiene and 5-methyl-2-cyclohexenone gave mainly a *trans*-octalone due to epimerization of the *cis* adduct in situ. Fortunately, the *cis*-oxime could be obtained by reaction of the *trans*-ketone with hydroxylamine under equilibration conditions.

A method for the synthesis of epibatidine (**9-46**)[40] according to a retrosynthetic analysis shown below has, a priori, a pitfall in that the hydrogenation of the 7-azabicyclo[2.2.1]hept-2-ene would occur from the *exo*-side, leading to a product possessing an *endo*-substituent. The synthetic scheme must rely on epimerization of

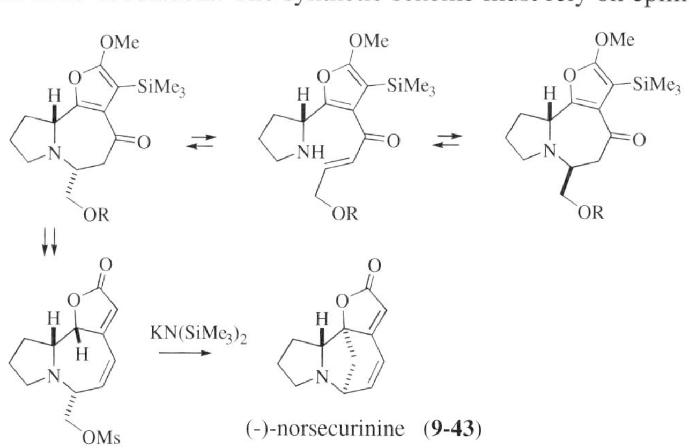

(-)-norsecurinine (**9-43**)

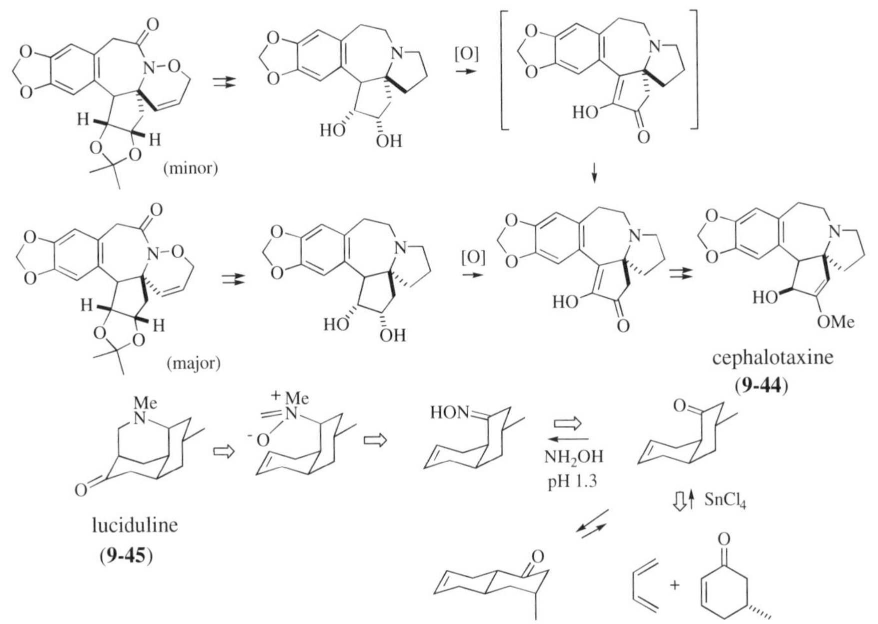

cephalotaxine (**9-44**)

luciduline (**9-45**)

the latter compound to become viable. The high acidity of the picolyl proton contributed to the successful conclusion of this synthetic endeavor. The crucial transformation was accomplished by heating the *endo* compound with potassium *t*-butoxide.

An early synthesis of dendrobine (**9-47**)[41] featured the formation of a hydrindan ester by an intramolecular Diels–Alder reaction. The major *endo* adduct was converted into a nitrile by homologation at the cyclopentanone moiety as a prelude to

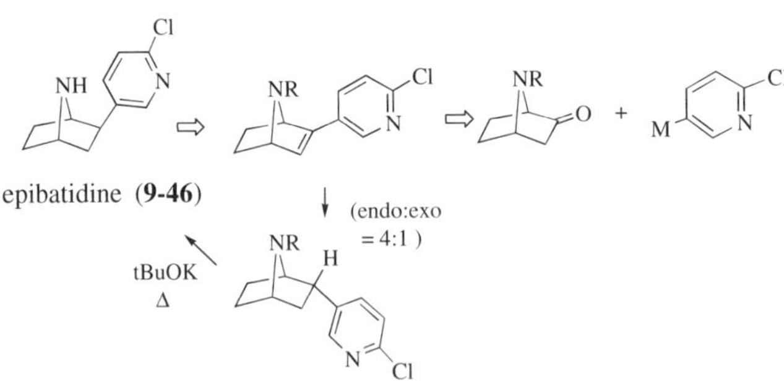

epibatidine (**9-46**)

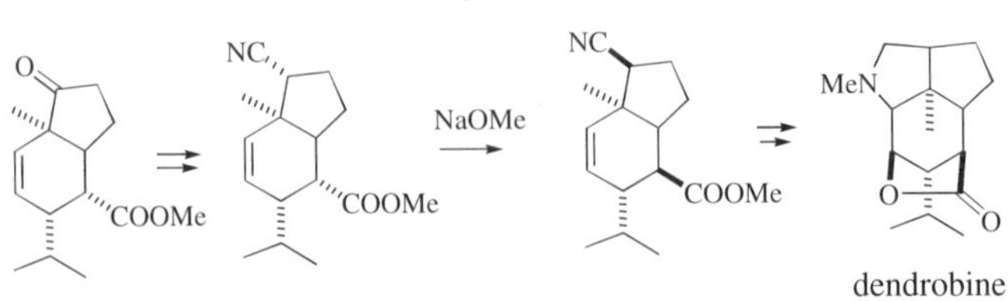

dendrobine (**9-47**)

the formation of the pyrrolidine ring. Treatment of the cyano ester brought about epimerization at two sites, thereby establishing two more stereocenters corresponding to those in dendrobine.

An access to (*R*)-1,7-dioxaspiro[5.5]undecane (**9-48**) initiated by cyclization of a chiral sulfinyldihydropyran occurred under kinetic control, apparently involving a metal chelate and the equatorial protonation of the α-sulfinyl carbanion. The sulfinyl spiroacetal could be isomerized by treatment with TsOH in methanol to gain entry into the (*S*)-series (**9-49**).[42] The process was thermodynamically driven as the generation of a monocyclic oxonium intermediate permitted a conformational change such that spiroacetal reclosure did give an isomer with an equatorial sulfinyl substituent.

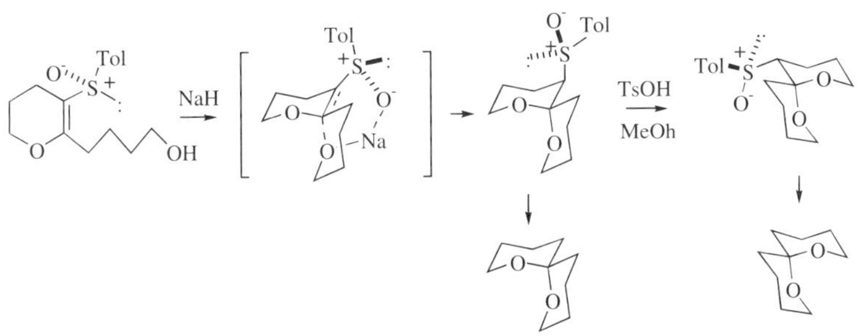

(**9-48**) (*R*)- (**9-49**) (*S*)-

During a synthesis of reserpine[43] the reduction after Bischler–Napieralski cyclization led to the pentacycle (**9-50**) having a 3α-H. The product is stable because all three substituents in the E-ring occupy equatorial positions, and with a *trans*-quinolizidine system for the CD-ring segment, the sterically demanding indole ring does not interact unfavorably with other bulky groups. In considering the structure (**9-50**), the prospect of achieving counterthermodynamic inversion of the configuration of $C_{(3)}$ was not encouraging. However, an ingenious plan to rescue the brilliant effort consisted of lactonization across the E-ring, involving substituents at $C_{(16)}$ and $C_{(18)}$, to lock all of the E-ring substituents in the axial positions, and also changed the 3α-H into an equatorial position in the quinolizidine system which must

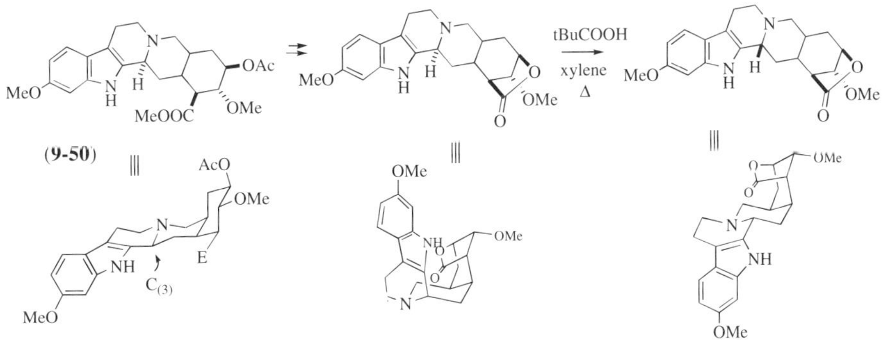

now be *cis* fused. Consequently, the indole nucleus was forced to assume an axial orientation and crash into the lactone carbonyl group if each of the C-, D-, and E-rings were to maintain a chair conformation. Under such conditions the molecule was liable to seek any relief and heating the lactone with pivalic acid in refluxing xylene gave rise to the 3β-H isomer. Essentially quantitative isomerization to the reserpic acid lactone was observed.

Spontaneous deracemization of a tricyclic ketone by heating with (–)-10-camphorsulfonic acid in ethyl acetate greatly facilitated a synthetic approach to chiral emetine (**9-51**).[44] While the salt of (–)-ketone deposited from the solution, the (+)-isomer underwent racemization through a retro-Mannich reaction. Thus, inversion of two stereocenters was involved.

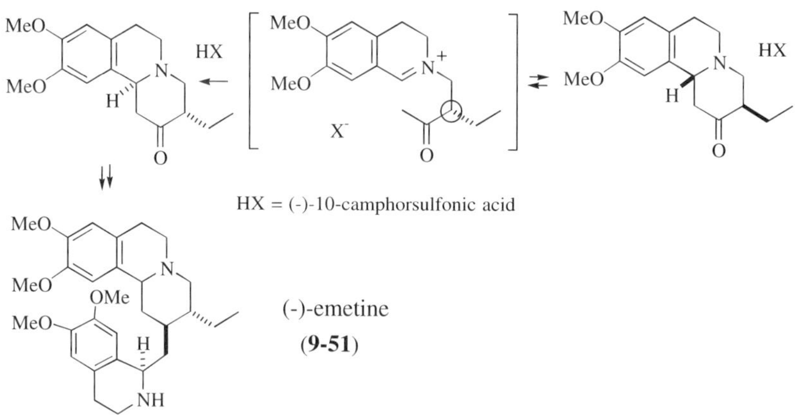

(-)-emetine
(**9-51**)

A vincamine (**9-52**) synthesis[45] consisted of preparation of a tetracyclic aldehyde intermediate. The equimolar mixture of *cis*- and *trans*-isomers was produced, but the undesirable *trans*-racemate could be converted into the *cis*-racemate by heating its *p*-toluenesulfonic acid salt in dioxane, the salt of the *cis*-aldehyde being more soluble.

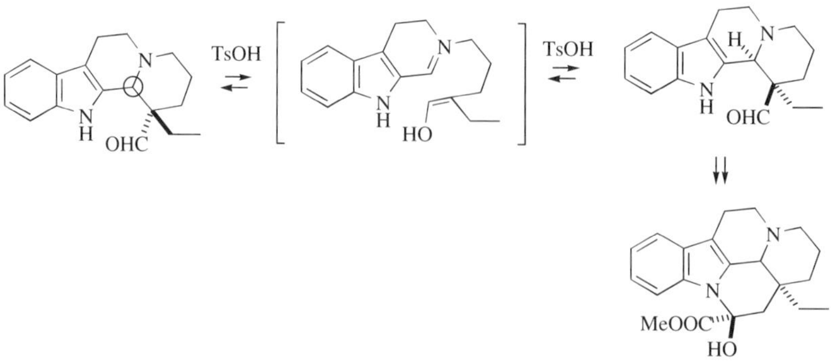

vincamine (**9-52**)

9.2. CONFIGURATION INVERSION BY OTHER METHODS

It is common practice to invert secondary alcohols by the Mitsunobu reaction. Except for one example, this and other typical S_N2 reactions will not be examined here because of their conceptual simplicity. The example concerns a synthetic approach to the tetronic acid fragment (**9-53**) of kijanolide and chlorothricolide.[46]

A promising plan would be that involving a Diels–Alder reaction because the cyclohexene subunit contains *cis* substituents at $C_{(3)}$ and $C_{(6)}$ which are readily preinstalled in the diene component. On the other hand, the choice of the dienophile is not that apparent owing to the fact that an α-oxygenated acrolein or acrylic acid derivative is not applicable if the Diels–Alder reaction goes through an *endo* transition state. A satisfying solution involved the use of α-bromoacrolein and displacement of the bromine atom in the cycloadduct by an oxynucleophile with inversion of configuration. An S_N2 reaction was precluded at the quaternary stereocenter except for an intramolecular version which traversed an epoxide intermediate. The addition of the Corey oxazaborolidine template/catalyst enforced an *endo* transition state for the Diels–Alder reaction.

These key steps were patterned after those delineated in a synthesis of fumagillin (**9-54**)[47] which has a spiroepoxide moiety.

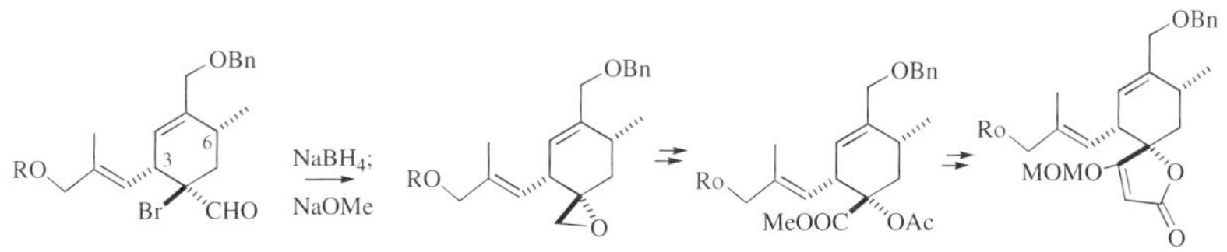

(**9-53**)

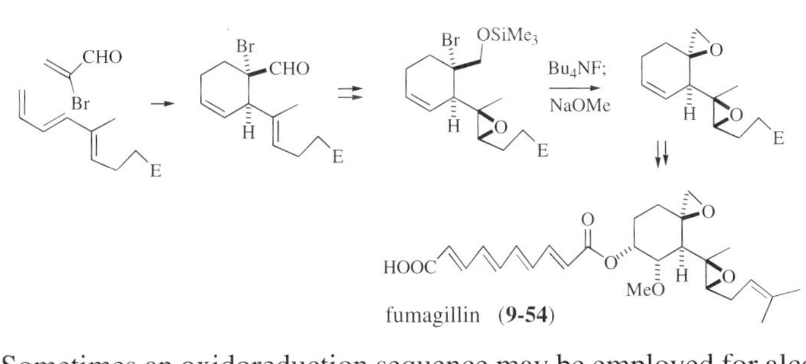

fumagillin (**9-54**)

Sometimes an oxidoreduction sequence may be employed for alcohol inversion. It is applicable to cases where an epimeric but separable alcohol mixture is formed in a reaction as well as when a single but undesirable diastereomer is generated. Such a tactic is important in a synthesis of slaframine (**9-55**).[48]

A key step in a synthesis of veatchine (**9-56**)[49] was the [2+2]-photocycloaddition of allene to a tetracyclic enone, placing the small addend on the same side as the piperidine ring onto the hydrophenanthrene framework. This stereochemical outcome required inversion of the two-carbon side chain derived from oxidative degradation of the four-membered ring before formation of the bicyclo[3.2.1]octane unit. This

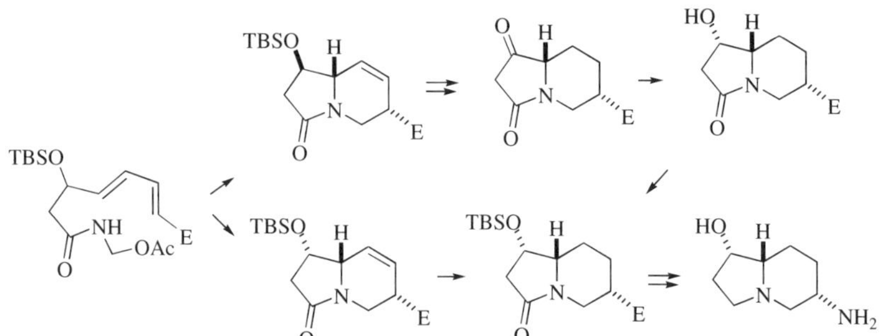

slaframine
(**9-55**)

intriguing transformation was achieved by a Wolff rearrangement which effected a formal insertion of a carbene to the ketone.

The generation of a diastereomer of isovalleral (**9-57**)[50] in which the cyclopropane is *trans* to the methine hydrogen atoms at the hydrindan ring juncture was not a disaster with regard to a synthesis of the natural terpene molecule. On heating in mesitylene a thermal equilibrium occurred and half of the material was converted into isovalleral through a retroene–ene reaction sequence.

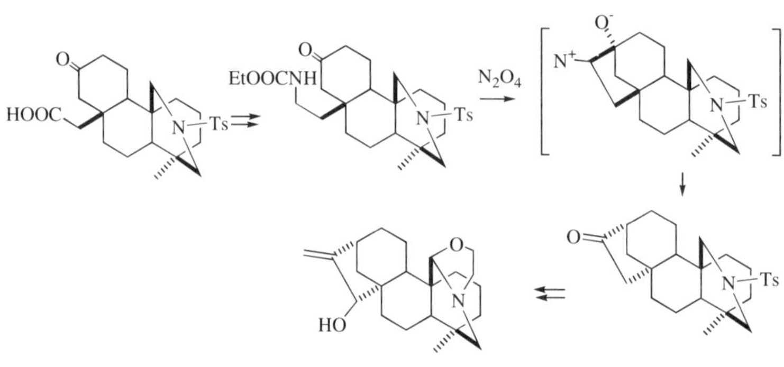

veatchine (**9-56**)

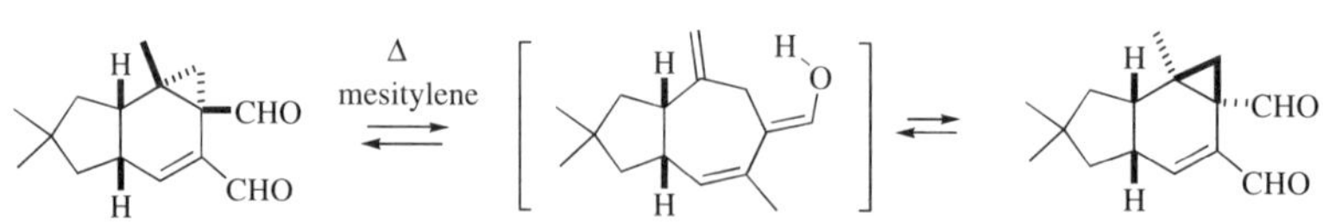

(+)-isovalleral
(**9-57**)

9.3. KINETIC TRAPPING

It is always challenging to navigate through an unstable intermediate in a synthetic endeavor. Perhaps the most reliable way to achieve such a task is by trapping, particularly with a functional group in the same molecule.

The simplest form of kinetic trapping involves deprotonation–protonation of ketones and esters. To transform a readily available norbornenecarboxylic ester to a fused lactone in an approach to iridoids such as boschnialactone (**9-58**), iridomyrmecin, and isoiridomyrmecin,[51] epimerization of the ester via protonation of the enolate from the less hindered side was effected.

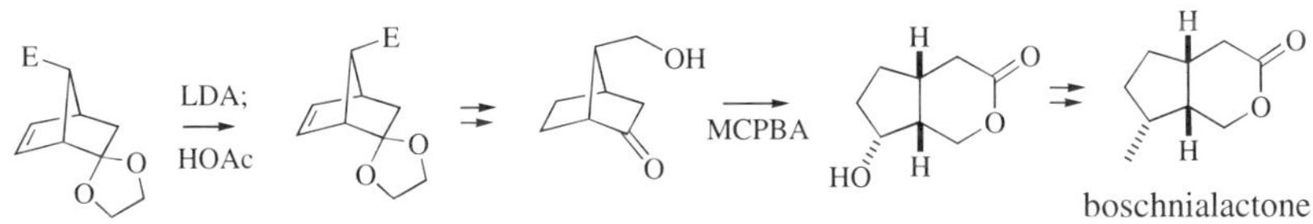

boschnialactone
(**9-58**)

Tetronomycin embodies a 1,2,3-trisubstituted cyclohexane unit in which the central side chain is branched. It has been demonstrated that such a synthon is accessible from a fused lactone, and the inappropriate orientation of the α-methylated product was corrected (to favor the desired *endo*-methyl isomer by 11:1) by a deprotonation–kinetic protonation maneuver.[52]

A method for the synthesis of the prostaglandin PGE_1 methyl ester[53] consisted of conjugate addition of a di-(*Z*)-alkenylcuprate reagent to a cyclopentenone. The product was transformed into the (*E*)-allylic alcohol with an (*S*)- configuration by thermal rearrangement of the derived *p*-toluenesulfenate to the 1,3-transposed sulfoxide and treatment of the latter species with trimethyl phosphite. Only the sulfenate in equilibrium with the sulfoxide was reactive toward the phosphite reagent and it suffered O–S bond cleavage. The most important aspect is that the [2,3]-sigmatropic rearrangement effected both inversion of a stereogenic center as well as the configuration of the double bond.

An oxy-Cope rearrangement approach to generate the tricyclic skeleton of pleuromutilin (**9-59**)[54] implies that the substrate must contain a vinyl substituent in an *exo* configuration with respect to the bicyclo[3.2.1]octane substructure. The preparation of this compound by a direct Grignard reaction was problematical because the tricyclic ketone favored equatorial attack. An inversion of configuration at the carbinyl center was required.

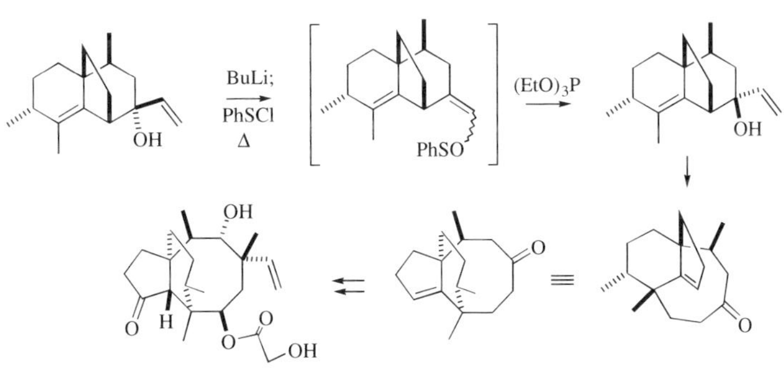

pleuromutilin (**9-59**)

The quaternary nature of the carbinyl center precluded equilibration processes in the classical sense. Yet failure to epimerize the alcohol would have condemned the overall scheme to oblivion. It is fortunate that the alcohol is allylic and subject to isomerization via the allyl sulfenate/allyl sulfoxide duad. The equilibrated sulfenates could be cleaved with triethyl phosphite. In the present case the undesired isomer was slightly favored (ratio 1.2:1) but it could be separated and recycled. This process at least made the best of an unfavorable situation.

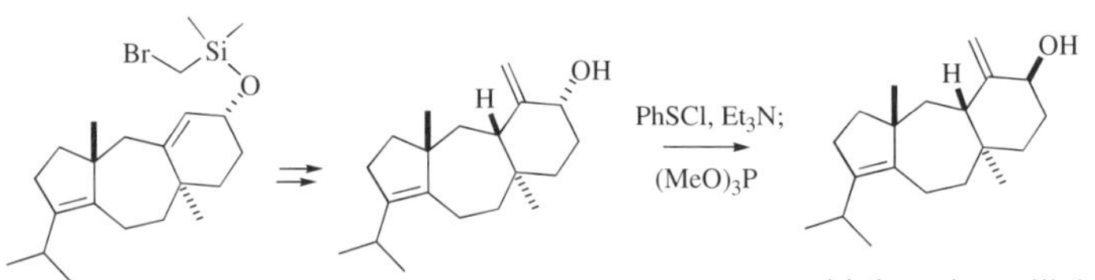

14-deoxyisoamijiol
(**9-60**)

The access to a tricyclic allylic alcohol related to isoamijiol prompted further conversion into a natural product. The effort which led to 14-deoxyisoamijiol (**9-60**)[55] involved sulfenate–sulfoxide equilibration. Apparently, the α-oriented sulfenate was less stable and present in small amounts, while most of the trappable sulfenate was the β-isomer. Hence, 14-deoxyisoamijiol was obtained in a 77% overall yield when the transposed allylic sulfoxide was treated with trimethyl phosphite. The Mitsunobu inversion[56] of the allylic alcohol gave a 30% yield only.

Very serious steric compression exists in annotinine (**9-61**) between the γ-lactone and the secondary methyl group, and it is therefore not surprising that in the open form the carboxyl group tends to epimerize. This stereochemical quandary during reconstitution of annotinine is fortunately not difficult to solve because lactone formation can be effected under equilibrating conditions.[57] The same tactic was

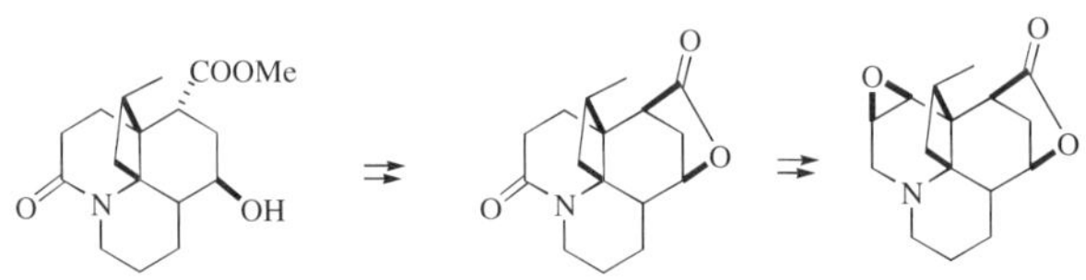

annotinine (**9-61**)

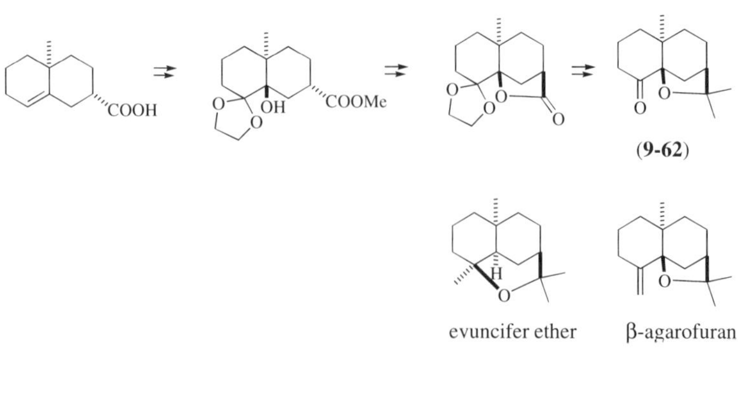

(**9-62**)

evuncifer ether β-agarofuran

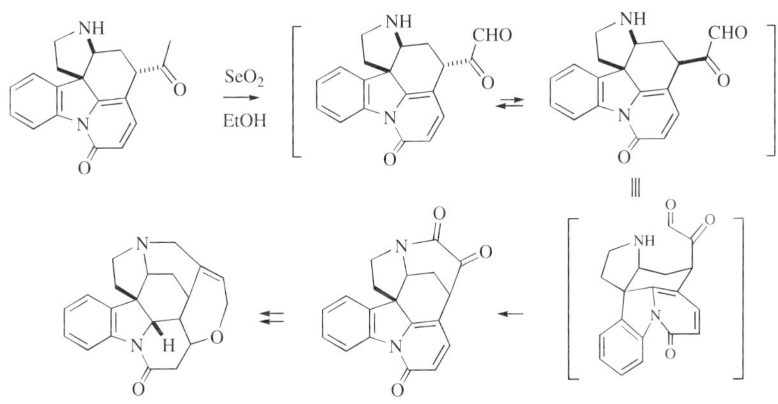

strychnine (**9-63**)

employed in a synthesis of the norketone precursor (**9-62**) of β-agarofuran and evuncifer ether.[58,59]

The bridging of the nitrogen atom to an acetyl side chain to form a new ring was a critical step in a synthesis of strychnine (**9-63**).[60] The acetyl group was pseudoequatorial and a configurational change must precede the desired transformation. The success of this reaction by selenium dioxide oxidation was due to the absence of excessive nonbonding interferences in the axial epimer, and only the amino group provided a 1,3-diaxial interaction. Perhaps the axial orientation of the acetyl or the glyoxal chain is not much more unfavorable because $A^{1,3}$-strain could be avoided. Another factor could be that once the carbinolamine was formed it would undergo rapid oxidation to give the α-ketolactam. The reverse process was suppressed.

It is instructive to compare the cyclization with the enforced abortion of a synthesis of akuammicine (**9-64**)[61] which was based on the cationic aza-Cope rearrangement/Mannich cyclization to construct the tetracyclic intermediate.

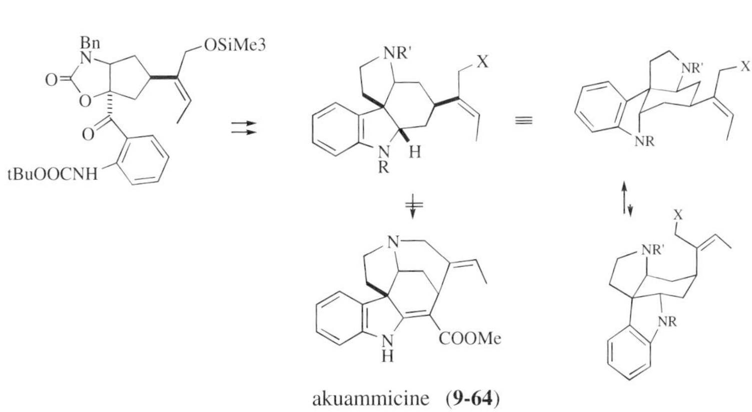

akuammicine (**9-64**)

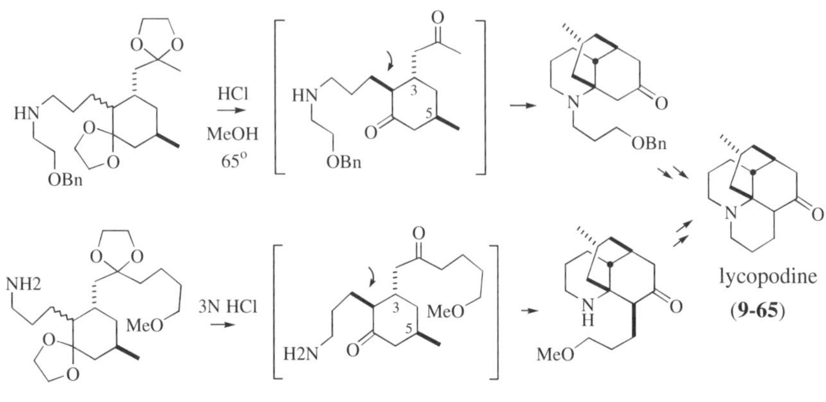

The benefit of stereochemical consequences from reaction under equilibration conditions was clearly evident in two versions of an approach to lycopodine (**9-65**).[62] During establishment of two stereogenic centers ($C_{(3)}$ and $C_{(5)}$) in a cyclohexanone, one of which induced the other by directing entry of an electrophile in a conjugate addition–trapping protocol. The third center which is α to the carbonyl group was subjected to equilibration, and when the tricyclic framework was created by an intramolecular Mannich reaction there was no doubt that the correct configuration would emerge.

A more detailed analysis of the synthesis plan is as follows. In the chair cyclohexanone, either the methyl group or the acetonyl unit can be equatorially oriented, but the acetonyl chain must be axial to participate in cyclization. Thus, of the two possible bicyclic iminium species, only the one formally derived from the *cis,trans*-2,3,5-trisubstituted cyclohexanone would be active toward cyclization. In other words, no other side products would be generated.

The interesting alkaloid porantherine (**9-66**) was the object of at least two syntheses. One approach[63] involved treatment of a tricyclic enamine with acid to bring about closure of the fourth ring. This cyclization step required isomerization and enolization of the acetyl chain to place it in an axial position to intercept the iminium ion.

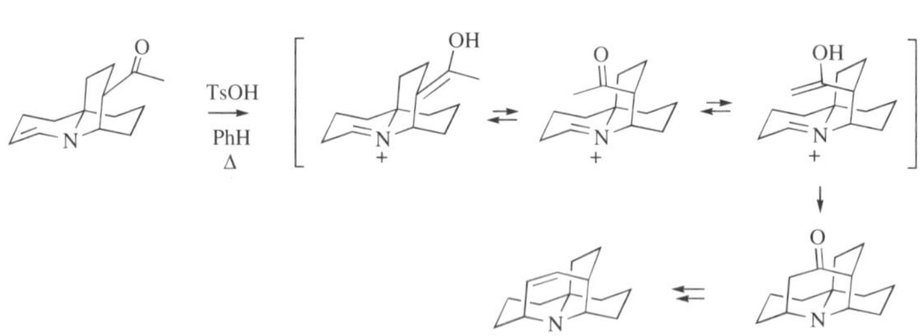

porantherine
(**9-66**)

A key step in a synthesis of longifolene (**9-67**)[64] was the formation of the bridged ring system by an intramolecular Michael reaction. Although the transformation proceeded in low yield, due to many available pathways for the product to undergo decomposition, there was no need for concern about the stereochemical propriety of the substrate. Whereas only the *cis*-homo-octalinone can cyclize, it was also accessible by equilibration.

From norcedrenedicarboxylic acid, the reconstitution of cedrol (**9-68**)[65] passed through a methyl ketone. An intramolecular Claisen condensation served to establish the tricyclic skeleton without complication, even though the methyl ketone chain is subject to epimerization. There is no doubt that kinetic trapping led to a stabilized β-diketone enolate, but the preference for the acetyl group to assume an *exo* configuration in a diquinane framework should not be ignored. Thus, in this case the cyclization substrate already existed in the most favorable situation.

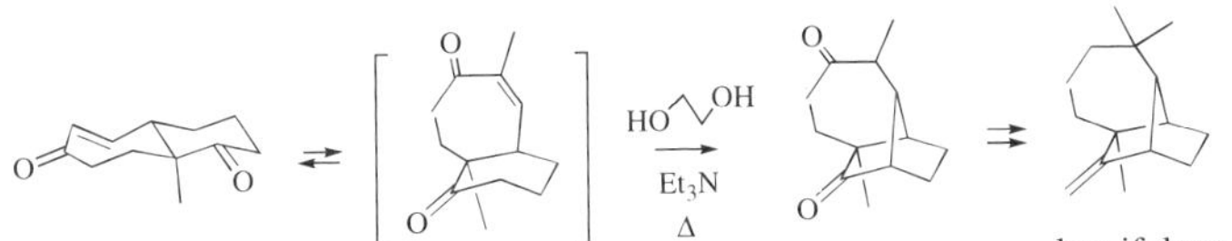

longifolene
(**9-67**)

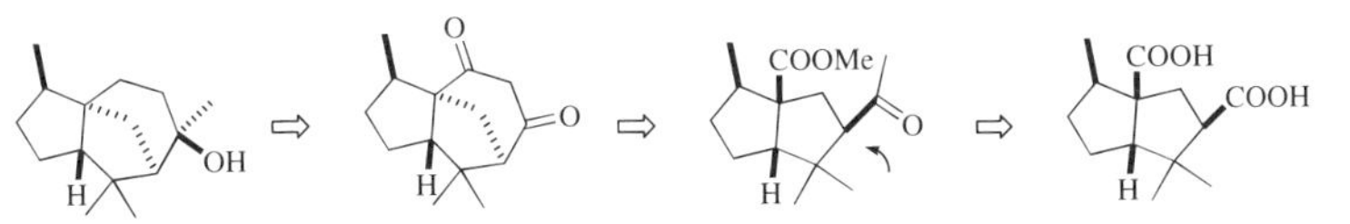

cedrol (**9-68**)

norcedrenedicarboxylic acid

Six stereocenters are present in ormosanine with one pair locked in a bridged system. In any synthetic effort, attention must be paid to the relative configuration of the bridgehead and the angular carbon atom of the quinolizidine moiety. Perhaps it is fortunate that a *trans*-quinolizidine is epimerizable through lone-pair electron inversion, and such a structural characteristic can be exploited.

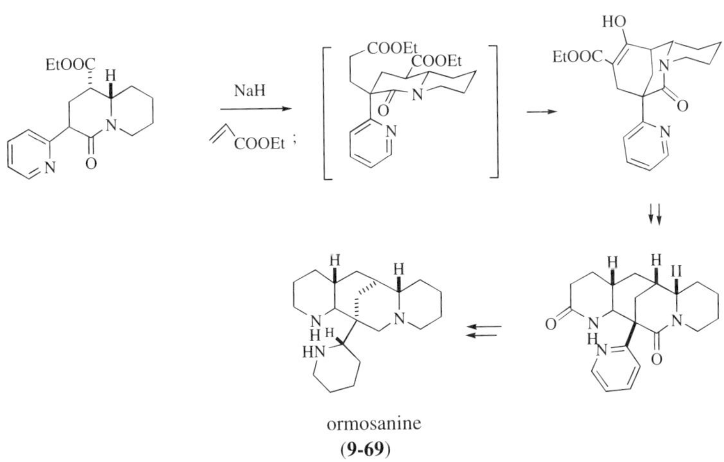

ormosanine
(**9-69**)

A synthetic route to ormosanine (**9-69**)[66] featuring construction of a bridged cyclohexanone by a Michael–Dieckmann reaction tandem relied on the first step to occur from the more exposed (equatorial) activated corner of an oxoquinolizidine. The steric constraint for the pimelic ester segment to undergo Dieckmann condensation is the equatorial disposition. Usually the six-membered ring containing part of the diester is required to adopt a boat conformation so as to bring the two reactive termini to within bonding distance. The situation is somewhat different in this instance as the piperidone portion is a flattened chair (lactam ring). Thus, the lone-pair electron inversion at the nitrogen atom does not convert it into a boat form. The alternative chair form is not severely destabilizing due to compensation of the bulky pyridine ring in becoming pseudoequatorial as the two other side chains are placed in pseudoaxial positions. More importantly, cyclization of the "active" conformer shifted the equilibrium and eventually rendered all of the diester into the desired product.

The Dieckmann condensation of a tetrahydro-β-carbolinedicarboxylic ester derived from (−)-tryptophan constituted a crucial step in a synthesis of koumine.[67] The α-ester group must undergo epimerization prior to cyclization. [The use of this major diester in the synthesis led to (+)-koumine (**9-70**) which is enantiomeric to the natural alkaloid.]

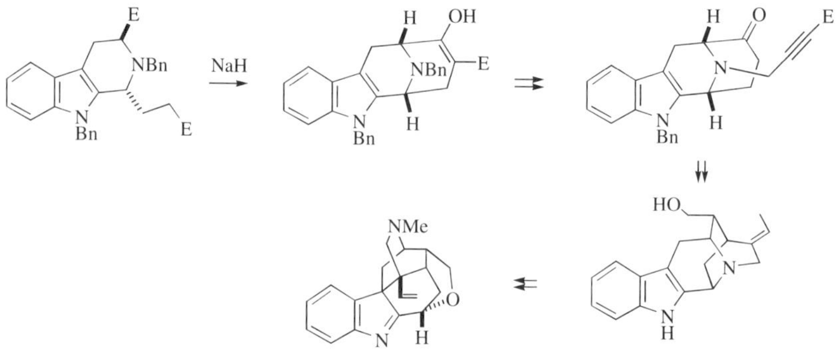

(+)-koumine (**9-70**)

A legitimate question concerning a stereoselective elaboration of β-vetivone (**9-71**)[68] is whether a precursor in which the secondary methyl group is delitescent in a functionalized form can be prepared and used to advantage for directing the installation of the isopropylidene side chain?

A positive answer emerged when a cyclohexenone containing a carboxyl group and a protected butanal chain was treated with acid. Intramolecular aldol condensation occurred and the secondary alcohol was trapped in situ as a γ-lactone. Equilibrium conditions ensured that the epimeric alcohol was channeled to the same product by way of a retroaldol fission to regenerate the monocarbocyclic ketone.

Structural correlation of the oxygenated position in the cyclopentane unit with that of the isopropylidene group in β-vetivone enabled the proper transformations to be carried out. In fact, after protection of the enone and reduction of the lactone to the diol, all the stereochemical issues were settled.

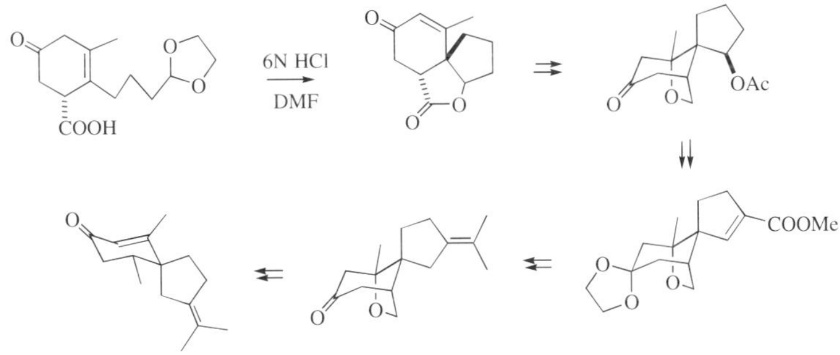

β-vetivone (**9-71**)

A diquinane containing vicinal vinyl and formyl groups in unspecified (mixture?) configurations was formed and further elaborated into hirsutene (**9-72**).[69] Advantage was taken of the self-adjusting stereochemistry in the intramolecular aldol condensation of the keto aldehyde for the closure of the third ring. Preference for an *exo* orientation in a diquinane skeleton by both functional chains is expected, any of the chains in an *endo* position would epimerize and participate in the cyclization.

By means of an intramolecular 1,3-dipolar cycloaddition of a derived nitrone, a useful intermediate (**9-73**) for the synthesis of hirsutene[70] was obtained. Notwithstanding a complete neglect of the stereochemical arrangement of the unsaturated ketone, this route automatically led to a *cis-anti-cis*-triquinane.

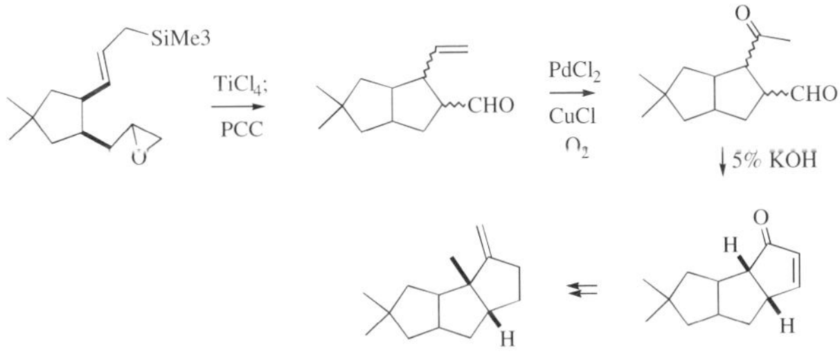

hirsutene (**9-72**)

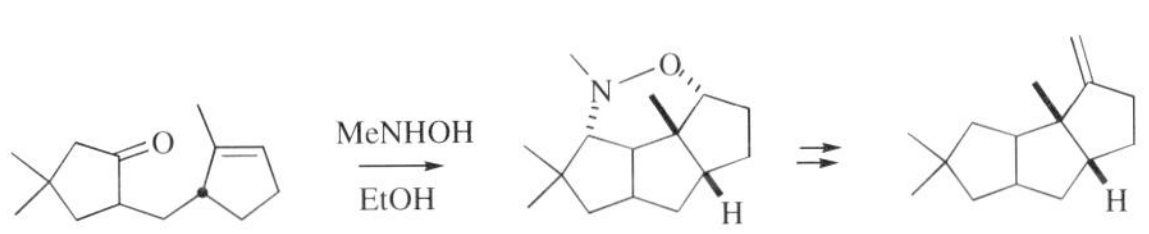

(**9-73**) hirsutene

Thermolysis of an *exo*-6-alkenylbicyclo[3.1.0]hex-2-ene successfully produced a precursor of sinularene (**9-74**).[71] Apparently, epimerization of the *exo*-substituent occurred (via a diradical?) to the *endo* isomer which then underwent a rapid Cope rearrangement. By this process the difficult task of synthesizing the *endo* isomer for the transformation was obviated.

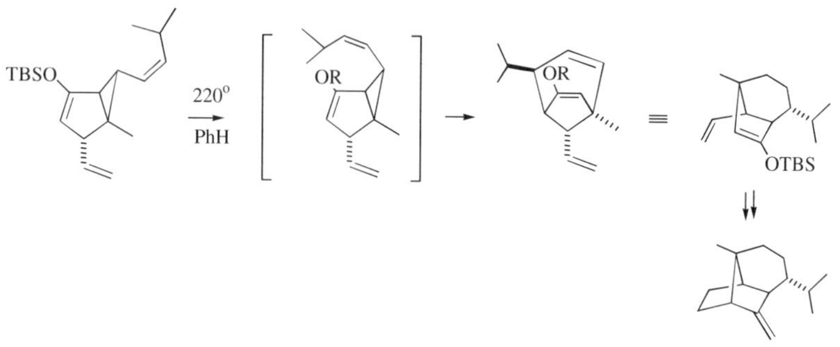

sinularene (**9-74**)

9.4. SELF-REGULATION OF CONFIGURATIONS

Certain reaction conditions are conducive to generation of the most stable products. Reduction of cyclic ketones with alkali metals in an alcoholic solvent or liquid ammonia is representative. The reductive cleavage of a cyano group in conjunction with α-alkylation of the nitrile constitutes a useful protocol for stereoselective C–C bond formation, as illustrated by the examples given below.

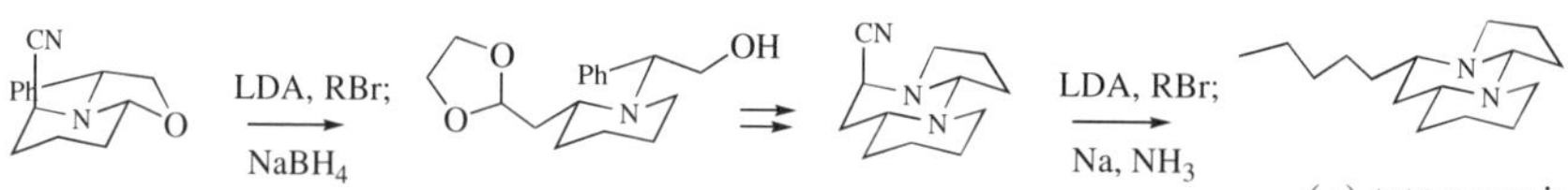

(+)-tetraponerine-8 (**9-75**)

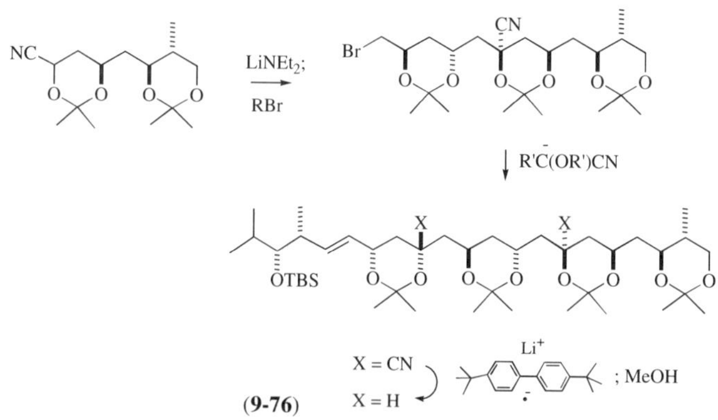

(**9-76**)

The method is limited to α-heterosubstituted nitriles. Nevertheless, the alkylation–cleavage sequence was exploited on two separate occasions in a synthesis of (+)-tetraponerine-8 (**9-75**).[72] Initially, the chiral bicyclic oxazolidine, which is readily prepared from glutaraldehyde, underwent alkylation and reduction. In the second step, both C–O bond cleavage and decyanation occurred, resulting in a piperidine with an equatorial substituent at $C_{(2)}$. Then, the most stable tricyclic intermediate containing a new α-aminonitrile function was installed to enable pentylation. As expected, the final step of reductive decyanation left the pentyl group in an equatorial position.

The alkylation of a 6-substituted 3-cyano-1,3-dioxane unit showed preference for equatorial entry to give the *cis*-3,6-dialkyl-1,3-dioxane. Reductive removal of the axial cyano group with retention of configuration by treatment with lithium di-*t*-butylbiphenylide completed the assembly of a masked *syn*-1,3-diol. An application of this method to synthesis of the polyol chain (**9-76**) of (–)-roxaticin is on record.[73]

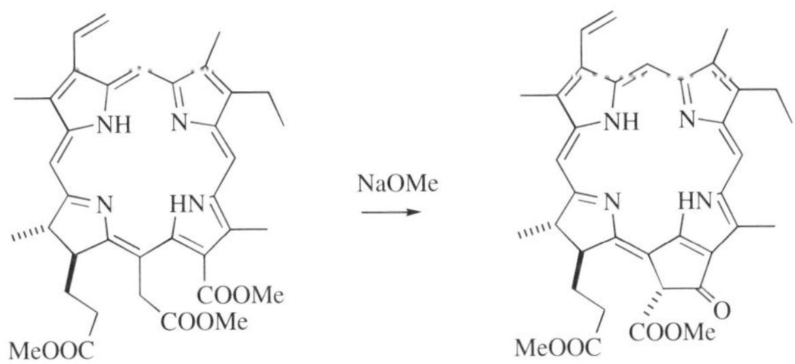

chlorin-e_6 trimethyl ester

methyl phaeophorbide-*a* (**9-77**)

In the reconstitution of methyl pheophorbide-*a* (**9-77**) from chlorin-e_6 trimethyl ester[74] by a Dieckmann condensation, the methoxycarbonyl group appeared in a sterically more favorable *trans*-relationship with the propyl ester chain at $C_{(17)}$. This result constituted the prior art for the last stage of the total synthesis of chlorophyll-*a* by Woodward et al.

Another eminent case is embodied in a synthesis of aspidospermine (**9-78**).[75] Four contiguous stereocenters are present in the alkaloid and on that basis its synthesis

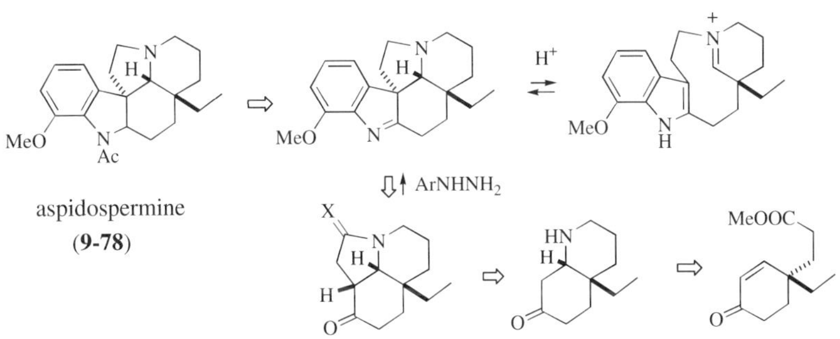

aspidospermine (**9-78**)

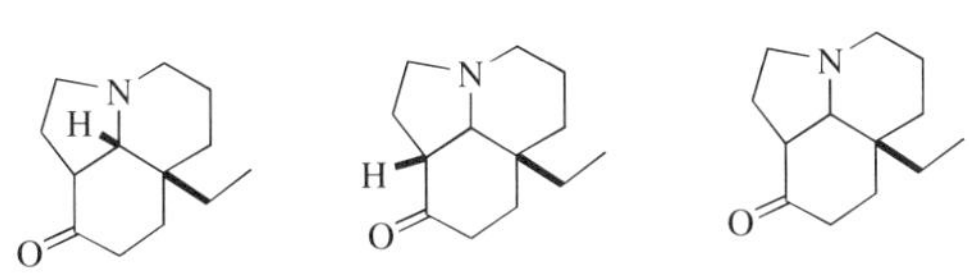

(**9-79**) (**9-80**) (**9-81**)

is quite challenging. Actually, the apparent problem vanishes should one choose Fischer indolization of a tricyclic ketone to assemble the skeleton. The pentacyclic product is the stabilomer with configuration of its three stereocenters corresponding to those of aspidospermine. Any isomer produced in the indolization is subject to equilibration via a seco iminium ion generated by a retro-Mannich fragmentation.

Support for this notion is the fact that the same product has been obtained from other tricyclic ketone stereomers (**9-79**, **9-80**, **9-81**).[76]

There is little doubt that the bisannulation featured in another route to aspidosperma alkaloids such as vindorosine (**9-82**)[77] is a process under thermodynamic control.

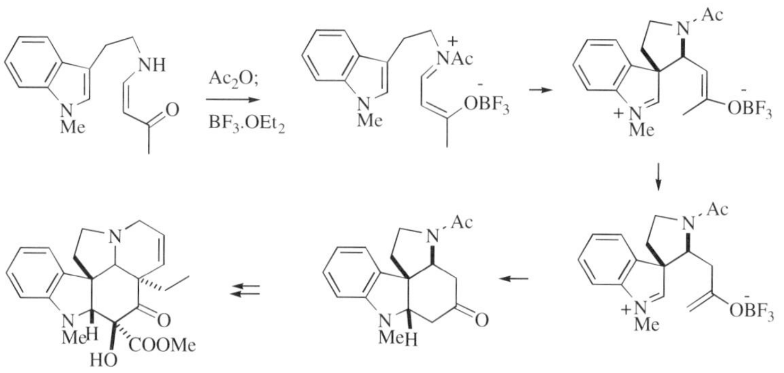

vindorosine (**9-82**)

The key step for an approach to the uleine group of alkaloids [e.g., dasycarpidone (**9-83**)] featured Mannich cyclization of a *cis*-disubstituted tetrahydropyridinium species[78] which disposed the indolylmethyl group in an axial orientation. The reversible reaction favored the product with an equatorial ethyl group. The ethyl group was susceptible to epimerization via the enamine–iminium equilibrium.

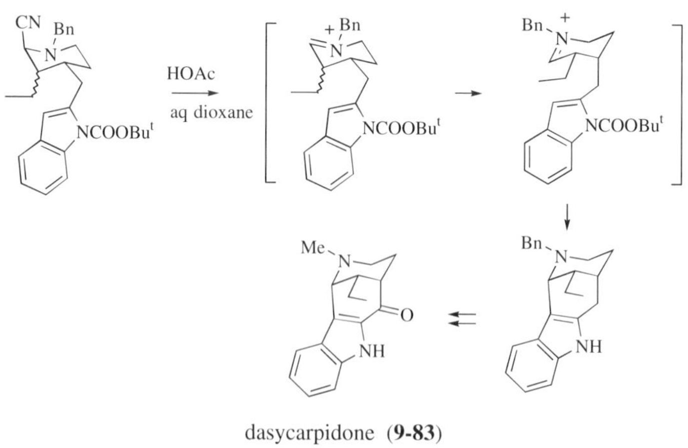

dasycarpidone (**9-83**)

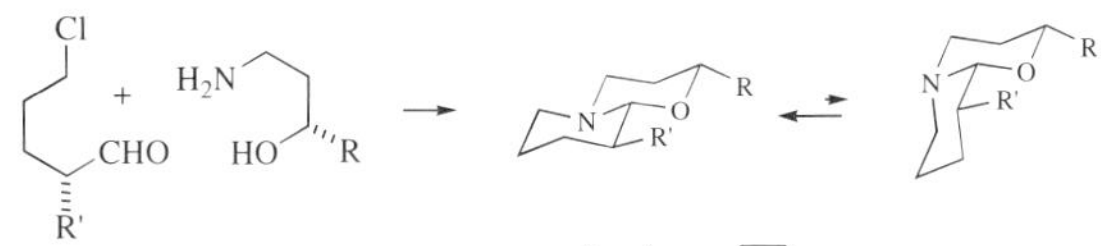

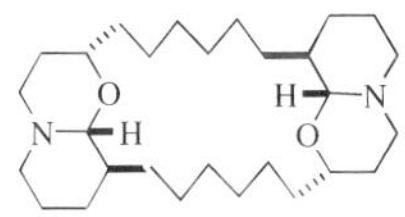

xestospongin-A
(**9-84**)

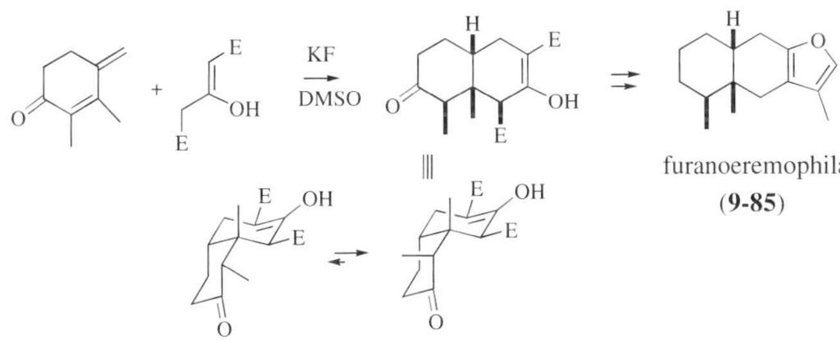

furanoeremophilane
(**9-85**)

The configuration of oxaquinolizidine units such as those existing in xestospongin-A (**9-84**) is self-adjustable through ring-chain equilibration. Accordingly, advantage has been taken of this characteristic in its synthesis.[79]

The three contiguous stereocenters of furanoeremophilane (**9-85**) would command uppermost attention in any attempt at its synthesis.[80] Yet an annulation by a double Michael reaction between 2,3-dimethyl-4-methylene-2-cyclohexenone and dimethyl 3-oxoglutarate was all it needed to construct the dicarbocyclic framework.[80] The most stable isomer of the decalindione product possesses the desired stereochemistry.

A *cis*-decalindione monoethyleneacetal used in a synthesis of luciduline (**9-86**)[81] contains an epimerizable methyl group. However, its *cis*-relationship to the angular hydrogen atoms was ensured by conformational bias of the decalin skeleton due to the acetal group. Actually the required isomer was obtained by equilibration of the monoketone.

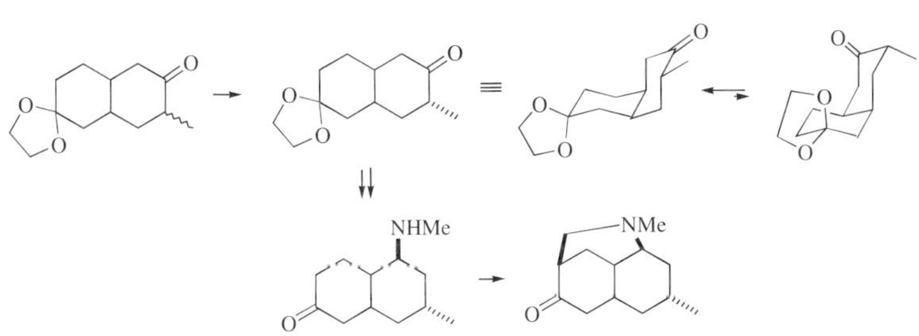

luciduline (**9-86**)

A retrosynthetic analysis of culmorin (**9-87**), particularly the corresponding diketone, indicates tetrahydroeucarvone to be an appropriate starting material. Thus, a rather straightforward construction scheme involving C–C bond formation at the α-

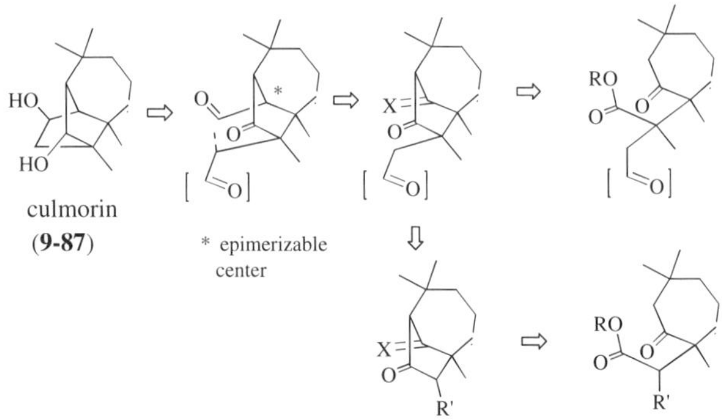

culmorin (**9-87**)

* epimerizable center

and α′-positions of the ketone by alkylation, Claisen condensation, and homologation at the ipso (C=O) carbon to afford a keto diester for Dieckmann cyclization was readily conceived.[82] The configuration of the rehybridized ipso carbon atom was impertinent.

It is of interest to note that the three-carbon unit constituting a portion of the D-ring and the angular aldehyde of aldosterone (**9-88**)[83] were each derived from a methallyl group in a synthesis from the tricarbocyclic ketone previously developed for a cortisone synthesis.[84] One (axial) of the two derived acetonyl units underwent cyclization to furnish a bridged dihydropyran, leaving the other for D-ring construction by an aldol condensation. The dihydropyran was later degraded to reveal a formyl substituent and a hydroxyl group (combined in a lactol form) as required.

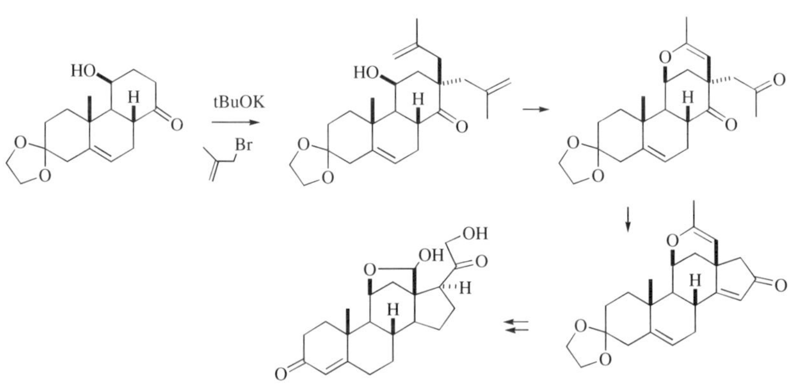

aldosterone (**9-88**)

A tandem aldol condensation-intramolecular Diels–Alder reaction served to construct a precursor (**9-88a**) of cytochalasan-K.[84a] Notwithstanding the incorrect configuration of the quaternary carbon center, the usefulness of this method was ascertained by the interchangeability of the lactone-lactam pair.

The occurrence of many structurally intriguing antibiotics which contain a spiroacetal subunit has impacted on synthetic studies of this class of compounds. Two attributes facilitate such efforts. The tendency (anomeric effect) to dispose the other

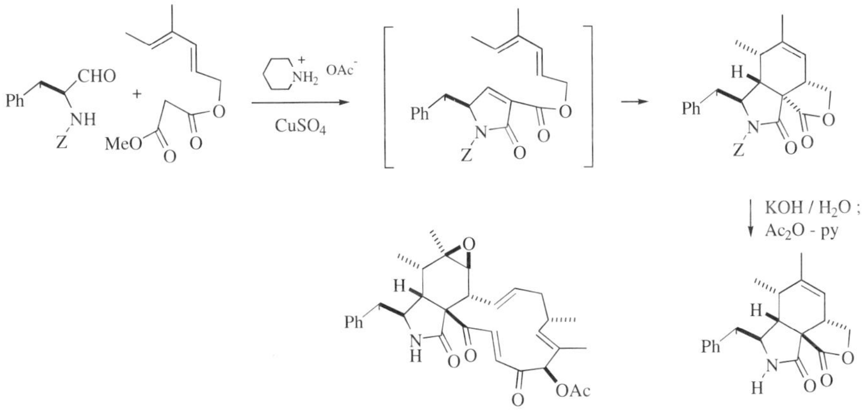

cytochalasan-K

(9-88a)

oxygen atom axially to each of the tetrahydropyran rings solves an important stereochemical problem; it is also often possible to manipulate stereocenters adjacent to the spirocyclic center as well as those bearing the cyclic oxygen atoms, for example by forming the spiroacetal subunit on the basis of a Michael reaction in a synthesis of milbemycin-β_3 (**9-89**).[85]

Analogous applications of the Michael reaction to construct oxacycles under thermodynamic control are also found in a synthesis of palytoxin.[86]

The recognition of hidden local symmetry in the quinuclidine portion of ajmaline (**9-90**) led to a synthesis design[87] based on elaboration of a cyclopentene derivative. One of the two equivalent acetaldehyde arms of the severed ring can participate in the formation of a cyclic carbinolamine, with a free choice to afford the most stable product. The formation of the C-ring is also self-regulated. The configuration of the benzoyloxymethyl group was of no importance because ample opportunity presented itself for stereochemical readjustment during the process for transformation into an aldehyde and eventual interception by the β-carbon of the indole nucleus. (For a very similar strategy, see ref. 88)

The autonomic selection of reaction partners is often witnessed in the formation of bridged ring systems. Of synthetic significance are intermediates prepared for gibberellic acid (**9-91**)[89] and cedrol (**9-68**).[90,91] It should be emphasized that the stereochemical outcome might not always be agreeable, as an attempted synthesis of a clovane framework failed.[92]

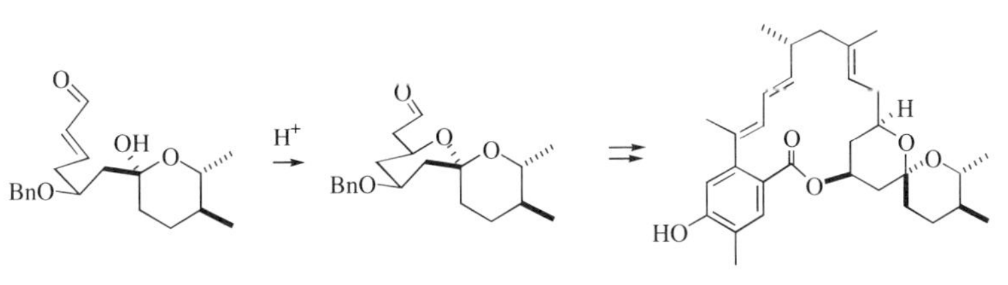

milbemycin-β_3 (**9-89**)

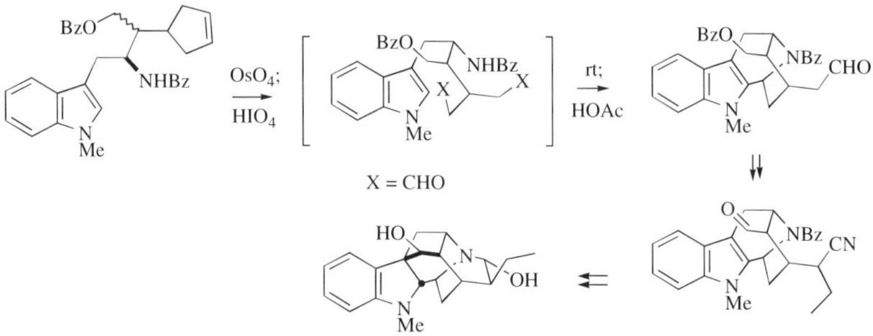

ajmaline (**9-90**)

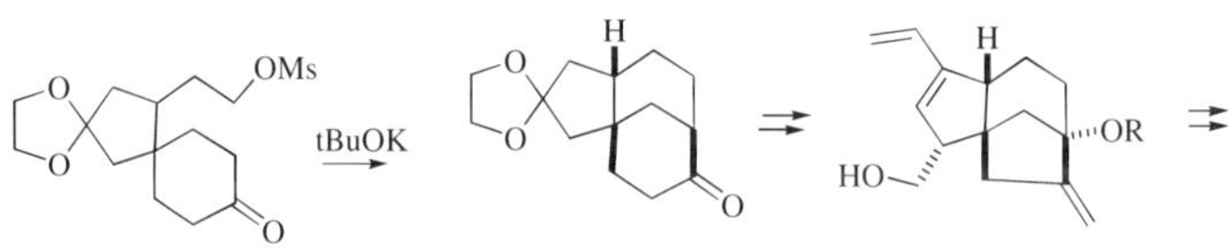

gibberellic acid (**9-91**)

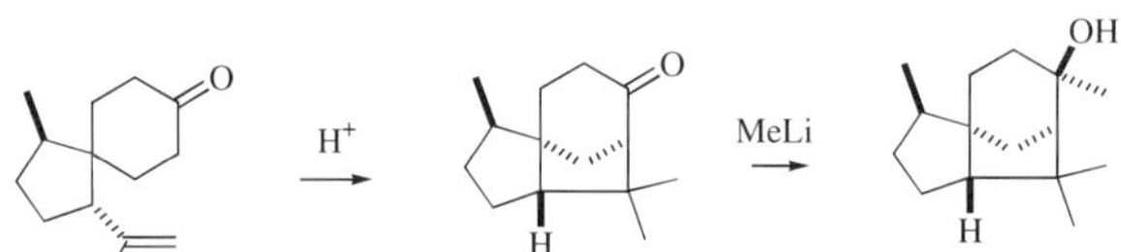

cedrol (**9-68**)

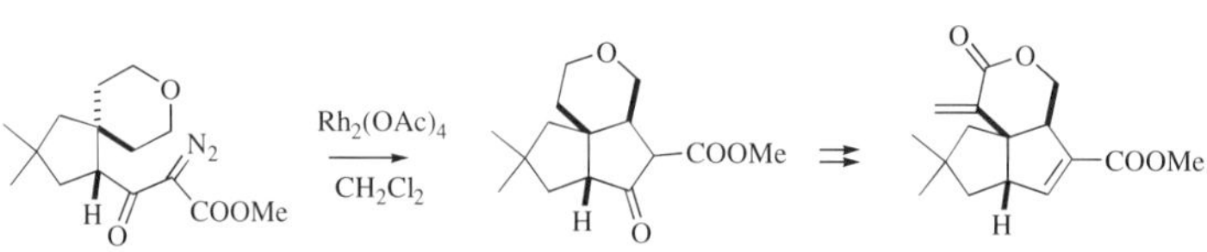

pentalenolactone-E methyl ester (**9-92**)

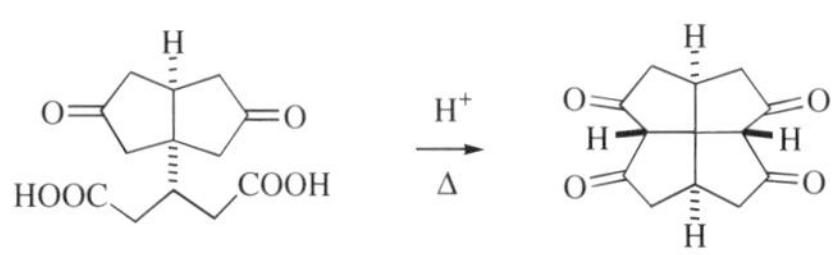

(**9-93**)

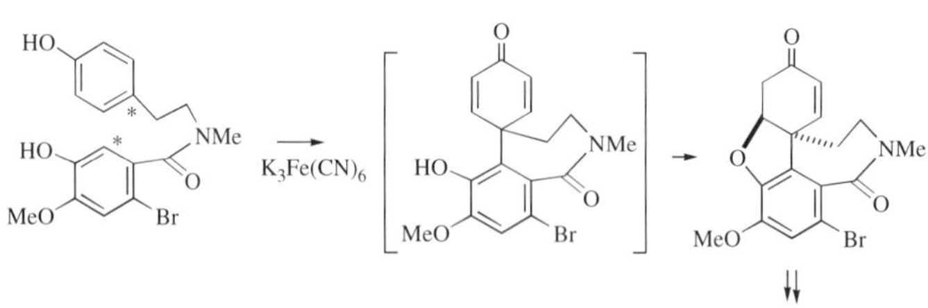

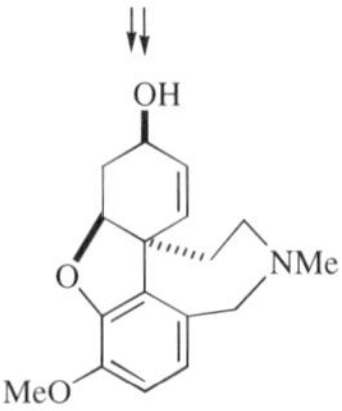

galanthamine (**9-94**)

Elaboration of certain condensed ring systems also benefited by the autoselection imposed by predisposed functional groups. Examples include an intramolecular carbenoid insertion to a C_δ–H bond in a pentalenolactone-E (**9-92**) synthesis,[93] an acylation process to give a stauranetetraone (**9-93**) with well-defined configuration at the ring junctions,[94] and formation of the furano ring leading to the skeleton of galanthamine (**9-94**).[95]

Interestingly, spontaneous deracemization of narwedine (**9-95**) has been reported.[96] Either enantiomer may be secured in high yield and excellent purity by seeding a supersaturated solution of the racemate with one of the enantiomers.

An unexpected stereorandom epoxidation of a bridged alkene was observed during synthesis of actinobolin (**9-96**)[97] It was fortunate that diastrous consequences were absolved immediately by formolysis and hydrolysis. Diaxial opening led to two formates which converged to the same diol.

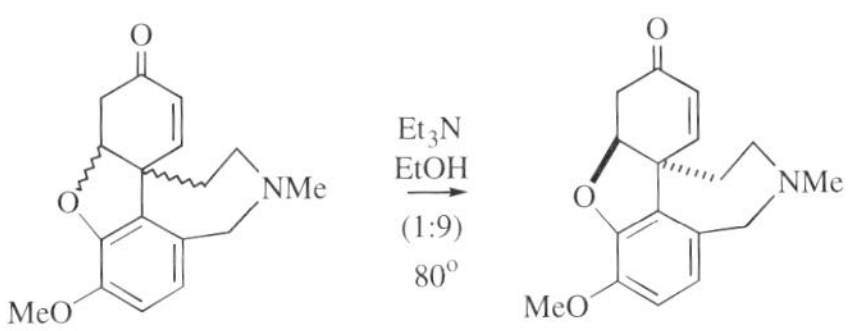

(-)-narwedine (**9-95**)

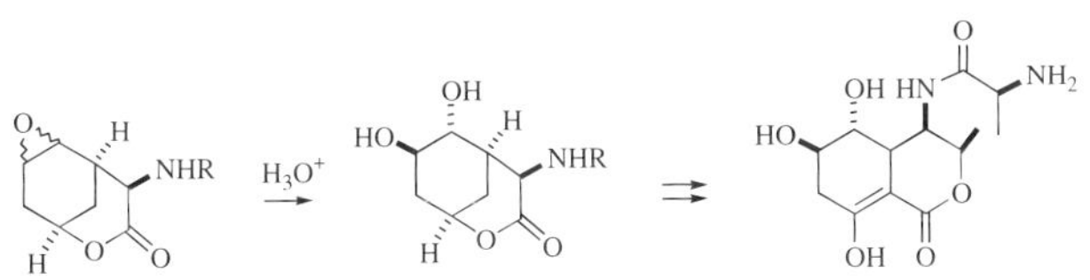

actinobolin (**9-96**)

9.5. ALKENE ISOMERIZATION

It is not always possible to obtain directly an alkene of the desired configuration in synthesis. Compromise is often made to acquire the isomeric structure and rely on a subsequent isomerization maneuver to reach the target.

Gadain was formed by a Pd(0)-catalyzed coupling reaction. To obtain savinin (**9-97**), gadain was subjected to isomerization by a free radical process.[98]

The Wittig reaction of aldehydes with unstabilized ylides usually leads to (*Z*)-alkenes as the predominant products. A synthesis of humulene[99] employing a Wittig reaction for assembly was destined to give an (*E,Z,E*)-cycloundecatriene on cyclization. Humulene (**9-98**) was obtained by a photoisomerization process in the presence of diphenyl disulfide.

A synthetic route to caryophyllene featuring an intramolecular reductive coupling of a keto ester as the key step for the nine-membered ring formation was diverted to the norketone of isocaryophyllene,[100] due to (*E-Z*)-isomerization of the substrate. A recent observation on the sensitized photoisomerization (with methyl

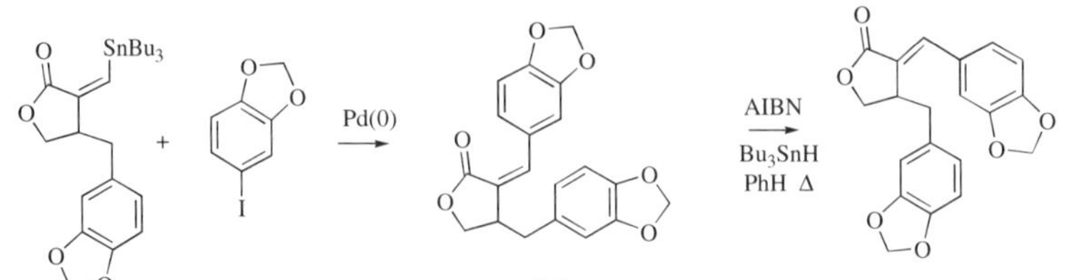

gadain savinin (**9-97**)

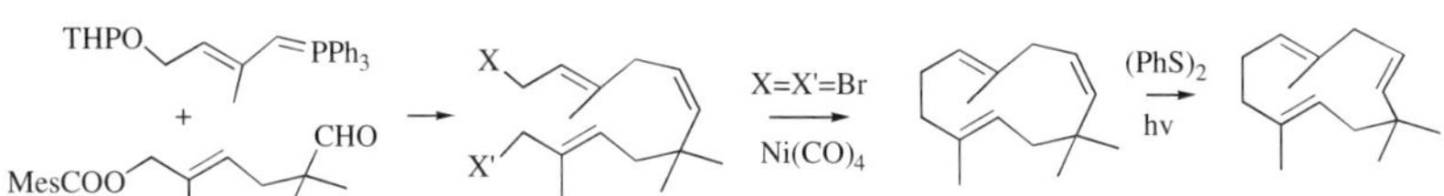

humulene (**9-98**)

benzoate) of the more stable isocaryophylene (**9-3**) to caryophyllene (**9-2**)[101] served to reinstate the original goal.

The expediency of a forskolin synthesis[102] based on an intramolecular Diels–Alder reaction (see **8-9**) has been greatly increased by using the much more readily available (*Z,E*)-isomer which underwent isomerization in situ to the desired (*E,E*)-isomer in the presence of benzenethiol. The former compound could be prepared directly by a Wittig reaction in 76% yield, whereas the latter compound in only 11% yield by a four-step sequence involving reaction with 3-methoxy-propynyllithium, Lindlar hydrogenation, allylic carbonate formation, and Pd(0)-catalyzed elimination.

Thermal isomerization of a chiral *N*-lithioenamine has been demonstrated.[103] Enantiomeric 2-methylcyclododecanones (**9-99, 9-100**) were obtained after methylation of the pre- and post-equilibrated metalloenamines.

An access to cyclopenin (**9-101**)[104] depended on partial isomerization of the (*E*)-benzylidene dilactam intermediate on heating with piperidine in methanol.

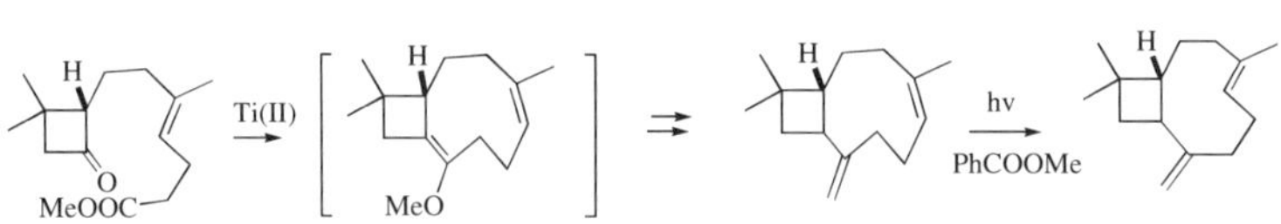

isocaryophyllene (**9-3**) caryophyllene (**9-2**)

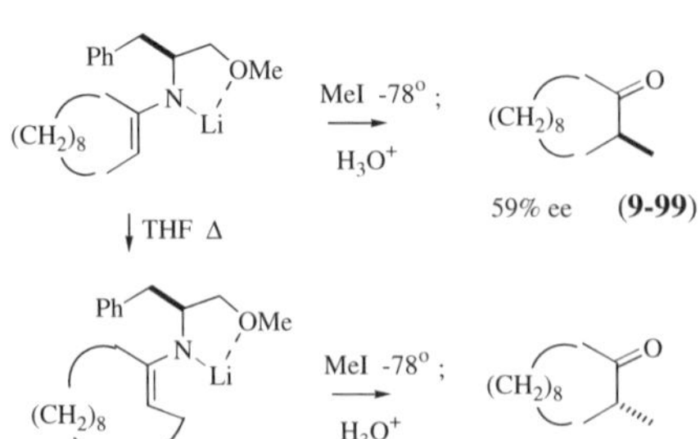

59% ee (**9-99**)

81% ee (**9-100**)

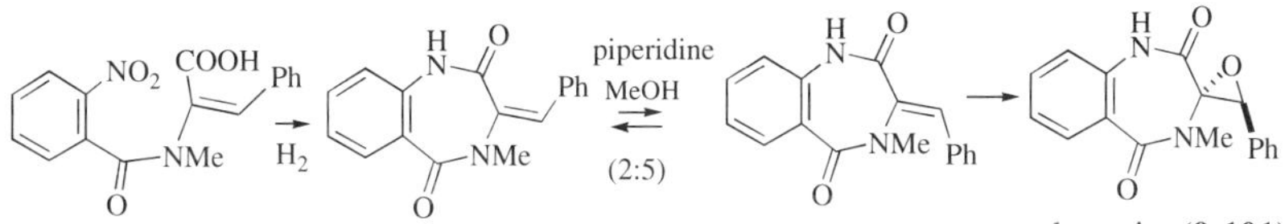

cyclopenin (**9-101**)

REFERENCES

1. Mehta, G.; Rao, K.S. *J. Am. Chem. Soc.* **1986**, *108*, 8015.
2. Corey, E.J.; Mitra, R.B.; Uda, H. *J. Am. Chem. Soc.* **1964**, *86*, 485.
3. Oppolzer, W.; Fehr, C.; Warneke, J. *Helv. Chim. Acta* **1977**, *60*, 48.
4. Ihara, M.; Suzuki, T.; Katogi, M.; Taniguchi, N.; Fukumoto, K. *Chem. Commun.* **1991**, 646.
5. De Clercq, P.; Vandewalle, M. *J. Org. Chem.* **1977**, *42*, 3447.
6. Termont, D.; De Clercq, P.; De Keukeleire, D.; Vandewalle, M. *Synthesis* **1977**, 46.
7. VanBever, W.; Kohen, F.; Ranade, V.V.; Counsell, R.E. *Chem. Commun.* **1970**, 758.
8. Woodward, R.B.; Sondheimer, F.; Taub, D.; Heusler, K.; McLamore, W.M. *J. Am. Chem. Soc.* **1952**, *74*, 4223.
9. van Tamelen, E.E.; Shamma, M.; Burgstahler, A.W.; Wolinsky, J.; Tamm, R.; Aldrich, P.E. *J. Am. Chem. Soc.* **1958**, *80*, 5006.
10. Goldberg, S.I.; Lipkin, A.H. *J. Org. Chem.* **1970**, *35*, 242.
11. Magnus, P.; Giles, M.; Bonnert, R.; Kim, C.S.; McQuire, L.; Merrit, A.; Vicker, N. *J. Am. Chem. Soc.* **1992**, *114*, 4403.
12. Liu, G.; Smith, T.C.; Pfander, H. *Tetrahedron Lett.* **1995**, *36*, 4979.
13. Jacobi, P.A.; Martinelli, M.J.; Polanc, S. *J. Am. Chem. Soc.* **1984**, *106*, 5594.
14. Deighton, M.; Hughes, C.R.; Ramage, R. *Chem. Commun.* **1975**, 662.
15. Corey, E.J.; Nozoe, S. *J. Am. Chem. Soc.* **1965**, *87*, 5728.
16. Piers, E.; Geraghty, M.B.; Smillie, R.D.; Soucy, M. *Can. J. Chem.* **1975**, *53*, 2849.
17. Grieco, P.A.; Williams, E.; Tanaka, H.; Gilman, S. *J. Org. Chem.* **1980**, *45*, 3537.
18. Nakata, T.; Nagao, S.; Mori, N.; Oishi, T. *Tetrahedron Lett.* **1985**, *26*, 6461.
19. Miyaoka, H.; Saka, Y.; Miura, S.; Yamada, Y. *Tetrahedron Lett.* **1996**, *37*, 7107.
20. Corey, E.J.; Danheiser, R.L.; Chandrasekaran, S.; Keck, G.E.; Gopalan, B.; Larsen, S.D.; Siret, P.; Gras, J.-L. *J. Am. Chem. Soc.* **1978**, *100*, 8034; Corey, E.J.; Danheiser, R.L.; Chandrasekaran, S.; Siret, P.; Keck, G.E.; Gras, J.-L. *J. Am. Chem. Soc.* **1978**, *100*, 8031.
21. Stork, G.; Clark, G.; Shiner, C.S. *J. Am. Chem. Soc.* **1981**, *103*, 4948.
22. Ohta, S.; Kimoto, S. *Tetrahedron Lett.* **1975**, 2279.
23. Irie, H.; Nagai, Y.; Tamoto, K.; Tanaka, H. *Chem. Commun.* **1973**, 302.
24. Torii, S.; Inokuchi, T.; Kawai, K. *Bull. Chem. Soc. Jpn.* **1979**, *52*, 861.
25. Nagakura, I.; Maeda, S.; Ueno, M.; Funamizu, M.; Kitahara, Y. *Chem. Lett.* **1975**, 1143.
26. Hortmann, A.G.; Daniel, D.S.; Martinelli, J.E. *J. Org. Chem.* **1973**, *38*, 728.
27. Näf, F.; Decorzant, R.; Thommen, W. *Helv. Chim. Acta* **1982**, *65*, 2212.
28. Solladié, G.; Hutt, J. *Tetrahedron Lett.* **1987**, *28*, 797.
29. Oppolzer, W.; Strauss, H.F.; Simmons, D.P. *Tetrahedron Lett.* **1982**, *23*, 4673.

30. Oppolzer, W.; Strauss, H.F.; Simmons, D.P. *Tetrahedron Lett.* **1984**, *25*, 825.
31. Corey, E.J.; Carpino, P. *J. Am. Chem. Soc.* **1989**, *111*, 5472.
32. Niwa, H.; Wakamatsu, K.; Hida, T.; Niiyama, K.; Kigoshi, H.; Yamada, M.; Nagase, H.; Suzuki, M.; Yamada, K. *J. Am. Chem. Soc.* **1984**, *106*, 4547.
33. Lansbury, P.T.; Serelis, A.K. *Tetrahedron Lett.* **1978**, 1909.
34. Lansbury, P.T.; Hangauer, D.G.; Vacca, J.P. *J. Am. Chem. Soc.* **1980**, *102*, 3964.
35. Stork, G.; Darling, S.D.; Harrison, I.T.; Wharton, P.S. *J. Am. Chem. Soc.* **1962**, *84*, 2018.
36. Hamelin, O.; Depres, J.-P.; Greene, A.E.; Tinant, B.; Declercq, J.-P. *J. Am. Chem. Soc.* **1996**, *118*, 9992.
37. Jacobi, P.A.; DeSimone, R.W.; Udodong, U.E.S.; Blum, C.A. *Tetrahedron Lett.* **1989**, *30*, 7173.
38. Burkholder, T.P.; Fuchs, P.L. *J. Am. Chem. Soc.* **1990**, *112*, 9601.
39. Oppolzer, W.; Petrzilka, M. *Helv. Chim. Acta* **1978**, *61*, 2755.
40. Fletcher, S.R.; Baker, R.; Chambers, M.S.; Herbert, R.H.; Hobbs, S.C.; Thomas, S.R.; Verrier, H.M.; Watt, A.P.; Ball, R.G. *J. Org. Chem.* **1994**, *59*, 1771.
41. Roush, W.R. *J. Am. Chem. Soc.* **1980**, *102*, 1390.
42. Iwata, C.; Fujita, M.; Hattori, K.; Uchida, S.; Imanishi, T. *Tetrahedron Lett.* **1985**, *26*, 2221.
43. Woodward, R.B.; Bader, F.E.; Bickel, H.; Frey, A.J.; Kierstead, R.W. *Tetrahedron* **1958**, *2*, 1.
44. Openshaw, H.T.; Whittaker, N. *J. Chem. Soc.* **1963**, 1461.
45. Oppolzer, W.; Hauth, H.; Pfäffli, P.; Wenger, R. *Helv. Chim. Acta* **1977**, *60*, 1801.
46. Marshall, J.A.; Xie, S. *J. Org. Chem.* **1992**, *57*, 2987.
47. Corey, E.J.; Snider, B.B. *J. Am. Chem. Soc.* **1972**, *94*, 2549.
48. Gobao, R.A.; Bremmer, M.L.; Weinreb, S.M. *J. Am. Chem. Soc.* **1982**, *104*, 7065.
49. Wiesner, K.; Uyeo, S.; Philipp, A.; Valenta, Z. *Tetrahedron Lett.* **1968**, 6279.
50. Bergman, R.; Hansson, T.; Sterner, O.; Wickberg, B. *Chem. Commun.* **1990**, 865.
51. Wang, T.-F.; Yang, C.-F. *Chem. Commun.* **1989**, 1876.
52. Semmelhack, M.F.; Epa, W.R.; Cheung, A. W.-H.; Gu, Y.; Kim, C.; Zhang, N.; Lew, W. *J. Am. Chem. Soc.* **1994**, *116*, 7455.
53. Miller, J.G.; Kurz, W.; Untch, K.G.; Stork, G. *J. Am. Chem. Soc.* **1974**, *96*, 6774.
54. Boeckman, R.K.; Springer, D.M.; Alessi, T.R. *J. Am. Chem. Soc.* **1989**, *111*, 8284.
55. Majetich, G.; Song, J.-S.; Ringold, C.; Nemeth, G.A.; Newton, M.G. *J. Org. Chem.* **1991**, *56*, 3973.
56. Mitsunobu, O. *Synthesis* **1981**, 1.
57. Wiesner, K.; Poon, L.; Jirkovsky, I.; Fishman, M. *Can. J. Chem.* **1969**, *47*, 433.
58. Heathcock, C.H.; Kelly, T.R. *Chem. Commun.* **1968**, 267.
59. Kelly, T.R. *J. Org. Chem.* **1972**, *37*, 3393.
60. Woodward, R.B.; Cava, M.P.; Oliis, W.D.; Hunger, A.; Daeniker, H.U.; Schenker, K. *Tetrahedron* **1963**, *19*, 247.
61. Overman, L.E.; Angle, S.R. *J. Org. Chem.* **1985**, *50*, 4021.
62. Heathcock, C.H.; Kleinman, E.F.; Binkley, E.S. *J. Am. Chem. Soc.* **1982**, *104*, 1054.
63. Ryckman, D.M.; Stevens, R.V. *J. Am. Chem. Soc.* **1987**, *109*, 4940.

64. Corey, E.J.; Ohno, M.; Mitra, R.B.; Vatakencherry, P.A. *J. Am. Chem. Soc.* **1964**, *86*, 478.

65. Stork, G.; Clarke, F.H. *J. Am. Chem. Soc.* **1961**, *83*, 3114.

66. Liu, H.-J.; Sato, Y.; Valenta, Z.; Wilson, J.S.; Yu, T.T.-J. *Can. J. Chem.* **1976**, *54*, 97.

67. Magnus, P.; Mugrage, B.; DeLuca, M.R.; Cain, G.A. *J. Am. Chem. Soc.* **1990**, *112*, 5220.

68. Yamada, K.; Nagase, H.; Hayakawa, Y.; Aoki, K.; Hirata, Y. *Tetrahedron Lett.* **1973**, 4963.

69. Sarkar, T.K.; Gosh, S.K.; Subba Rao, P.S.V.; Mamdapur, V.R. *Tetrahedron Lett.* **1990**, *31*, 3465.

70. Funk, R.L.; Bolton, G.L.; Daggett, J.U.; Hansen, M.M.; Horcher, L.H.M. *Tetrahedron* **1985**, *41*, 3479.

71. Piers, E.; Jung, G.L. *Can. J. Chem.* **1987**, *65*, 1668.

72. Yue, C.; Royer, J.; Husson, H.-P. *J. Org. Chem.* **1990**, *55*, 1140.

73. Rychnovsky, S.D.; Rodriguez, C. *J. Org. Chem.* **1992**, *57*, 4793.

74. Fischer, H.; Oestreicher, A. *Liebigs Ann. Chem.* **1941**, *546*, 49.

75. Stork, G.; Dolfini, J.E. *J. Am. Chem. Soc.* **1963**, *85*, 2872.

76. Ban, Y.; Sato, Y.; Inoue, I.; Nagai, M.; Oishi, T.; Terashima, M.; Yonemitsu, O.; Kanaoka, Y. *Tetrahedron Lett.* **1965**, 2261; Kuehne, M.E.; Bayha, C. *Tetrahedron Lett.* **1966**, 1311.

77. Büchi, G.; Matsumoto, K.E.; Nishimura, H. *J. Am. Chem. Soc.* **1971**, *93*, 3299.

78. Gracia, J.; Casamitjana, N.; Bonjoch, J.; Bosch, J. *J. Org. Chem.* **1994**, *59*, 3939.

79. Hoye, T.R.; North, J.T.; Yao, L.J. *J. Am. Chem. Soc.* **1994**, *116*, 2617.

80. Irie, H.; Katakawa, J.; Mizuno, Y.; Udaka, S.; Taga, T.; Osaki, K. *Chem. Commun.* **1978**, 717.

81. Scott, W.L.; Evans, D.A. *J. Am. Chem. Soc.* **1972**, *94*, 4779.

82. Roberts, B.W.; Poonian, M.S.; Welch, S.C. *J. Am. Chem. Soc.* **1969**, *91*, 3400.

83. Heusler, K.; Wieland, P.; Wettstein, A. *Helv. Chim. Acta* **1959**, *42*, 1586.

84. Sarett, L.H.; Johns, W.F.; Beyler, R.E.; Lukes, R.M.; Poos, G.I.; Arth, G.E. *J. Am. Chem. Soc.* **1953**, *75*, 2112.

84a. Schmidlin, T.; Zürcher, W.; Tamm, C. *Helv. Chim. Acta* **1981**, *64*, 235; Zehnder, M. *Helv. Chim. Acta* **1981**, *64*, 251.

85. Smith, A.B., III; Schow, S.R.; Bloom, J.D.; Thompson, A.S.; Winzenberg, K.N. *J. Am. Chem. Soc.* **1982**, *104*, 4015.

86. Ko, S.S.; Klein, L.L.; Pfaff, K.-P.; Kishi, Y. *Tetrahedron Lett.* **1982**, *23*, 4415.

87. Masamune, S.; Ang, S.K.; Egli, C.; Nakatsuka, N.; Sarkar, S.K.; Yasunari, Y. *J. Am. Chem. Soc.* **1967**, *89*, 2506.

88. van Tamelen, E.E.; Oliver, L.K. *J. Am. Chem. Soc.* **1970**, *92*, 2136.

89. Corey, E.J.; Smith, J.G. *J. Am. Chem. Soc.* **1979**, *101*, 1038.

90. Corey, E.J.; Girotra, N.N.; Mathew, C.T. *J. Am. Chem. Soc.* **1969**, *91*, 1557.

91. Crandall, T.G.; Lawton, R.G. *J. Am. Chem. Soc.* **1969**, *91*, 2127.

92. Danishefsky, S.; Hatch, W.E.; Sax, M.; Abola, E.; Pletcher, J. *J. Am. Chem. Soc.* **1973**, *95*, 2410.

93. Taber, D.F.; Schuchardt, J.L. *Tetrahedron* **1987**, *43*, 5677.

94. Deshpande, M.N.; Jawdosiuk, M.; Kubiak, G.; Venkatachalam, M.; Weiss, U.; Cook, J.M. *J. Am. Chem. Soc.* **1985**, *107*, 4786.

95. Kametani, T.; Yamaki, K.; Yagi, H.; Fukumoto, K. *J. Chem. Soc. C* **1969**, 2602.

96. Shieh, W.-C.; Carlson, J.A. *J. Org. Chem.* **1994**, *59*, 5463.

97. Garigipati, R.S.; Tschaen, D.M.; Weinreb, S.M. *J. Am. Chem. Soc.* **1985**, *107*, 7790.

98. Lee, E.; Hur, C.U.; Jeong, Y.C.; Rhee, Y.H.; Chang, M.H. *Chem. Commun.* **1991**, 1314.

99. Corey, E.J.; Hamanaka, E. *J. Am. Chem. Soc.* **1967**, *89*, 2758.

100. McMurry, J.E.; Miller, D.D. *Tetrahedron Lett.* **1983**, *24*, 1885.

101. Suginome, H.; Kondoh, T.; Gogonea, C.; Singh, V.; Goto, H.; Osawa, E. *J. Chem. Soc. Perkin Trans. 1* **1995**, 69.

102. Hashimoto, S.; Sakata, S.; Sonegawa, M.; Ikegami, S. *J. Am. Chem. Soc.* **1988**, *110*, 3670.

103. Meyers, A.I.; Williams, D.R.; White, S.; Erickson, G.W. *J. Am. Chem. Soc.* **1981**, *103*, 3088.

104. White, J.D.; Haefliger, W.E.; Dimsdale, M.J. *Tetrahedron* **1970**, *26*, 233.

INDEX